Environmental Issues and Waste Management Technologies in the Ceramic and Nuclear Industries IV

Related titles published by the American Ceramic Society:

Environmental Issues and Waste Management Technologies in the Ceramic and Nuclear Industries III
Edited by David K. Peeler and James C. Marra
© 1998, ISBN 1-57498-035-1

Environmental Issues and Waste Management Technologies in the Ceramic and Nuclear Industries II
Edited by Vijay Jain and David Peeler
© 1996, ISBN 1-57498-023-8

Environmental Issues and Waste Management Technologies in the Ceramic and Nuclear Industries
Edited by Vijay Jain and Ron Palmer
© 1995, ISBN 1-57498-004-1

Environmental and Waste Management Issues in the Ceramic Industry II
Edited by Dennis Bickford, Steven Bates, Vijay Jain, and Gary Smith
© 1994, ISBN 0-944904-79-3

Environmental and Waste Management Issues in the Ceramic Industry
Edited by George B. Mellinger
© 1994, ISBN 0-944904-71-8

For information on ordering titles published by The American Ceramic Society, or to request a publications catalog, please contact our Customer Service Department at 614-794-5890 (phone), 614-794-5892 (fax),<customersrvc@acers.org> (e-mail), or write to Customer Service Department, 735 Ceramic Place, Westerville, OH 43081, USA.

Visit our on-line book catalog at <www.acers.org>.

Ceramic Transactions
Volume 93

Environmental Issues and Waste Management Technologies in the Ceramic and Nuclear Industries IV

Edited by
James C. Marra
Westinghouse Savannah River Company

Gregory T. Chandler
Westinghouse Savannah River Company

Published by
The American Ceramic Society
735 Ceramic Place
Westerville, Ohio 43081

Proceedings of the Science and Technology in Addressing Environmental Issues in the Ceramic Industry Symposium and the International Symposium on Waste Management Technologies in Ceramic and Nuclear Industries, held at the 100th Annual Meeting of The American Ceramic Society in Cincinnati, Ohio, May 3-6, 1998.

COVER PHOTO: "UO_2 (ex-AUC) powder," is courtesy of M. De Vleeschhauwer, M. Lippens, F. Eeckhout, and E. Pelckmans, and appears as figure 2 in their paper "Powder and Pellet Characterization in the MIMAS-Process," which begins on page 79.

Library of Congress Cataloging-in-Publication Data
A CIP record for this book is available from the Library of Congress.

For information on ordering titles published by The American Ceramic Society, or to request a publications catalog, please call 614-794-5890.

Printed in the United States of America.

1 2 3 4–02 01 00 99

ISSN 1042-1122
ISBN 1-57498-057-2

Contents

Nuclear Fuel Development

Stabilization of Low-level Radioactive and Hazardous Wastes

High-Level Radioactive Waste Management

Waste Form Leaching and Property Measurement

Actinide Material Treatment and Stabilization

Preface

As we approach the new millenium, it is increasingly important that we identify and adequately address environmental issues. This includes finding ways to reduce waste and inefficiency in manufacturing processes, recycle or stabilize waste materials, and remediate legacy environmental conditions.

Crystalline ceramics and glasses are presently being used to tackle many of these challenging environmental issues. These ceramic materials are frequently used as immobilization forms for both radioactive and hazardous wastes. There continues to be significant technological growth and scientific interest in the development of immobilization forms and processes, in the understanding of product performance and in the recycling of waste stream materials. In addition, the ceramic manufacturing industry is taking an increased interest in applying new environmental practices. Implementing the ISO14001 environmental program, performing life-cycle analyses and closely monitoring emissions are just a few of the many practices being used to foster an environmental awareness in the ceramic industry.

The Nuclear and Environmental Technology (NET) Division of The American Ceramic Society recently sponsored two symposia at the 100th Annual Meeting of The American Ceramic Society focusing on these environmental issues. The symposium focusing on the science and technologies associated with addressing environmental issues in the ceramic industry was sponsored by the NET division in conjunction with The American Ceramic Society's Basic Science, Glass and Optical Materials and Cements divisions. The symposium on waste management technologies in the nuclear and ceramic industries was co-sponsored by the Glass and Optical Materials and Cements divisions.

This book is the sixth in the series of The American Ceramic Society's *Ceramic Transactions* volumes on the subjects of waste management and environmental issues. Previous proceedings include *Ceramic Transactions* volumes 39, 45, 61, 72, and 87.

The editors of these proceedings are grateful to all those who contributed to the success of the symposia. We are especially indebted to the authors and reviewers who helped deliver a quality product. We would also like to acknowledge the contributions of the session chairs: Dane Spearing, Xiangdong Feng, Gary Smith, Tom Rankin, Vijay Jain, David Peeler, and Sharon Marra. In addition to chairing their sessions and facilitating many interesting discussions, these individuals also assisted in ensuring the timely review of the manuscripts. The editors would finally like to thank the book publishing team at The American Ceramic Society: Mary Cassells, Sarah Godby, and Natalee Sperry. Their efforts were instrumental in the publication of these proceedings.

James C. Marra

Gregory T. Chandler

Environmental Strategies and Waste Treatment/Recycling Technologies

STRATEGY FOR ENHANCING THE MATERIALS ROLE IN ENVIRONMENTAL TECHNOLOGY AND SUSTAINABILITY: ASSESSMENT AND LINKAGES

Kelly S. Kirkpatrick and Robert J. Eagan
Sandia National Laboratories
P.O. Box 5800
Albuquerque, NM 87185-0513

David G. Howden
Ohio State University
College Of Engineering
1248 Adams Drive
Columbus, OH 43221

Toni Grobstein Marechaux
U.S. Department of Energy
Energy Efficiency and Renewables
1000 Independence Avenue SW
Washington, DC 20585-0121

Samuel J. Schneider, Jr.
National Institute of Standards and Technology
Materials Science and Engineering Laboratory Office
MATLS B309
Gaithersburg, MD 20899

Jorn Larsen-Basse
National Science Foundation
4201 Wilson Blvd.
Arlington, VA 22230

Materials are essential ingredients of industrial production and represent an important element of the U.S. economy. How we produce, use and dispose of materials are at the heart of our ability to achieve a sustainable future. Strategies that promote industrial ecology and sustainable development inevitably focus on material and energy flows to help explain and optimize modern production systems, consumption and use, and national economies. While materials clearly play a central role in environmental technology and sustainable development, scientists and policymakers have come to appreciate the complexity of material flows through the economy and the difficulty in developing clear strategies and policies for materials production and use.

The Federation of Materials Societies (FMS) and the MatTec Environmentally Conscious Materials Working Group (EMAT) sponsored a workshop on December 17-18, 1997, entitled, **Strategy For Enhancing the Materials Role in Environmental Technology and Sustainability: Assessment and Linkages**. Taking the foundation of knowledge laid out from the 1996 FMS-EMAT workshop that identified the key environmental technology research and development funding areas and activities, this workshop covered the following key areas: assessment of materials research-related proposals in the environmental technologies and sustainability arena; examination of effective ways to develop collaborations among academia, Federal government and industry; and initiation of the development of R&D long-range strategies leading to pollution avoidance in the industrial sectors that are most energy and pollution intensive industries. This report summarizes the key points drawn from the workshop.

The purpose of this meeting was to convene high-level government officials, policymakers and scientists who assessed strategies that can improve the efficiency of materials flows in the U.S. economy and contribute to environmental and sustainable development goals. The workshop was successful in improving the understanding of materials cycles and the R&D strategies being pursued by government and industry to produce and use materials in an environmentally responsible way. Key findings of the workshop were:

- Industrial ecology is a useful framework for viewing the relationship among industrial processes, products, and the environment. Its goal is to make industrial production system sustainable within the economy and the environment. Industrial ecology is beginning to shift from being a separate, post-design activity to a strategic one in which environmental issues are integrated into the entire design process.

- Sustainable development may be an inadequate goal for society to focus on because it cannot be clearly defined and it is difficult to benchmark.

- Unique and valuable solutions to the complex problems of dematerialization, decarbonization and toxic use often emerge when they are analyzed using a systems approach.

- DOE/NSF workshops on sustainability determined that R&D should be organized to address system changes, institutional changes, and cross-disciplinary strategies.

- The Partnership for the New Generation of Vehicles may serve as a good model for government-industry collaboration but more must be done to replicate successes and avoid known pitfalls.

- Effective materials research and engineering design can accomplish pollution prevention at various levels of detail: macro-level (industrial), meso-level (unit operations), and micro-level (molecular). Setting goals at the macro-level can help to prioritize research needs at the meso-and micro-level.

- Successful R&D strategies for sustainable materials invariably benefit from collaboration among industry, government and academia. Examples of successful partnerships can be found in DOE's Industries of the Future Strategy, the aluminum industry partnership, DOE's environmental programs, EPA's materials selection approaches, and NIST's activities in advanced building materials. Each partnership is driven by different needs but all benefit from the convergence of diverse technical expertise and perspectives to solve complex materials issues.

The materials societies that comprise FMS, including the American Ceramic Society, and the members of EMAT face an important challenge: to draw upon a new understanding of materials flows and the successful experiences of collaborative partnerships to recommend effective technology policies and develop clear strategies for materials R&D. Four key partners are needed to help the materials community elevate materials issues to a national level.

- The <u>Administration</u> must provide a vision for materials at the highest levels.

- <u>Government agencies</u> must devise and enforce regulations, develop technologies, and provide the data framework for materials studies through collaborative partnerships.

- Trade associations must focus industry needs, advocate the vision within their respective industries, and inform companies of partnership opportunities with the government.

- Professional societies must educate their members about the critical issues of environmental technology and sustainability, broker partnerships between the societies and government, and advise the research community on materials and sustainability issues.

All partners need to make a proactive effort to keep each other updates on new developments to foster the flow of information and new developments in environmental technology and sustainability amongst them. Working together and maintaining open lines of communication will catalyze the comprehensive understanding of what steps need to be taken to solidify the vision that will be created in this area. By working together, professional societies can identify and address interdisciplinary needs. Getting professional society members to address problems at a local level and to develop mechanisms to drive the cost of waste minimization down at the national meetings would allow local issues to be translated through the society leadership to our political leaders in Washington.

Each FMS member society represented at the meeting provided thoughts on the meeting and ideas for the next steps to take beyond this workshop. The American Ceramic Society will consider actively work with its industrial members to help identify effective mechanisms for becoming more environmentally responsible by organizing panels and workshops. Currently, three committees in the American Ceramic Society are concerned with environmental technology-related issues. In general, the Ceramics community needs to identify strategies for becoming more proactive when addressing environmental issues.

Additional suggestions for action items include the following:

- FMS gathers societies at the executive level. There should be a technical liaison committee comprised on societies working to determine where materials fit into the various industrial sectors.

- It must be recognized that societies serve their membership and their needs vary among the societies. This can lead to a different agenda for each society and communication breakdowns. A format for useful discussions is needed. One common goal among all societies is environmental improvement; member societies need to be more actively involved to work towards this end. Organizing workshops and symposia amongst professional societies will help to develop an understanding cross-cut areas of mutual concern in environmental technology.

- A vision and roadmap for environmental technology for given materials systems should be jointly composed among member societies.

- The professional societies' role is to serve as an engine for communication and education. By distilling the large amounts of information that exist in the areas of environmental technology and sustainability to be member specific, the professional society will provide a tool for its members to use in communicating with their colleagues and to help them understand more effective ways to address environmental technology and sustainability in a proactive way and to create programmatic priorities in R&D.

The NSF/DOE workshops on new themes for basic research: **Vehicles for the Future**, **Environmentally Responsive Technologies of the Future**, and **Organizing for the R&D in the 21st Century**, have taken this approach for sustainability. The workshop participants were roughly 40% academia, 30% industry and 30% government agencies and national laboratories. These workshops have resulted in national goals to which the materials community must link its initiatives in order to meet with future support. The rationale for the workshops was to conceive of a new mission challenge for the national R&D infrastructure to pursue. Dr. Michael Knotek and Dr. Peter Eisenberger, co-chairs of these workshops, are now planning the next workshop in this series entitled, Science Needs for Sustainability, that is scheduled for October 1998. An FMS-EMAT representative is working in their organizing committee to share insights gained from the workshop discussed in this paper, **Strategy For Enhancing the Materials Role in Environmental Technology and Sustainability: Assessment and Linkages**.

It is important that the materials community identify some common ground that can provide the basis for moving forward with a clear, crisp message on environmental technology strategies. The professional societies represented at the workshop agreed that a greater awareness is needed of the role of materials in environmental technology and sustainability. They also indicated that despite a variety of initiatives within their societies, more must be done to incorporate industrial ecology and sustainability strategies into their committees and activities. Focused collaboration among the professional societies will be important in making this happen. Strategies must then quickly translate into specific R&D initiatives and technology implementation. Only then will we begin to see change.

To obtain the full report of the Federation of Materials Societies (FMS) and the MatTec Environmentally Conscious Materials Working Group (EMAT) workshop entitled, **Strategy For Enhancing the Materials Role in Environmental Technology and Sustainability: Assessment and Linkages**, contact Betsy Houston, Executive Director of FMS (phone: 202/296-9292; email: betsyhou@ix.netcom.com).

IMPLEMENTATION OF ISO14001 AT THE SAVANNAH RIVER SITE

Sharon L. Marra and Ronald D. Reeves
Westinghouse Savannah River Company
Savannah River Site
Aiken, SC 29808

ABSTRACT

The Department of Energy's Savannah River Site (SRS) in Aiken, SC recently received ISO14001 certification. ISO14001 is an internationally recognized standard that delineates the elements of an effective environmental management system (EMS) and enhances environmental stewardship. SRS preparations for ISO14001 certification involved a comparison of existing programs to the requirements of the standard. Gaps in the program were identified and work initiated to fill those gaps. Primarily, these deficiencies were related to documentation of the SRS EMS and employee training. Certification was granted after an extensive review by a team of independent auditors. The review included personnel interviews, documentation reviews, and work practice observations. An overview of the preparation process as well as the independent review will be presented.

INTRODUCTION

The Savannah River Site (SRS) is a 310 square mile U.S. Department of Energy (DOE) facility located in Aiken, SC bordering the Savannah River. SRS was constructed in the early 1950's to produce tritium and plutonium, basic materials used in nuclear weapons. The last of five reactors was shut down in the late 1980's. However, other facilities are operating in support of the current SRS mission to reduce nuclear dangers, clean up the site and manage the wastes SRS has produced. The environmental restoration mission of the site is to clean up 430 waste sites.

One of the largest efforts at SRS is the management and treatment of 30 million gallons of liquid high level radioactive waste. The high level waste system treats the waste to obtain a low and high level waste fraction. The high level waste fraction is then immobilized in borosilicate glass which is poured into stainless steel canisters in the Defense Waste Processing Facility (DWPF). The canisters are ten feet high and two feet in diameter. The canisters are temporarily stored at SRS and eventually will be sent to a federal repository.

In 1995, SRS senior management voluntarily chose to pursue certification to the new ISO14001 environmental management standard. This decision was based on the belief that implementing the elements of ISO14001 will increase cost effectiveness and environmental compliance efficiency. In addition, ISO 14001 certification provides evidence, by an independent third party, to stakeholders and regulators that SRS is committed to an environmentally safe site, pollution prevention, environmental compliance, and continuous improvement.

ISO14001

The International Organization for Standardization (ISO) is comprised of approximately 120 member countries. ISO is an organization that promotes the development of international manufacturing, trade, and communication standards.

ISO14000 is a series of environmental management standards and guidelines. United States agencies that favor ISO14000 as a voluntary standard include the Environmental Protection Agency (EPA), Department of Energy (DOE), Department of Commerce, and Department of Justice. In fact, EPA participated on the team that drafted the standards.

ISO14001 is the Environmental Management System Standard.[1] It is a voluntary standard containing elements of environmental management systems based on the implementation of good business practices rather than command-and-control compliance.[2] The basic environmental management system principles or elements are:

- Policy - The organization must define a policy and ensure commitment.
- Planning - The organization must formulate a plan for fulfilling the environmental policy.
- Implementation and Operation - The organization must develop capabilities and support mechanisms necessary to achieve its environmental policy, objectives and targets.
- Checking and Corrective Action - The organization must measure, monitor, and evaluate environmental performance.
- Management Review - The organization must review and continually improve the environmental management system with the objective of improving overall environmental performance.

Once an organization has implemented the elements of ISO14001, self declaration or third party certification can follow. Self declaration means an organization has determined by some evaluation process, other than a registration audit, that it is conforming with all of the requirements of ISO14001. This can be achieved by completing a gap analysis and implementing corrective actions. The gap analysis and corrective actions can be performed by outside consultants or employees with appropriate knowledge.

As an alternative to self-declaration, a registrar can certify an environmental management system after an independent team certified by the American National Standards Institute/Registrar Accreditation Board performs an ISO14001 audit and recommends certification. This option would add more credibility than self-declaration.

IMPLEMENTATION AT SRS

Prior to initiating efforts for ISO14001 certification, SRS management needed to define the scope of the certification. SRS could have pursued certification for the entire site or just for individual facilities. Due to the existing programs already being used and the site organizational structure the decision was made to pursue certification of the entire site. The certificate would then include the following organizations: DOE Savannah River, Westinghouse Savannah River Company (WSRC), Bechtel Savannah River Incorporated (BSRI), Babcock & Wilcox (B&W) at SRS, British Nuclear Fuels Limited (BNFL) at SRS, University of Georgia, Savannah River Natural Resources Management and Research Institute, and Wackenhut Services Incorporated (WSI) at SRS. This management commitment and decision on scope was made in December 1995.

See Figure 1 for a timeline of ISO14001 implementation/certification activities.

Gap Analysis and Correction of Deficiencies

Once the decision was made to pursue ISO14001 certification a 15 member team representing all affected divisions on site was organized in January 1996. A gap analysis was performed over a four month period (see Figure 1) to determine how well existing programs met the ISO14001 standard. The gap analysis showed that SRS already had good environmental stewardship programs in place but enhancements were needed to fully conform to the standard. A

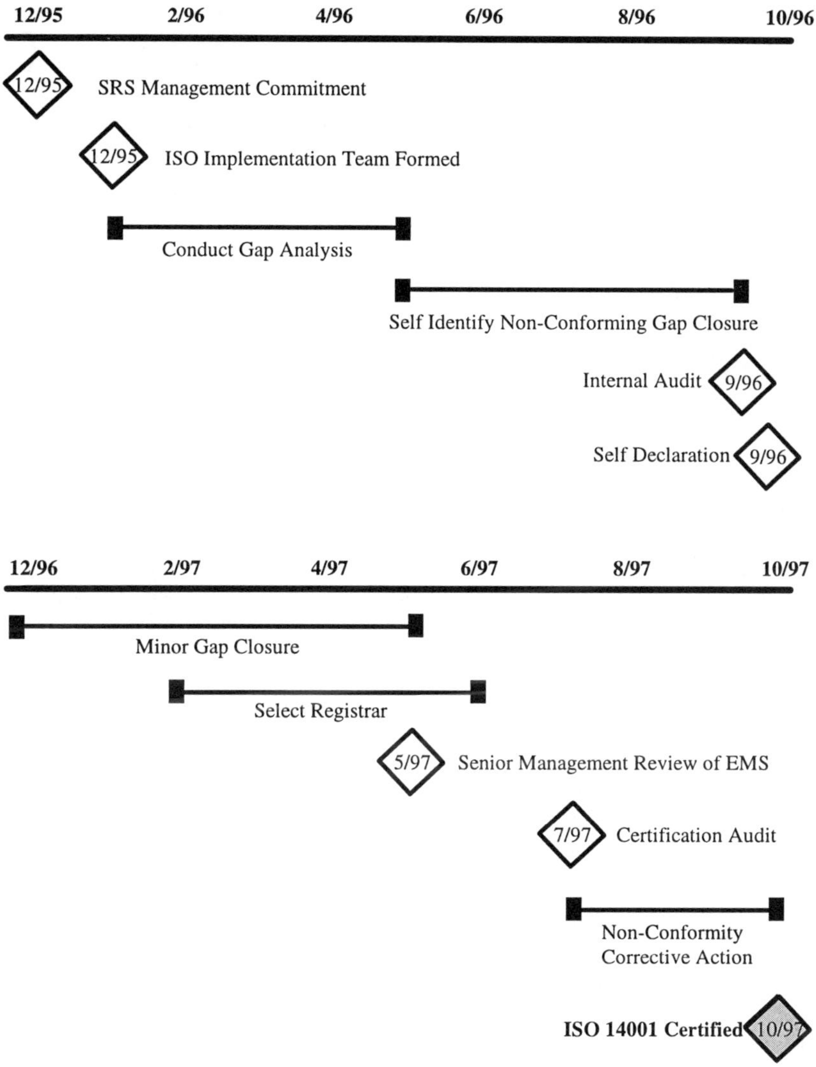

Figure 1. Timeline for Implementation of ISO14001 at the Savannah River Site.

conformance plan was put together which identified the actions needed to come into conformance. The following actions were identified:

- Policy - A written environmental policy statement signed by SRS senior management was needed.
- Planning - A new chapter was needed in the Environmental Implementation Plan which summarized environmental aspects of activities. Environmental review aspects needed to be broadened to include facility activities not just project activities. An introduction by the site manager, emphasizing commitment, was needed to enhance the already existing Waste Minimization and Pollution Prevention Plan.
- Implementation and Operation - The Environmental Management System (EMS) needed to be communicated to all site personnel and ISO14001 awareness training was needed. A "road map" describing how documents and programs fit into the EMS was needed.
- Checking and Corrective Action - Development/implementation of procedures for identifying and handling key environmental records was needed. A revision to the existing self assessment program was also needed to assess the environmental management system.
- Management Review - Development of procedures that establish routine periodic senior management review of the environmental management system was needed.

From June until September 1996, SRS worked to complete the above actions to conform to the ISO14001 standard. The existing ISO14001 team lead the corrective action implementation. Subgroups were formed to work on individual items such as policy and training.

A site environmental policy statement was adopted and signed by SRS senior management. This policy statement was distributed to all personnel. The policy commits SRS to pollution prevention, regulatory compliance, and continuous improvement. The objective of the policy is as follows:

"The objective of this policy is to ensure every employee of the DOE Savannah River Operations Office (SR), all contractors, subcontractors, and other agencies performing work at the Savannah River Site (SRS) do so in accordance with DOE order 5400.1 and the mission, the vision, the core values, and the environmental goals and objectives of the Savannah River Strategic Plan."

A large effort was initiated to provide a roadmap to describe how the documents and programs fit into the EMS. This document includes sections for each of the elements of the EMS. Within each section, key reference and implementing documents are listed. The SRS policy statement is included along with a matrix of SRS organizational missions and environmental activities and responsibilities. This document was provided to all SRS managers prior to the certification audit (see below).

The most time consuming and intensive effort for implementation was employee awareness training. While SRS employees certainly consider environmental stewardship in their day to day activities, an understanding of the relationship of their activities to the environmental management system (EMS) and to ISO14001 was needed. Each employee was expected to know that site management had committed SRS to good environmental stewardship and to know the environmental aspects of their job. All personnel received a briefing on ISO14001 and received a copy of the SRS policy statement that also included questions and answers on ISO14001. Job specific training was also provided that was intended to make personnel think about the environmental aspects of their job. A matrix similar to that shown in Figure 2 was used to initiate discussion. Personnel were asked to list the activities they perform or the service they provide. The potential environmental aspect of that activity and then the potential environmental impact of that activity were then also listed. This exercise allowed employees to think about their jobs and

Activity/Product/ Service	Potential Environmental Aspect	Potential Environmental Impact
Shipping crude oil in tanker	Oil spill	Water pollution, dead birds, fish, etc.
Manage high-level waste	Release high-level waste to the environment	Soil, air and water contamination. Adverse health effects.
Equipment decontamination	1) Waste generation 2) Reuse of equipment	1) Waste disposal 2) Reduce disposal, save resources

Figure 2. Example of Environmental Aspects Matrix.

how their jobs may have an impact to the environment even if they do not perform field activities (e.g. engineering, budget planning, etc.).

CERTIFICATION PROCESS

In September 1996, SRS made a self declaration that they were in compliance with the ISO14001 standard. An internal audit was performed with assistance from personnel from other government owned, contractor operated (GOCO) DOE sites to provide an independent review of the SRS program prior to self declaration.

During the first several months of 1997, activities were completed to close the gap on ISO14001 conformance. This included completion of employee awareness training and distribution of the EMS description document. Also, in the first part of 1997, the process for selection of a registrar for certification was initiated. Prior to selection of the registrar, all bidders visited SRS to discuss the scope of certification, receive a tour of the site, and review procedures for submitting bids to SRS.

In May 1997, SRS senior management performed a review of the EMS to ensure the current policy was still applicable and that selected targets and objectives were adequate for an effective EMS. A registrar was selected in June 1997. These independent certified auditors then visited SRS to receive documentation describing site activities and to conduct a pre-audit review to allow SRS personnel to become familiar with the process. One of the facilities selected for this pre-audit review was the Defense Waste Processing Facility (DWPF).

In July 1997, the certification audit was conducted. Facility contacts were assigned to accompany auditors to the field and assist with the logistics of arranging interviews, documentation reviews, etc. The pre-audit review conducted in June was extremely valuable in allowing SRS to prepare for the certification audit in these areas. The auditors reviewed documentation and visited several facilities on site to observe activities and conduct interviews.

Due to an identified non-conformity, SRS didn't receive certification immediately following the audit. The major non-conformity found was that SRS didn't have an audit program that adequately defined all elements of ISO14001. Several observations were also noted. SRS followed up on the non-conformity by incorporating an audit program for the EMS into the existing self assessment program. New criteria were added to the program to evaluate the EMS and compliance with ISO14001. Following the completion of this action SRS received the ISO14001 certificate in October 1997.

This certificate is good for three years, however, an annual surveillance audit is performed to ensure conformance. The first surveillance audit was performed in February 1998. This audit yielded 14 observations but no non-conformities. If the audit team finds a non-conformance the certificate can be revoked. SRS must also show progress in responding to any observations that are noted. SRS is currently working on the observations noted during the February 1998 surveillance.

CONCLUSIONS

SRS became one of the first sites in the DOE complex to receive ISO14001 certification. The receipt of this certification, in the Fall of 1997, validated an almost two year effort. This effort involved an evaluation of existing programs against the standard followed by actions which were needed to bring SRS into conformance. SRS will continue to monitor its performance as an indication of the value of ISO14001 implementation.

ACKNOWLEDGMENT

This paper was prepared in connection with work done under Contract No. DE-AC09-96SR18500 with the U.S. Department of Energy.

REFERENCES

1. ISO14001: 199X, Environmental Management Systems - Specification with Guidance on Use, Committee Draft Prepared by ISO/TC207/SC1/WG1: February 1995.

2. D. Sayre, **Inside ISO14000 - The Competitive Advantage of Environmental Management**, St. Lucie Press, Delray Beach, FL, 1996.

FIRST COMMERCIAL U.S. MIXED WASTE VITRIFICATION FACILITY: PERMITS, READINESS REVIEWS, & DELISTING of FINAL WASTEFORM

John B. Pickett and Stephen W. Norford
Westinghouse Savannah River Co., Aiken, SC 29808

Glenn A. Diener
GTS Duratek, Columbia, MD 21046

ABSTRACT

Westinghouse Savannah River Co. (WSRC) contracted GTS Duratek (Duratek) to construct and operate the first commercial vitrification facility to treat an F-006 mixed (radioactive/hazardous) waste in the United States. The permits were prepared and submitted to the South Carolina state regulators by WSRC - based on a detailed design by Duratek. Readiness Assessments were conducted by WSRC and Duratek at each major phase of the operation (sludge transfer, construction, cold and radioactive operations, and a major restart) and approved by the Savannah River Department of Energy prior to proceeding. WSRC prepared the first "Upfront Delisting" petition for a vitrified mixed waste. Lessons learned with respect to the permit strategy, operational assessments, and delisting from this "privatization" project will be discussed.

INTRODUCTION

The Savannah River Site (SRS) is a 300 square mile complex which has been dedicated to the production of national defense materials since 1952. The SRS is operated by the Westinghouse Savannah River Co. (WSRC) for the U.S. Department of Energy (DOE). One of the legacies of the production activities was a mixed waste plating line sludge, generated in the Reactor Materials Area (M-Area). This plating line sludge was generated during the wastewater treatment of effluents from the nickel plating of depleted uranium cores, which were subsequently irradiated in the site's reactors to form plutonium-239.

The M-Area plating line sludge is a hazardous waste (F-006 listed waste) and since it contains a significant amount of depleted uranium, it is a "mixed" waste as defined by the Federal Facilities Compliance Act of 1992. In order to comply with the RCRA (Resource Conservation and Recovery Act) laws, the waste has to be treated to meet the Land Disposal Restriction (LDR) regulations [1]. In addition, since the sludge had been stored in on-site tanks longer than allowed by the LDR's, the sludge was subject to a 1991 Federal Facilities Compliance Agreement (FFCA) between the US Environmental Protection Agency (EPA), Region IV and the DOE-Savannah River.

The original concept to stabilize this waste was to construct a small, permanent (75,000 gallon/year) cementitious treatment facility in M-Area. This facility would have treated the on-going generation of sludges from the M-Area facilities, and worked off the stored waste over a 10 year period. However, in 1991, the DOE Savannah River (DOE-SR) directed WSRC to place the M-Area production facilities in standby. Without the on-going generation of new wastes, the construction and operation of a permanent, 10 year life facility was not warranted. WSRC proposed to the DOE-SR and the EPA, Region IV that a temporary facility, owned and operated by an independent sub-contractor, be utilized to treat the stored sludge. This approach was accepted in early 1992, and a Request For Proposals (RFP) was issued June 2, 1992. The technology to be used was not specified, just that the final wasteform had to meet the LDR treatment standards. All of the cementitious and vitrification treatability test results [2] that had been conducted by the Savannah River Technology Center (SRTC) were included with the RFP. A contract was awarded to GTS Duratek of Columbia, MD in November, 1993 - who proposed to treat the M-Area sludge using vitrification in a borosilicate matrix. The was/is the first commercial vitrification of a mixed waste in the United States.

The M-Area Vendor Treatment Facility (VTF) was selected by the DOE as one of five privatization "successes", and presented as such to the US Congress in the spring of 1996. "Privatization" is usually defined as an operation in which a contractor has a direct contract with the DOE, furnishes their own capital funds, and builds and operates the facility. The M-Area VTF has been included in this definition, since although Duratek was sub-contracted by WSRC (instead of DOE), GTS Duratek designed, used its own funds to construct, and has and is operating the VTF. This paper describes the complex interactions between the contractor (WSRC), the DOE-Savannah River (SR) and DOE-Headquarters (HQ), and the sub-contractor that are required to conduct such a "privatized" operation.

NATIONAL ENVIRONMENTAL POLICY ACT - NEPA

During the time the contract was being negotiated and awarded, WSRC prepared an Environmental Assessment (EA), to meet the requirements of the National Environmental Policy Act (NEPA). Since the qualified offerors had proposed both cementitious and vitrification stabilization, both of those options were included in the EA. The EA was initiated in November of 1992, and submitted to the DOE-HQ for approval in October of 1993. The EA included all of the proposed options; a permanent cement facility in M-Area, a permanent centralized cement facility at the SRS, temporary cement or glass treatment a by sub-contractor in M-Area, and a "No Action" option. The total Life Cycle Cost estimates for all of the options were prepared by WSRC Systems Engineering. The preferred option in the EA was treatment by a subcontractor, using either a temporary cement or vitrification facility.

The EA utilized the generic descriptions provided by the offerors to calculate the on and off-site hazardous chemical and radiological risks via the preparation of a Preliminary Hazards Assessment (PHA) by Savannah River Technology Center (SRTC). The radiological and chemical air emission concentrations from a worst case release of the waste sludge were modeled by SRTC, based on the detailed analyses of the wastes that WSRC had completed prior to the Request for

Proposals. Other exposure hazards were based on the chemicals proposed for the treatment technologies. The radiological components in the waste sludge resulted in the VTF being classified as a Low Hazard Nuclear facility. The worst case predicted airborne hazard, due to any chemical constituent present in the sludge or used in the process, resulted from a spill of a process chemical; sodium hydroxide.

When the contract was awarded to Duratek (who proposed a vitrification process) on November 4, 1993, the DOE-HQ requested that the EA be revised to indicate that vitrification by a sub-contractor was the preferred alternative. A revised EA was transmitted from the DOE-SR to DOE-HQ on December 16, 1993. The DOE-HQ was also requested to issue an exemption to the NEPA regulations to allow Title II design to start, prior to the approval of the EA. The DOE-HQ granted the exemption request on January 12, 1994, and Duratek immediately initiated preparation of the Detailed Design. The EA design exemption by DOE-HQ was critical, because WSRC/ DOE-SR had an FFCA commitment to submit the required permits to the regulatory authorities by June 30, 1994.

The revised EA [3] was approved and a Finding of No Significant Impact (FONSI) was issued by DOE-HQ on August 1, 1994. Overall, this took almost 2 years to prepare and approve. The August, 1994 approval timing of the FONSI was also a critical path item, as Duratek had to initiate the procurement of long lead time equipment to meet the contractual construction requirements, and this was not allowed until the EA/FONSI was approved.

NATIONAL EMISSION STANDARDS FOR HAZARDOUS AIR POLLUTANTS (NESHAPS)

Under the federal NESHAPS regulations, the EPA must issue a NESAHPS permit, including an "approval to construct", if the hazardous emissions from a proposed facility managing radioactive materials could exceed an annual dose of 0.1 mrem/yr. to the affected public. This approval must be issued before construction of such a facility may be initiated. Initial calculations prepared by WSRC indicated the vitrification facility would exceed the 0.1 mrem/yr. limit. Therefore, WSRC transmitted an "alternative source term calculation" to the EPA, Region IV, which demonstrated that the uranium compounds in the M-Area waste would not volatilize under the vitrification conditions, and that the annual dose would not exceed the 0.1 mrem/yr. limit. Approval of the federal NESHAPS exemption by the EPA was a another critical path item. The "alternative source term calculation" request was granted by the EPA June 22, 1994 - which in conjunction with the FONSIE issuance in August of 1994 - allowed Duratek to commence procurement of long-lead time items, and keep the construction schedule on track to meet the FFCA construction commitments.

The calculated potential radioactive emissions were 0.22 mrem/yr., if the baghouse and HEPA filtration systems were not included in the calculations. This exceeded the 0.1 mrem/yr. limit, so Continuous Air Monitoring (CAM) equipment was required. WSRC utilized existing CAM equipment to conduct this monitoring. The CAM had been previously procured for one of the shut-down operational M-Area buildings, and it was modified to meet the vitrification off-gas conditions. Duratek was requested to provide the off-gas connection piping. WSRC installed the CAM, operated and maintained it, and provided the radiological results to the regulator. WSRC and Duratek had to coordinate closely to maintain the correct humidity and temperature of the off-gas feed to the CAM.

PERMITTING

All of the permits required for operation of the VTF are the responsibility of WSRC. They include an Industrial Wastewater Treatment permit, the NPDES permitted outfall to which the wastewater effluents discharge, and an Air Quality permit; for which the regulator was the South Carolina Department of Health and Environmental Control (SCDHEC). The permits were prepared by WSRC personnel based on the "Detailed Design" (DD) provided by Duratek. The DD was the first deliverable under the contract, and was intended to be as detailed as a "Title II" design for a line-item capital project. As mentioned previously, Duratek could not initiate the DD until the NEPA exemption for design was approved by DOE-HQ on January 12, 1994, and they had 90 days to submit a draft DD, and a total of 150 days to submit the final DD to WSRC. They submitted both draft and final DD's on schedule. WSRC initiated preparation of the Wastewater and Air Permit applications based on the draft, but WSRC had only 2 weeks (June 15 to June 30, 1994) to prepare the final permit applications, obtain WSRC and DOE-SR site approvals, and submit to SCDHEC to meet the FFCA commitment. The permits also had to be signed/certified by a Professional Engineer (PE) licensed in SC. This was provided by Duratek, as the permits were based on Duratek's DD.

The next key contract deliverable was to initiate construction within 15 days after all of the permits were approved by SCDHEC. This was a "floating" date, since the time for regulator's approval could only be roughly estimated. This turned out to be an important "float", since a few months after the Wastewater (WW) and Air permit applications were submitted, Duratek discovered technical problems with the proposed technique to control NO_x emissions. WSRC requested SCDHEC to delay consideration of the WW permit, while Duratek and Catholic University of America (CUA) conducted additional bench and full scale off-gas treatment tests.

WSRC and SRTC also conducted an independent technical review of the proposed new off-gas treatment process, and WSRC participated in a full scale demonstration of the revised process at CUA's 1 ton/day melter in December, 1994. The bench scale and full scale results were incorporated into a revised DD by Duratek; the WW and Air permit applications were revised by WSRC; and the permits were re-submitted to SCDHEC March 27, 1995. The permits were approved by SCDHEC July 13, 1995.

Duratek initiated construction of the VTF one day later by installing the melter support frame into a previously constructed "Butler type" building. Construction of the entire VTF was completed January 22, 1996 (approximately 10 days later than the 180 day contract deliverable). It should be noted that the contract assumed/suggested that a "temporary" facility would be provided by the sub-contractor. However, Duratek decided to construct a much more permanent facility in the expectation of treating additional mixed waste streams after completing treatment of the M-Area sludge. Construction of the VTF in the 180+ day time-frame was a very significant accomplishment.

READINESS ASSESSMENTS

The startup requirements for the VTF were based on the DOE Order 5480.31, "Startup and Restart of Nuclear Facilities". The VTF was defined to be a non-reactor nuclear facility and had been determined to be a "Radiological Facility" by the Hazards Assessment. The VTF was a new facility, but since a "Radiological Facility" is less than Hazards Category 3, only a Readiness Assessment (RA) was required, vs. a full Operational Readiness Review (ORR).

In order to conduct the Readiness Assessment, WSRC had to define exactly how the VTF construction and operation would be assessed vs. the applicable DOE orders and Westinghouse procedures. This was accomplished by the preparation of a Readiness Assessment Action Plan (RAAP). WSRC used and complied with the DOE Order 5480.31, and the WSRC "Startup and Operational Readiness Manual, 12Q" to prepare the RAAP. The RAAP was based on the detailed design of the VTF, the Safety and Health Plan for the VTF (prepared by Duratek, and approved by WSRC Procurement Safety oversight), and all of the operational and maintenance procedures for the VTF (supplied by Duratek). Preparation of the RAAP took approximately one year, with the total dedication of one of the authors of this paper (GAD). The RAAP was approved in March, 1996 by WSRC management (Excess Facilities and Spent Fuel Division), with concurrence by DOE-SR (Reactor Division).

Two of the main conclusions of the RAAP were:

1. The VTF could be constructed under the design and QA requirements for commercial chemical facilities, rather than the requirements of DOE Order 6430.1A, "General Design Criteria for Reactors or Non-Reactor Nuclear Facilities". This was due to the fact that since Duratek owned the completed facility, it was not a Government facility, and as such, 6430.1A did not apply. The design requirement for nuclear facilities are EXTENSIVE, and in the authors' opinion, would have increased the VTF design and construction costs 2-3X.

2. The VTF was constructed under a Purchase Requisition, and the requirements of DOE Order 4700.1 for a DOE cost or capital project construction were not applicable. Again, the authors' estimate the cost savings at an additional 2-3X of Duratek's cost of construction.

The Readiness Assessment was divided into phases, each of which were conducted jointly by WSRC and Duratek, before that operational phase began.

Phase I, Sludge Transfer

Duratek had decided to modify six existing 35,000 gallon tanks, for their use as sludge feed make-up, feed, and storage of off-gas condensate solutions. In order to modify the tanks, the tanks had to be emptied of the mixed waste sludge. The first Readiness Assessment (RA) phase was conducted on all of the Radiological and Operational Procedures that Duratek had developed to remove and transfer the sludge to one of the nearby 500,000 gallon tanks. The RA was approved May 9, 1995 by WSRC, and Duratek immediately started the sludge transfer activities. It should be noted that these activities did not involve any "construction", or "major modifications" to existing equipment, as the Wastewater Construction Permit had not yet been approved.

Phase II, Construction

The RA phase allowing Duratek to "Start Construction" was approved by WSRC management July 10, 1995. The approved WW permit from SCDHEC was received July 13, 1995. Duratek started construction July 14, 1995.

Phase III, Cold Chemical Operations, and Preparation of First Feed Batch

When construction was completed January 22, 1996 the major phase of the RA was initiated. This phase considered all aspects of the melter and off-gas treatment operations, and transfer of the radioactive feed batches to the melter. Three months were allowed for WSRC to conduct the RA and obtain DOE approval, although it was hoped that it could be completed sooner. Duratek was in a "stand-by" mode during this period, assisting WSRC in the RA, and maintaining the operational readiness and training of their operations personnel. The Readiness Assessment was approved by DOE-SR April 17, 1996. Duratek immediately started preparation of the first radioactive sludge feed batch and initiated cold chemical feeding to the melter, using a temporary feed tank. One additional hold was built into the Readiness Assessment approval to begin full radioactive operations, in that WSRC planned to conduct a Conduct of Operations assessment prior to radioactive feed to the melter.

Phase IV, Conduct of Operations Self Assessment Prior to Radioactive Feed

WSRC completed a Conduct of Operations Self Assessment in July, 1996, which confirmed the readiness to commence radioactive operations. This allowed Duratek full authority to start vitrifying the M-Area sludge.

Phase V, Addendum Self Assessment

During the cold chemical check-out of the melter and ancillary equipment (April to September 1996) Duratek experienced significant equipment problems, mainly associated with the Duramelter™ 5000. In order to ensure that the design and procedural modifications that Duratek implemented during the "Cold Feed" period - to address those problems - WSRC conducted an "Addendum Self-Assessment" prior to radioactive feed to the melter. This was approved by WSRC October 15, 1996, and Duratek started radioactive feed to the melter 3 days later.

It should be noted that while WSRC conducted the self-assessments, Duratek personnel were intimately involved in all aspects of the assessments. In some cases this involved responding to comments on operations procedures, and modifying the procedures; in other cases it involved in-the-field observations of Duratek personnel in all operational activities.

Phase VI, Restart Readiness Review

After operating for approximately 5 months on radioactive feed, Duratek determined that the original melter was no longer reliable and ceased operations March 27, 1997. Duratek removed the original melter and installed a re-designed melter. WSRC then conducted a "re-start" Readiness Review (RR) to verify satisfactory replacement of the melter, and that the operations personnel had maintained their training and professionalism during the outage. The Restart Review was completed November 11, 1997, Duratek initiated cold chemical feed in late November 1997, and started radiological feed December 10, 1997.

DELISTING PETITION

WSRC has also prepared an "Upfront Delisting Petition" [4] for the vitrified M-Area waste. This is first delisting petition submitted to the EPA for a vitrified mixed waste. Delisting should result in significant cost savings.

Cost Savings — Disposal Cost, $/cu. Ft.

SRS On-Site Disposal LLW	SRS On-Site Disposal Mixed Waste	Off-Site Disposal (Envirocare) LLW	Off-Site Disposal (Envirocare) Mixed Waste
$30-50	$700*	$10	$40 - 100**

* Assumes a SRS disposal vault constructed; ** volume dependent

Assuming that vitrification of the M-Area sludge will result in ~150,000 gallons, or ~20,000 cu. ft., the disposal cost savings range from approximately $13,000,000 to $1,800,000. Preparation of, and analyzing the final vitrified waste for the delisting petition, will cost $150,000 to $250,000. The cost savings for on-site disposal are enormous, but not realistic, as SRS has no plans to build a mixed waste disposal vault on-site. However, if delisted, the glass could be disposed on-site, which is the preferred option. Final disposal at the SRS LLW disposal vault will save approximately $750,000 (a delta cost of $50/cu.ft. x 20,000 cu. ft. - $250,000 to delist) vs. off-site mixed waste disposal.

Constituents of Concern, and Delisting Levels

The constituents of concern in the final vitrified waste were based on all of the total constituent analyses conducted on the sludge and vitrified sludge, and Toxicity Characteristic Leaching Procedure (TCLP) analyses conducted on the glass. No organic constituents were detected in the glass (as expected). Those inorganic constituents which were detected above the analytical detection limit were designated as "Constituents of Concern".

Constituents of Concern	Delisting Level # TCLP, mg/L	Constituents of Concern	Delisting Level # TCLP, mg/L
NH_3	100	F	400
As	5	Pb	1.5
Ba	200	Hg	0.2
Be	0.4	Ni	10
Cd	0.5	Se	5
Cr	10	Ag	20
Zn	1000		

The delisting level assumes 20,000 cu. ft (~750 cu. yd.) of glass to be disposed in one year, which provides a Dilution-Attenuation Factor (DAF) of 100.

The EPA maximum contaminant Drinking Water Standards (DWS) is multiplied by the DAF of 100 to obtain the TCLP concentration limit for the final waste. For example, the DWS for nickel is 0.100 mg/L times a DAF of 100, which gives a TCLP concentration limit of 10 mg/L for the final glass. It should be noted that all of the TCLP analyses conducted on the vitrified M-Area waste to date have resulted in TCLP levels less than 0.5 mg/L for nickel. This level is 20 times lower than the level needed to delist.

CONCLUSIONS

Westinghouse Savannah River Co., the Department of Energy, and GTS Duratek have worked closely together to demonstrate the first commercial vitrification of a mixed waste plating line sludge in the United States. The keys to a successful, cost effective "privatization" project are:

- preparing a very tight contract Specification and Scope of Work,
- utilizing commercial construction standards, when appropriate for Low Level radioactive waste treatment, and
- eliminating the very costly government project oversight, which can reduce final design and construction costs by 70-80%.

It should be noted that the day-to-day operations costs for such a facility will not differ significantly for a private sub-contractor vs. operation by a Government Owned/Government Operator (GOCO). This is due to the fact that a subcontractor operating on or at a DOE site will have to follow most, it not all, of the DOE operational Orders. The M-Area vitrification project has demonstrated that DOE, WSRC (as the GOCO), and a private sub-contractor can successfully treat a complex mixed waste, using state-of-the-art vitrification technology.

REFERENCES

1. U. S. Environmental Protection Agency, "Land Disposal Restrictions for Third Scheduled Wastes, Final Rule", *55 Federal Register 22627* (1990).

2. J. B. Pickett, J. C. Musall, and H. L. Martin, "Treatment and Disposal Of A Mixed F006 Plating Line Sludge At the Savannah River Site", *Proceedings of the Second International Mixed Waste Symposium, Baltimore, MD,* 1.3.1 (1993).

3. U. S. Department of Energy (DOE). "Environmental Assessment, Treatment of M-Area Mixed Wastes at the Savannah River Site", DOE/EA-0918, June, 1994, Savannah River Operations Office, Savannah River Site, Aiken, SC 29808.

4. U. S. Department of Energy (DOE). "Upfront Delisting Petition for Vitrified M-Area Plating Line Wastes", WSRC-TR-96-0244, September, 1996, Savannah River Operations Office, Savannah River Site, Aiken, SC 29808.

Volatilization Studies of a Lanthanide Lead Borosilicate Glass

R.F. Schumacher, D.S. McIntyre, D.K. Peeler, and J.M. Pareizs
Westinghouse Savannah River Company
Savannah River Technology Center
Aiken, SC 29808

ABSTRACT

The gravimetric weight loss of lead oxide and boric oxide from lead lanthanide borosilicate type glasses was investigated by remelting glass grains in small platinum crucibles at varying temperatures and times. A simple linear equation was developed to describe the volatilization rate of the glass in the crucibles with changes in temperature. The holding time at temperature does not have a very strong impact on the calculated volatilization rate when the time is within a two day period. The volatilization rate was calculated and compared to values obtained from the pilot facility bushing melter off-gas studies at the same temperatures. The calculated volatilization rate from crucibles was approximately twice that measured in the bushing melter off-gas which indicates that either the bushing may have a colder air/glass interface temperature or possibly the off-gas investigations did not capture all of the volatile material.

INTRODUCTION

Special glass compositions and melter systems are being developed by the Savannah River Technology Center (SRTC) to immobilize a stored nitric acid solution containing relatively large quantities of lanthanides and the radioactive isotopes: americium-243 and curium-244. These materials are presently stored in Tank 17.1 at the Separations Canyon in F Area at the Savannah River Site. Immobilization of this material will permit shipment of the material to the Oak Ridge National Laboratory - Heavy Element and Advanced Neutron Facility where it will be transformed by nuclear reactions into other useful transplutonium isotopes. The glass compositions for this program were based on a lead borosilicate optical glass composition developed during World War II by Loeffler.[1] This glass system was initially selected based on its ability to incorporate large quantities of lanthanides and potentially actinides into the glass network structure.

These glasses melt sharply at normal glass melting temperatures and have unusually low viscosity making them desirable from a vitrification processing standpoint. Research has shown that increasing levels of lanthanide oxides acts as a flux or softening agent and reduces the melt viscosity.

One of the melter systems under development at SRTC is a closed, platinum rhodium alloy box which is heated by passing large electrical currents through the platinum alloy (resistance heat). This type of system was initially developed for use in remelting glass marbles for the manufacture of continuous glass fibers[2] and an extensive body of patent and journal literature exists for this area of glass manufacture. This commercial technology was modified and is being further developed for remote operation with the lead-lanthanide borosilicate glasses using a coupled liquid-frit feed stream.[3] This modified remelt bushing system may be employed for the remote melting and casting of the Am/Cm glass into small cans in the Separations Canyon in F Area.

The volatilization of lead oxide from the Loeffler glasses was considered early in the development of the laboratory process hazards review. One of the major concerns was the volatilization of lead compounds from the glass melt to the off-gas system and the hazards this might cause for personnel working with the system. The findings of J. Matousek [4] were employed to provide an initial estimate of the volatility of lead from a typical alkali lead silicate glass. A rough order of magnitude, estimate of 20 mg/sq.cm/hr for the volatilization at 1400°C of the lead oxide was chosen based on an alkali lead glass containing about 20 wt% lead oxide.

It has been shown [5-7] that the volatilization of lead oxide is significantly increased by increasing temperature, but the rate of volatilization decreased with increasing soak time. Most authors also agree that under equal conditions, the volatilization of lead increases with the fraction of lead in the glass or with reduced SiO_2 content in the glass. The volatilization products from lead borosilicate glasses consist mainly of lead oxide while other volatile constituents, e.g. alkali and boron compounds, begin to volatilize at temperatures above 1200°C but at lower rates than the lead oxide. Volatilization appears as a complex process of physical chemistry, in which three simultaneous sub-processes may take part in series:

a. the volatile component diffuses through the melt to the surface of the glass,
b. the volatile component evaporates from the surface of the melt, and
c. the diffusion of the volatile component through the gas phase from the melt surface into the adjacent gaseous phase.

Diffusion in the gaseous phase is usually assumed to proceed much faster than the other processes, a and b, which then places these processes (a & b) as the rate controlling processes. The total loss of the volatile component can be expressed as:

$$M(t) = 2\ Co \sqrt{D\,t/\pi} \qquad \text{Eq. 1}$$

where M(t) is the mass of the volatile component after time t, Co is the initial concentration of the volatile component, and D is the diffusion coefficient of the volatile component. When there is little convection in the glass and temperatures are low, diffusion to the surface should be rate controlling. As temperatures increase, surface evaporation will have more influence and may dominate at extremely high temperatures. In general, the amount of lead vaporized should therefore be proportional to the square root of time. The diffusion coefficient and evaporation coefficient should be related to the reciprocal of temperature expressed in degrees Kelvin. While these theoretical relationships deserve consideration, the experimental conditions found in the present study may be too complicated for such simple relationships. The work described in this report was generally considered scoping in nature due to the complexity of relating small-scale crucible melts to large-scale processing systems.

EXPERIMENTAL CONDITIONS

Glass Preparation

Approximately 1650 grams of simulated Tank 17.3 solids* prepared from reagent grade chemicals. The Tank 17.3 solids were then mixed with the B2000 glass frit** in the proportions of 40, 35, 30, 25, and 20 wt% Tank 17.3 solids. The one kilogram powder mixtures were mixed by thorough shaking and tumbling in 2 liter plastic containers for about ten minutes prior to melting. The calculated chemical compositions of the frit, the 17.3 solids, and the resulting glasses are presented in Table 1.

	B2000 Frit	Tank 17.3 Simulant	Tank 17.3 Loading 40	35	30	25	20
Oxides	Wt%	Wt%	Wt%	Wt%	Wt%	Wt%	Wt%
La2O3	7.12	11.23	8.76	8.56	8.35	8.15	7.94
Ce2O3		12.90	5.16	4.52	3.87	3.23	2.58
Pr2O3		12.90	5.16	4.52	3.87	3.23	2.58
Nd2O3		27.35	10.94	9.57	8.21	6.84	5.47
Sm2O3		6.34	2.54	2.22	1.90	1.59	1.27
Eu2O3		1.25	0.50	0.44	0.38	0.31	0.25
Gd2O3		3.22	1.29	1.13	0.97	0.81	0.64
Er2O3		12.90	5.16	4.52	3.87	3.23	2.58
Al2O3	6.04	0.64	3.88	4.15	4.42	4.69	4.96
B2O3	9.20		5.52	5.98	6.44	6.90	7.36
CaO		0.03	0.01	0.01	0.01	0.01	0.01
Cr2O3		0.26	0.10	0.09	0.08	0.07	0.05
Fe2O3		2.19	0.88	0.77	0.66	0.55	0.44
K2O		0.08	0.03	0.03	0.02	0.02	0.02
MnO		8.34	3.34	2.92	2.50	2.09	1.67
Na2O		0.28	0.11	0.10	0.08	0.07	0.06
NiO		0.09	0.04	0.03	0.03	0.02	0.02
SiO2	44.23		26.54	28.75	30.96	33.17	35.38
ZnO		0.01	0.00	0.00	0.00	0.00	0.00
PbO	24.38		14.63	15.85	17.07	18.29	19.50
BaO	9.02		5.41	5.86	6.31	6.77	7.22
Total	99.99	100.01	100.00	100.00	100.00	100.00	99.99

Table 1. Calculated Chemical Composition of B2000 Glass Frit, Tank 17.3 Solids and Resulting Glasses .

* Feed Material currently in Tank 17.3 will be retrieved and washed according to a specific pretreatment option. The pretreated feed is to be stored in Tank 17.3

** Frit B2000 was supplied by Ferro Corporation - Cleveland, Ohio.

The materials from Table 1 were melted in a small platinum-rhodium bushing melter. The mixtures were ramped up to 1475°C and additional batch added to bring up the glass level. Problems were encountered with melting and refining of the 20 and 25 wt% Tank 17.3 loaded glasses and only half of the one kilogram batch was melted. All of the available batch for the higher loaded glasses was melted. After melting for two hours at 1475°C the bushing temperature was dropped to about 1400°C and the glass drained.

The glasses were observed to be quite fluid when they were drained into water. Samples of the "as fabricated" glass were submitted to an analytic laboratory for determination of the chemical composition by Inductively Coupled Plasma (ICP) chemical analysis.

Volatility Studies

Two distinct approaches were taken to investigate volatilization of the Tank 17.3 - B2000 system. The first approach involved simple gravimetric studies of all of the glasses (i.e., various 17.3 loadings) held under isothermal conditions for 48 hours (in addition to the 2 hours at 1475°C during initial fabrication) at temperatures ranging between 1200 to 1450°C. Based on mass loss alone, one would not be able to tell the volatile species in this system. Therefore, the heat treated samples were submitted for chemical analysis by ICP and were compared to the "as-fabricated" compositions. The second approach involved heat treating only the 40 and 35 wt% Tank 17.3 loaded glasses at 1300 and 1400°C for varying lengths of time. The objective and general experimental technique for each approach follows.

Approach No.1

The objective of this initial task was to gain insight into the volatilization as a function of temperature for the Tank 17.3 - B2000 system for a given time period. Temperatures ranging from 1450°C (the nominal operating temperature of the bushing melter for the B2000 system) to 1200°C were evaluated. The lower limit was chosen based on volatilization data from the literature dealing with lead-silicate glasses. The amount of dry glass grains placed in the weighed platinum crucible was between 40 and 50 grams. The degree of volatilization was measured by weight loss for the Tank 17.3 series of glasses. An experimental design, as shown in Table 2, was selected for this temperature - waste loading investigation.

Table 2. Experimental Design No.1 - Volatility After 48 Hours

	Tank 17.3 Solids Loading				
Temperature	40%	35%	30%	25%	20%
1450°C	X	X	X	X	X
1400°C	X	X	X	X	X
1350°C	X	X	X	X	X
1300°C	X	X	X	X	X
1250°C	X	X	X	X	X
1200°C	X	X	X	X	X

The open surface area in the platinum crucibles was calculated to be about 12.57 square centimeters. Upon removal from the furnace the samples were cooled to room temperature in a dessicator and weighed again. Total mass loss was calculated as a measure of the degree of volatilization. To obtain semi-quantitative analysis on the volatile species, heat treated samples were submitted for chemical analysis and compared to the "as fabricated" sample compositions.

It should be noted that the compositional comparison of the "as fabricated" glasses (1475°C for 2 hours) to the "heat treated" glasses will underestimate the total degree of volatilization. That is, during initial fabrication (melting) of the glasses, some degree of volatilization occurs. The volatilization during fabrication is not accounted for by these comparisons.

Approach No. 2

Since the initial baseline flow sheet targets a 35 wt% Tank 17.3 loading - B2000 glass, a more detailed evaluation of the effects of time and temperature on volatilization was undertaken.

A second glass (40 wt% Tank 17.3 loading) was also evaluated due to the potential for higher waste loadings and because it appears that the lower waste loading glasses were rather difficult to melt at 1450°C. The "as fabricated" glasses were heat treated for various times and temperatures. The degree of volatilization was determined by gravimetric analysis as in the initial study (approach No.1). The same crucibles were employed for these measurements. This will allow for the compilation of data between the two approaches. Table 3 summarizes the experimental design for this approach.

Table 3. Experimental Design No.2 - Volatility at 1400 and 1300°C with Increasing Time.

	Tank 17.3 Solids Loading at Temperature			
	1400°C		1300°C	
Time (hours)	40%	35%	40%	35%
2		Y	Y	Y
4	Y	Y	Y	Y
8	Y	Y	Y	Y
24	Y	Y	Y	Y
48 [f]	X	X	X	X

[f] Results from Approach No. 1 will be used to supplement these tests.

RESULTS AND DISCUSSION

Initial Glass Fabrication

The glasses after melting were chemically analyzed by ICP and compared to the targeted calculated compositions. The B_2O_3 and PbO were found to be consistently lower than the calculated values while the other oxides were similar. It is expected that some of the lead and boron volatilized during the melt preparation of the glasses prior to any gravimetric testing. Closer inspection of the lead oxide analyses indicated that the loss during melting was usually less than one wt%.

Approach No. 1

Samples of all the glasses were dried and weighed and placed into weighed platinum crucibles in a furnace at the selected temperatures (see Table 2). After 48 hours at the temperature, the crucibles were removed, cooled and weighed again. The weight lost due to volatilization was obtained and plotted in Figure 1.

This figure is a bar graph showing the average volatilization rate per hour from each of the five glasses with Tank 17.3 loadings at the chosen temperatures. The data point for the 20 wt% loaded glass at 1250°C appears suspect, but all the other data points appear to be acceptable within the limits of this test.

The glass volatilization did not appear to be greatly influenced by the level of the Tank 17.3 loading. Accepted theory would lead to the expectation that the highly loaded Tank 17.3 glasses would be low in volatility as they contain less lead oxide. While the glasses melted at 1400 and 1450°C may show some variation due to the content of lead and boron, it is doubtful that these differences would be statistically significant. It should also be noted that three of the glasses with the lower levels of Tank 17.3 loadings were different in appearance after the heat treatment at 1450°C. There was a slight milky opalescence within the glasses. The high level of PbO volatilization, greater than 10 wt%, may have unstabilized the glass, leading to phase separation or devitrification. The glasses treated at 1200 and 1250°C were examined by x-ray diffraction and some small level of crystallization was detected. The phase identification was extremely difficult but may have been rare earth oxides and lead silicate.

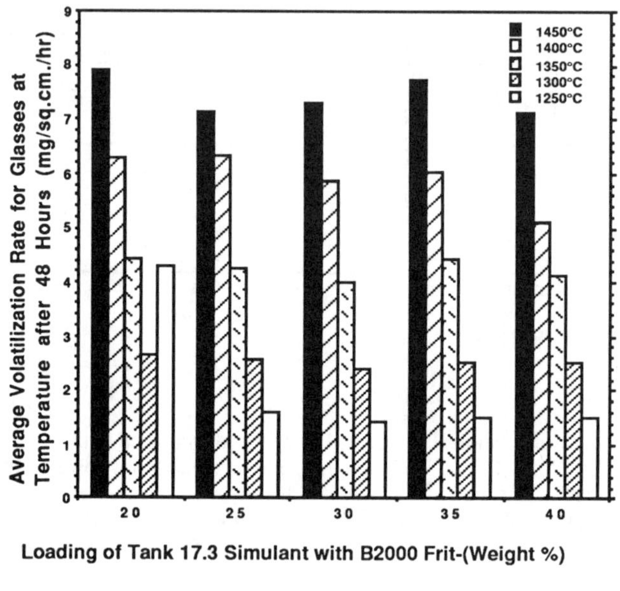

Figure 1. Measured Volatilization Rate from Tank 17.3 Glasses After 48 Hours at Temperature.

The glasses heated at 1400°C for 48 hours were analyzed chemically by ICP and compared to the "as fabricated" glass composition. A comparison was made by a mass balance calculation and this data is presented in Table 4. The numbers presented are the product of the composition in weight percent and the weight of glass used in the test.

This calculation clearly indicates that the lead and boron losses were the principle cause for the measured weight loss for the 1400°C samples. Based on the mass balance calculation, barium (the other toxic element of B2000 frit) was not volatile. Further calculations showed that the lead loss was approximately 70 to 80% of the total loss and boron was responsible for perhaps 10 to 20% of the loss. This finding will later be shown to be consistent with off-gas measurements on pilot system.

Table 4. Glass Compositions as Fabricated and after 1400°C for 48 Hours

Oxides	40Wt%		35Wt%		30Wt%		25Wt%		20Wt%	
La2O3	3.60	3.64	3.89	3.94	3.46	3.43	3.76	3.74	3.46	3.39
Ce2O3	1.02	1.04	1.05	0.99	0.76	0.78	0.72	0.74	0.53	0.52
Pr2O3	1.86	1.88	1.81	1.78	1.31	1.37	1.29	1.30	0.93	0.92
Nd2O3	4.04	4.25	3.95	3.81	2.91	2.95	2.86	2.79	2.07	2.04
Sm2O3	1.00	0.99	0.95	0.93	0.70	0.73	0.69	0.66	0.50	0.48
Eu2O3	0.19	0.20	0.19	0.19	0.14	0.14	0.14	0.14	0.10	0.10
Gd2O3	0.48	0.50	0.48	0.49	0.41	0.42	0.35	0.37	0.32	0.30
Er2O3	2.08	2.13	2.06	2.07	1.54	1.57	1.48	1.48	1.09	1.04
Al2O3	2.16	2.15	2.36	2.31	2.20	2.20	2.33	2.38	2.42	2.32
B2O3	1.67	1.34	2.18	1.57	2.06	1.40	2.39	1.69	2.35	1.66
CaO	0.12	0.12	0.12	0.12	0.12	0.12	0.09	0.08	0.10	0.09
Cr2O3	0.04	0.03	0.04	0.03	0.03	0.03	0.03	0.02	0.02	0.01
Fe2O3	0.41	0.40	0.38	0.38	0.30	0.31	0.24	0.27	0.21	0.21
MnO	1.49	1.48	1.43	1.46	1.12	1.16	1.08	1.10	0.82	0.84
Na2O	0.12	0.09	0.12	0.09	0.12	0.09	0.10	0.09	0.12	0.10
SiO2	11.87	11.68	13.44	13.09	12.59	12.81	14.96	14.81	14.55	15.23
ZnO	0.41	0.42	0.43	0.41	0.33	0.33	0.33	0.31	0.27	0.25
PbO	5.16	2.94	6.68	3.78	6.41	3.65	7.50	4.82	7.27	4.49
BaO	2.43	2.41	2.83	2.79	2.66	2.71	3.27	3.15	3.21	3.06
Weight	Loss:	3.08		3.65		3.53		3.82		3.80
PbO		2.22		2.90		2.76		2.68		2.79
(Wt.%)		(72.1)		(79.3)		(78.2)		(70.2)		(73.4)
B2O3		0.33		0.61		0.66		0.70		0.70
(Wt%)		(10.7)		(16.6)		(18.8)		(18.4)		(18.4)
Total	Wt%	(82.8)		(96.0)		(97.0)		(88.5)		(91.8)

Attempts to plot this data compared to 1/T (Kelvin) did not exactly meet the expected linear relationships predicted by simple theory. See Figure 2. There was a slight curvature to the combined data. However, when the data was plotted with a "best fit" line the following simple linear equation was obtained:

Average Volatility Rate (mg/sq.cm/hr) = 50.17 - (73,630/Temperature - K) Eq. 2

This equation provides an approximation of the average volatility in mg/sq.cm/hr after a significant time period (48 hours) at the selected temperature in Kelvin. This is the equation for small scale, crucible melts. Use of this equation to estimate the volatility from large systems should be used with caution due to the complexities associated with volatilization.

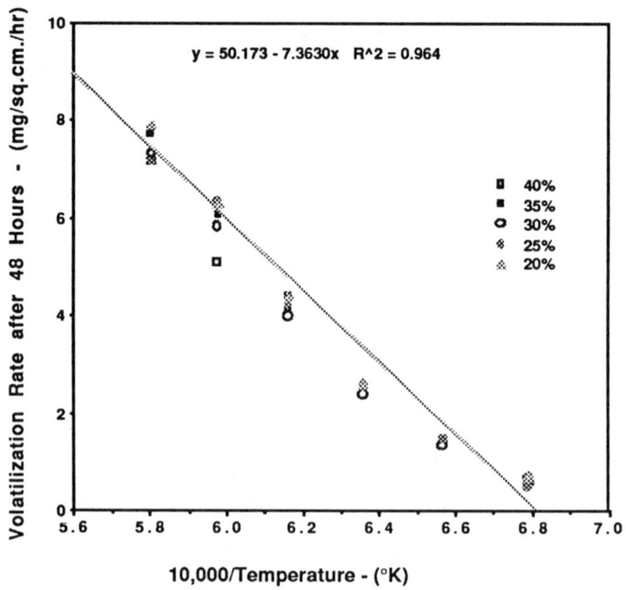

Figure 2. Volatilization Rate of all Glasses as a Function of Temperature.

Approach No. 2

The glasses with 35 and 40% Tank 17.3 loading were heat treated for time periods between two and twenty-four hours at 1400°C and 1300°C. The resulting weight loss data was combined with the appropriate 48 hour data from Approach No. 1 and is plotted as volatilization weight loss per square centimeter versus time in Figure 3.

A continuous almost linear relationship was obtained as shown. It appears that the volatilization at 1400°C is approximately double the volatilization at 1300°C and this is generally supported utilizing Equation 2. This data was also plotted versus the square root of time. The data did not support the square root of time linear relationship which would have identified diffusion as the rate controlling step. It is possible that the low viscosity glasses are experiencing convection at the higher temperatures or other experimental factors could be limiting the utilization of the simple theoretical relationships e.g. the vapor space above the melt is saturated with PbO suppressing volatilization. In addition, the fact that two elements, lead and boron, are both volatilizing may be interfering with the simple explanations of volatility. Again, Equation 2 should provide a reasonable estimate of the volatilization within the temperature range of interest for crucible tests.

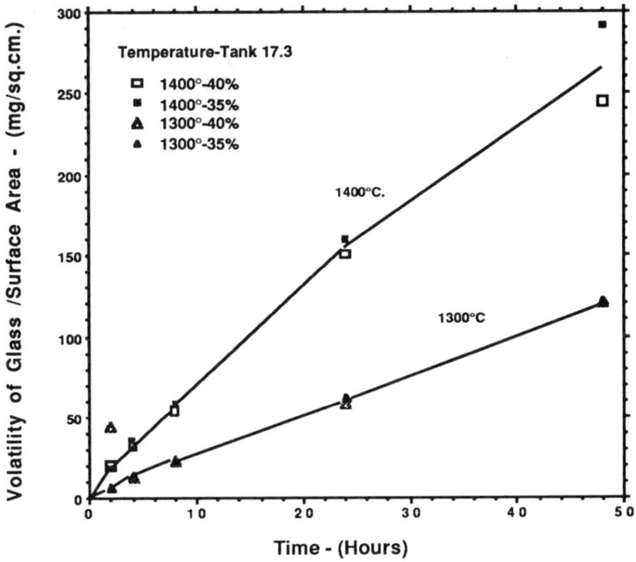

Figure 3. Measured Volatilization per Area as a Function of Time for 35 and 40% Tank 17.3 Loading Glass.

Volatility Compared to Actual Am/Cm Bushing Testing

The off-gas system of the Am/Cm bushing melter was tested under a variety of processing conditions using the B2000 glass system.[10] Although six test runs were completed, only the runs where glass was idled at 1450 °C and 1150 °C are comparable to this volatility investigation. Frit and surrogate feed were not fed during these two runs. Additionally, only air was introduced into the off-gas film cooler. The off-gas stream was sampled between the control air inlet and the steam eductor. The sampled stream passed through a cascade impactor and then through an EPA Method 29 sampling train. Filter papers (retention to 0.18 µm) within the impactor were weighed before and after each test run to determine total particulate loading in the off-gas streams.

The compositions of impinger solutions and material retained on the impactor filters were determined by Inductively Coupled Plasma-Atomic Emission Spectroscopy (ICP-AES). The ICP-AES was used to measure Pb, Ag, Al, As, B, Ba, Ca, Cd, Ce, Cr, Cu, Er, Eu, Fe, Gd, La, Mn, Nd, Ni, Pr, Se, Sm, and Sr concentrations. Silicon was not measured because the filter papers within the impactor were primarily silicon. Boron results could not be utilized because of the probable boron contamination from the borosilicate glassware. Additionally, high background levels of aluminum and calcium made results for these elements unreliable.

The bushing melter has a glass/air surface area of approximately 177.8 sq. cm. The measured volatilization rates of total emission and PbO were 3.04 $mg/cm^2/hr$ and 1.91 $mg/cm^2/hr$ respectively while idling at 1450°C. Thus, PbO accounted for 63 percent of the material lost from the glass surface. Comparison with the crucible studies showed PbO and B_2O_3 were responsible for 70 to 80 and 10 to 20 percent, of the volatile losses at 1400°C (Approach #1). Considering the major differences in equipment and conditions, these results are very similar. However, the total rates of volatilization from the crucibles at 1450°C were somewhat different. The average total volatility rate at 1450°C (using Equation 2) was 7.44 mg/sq. cm/hr. The rate of volatilization in crucible melts at 1450°C was somewhat more than double the rate in the bushing melter at 1450 °C. Assuming that greater convection currents exist in the bushing melter (due to a larger thermal gradient and the use of forced convection (bubbler), the rate of PbO and B_2O_3 volatilization should be greater. One possible explanation is the difference between the measured temperature of the bushing melter and the unknown true temperature of its glass surface. Another explanation may be that the off-gas test did not capture all of the volatile material. It is known that some amount of material did plate out on the off-gas film cooler and the off-gas line prior to sampling.

The bushing melter temperature is measured at the outside of the bushing wall. The surface temperature of the glass inside the bushing is not normally measured but is less than the control temperature. A lower measured volatilization rate from the melter may indicate glass surface temperatures below 1400 °C. Until the true temperature of the glass surface is determined, an absolute comparison between the bushing melter and crucible volatilization cannot be made. There was no measurable PbO emission in the bushing melter at the 1150 °C, and total emission was only 0.06 mg/cm^2 hr. Extrapolation of crucible results from Equation 2 also predicts a volatilization rate close to zero below 1200°C.

CONCLUSIONS

Examination of the data presented in this report provided the following general conclusions for the volatility of PbO and B_2O_3 from Tank 17.3 -B2000 glass melts:

- The level of waste loading (20 to 40 wt%) does not have a strong influence on volatility even though the level of PbO in the glasses decreases from 19.5 to 14.6 wt%.
- The average volatility rate in crucibles, after 48 hours, was described by the equation:
  Volatility Rate (mg/sq.cm/hr) = 50.17 - 73,630/(Temperature-°K).
- The total volatilization from crucibles after 48 hours at 1400°C appears to be mostly due to PbO loss (70 to 80 wt%) and B_2O_3 loss (10 to 20 wt%).
- The volatilization rate in crucibles is approximately linear with respect to time between 0 to 48 hours indicating that the volatilization is mass transfer controlled.
- Comparing crucible volatility to bushing melter off-gas measurements tends to suggest that the glass/air surface temperature in the bushing is considerably below the bushing control temperature.
- Any future volatility studies should strongly consider filling the crucibles with glass to reduce vapor phase considerations. The simple volatilization theories are very likely influenced by the presence in this system of two volatile oxides (PbO and B_2O_3) and the unusually low viscosity of these glass systems.

ADDENDUM

Due to various problems concerning glass devitrification with B2000, the composition of the frit for the americium/curium solidification program has been modified.

REFERENCES

1. M.B. Volf, "Chemical Approach to Glass," Series on Glass Science and Technology - 7, Chapter 27 - Lanthanoids, Elsevier, New York, 1984.
2. K.L. Lowenstein, "The Manufacturing Technology of Continuous Glass Fibers," Second Ed., p.118-168, Elsevier, New York, 1983.
3. R.F. Schumacher, W.G. Ramsey, F.M. Johnson, T.M. Jones, D.H. Miller, and B.J. Hardy, "Development of a Remote Bushing System for Americium-Curium Vitrification," Ceramic Transactions, Vol.87, pg 3-12, Amer. Ceram. Soc. Westerville, OH.
4. J.Matousek, "Volatilization of Lead Glasses within the 1200 to 1400°C Temperature Range," (English) Silikaty, 1 pg 1-12, 1972.
5. M.B. Volf, "Chemical Approach to Glass," Series on Glass Science and Technology - 7, Chapter 35 Lead, Elsevier, New York, 1984.
6. O. Andersen, "The Volatilization of Lead Oxide from Lead Silicate Melts," J. Amer. Ceram. Soc., [2], p.784, (1919).
7. E. Preston and W.E.S. Turner, "A Study of the Volatilization and Vapour Tension at High Temperatures of an Alkali-Lead Oxide-Silica Glass," J. Soc. Glass Technology, p. 219 (1932)
8. J.R. Zamecnik, Analysis of Cascade Impactor and EPS Method 29 Data from the Americium/Curium Pilot Melter System-(U)," WSRC-RP-97-00937, Rev. 0, November 19, 1997.

Acknowledgments

The information contained in this article was developed during the course of work under Contract No. DE-AC09-89SR18035 with the U.S. Department of Energy.

SELF-ASSEMBLED MONOLAYERS ON MESOPOROUS SILICA, A SUPER SPONGE FOR ACTINIDES

X. Feng*[1], L. Rao[1], T. R. Mohs[1], J. Xu[2], Y. Xia[1], G. E. Fryxell[1], J. Liu[1], and K. N. Raymond[2]
[1]Pacific Northwest National Laboratory, Box 999, P8-37, Richland, WA 99352;
[2]University of California, Berkeley, Department of Chemistry, Berkeley, CA 94720-1460

ABSTRACT

The combination of two frontier material sciences: mesoporous silica and self-assembled organic monolayers, resulted in a breakthrough material for actinide separation. Self-assembled monolayers such as hydroxypyridonates (HOPO) and propylthiol were installed on large surface areas of mesoporous silica, producing effective scavengers for radionuclides. The HOPO derived materials effectively removed Am(III) at initial 302000 cpm/ml to almost back ground level of 390 cpm/ml with a Kd value as large as $4.5x10^5$ in pH 0.88 to 6.80 solutions with 0.1 m $NaNO_3$. Large Kd values were also obtained for removing Np(V), Pu(IV), Th(IV), and U(VI). Different monolayers on the mesoporous materials have different selectivities for each actinide, leading to possible separation of different actinides. The thiol-based materials were shown previously as heavy metal-soakers such as mercury, lead, silver, cadmium, and copper with Kd values as high as $1x10^8$.

INTRODUCTION

Self-assembled monolayers on mesoporous silica (SAMMS) represents a new class of materials that integrate two frontiers of science: mesoporous oxides and self-assembled monolayers.[3] Mesoporous oxides[4,5] are a class of materials with large surface areas (~1000 m^2/g) in which the pore size, ceramic substrate, and geometry can be manipulated. Self-assembled organic monolayers bind to the ceramic substrate with prescribed densities and contain free functional groups that can selectively bond metals. The SAMMS technology has been shown to reduce Hg concentration in wastewater to below 10 ppt (part per trillion) with a K_d as high as 10^8.[6] The Hg-laden SAMMS was also shown to pass U.S. Environmental Protection Agency (EPA) Toxicity Characteristic Leach Procedure (TCLP) tests and has long-term chemical durability as a final waste form because of the stable ceramic matrix and covalent binding between metals and monolayers.[6]

Separation and removal of actinides at very low concentration have been a great challenge to separation chemists and are in great demand in our efforts to clean up nuclear weapon and nuclear power sites around the world. The development of SAMMS as an efficient "super sponge" for removing actinides is described in this communication. As discussed above, SAMMS materials consist of two distinct parts: mesoporous oxides and self-assembled organic monolayers. The mesoporous oxides provide the foundation and high surface area (1000 m^2/g) for SAMMS, which ensures a high metals-loading capacity of SAMMS materials. The molecular recognition, or the selectivity, is derived from the specific functional groups on the self-assembled monolayers.

Many radioactive liquid wastes stored within the U. S. Department of Energy repositories contain various actinides. Some of these wastes could be treated as low-level radioactive wastes if the actinides could be selectively removed. This would substantially reduce the volume of high level radioactive wastes, thereby significantly reducing handling and disposal costs. Separation of each actinide is also needed for risk assessment. Removing actinides from high level wastes is very challenging in that these wastes are usually complex salt solutions, containing high concentrations of nonradioactive cations that can compete with the actinides in binding to the sorbent materials and actinides are similar in chemicl behaviors. Therefore, potential sorbents must be extremely selective. Solvent extraction and ion exchange have been extensively studied for this purpose.[7-13] The objective of the current research is to demonstrate that SAMMS can be applied, in a highly selective fashion, for actinide separation in the presence of highly concentrated salts.

The high selectivity for SAMMS is achieved by incorporating the actinide-specific complexants developed by Raymond and his co-workers.[14-19] In nature, siderophores use Lewis basic chelating groups, including catechols (e.g., enterobactin) and hydroxamic acids (e.g., desferrioxamine B). These chelating groups are often combined in polydentate ligands to fully bind the metal in a six-coordination complex. Raymond and coworkers made actinide sequestering agents by incorporating these chelating groups into multidentate ligands. They have also attached these ligands onto insoluble polymers for use as solid/liquid extractants.[20] The distribution coefficients, Kd, reported for Th(IV) were as high as 70 for 1,2-HOPO and 250 for 3,4-HOPO resins at pH 5.8.

EXPERIMENTAL

In our studies, we synthesized mesoporous silica in cetyltrimethylammonium chloride/hydroxide, silicate, and mesitylene solutions.[4,5] The thiol-SAMMS was made by incorporating propylmercaptan onto the mesoporous materials as reported before.[3]

The HOPO-SAMMS were made by first incorporating a propylamine monolayer on the mesoporous materials through silinization methods similar to that of thiol-SAMMS. The analysis of the amino-SAMMS found C, 15.73; H, 4.08; N, 5.45; Al, 1.61; Si, 11.00.

Dimethylformamide was dried over 4 Å molecular sieves, and pyridine was dried over potassium hydroxide. All other reagents were purchased from commercial sources and used as received.

We dissolved 2.20 g (14.2 mmol) of 1,2-HOPO-6-Carboxylic Acid in 150 mL of dimethylformamide. To this was added 2.57 g (15.8 mmol) of 1,1'-carbonyldiimidazole under nitrogen, and the solution was stirred for 2 hours. We added 2 g of amino-SAMMS, and stirred the solution at room temperature for 4 days. The solution was filtered, washed with methanol, and dried in a vacuum at 60°C overnight to afford 2.43 g of an off-white powder. Analysis found C, 17.90; H, 3.91; N, 6.26; Si, 9.88; Al, 1.44. IR (KBr, cm^{-1}): 1622 (s, C=O).

We added 5.87 g (27.1 mmol) of 3-benzyloxy-4-HOPO-N-methylcarboxylic acid to 160 mL of pyridine. To this was added 2.45 g (21.3 mmol) of N-hydroxysuccinimide and 4.40 g (21.3 mmol) of 1,3-dicyclohexylcarbodiimide. The solution was stirred overnight at room temperature. We added 4 g of amino-SAMMS and stirred the solution at room temperature under nitrogen for 4 days. The solution was filtered, and the powders were washed with boiling tetrahydrofuran, distilled water, and methanol. The powder was stirred for 2 days in 150 mL of a mixture of 1:1 glacial acetic acid/concentrated hydrobromic acid. The powder was collected by filtration, washed with distilled water and methanol, and dried in a vacuum at 60°C overnight to obtain 5.29 g of an off-white powder. Analysis found C, 20.57; H, 4.75; N, 5.35; Al, 0.79; Si, 9.78. IR (KBr, cm^{-1}): 1627 (s, C=O).

Am(III), Th(IV), Pu(IV), Np(V), Pu(VI) and U(VI) were selected as the models for tri-, tetra-, penta- and hexa-valent actinides, respectively. $NaNO_3$ was added to give a 0.1 M of Na^+ concentration to ensure a rigorous test of selectivity. The solutions were adjusted to different pHs with final equilibrium pHs ranging from 0.84 to 6.8. The tests were performed in batch mode using 20 mg of SAMMS powders (except for experiments with Thiol-SAMMS) in 5.0 mL of actinide solutions. The mixture was shaken at room temperature for 4 hours and was then filtered through a 22-micrometer filter before the filtrate was analyzed. Liquid scintillation counting was used to measure ^{241}Am, ^{237}Np, and ^{239}Pu and inductively coupled plasma - mass spectrometry was used to analyze ^{238}U and ^{232}Th.

RESULTS AND DISCUSSIONS

The molecular structure for 1,2-HOPO-SAMMS, 3,4-HOPO-SAMMS, and thiol-SAMMS are shown in Figure 1. The actinide binding results are shown in Table 1. The data indicate the following:

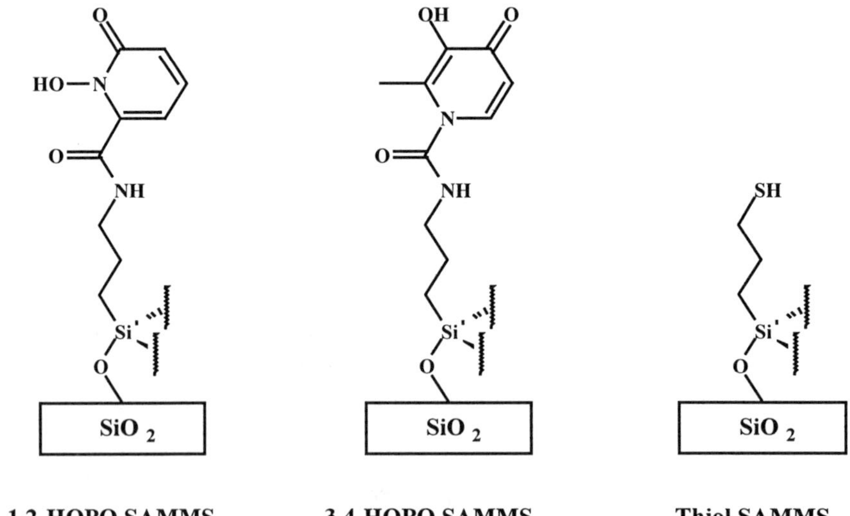

Figure 1. Molecular structure of 1,2-HOPO-SAMMS, 3,4-HOPO-SAMMS, and thiol-SAMMS.

Table 1. Actinide Absoption on SAMMS

Element	Solu mL	SAMMS g	pH	Final cpm/mL	Initial cpm/mL	Final ppb	Initial ppb	Kd
1,2-HOPO-SAMMS								
Am(III)	5	0.02	0.88	390	302000	0.051	39.8	1.93E+05
Am(III)	5	0.02	4.50	28	24200	0.004	3.2	2.14E+05
Am(III)	5	0.02	6.80	30	53500	0.004	7.1	4.46E+05
Np(V)	5	0.02	0.88	41300	73400	26993	48000	1.95E+02
Np(V)	5	0.02	3.20	27200	73400	17778	48000	4.25E+02
Np(V)	5	0.02	5.60	5140	72200	3359	47200	3.26E+03
Np(V)	5	0.02	6.30	4450	69000	2908	45100	3.63E+03
Th(IV)	5	0.02	0.82			86300	141400	1.60E+02
Th(IV)	5	0.02	3.40			30500	133100	8.41E+02
U(VI)	5	0.02	0.85			78500	135400	1.81E+02
U(VI)	5	0.02	3.00			39400	136800	6.18E+02
Thiol-SAMMS								
Am(III)	5	0.02	0.84	108000	302000	14.248	39.8	4.49E+02
Am(III)	5	0.02	3.10	17500	296000	2.309	39.1	3.98E+03
Am(III)	5	0.02	3.70	11300	291000	1.491	38.4	6.19E+03
Np(V)	5	0.02	0.84	67900	70500	44379	46100	9.69E+00
Np(V)	5	0.02	3.10	67900	70400	44379	46000	9.13E+00
Th(IV)	5	0.02	0.87			86800	141400	1.57E+02
Th(IV)	5	0.02	3.00			53800	133100	3.68E+02
U(VI)	5	0.02	0.85			132600	135400	5.28E+00
U(VI)	5	0.02	3.00			105700	136800	7.36E+01
Pu(IV)	5	0.10	1.00	350	12000			1.66E+03
Pu(VI)	5	0.10	3.00	backgd	8000			>10000
3,4-HOPO-SAMMS								
Am(III)	5	0.02	0.84	292000	302000	38.522	39.8	8.56E+00
Am(III)	5	0.02	2.60	288000	296000	37.995	39.1	6.94E+00
Np(V)	5	0.02	0.75	69700	70500	45556	46100	2.99E+00
Np(V)	5	0.02	2.50	69500	70400	45425	46000	3.16E+00
Element	Solu mL	SAMMS g	pH	Final cpm/mL	Initial cpm/mL	Final ppb	Initial ppb	Kd

Table 1. Actinide Absoption on SAMMS

Th(IV)	5	0.02	0.84			109000	141400	7.43E+01
Th(IV)	5	0.02	2.60			91500	133100	1.14E+02
U(VI)	5	0.02	0.86			130000	135400	1.04E+01
U(VI)	5	0.02	2.60			105400	136800	7.45E+01

- <u>For Trivalentactinides</u>
  The 1,2-HOPO-SAMMS are very effective in removing trivalent americium with Kd in the range of 1.9×10^5 at pH 0.88 to 4.5×10^5 at pH 6.8. The 1,2-HOPO-SAMMS reduced ^{241}Am concentration in solutions from 7.1 ppb to 4 ppt and radioactivity from 53,500 cpm/mL to 30 cpm/mL at pH 6.8. Thiol-SAMMS are also effective in removing Am (III) with Kd in the range of 449 to 6190. The 3,4-HOPO-SAMMS showed little binding for trivalent americium tested. Los Alamos National Laboratory performed a study of actinide-binding properties on 60 selected commercial and laboratory absorbers[20] and reported the highest Kd values of 9.4 in the acidic conditions similar to this study for total Am.

- <u>For Tetravalent Actinides</u>
  Thiol- SAMMS, 1,2-HOPO-SAMMS, and 3,4-HOPO-SAMMS are effective in removing tetravalent thorium with Kd of 160 at pH 0.82 to 841 at pH 3.4 and Kd of 74 at pH 0.84 to 114 at pH 2.6 respectively. Only thiol-SAMMS was tested for binding to Pu(IV), and it showed a good binding with a Kd of 1660.

- <u>For Pentavalent Actinides</u>
  Only 1,2-HOPO-SAMMS shows a good binding for the pentavalent Np(V) (in the form of NpO^{2+}) with Kd values of 195, 425, 3260, and 3630 at pH 0.88, 3.2, 5.6, and 6.30, respectively. The binding is much stronger as the solution pH increases.

- <u>For Hexavalent Actinides</u>
  Thiol-SAMMS showed a strong selectivity for Pu(VI), with Kd values larger than 10,000. The reported highest Kd values by Los Alamos National Laboratory[21] for the total Pu among the 60 commercial and laboratory absorbents are 2185 after 6 hours of equilibrium with the solution. The 1,2-HOPO-SAMMS showed moderate binding for U(VI) with Kd values of 180 at pH 0.88 to 620 at pH 3.0. The thiol-SAMMS and 3,4-HOPO-SAMMS did not show significant binding with the other

actinides under the acidic conditions tested.

The 1,2-HOPO-SAMMS showed a higher affinity for Th(IV) (Kd of 840) than the previous work with the HOPO resins (Kd of 70). Even the thiol-SAMMS showed an improvement over the resins with a Kd of 368. However, the 3,4-HOPO-SAMMS showed a lesser affinity (Kd of 114) than did the corresponding 3,4-HOPO resin (Kd of 260). At this point, it is still not clear why the 1,2-HOPO-SAMMS behave so differently than the 3,4-HOPO-SAMMS in their actinide affinities, even though there is no difference in actinide binding between 1,2,-HOPO-Resin and 3,4-HOPO-Resin. The complex structure in the resin and in the SAMMS will have to be studied in greater detail to understand the differences.

In summary, this preliminary study showed that by incorporating actinide-specific functional groups such as 1,2-hydroxypyridinone into SAMMS, an effective scavenger for actinides was produced. It is effective in removing tri-, tetra-, penta-, and hexa-valent actinides from solutions. The thiol-SAMMS is also effective in removing tri-, tetra-, and hexavalent actinides. The 3,4-HOPO-SAMMS is not as effective as 1,2-HOPO-SAMMS.

ACKNOWLEDGMENT

Pacific Northwest National Laboratory is operated for the U.S. Department of Energy by Battelle under Contract DE-AC06-76RLO 1830.

REFERENCES

1. Pacific Northwest National Laboratory, Box 999, P8-37, Richland, WA 99352.
2. University of California, Berkeley, Department of Chemistry, Berkeley, CA 94720-1460.
3. Feng, X.; Fryxell, G. E;, Wang, L-Q.; Kim, A. Y.; Liu, J.; Kemnerl, K. M. Science, 1997, 276, 913.
4. Beck, J. S.; Vartuli, J. C.; Roth, W. J.; Leonowicz, M. E.; Kresge, C. T.; Schmitt, K. D., Chu, C. T-W.; Olson, D. H.; Sheppard, E. W.; McCullen, S. B.; Higgins, J. B.; Schlenker, J. L. J. Am. Chem. Soc. 1992, 114, 10834.
5. Kresge, C. T.; Leonowicz, M. E.; Roth, W. J.; Vartuli, J. C., Beck, J. S. Nature, 1992, 359, 710.
6. Feng, X.; Liu, J.; Fryxell, G. E.; Gong; M.; Wang, Li-Q.; Chen, X.; Kurath, D. E.; Ghormley, C. S.; Klasson, K. T.; Kemner, K. M. Self-

Assembled Mercaptan on Mesoporous Silica (SAMMS) Technology for Mercury Removal and Stabilization, Pacific Northwest National Laboratory Report # PNNL-11691, Richland, Washington, 1997.

7. Horwitz, E. P., Chiarizia, R.; Diamond, H.; Gatrone, R. C. Sovent and Ion Exchange, 1993, 11(5), 943-966.
8. Horwitz, E. P.; Dietz, M. L.; Chiarizia, R.; Dimond, H. Analytica Chimica Acta, 1995, 310(1), 63-78.
9. Karalova, Z. K.; Lavrinovich, E. A.; Myasoedov, B. F.; Fedorov, L. A. Radiochemistry, 1994, 36(5), 441-445.
10. Rozen, A. M.; Nikiforov, A. S.; Zakharkin, B. S.; Nikolotova, Z. I. Atomnaya Energiya, 1992, 72(5), 453-459.
11. Condamines, N.; Musikas, C. Solv. Extr. Ion Exch., 1992, 10(1), 69-100.
12. Mathur, J .N.; Murali, M. S.; Rizvi, G. H.; Iyer, R. H. Solv. Extr. Ion Exch., 1994, 12(4), 745-763.
13. Maiti, T. C.; Kaye, J. H.; Kozelisky, A. E. J. Radioanaly. Nucl. Chem.-Art., 1992, 161(2) 533-540.
14. Harris, W. R.; Raymond, K. N.; Weitl, F. L. J. Am. Chem. Soc., 1981, 103, 2667.
15. Riley, P. E.; Abu-Dari, K.; Raymond, K. N. Inorg. Chem., 1983, 22, 3940.
16. Scarrow, R. C.; Riley, P. E.; Abu-Dari, K.; White, D. L.; Raymond, K. N. Inorg. Chem., 1985, 24, 954.
17. Scarrow, R. C., Riley, P. E.; White, D. L.; Raymond, K. N. J.Am.Chem.Soc., 1985, 107, 6540.
18. Scarrow, R. C.; Raymond, K. N. Inorg. Chem., 1988, 27, 4140.
19. Hou, Z.; Whisenhunt, Jr., D. W.; Xu, J.; Raymond, K. N. J.Am.Chem.Soc., 1994, 116, 840.
20. Whisenhunt, Jr., D. W. Synthesis and Evaluation of Chelating Agents and Resins Specific for the Actinides; Ph.D. Thesis, University of California at Berkeley, 1994.
21. Marsh, S. F.; Svitra, Z. V.; Bowen, S. M. Los Alamos National Laboratory Report LA-12654; Los Alamos, New Mexico, 1993.

REMOVAL OF SILICON FROM WASTE HF SOLUTION FOR GLASS CLEANING

Hee Chan Park*, Kwang Ho Kim, Bong Ki Ryu, and Seong Soo Park

Department of Inorganic Materials Engineering, Pusan National University, Pusan 609-735, Korea

Hack Gi Shin

Department of Industrial Chemistry, Kyungnam Technical College, Pusan 616-012, Korea

ABSTRACT

The process for removal of silicon from waste HF solution contaminated by HF attack of TV panel glass was described for use in cleaning of the glass. The removal of silicon was carried out by mixing HF solution containing silicon fluoride ion with sodium carbonate, whereby sodium silico-fluoride precipitated. This process made it possible to reuse the waste HF solution for cleaning of the glass. In an industrial practice, the process was applied and resulted in 90~95wt% silicon removal at reasonable cost.

INTRODUCTION

It has been known that TV panel glass must be of very high quality with respect to its surface. An appropriate spraying of HF solution on the TV panel glass improves its flatness and removes surface blemishes. However, the leaching of Ba, Ca, Si, Na, etc. ions from the glass leads to contamination of leaching solution. During the cleaning operation, the glass is dissolved by HF to form hydrogen silico-fluoride(H_2SiF_6).[1] The fluoride component in the leachate solution was believed to be comprised primarily of silico-fluoride anions(SiF_6^{2-}).[2,3] In the presence of HF or fluoride ions, silico-fluoride anions will be formed according to the following reaction:

$$SiO_2(glass) + 6HF(liq) \rightarrow SiF_6^{2-} + 2H_2O + 2H^+ \quad (1)$$
$$\Delta G° = -46.107 \text{ J/mol}$$

where, ΔG° is the standard free energy in the aqueous solution at 25°C. This thermodynamic prediction has been verified by Berry et al.[2] and Paul.[3] They demonstrated that almost quantitative reactions take place in all practically attainable acid and alkaline media. The very corrosive action of fluorides on silicate glass is well known. Hydrofluoric acid[4,5] also will attack the silica group in the glass, allowing the structure to break up and release network modifiers such as sodium, barium, etc. The H_2SiF_6 will react with Ba, Ca, Na, K etc. ions to form silico-fluoride salts and the salts will be precipitated in the leaching solution. Thus, the leaching solution with the salts prevents the HF solution from passing through pipes and nozzles for spraying. This solution can not be used for the cleaning solution and is treated as waste HF solution. The objective of this study was to reuse the waste HF solution as cleaning solution by the removal of dissolved ions released by HF attack of the TV panel glass.

EXPERIMENTAL

TV panel glass and waste HF solution were provided by LG Electronics Inc., Changwon, Korea. The chemical composition of the TV panel glass and the concentration of Si in the leachate solution(waste HF solution) are given in Tables I and II, respectively. The HF cleaning system at LG Electronics Inc., outlined in

Table I. Chemical composition of as-received panel glass

Constituents	wt%
SiO_2	61.30
Al_2O_3	3.85
MgO	1.17
CaO	1.89
BaO	14.70
Na_2O	10.20
K_2O	6.85
Fe_2O_3	0.04

Table II. The Si concentration in leachate solution as-received and after-treated by the addition of Na_2CO_3 to waste HF solution

Sample	Amount of Na_2CO_3 added to waste HF solution (wt%)	Si (ppm)	Si removal from waste HF solution (wt%)
As-received waste HF solution	—	116000	—
1	0.13	7260	93.7
2	0.25	6820	94.1
3	0.50	6280	94.6
4	1.00	4820	95.8
5	2.00	2340	98.0
6	2.30	1660	98.6
7	2.50	1500	98.7
8	2.75	872	99.2
9	3.00	770	99.3
10	3.25	534	99.5
11	3.50	440	99.6
12	3.75	310	99.7
13	4.00	250	99.8
14	5.00	200	99.8
15	6.00	200	99.8
16	7.00	200	99.8

Fig. 1, is a kind of recirculation system that consists of nozzles for spraying, solution tanks, and pumps. The HF solution is continuously pumped through a recirculating loop. It takes about 5sec for each panel glass to be cleaned. Subsequent cleaning proceeds until the concentration of leachate solution becomes about $1.16x10^5$ppm Si. Then, the leachate solution is treated as the waste HF solution and moved to another tank for disposal.

The waste HF solution was kept in a graduated polyethylene cylinder, with an inside diameter of 30mm at room temperature. Then, an appropriate amount of sodium carbonate was added and mixed. The waste HF solution in cylinder was hand-shaken forcefully to distribute the components uniformly. The volume of initial solution for each sample was 200ml. As shown in Table II, chemical compositions in leach liquor and precipitate were analyzed after visual observing a sharp interface between leach liquor and solid-like precipitate. The salts present in the precipitate were determined by the X-ray diffraction(XRD) technique. X-ray diffraction data were taken using a Philips powder diffractometer with $CuK\alpha$

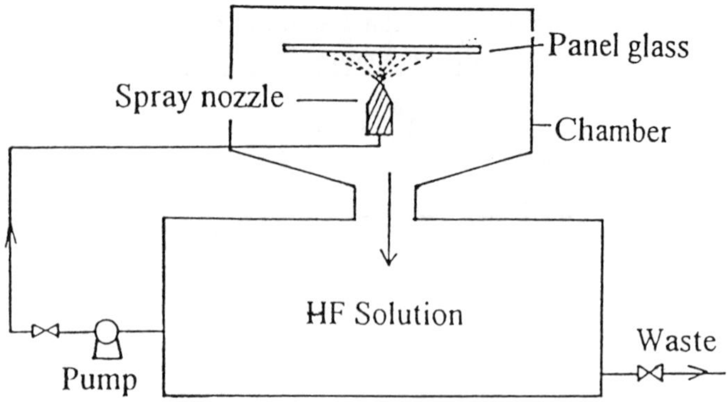

Fig. 1. Schematic representation of the recirculating flow loop for the cleaning process under the current production practice.

radiation operating at 40kV and 20mA. The chemical compositions of the HF solution were determined by chemical analysising using spectrochemical methods. The arithmetic mean as well as standard deviation were applied in the calculation of chemical compositions.

RESULTS AND DISCUSSION

(1) Waste HF Solution without Na_2CO_3 Addition

When no Na_2CO_3 was added to the waste HF solution in the cylinder, a precipitate was obtained within 5.0min after shaking the cylinder. The resultant leach liquor was decanted, and then transferred immediately into a plastic bottle. In X-ray diffraction analysis, the precipitate dried at 80°C for 2h was barium silico-fluoride($BaSiF_6$) as given in Table III. It further showed that the $BaSiF_6$ was preferably precipitated from the waste HF solution without Na_2CO_3 addition.

In this study, some of silico-fluoride anions in the waste HF solution were believed to be removed by reaction with Ba ion to form $BaSiF_6$. As given in Table IV, the solubility[6] of $BaSiF_6$ is much lower than that of other silico-fluoride salts with the exception of Na_2SiF_6, and calcium silico-fluoride($CaSiF_6$) is to be soluble in HF. This reflects the results that the precipitate mostly contains $BaSiF_6$. Fluorides or silicates other than $BaSiF_6$ might remain in the precipitate via the following processes, considering components in the TV panel glass although they

Table III. XRD data of $BaSiF_6$ from Joint Committee on Powder Diffraction Standards (JCPDS) and precipitate obtained from waste HF solution

JCPDS (No. 15-736)			Precipitate	
hkl	**d**(nm)	**I / I**$_{(max)}$	**d**(nm)	**I / I**$_{(max)}$
101	4.650	30	4.667	26
110	3.593	100	3.587	100
012	3.055	60	3.035	37
003	2.336	40	2.349	27
211	2.230	60	2.221	35
122	1.953	65	1.947	47
220	1.796	14	1.789	23

Table IV. Solubilities of silico-fluorides

Formular	Solubility (g/100ml)		
	Cold water	Hot water	Other solvents*
$MgSiF_6$	65.00		
$CaSiF_6$	sl s		s al, HF, HCl
Na_2SiF_6	0.0652	2.46	i al
$BaSiF_6$	0.026	0.09	i al, sl s a, NH_4Cl

*i, s, sl s, al, and a demote insoluble, soluble, slightly soluble, alcohol, and acid, respectively

were not detected by X-ray analysis.

$$CaO + 2HF \rightarrow CaF_2 + H_2O \quad (2)$$
$$BaO + 2HF \rightarrow BaF_2 + H_2O \quad (3)$$
$$SiO + 4HF \rightarrow SiF_4 + 2\,H_2O \quad (4)$$
$$3SiF_4 + 3H_2O \rightarrow H_2SiO_3 + 2H_2SiF_6 \quad (5)$$

(2) Waste HF Solution with Na_2CO_3 Addition

The leach liquor in a cylinder will still contain the state of SiF_6^{2-} and H^+ because major component of the TV panel glass is silica(SiO_2). Chemically, silico-fluoride anions can be removed from the waste HF solution by precipitation according to the following reaction:

$$H_2SiF_6 + Na_2CO_3 \rightarrow Na_2SiF_6 + H_2O + CO_2 \quad (6)$$

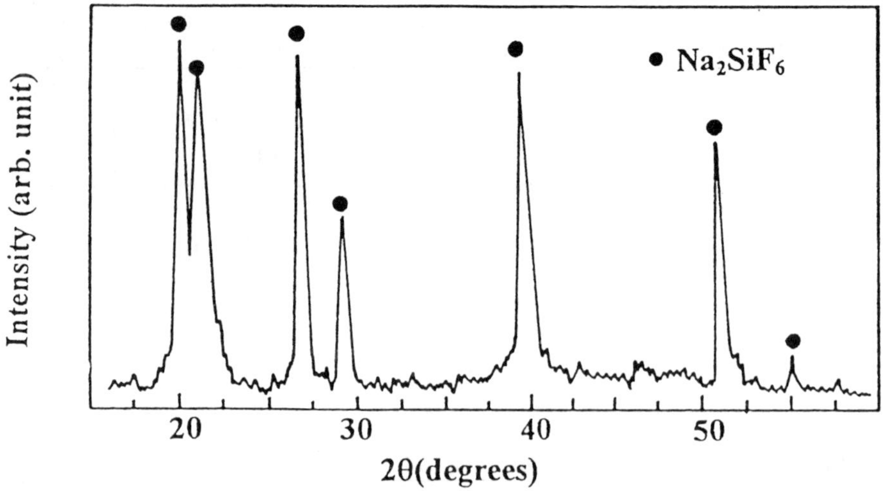

Fig. 2. XRD pattern of precipitate obtained by addition of Na_2CO_3 to leach liquor. Precipitate was dried at 80°C for 2h.

Sodium carbonate(Na_2CO_3) will be precipitant with the advantages of low cost and high solubility. Therefore, the effect of Na_2CO_3 addition on Si removal in the waste HF solution was mainly examined. The precipitate obtained by the addition of 1.0wt% Na_2CO_3 to the leach liquor was found to be Na_2SiF_6 as shown in Fig. 2, in agreement with Equation 6.

The data on the concentration of Si in the leach liquor separated from the precipitate by the addition of Na_2CO_3 to the waste HF solution are given in Table II. It was seen that increasing the Na_2CO_3 up to 2.0% decreased sharply in the concentration of Si. Increasing the Na_2CO_3 from 2.0 to 4.0% caused a slow decrease in the concentration of Si. However, with the addition of more than 4.0%, the concentration of Si was not changed. Table II also indicates that Si is mostly removed in the form of Na_2SiF_6, considering Equation 6 and X-ray analysis. The consumed fluoride ions by forming sodium silico-fluoride were supplied by adding an appropriate amount of new HF solution. The precipitation process for recycling of waste HF solution with $1.16x10^5$ppm Si was applied at LG Electronics Inc. This process was capable of at least 90% silicon removal at reasonable cost with only 1.0% Na_2CO_3 addition, without detrimental effect on the cleaning operation of the TV panel glass in commercial production.

CONCLUSIONS

This work has been demonstrated that waste HF solution contaminated by HF attack of TV panel glass can be recycled for cleaning of its glass, using sodium carbonate as a precipitant. The waste HF solution containing $1.16x10^5$ppm silicon, without sodium carbonate addition was partially precipitated. The precipitate was found to be barium silico-fluoride. The removal of silicon from the waste HF solution can be carried out by mixing the HF solution with sodium carbonate, whereby sodium silico-fluoride precipitates, indicating that the solution can be recycled for use in cleaning of silicate glass other than the TV panel glass.

ACKNOWLEDGMENTS

We thank LG Electronics Inc. and Research Institute of Industrial Technology, Pusan National University for financial support. We also thank Chapman & Hall Ltd. for permission to use the materials published in J. Mater. Sci. Lett., **17** [1] 11-13 (1998) by Hee C. Park et al., with title: Recycling of waste HF solution by Na_2CO_3 treatment for use in cleaning of TV panel glass.

REFERENCES

[1]K. W. Boer, *Survey of Semiconductor Physics*, Vol. II; p. 294, Van Nostrand Reinhold, New York, 1992.

[2]W. W. Berry and G. J. Rossister, "Process for Removing Fluoride from a Wastewater and Producing Hydrofluoric Acid Therefrom," U.S. Pat. No. 4 965 061, October, 1990.

[3]A. Paul, *Chemistry of Glasses*; pp. 121, Chapman & Hall, New York, 1982.

[4]H. Scholze, *Glass-Nature, Structure and Properties*; pp. 328, Springer-Verlag, New York, 1980.

[5]N. B. Kirk, "Study of Commonly Observed Defects Formed during the Process of Acid Polishing of Lead Crystal Glass," *Glass Technol.*, **37** [2] 52-56 (1996).

[6]D. R. Lide (Ed.), *CRC Handbook of Chemistry and Physics*, CRC Press, London, 1994.

RECYCLING OF SILICON WAFER CUTTING SLUDGE TO FORM SILICON NITRIDE CERAMICS

Soo young Lee and Byung teak Lee

Korea Institute for Machinery and Materials, Changwon, Korea, Kongju National University, Kongju, Korea

Silicon cutting sludge comes from cutting silicon single crystals to make silicon wafers. It consists mainly of silicon debris and organic lubricant. The waste silicon sludge was purified by washing with water and nitrided to form reaction-bonded silicon nitride (RBSN). The higher contents of oxygen in the purified silicon powder gave rise to the difficulty of nitridation. The addition of an Fe containing compound accelerated the nitridation. Two step gas pressure sintering resulted in fully densified sintered reaction-bonded silicon nitride (SRBSN). The reaction mechanism and mechanical properties of sintered RBSN are discussed.

INTRODUCTION

When silicon wafers are cut from a silicon ingot single crystal. with a diamond saw, a lot of silicon loss occurs. Silicon debris from the cutting process are usually mixed with organic agents. This mixture, as a sludge, is used to promote the cutting ability. Tons of silicon waste is produced every year in Korea. However, most of the silicon waste is discarded into landfills. This results in the loss of natural resources and causes pollution of the environment. In this work, low-cost reaction bonded silicon nitride (Si_3N_4) has been developed using waste silicon sludge. The purity of the ingots used for silicon substrate fabrication in the semiconductor industry is very high. It is, thus, possible to obtain silicon powder with high quality if the organic agents are separated effectively from the waste silicon sludge. The reaction mechanism and the properties of the reaction bonded silicon nitride (Si_3N_4) will be discussed.

EXPERIMENTS

The silicon sludge produced from cutting of a silicon ingot was rinsed with water three times to remove the organic agents, followed by drying on a hot plate.

Sieving was carried out to eliminate the coarse silicon particles. The average size of the sieved silicon particles was 8 μm, which was very fine and homogeneous as shown in Fig. 1. The chemical analysis of the silicon powder was performed by inductively coupled plasma spectrometry (ICP). Oxygen was analyzed by an oxygen analyzer. The sieved silicon powder and sintering additives (6 wt.% Y_2O_3 and 2 wt.% Al_2O_3) were mixed in ethanol and milled with silicon nitride (Si_3N_4) ball media. SiC at 30 wt.% was also mixed with the starting powder as a second phase. The powder mixture was pressed into pellets at 250 MPa by cold isostatic pressing. The pellets, 40 mm in diameter, were nitrided in flowing nitrogen/10% hydrogen gas mixture at 1400 °C for 20 hours. Post-sintering was performed at 1950°C for 5 hours under 1050 psi nitrogen gas pressure. The fracture strength was measured by the three point bending method using specimens measuring 3x4x30 mm. The fracture toughness, K_{IC}, was measured by the indentation method with a load of 30 kgf. X-ray diffraction (XRD) was used for phase determination. The microstructures were examined using optical microscopy, scanning electron microscopy (SEM) and transmission electron microscopy (TEM).

RESULTS AND DISCUSSION

The oxygen content of silicon powder after rinsing with water three times and drying was analyzed and compared with commercial silicon powder (Type C, Permascand, Sweden). Silicon powder obtained from silicon waste sludge after rinsing with water is designated as silicon(w) and the Permascand's type C silicon powder is designated as silicon(c).

The oxygen content and the amounts of SiO_2 in each powder is shown in Table 1. The oxygen content in the silicon powder obtained from silicon waste is very high and may affect the nitridation behavior of the silicon powder. Figure 2 is a TEM micrograph showing the SiO_2 layer on the silicon(w) powder.

The rate of nitridation of several samples of the starting materials is shown in Table 2. The system(a) contains silicon powder (silicon(w) or silicon(c)) and sintering aids (6 wt.% Y_2O_3, 2 wt.% Al_2O_3). The system(b) contains sintering aids and 30 wt.% SiC as a second phase. The system(c) contains sintering aids and 1 wt.% Fe. The system(d) contains HF treated silicon powder and sintering aids. The silicon(w) and silicon(c) powder were immersed in 50% HF solution for 24 hours before nitridation in order to remove the oxygen containing layer. The rate of nitridation of silicon(w) was generally lower than silicon(c) in all of the systems. This is due to the presence of large amounts of SiO_2 in the silicon(w). The SiO_2 on the silicon particles should be destroyed in order to

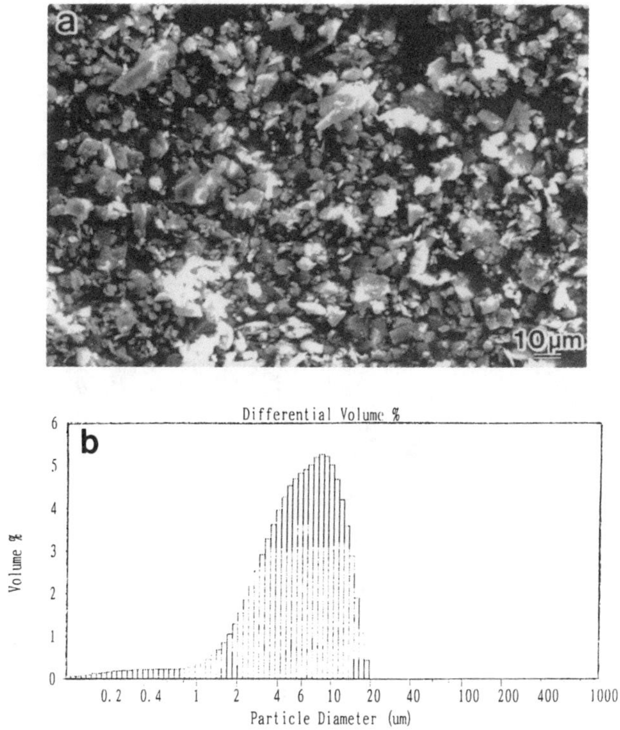

Figure 1. SEM micrograph (a) and particle size analysis (b) of Si(w) powder.

facilitate nitriding. Silicon monoxide gas will be produced by decomposition of the SiO_2 layer on silicon powder [1]. This decomposition will control the rate of nitridation, resulting in slow nitridation of silicon(w) powder. For this reason, the system(d) containing HF treated silicon powder showed higher nitridation than the system(a). The system(c) showed a higher nitridation rate than the system(a). It has been reported that the presence of Fe enhances the nitridation of silicon powder [2-3].

The fracture strength of the nitrided samples is shown in Table 3. The specimens with higher nitridation rates show higher fracture strengths. The fracture strength of silicon(w) is, thus, lower than silicon(c).

The nitridation temperature for silicon(w) was increased up to 1470°C in order to increase the nitridation rate. As shown in Figure 4, the specimen was nitrided

Table 1. Oxygen Content of Si Powders

	treatment	Oxygen content(wt.%)	Amounts of SiO_2(wt.%)
Si(c)	commercial Si	0.63	1.18
Si(w)	Si waste rinsed with water	6.68	12.3

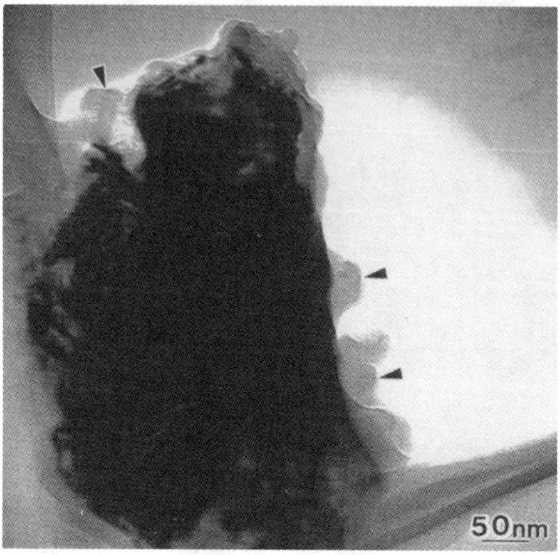

Figure 2. TEM micrograph of Si(w) powder. (Arrows indicate SiO_2 layer)

Table 2. Nitridation Levels of Several Systems

System	System(a) Si 6 wt.% Y_2O_3, 2 wt.% Al_2O_3	System(b) Si 6 wt.% Y_2O_3, 2 wt.% Al_2O_3 30 wt.% SiC	System(c) Si 6 wt.% Y_2O_3, 2 wt.% Al_2O_3 1 wt.% Fe	System(d) Si treated in HF 6 wt.% Y_2O_3, 2 wt.% Al_2O_3
Si(c)	84.7 %	98.4 %	94 %	98.7 %
Si(w)	53.4 %	78.4	80.7	95 %

fully at 1470°C. The nitrided samples were sintered again at 1950°C for 5 hours under 7 MPa nitrogen pressure. Before heating to the 1950°C sintering temperature, the RBSN was further nitrided by holding at 1500°C for 5 hours. The sample nitrided with silicon(c) achieved better than 99% of theoretical density and the sample nitrided with silicon(w) achieved 97% of theoretical density after sintering at 1950°C for 5 hours under 7 MPa nitrogen pressure.

The mechanical properties of SRBSN after sintering at 1950°C for 5 hours under 7 MPa nitrogen pressure is shown in Table 3. The mechanical property values of silicon(w) specimens were lower than that of silicon(c). The sample nitrided with silicon(w) was not densified effectively by post-sintering compared to the sample nitrided with silicon(c), resulting in lower mechanical property values. The sample nitrided with silicon(c) in the system(b) showed the highest fracture strength. The high strength associated with the SiC additive is due to the resulting fine and homogeneous microstructure as shown by the SEM micrographs in Figure 5. The addition of SiC also enhanced the nitridation rate and controlled pore formation effectively during the nitridation process, resulting in an increase in mechanical property values [4]. On the other hand, the addition of SiC to the system with silicon(w) did not improve the mechanical properties of the SRBSN because the sinterability of this system was inhibited by the presence of large amounts of residual silicon.

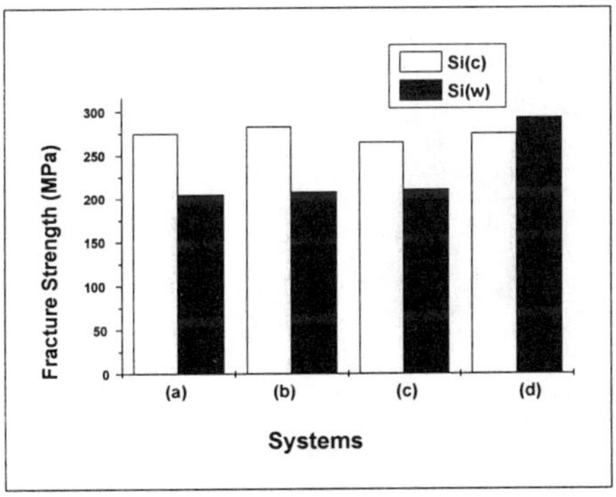

Figure 3. Fracture strength after Nitridation at 1400° C.

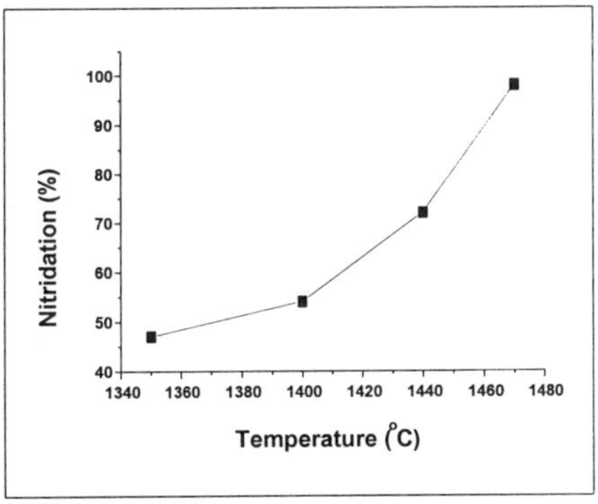

Figure 4. Nitridation level with reaction reaction temperature.

Table 3. Mechanical Properties of SRBSN

	Si(c)		**Si(w)**	
	System(a)	System(b)	System(a)	System(b)
Fracture Strength (MPa)	792 ± 43	1078 ± 52	698 ± 34	652 ± 41
Fracture Toughness (K_{IC})	7.2 ± 1.2	6.9 ± 0.9	5.2 ± 0.5	4.9 ± 0.4

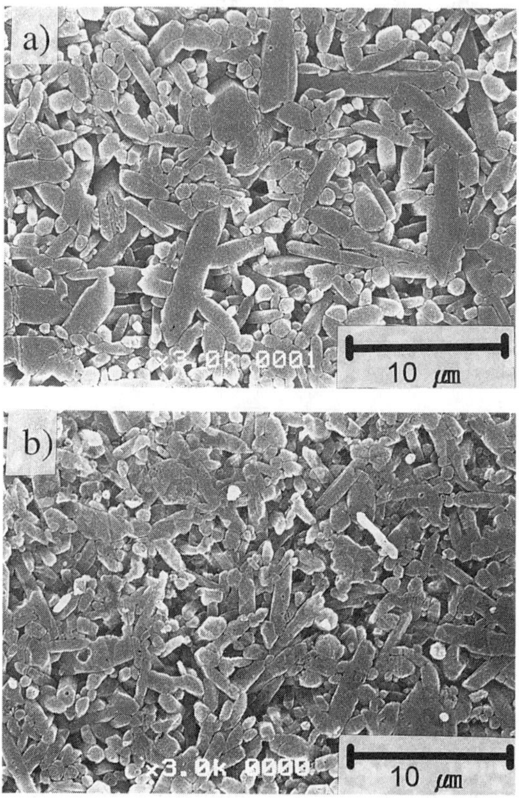

Figure 5. SEM micrographs of SRBSN fabricated with Si(w). ((a) No SiC and (b) 30 wt.% SiC).

CONCLUSIONS

The fabrication process of reaction-bonded silicon nitride (Si_3N_4) using waste silicon sludge has been developed and the following results were obtained:

1. Waste silicon sludge was treated to remove organic agents by rinsing with water. The average particle size of silicon powder was about 8 μm with irregular shape.
2. The rate of nitridation for silicon powder from waste is lower than the rate of nitridation for commercial silicon powder. Silicon powder from waste contains about 10 wt.% SiO_2 which inhibits the nitridation of silicon.
3. The presence of large amounts of SiO_2 in this system gave rise to formation of a Sialon phase after nitridation.
4. The removal of the SiO_2 layer on the silicon powder by HF treatment improved the rate of nitridation.
5. The nitrided body consisting of alpha and beta silicon nitride (Si_3N_4) was sintered by gas pressure sintering and was densified up to 97% of the theoretical density.
6. The highest values of fracture strength and fracture toughness of sintered RBSN fabricated from silicon was powder were 698 MPa and 5.2 MPa-$m^{1/2}$, respectively.

REFERENCES

1. K. Niihara and T. Hirai, "Chemical Vapor-Deposited Silicon Nitride," J. Materials Science, 11, 593-603 (1976).
2. S. Shinozaki and M.E. Milberg, "Electron Microscopy Study of the Effect of Iron in Reaction-Sintered Silicon Nitride," J. American Ceramic Society, 64, 383-386 (1981).
3. J.A. Mangeis and G.J. Tennenhouse, "Densification of Reaction-Bonded Silicon Nitride," Ceramic Bulletin, 59, 1216-1219 (1980).
4. S.Y. Lee, "The Fabrication of Silicon Nitride (Si_3N_4)/SiC Composite by Reaction-Bonding and Gas Pressure Sintering," J. American Ceramic Society, 81, 1262-1268 (1998).

EFFECT OF WASTEWATER REUSE IN GLAZED TILE MANUFACTURE ON GAS EMISSIONS AND SOLID WASTE

J.E. Enrique, E. Monfort, M. Monzó, E. Sánchez
Instituto de Tecnología Cerámica.
Asociación de Investigación de las Industrias Cerámicas.
Universitat Jaume I. Castellón. (Spain)

KEYWORDS: Wastewater reuse, glazed tile, gas emissions, solid waste

1 WATER IN THE TILE MANUFACTURING PROCESS

Water plays an important part in the ceramic tile manufacturing process, fulfilling specifically technological functions on being used as a raw material or material with auxiliary functions. The manufacturing process was therefore analyzed to determine in which process stages water was needed, establishing required water characteristics, and which process stages produced wastewater.

Wastewater flow rates and characteristics can exhibit very significant variations across time as most wastewater comes from washing operations in the glazing lines.

From a chemical point of view, the wastewater involved is characterized by the presence of:

- Suspended solids: clays, frit rests, insoluble silicates in general.
- Dissolved anions: sulfates, chlorides, fluorides, etc.
- Dissolved or suspended heavy metals.
- More or less variable quantities of boron.
- Traces of organic matter: screen-printing vehicles, glues and other organic compounds used in glazing operations.

Table I presents a standard wastewater composition produced in glazed tile manufacture.

Table I.- Average chemical analysis of non-treated wastewater arising in glazed tile manufacture.

Substance	Variation interval	Substance	Variation interval
pH	7-9	Boron (mg/L)	1-10
Suspended matter(mg/L)	1000-20000	Calcium (mg/L)	5-500
Settleable matter (mg/L)	5-30	Magnesium (mg/L)	10-100
Chlorides (mg/L)	100-700	Aluminium (mg/L)	< 2
Sulfates (mg/L)	100-500	Iron (mg/L)	< 0.5
Fluorides (mg/L)	< 2	Zinc (mg/L)	< 2
COD (mg/L)	150-1000	Lead (mg/L)	< 5
BOD5 (mg/L)	50-400		

2 IMPLEMENTATION OF WASTEWATER REUSE SYSTEMS

Growing water demands for industrial use and stricter pollution control regulations have led to a constant rise in industrial water supply and effluent treatment costs. This led to considering the possibility of wastewater reuse in the same manufacturing process.

The amounts of reusable wastewater were determined for a standard glazed tile manufacturing facility, as well as the process stages in which wastewater could be reused. Wastewater reuse in the same manufacturing process poses various problems that need to be adequately solved. However, wastewater reuse also entails significant advantages:

- Savings in water consumption.
- Less waste to be disposed of.
- Lower wastewater treatment costs, as it is usually cheaper to treat water for reuse than for disposal, since the limits of suspended and/or dissolved elements are fixed by the alteration to the process stage in which reuse occurs, and not by legislation requirements, which tend to be much stricter.

It is therefore important to identify the qualitative requirements of the water used throughout the process in order to know what wastewater treatment is needed prior to possible reuse.

The qualitative requirements of the water used in each tile manufacturing stage were determined. It was found that in the wet milling stage large quantities of wastewater can be used instead of industrial water, if suitable controls are implemented to monitor possible changes in the rheology of the body slip and appearance of efflorescence.

Having established qualitative water requirements in each process stage it was found that if wastewater were to be reused in body preparation, in principle no pretreatment would be needed other than a homogenization tank to ensure the most consistent possible water characteristics for body preparation. If the wastewater were to be reused for washing facilities however, water quality would need to be higher, so that a settler would be required, followed by aeration with or without subsequent chemical treatment for suppressing smells.

3 EFFECTS OF WASTEWATER REUSE ON AIR EMISSIONS AND SOLID WASTE

As indicated above, in wastewater reuse it is very important to take into account the effects that such reuse can have on final product quality and on the environment.

Studies performed by the Instituto de Tecnología Cerámica have shown the feasibility of reusing wastewater as a raw material in body preparation in the same manufacturing process. The controlled application of these recycling systems was found to produce no significant changes in process variables or in the quality of the final product.

In these recycling systems, most of the wastewater is eliminated by evaporation in spray drying, and the suspended and dissolved materials contained in the wastewater are assumed to be incorporated in the final product.

Fig. 1 depicts the schematic of an industrial wastewater treatment system in which zero wastewater emission can be achieved on reusing the arising wastewater as a raw material in body preparation.

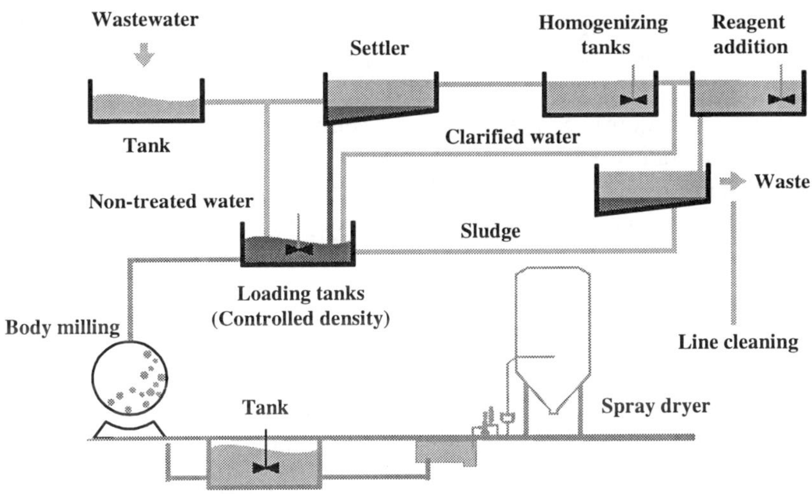

Figure 1. Wastewater treatment and reuse system.

In this figure it can be observed that the wastewater is reused directly in body preparation without any type of pretreatment. This is the case in most facilities in which the wastewater is simply homogenized, controlling density so as not to exceed the admissible solids contents determined in preceding tests. With this system it is necessary to slightly modify the deflocculant proportion to be added to the suspension to achieve the desired rheological behaviour.

On the other hand, it can also be observed in Fig. 1 that when the arising wastewater exceeds water consumption in body preparation, a part of the wastewater can be reused after chemical treatment in other operations: for washing, gas scrubbing in wet filters, etc. After such recycling, the arising wastewater can again be used in body preparation and zero wastewater emission can readily be achieved.

However, one of the issues as yet not addressed, and nonetheless of great importance is the assessment of the possible increased environmental impact on reusing wastewater in body preparation. It is assumed that this would mainly affect the resulting air emissions and composition of the solid waste (green and fired tile scrap) arising in the manufacturing process.

The following sections will separately assess the possible effects of wastewater reuse on air emissions and on the composition of the solid waste produced in glazed tile manufacture.

The experimental study was conducted at a porous single-fire wall tile manufacturing facility, at which wastewater and sludge were recycled according to the schematic depicted in Fig. 1, while the arising green scrap was also reused in the same process by being added to the body composition.

3.1 Effect of industrial wastewater reuse on air emissions

The study first addressed the pollutants present in the gas emissions produced in spray drying and firing, as these are the key tile manufacturing processes.

Boron and lead, which are not present in the body raw materials, were used as tracing elements of the reused wastewater.

These elements (boron and lead) come from the water used to wash the glaze preparation and application facilities. The greater part of these elements is retained in the spray-dried granules and the remainder, which is detected in the air emissions, is borne along with the evaporating water in spray drying.

These elements were chosen as tracing elements since finding them in the spray-drying emissions would indicate whether the compounds are incorporated into the body or are directly released into the air in wastewater reuse.

Thus the most important effect in this case was the possible increase in boron and lead concentrations in spray-drying emissions.

Table II details the variation intervals for the element concentrations found.

Table II.- Concentrations of substances present in spray drying air emissions.

Substance	Variation interval	Substance	Variation interval
Ps(mg/Nm^3)(*)	150 - 1500	B (mg/Nm^3)	< 0.3
NO_x (ppm)	3 - 15	Cl (mg/Nm^3)	1 – 5
CO (ppm)	1 - 15	Pb (mg/Nm^3)	< 0.15
CO_2 (%)	1.5 - 4		

(*) Ps: solid particles ; Nm^3: normal cubic meters

Boron and lead concentrations are low, as their concentrations in the cleaning water used for preparing the slip are low. Their presence stems from the dissolution of the boron and lead that is present in glaze rests in the cleaning water.

With regard to the study of air emissions produced in firing, the data on the nature and concentrations of these substances correspond to the exhaust stack, as this the only stack at which pollutant cleaning might be needed.

Table III lists the substances studied and determined in kiln air emissions. In this process stage the boron present in the emissions comes mainly from the water contained in the tile on entering the kiln, which is evaporated in the first firing stages. This water has remained in the tile after drying or has been incorporated into the body and/or glaze during the glazing operation.

Table III.- Concentrations of substances present in the exhaust stack

Substance	Variation interval	Substance	Variation interval
Ps (mg/Nm^3)	5 - 50	B (mg/Nm^3)	< 0.5
NO_x (ppm)	15 - 60	Cl (mg/Nm^3)	20 - 90
CO (ppm)	1 - 15	Pb (mg/Nm^3)	< 0.15
CO_2 (%)	1.5 - 4	F (mg/Nm^3)	5 - 40
SO_2 (mg/Nm^3)	< 10		

(*) Ps: solid particles ; Nm^3: normal cubic meters

Boron concentrations are very low, as the boron concentration in the cleaning water is very low.

The stream's lead content is quite low and mainly comes from the vaporization of glazes containing this element in their composition, currently quite a minor group of glazes.

3.2 Effect of wastewater reuse on the arising solid waste.

The study also addressed the possible effects of wastewater reuse on the arising solid waste in the manufacturing process.

Two types of solid waste are basically produced in the process: green scrap (broken pieces from pressing and drying), and fired scrap (defective or broken tiles etc.).

Management of these two types of waste is different. Green scrap is reused directly in the body preparation process without pretreatment, whereas fired scrap is disposed of as inert waste or used by third parties as a filler in building construction.

In order to carry out a study on the effects of wastewater reuse on the firing stage, boron was used as a tracing element on being one of the most significant elements, owing to its high water solubility and possible accumulation in the wastewater reuse process. Boron is moreover an element that is found in sufficient quantities to allow quantifying its presence in all the studied streams by chemical analysis.

Boron contents were first determined in the glaze raw materials, subsequently sampling during two weeks of standard production.

Using the values obtained in the various chemical analyses, and drawing up a mass balance of the process, the mass flow rates of the boron compounds were found for all the streams.

Mass flow rates are expressed in mg boron/kg processed body raw material.

The resulting boron mass flow rates, obtained by using the calculation indicated with experimental data and relevant calculations, are set out in Table IV.

Table IV.- Boron mass flow rates in different streams.

Stream	Boron (mg/kg body)
Waste reused in body preparation (wastewater+sludge+glaze rests+green scrap)	225
Spray-dried powder and unglazed green tile	224.7
Spray dryer emissions	0.3
Glaze raw materials	740
Unfired glazed tiles	739.7
Fired glazed tiles	739.1
Firing kiln emissions	0.6

The results show that the dissolved or suspended elements present in the reused wastewater are mainly incorporated into the fired product where they are immobilized. The only possible environmental impact would be the possible solubility of the resulting fired scrap (non-conforming to standard or broken tiles, etc.), but in this process such waste can also be reused in the body preparation stage as grog, so that the environmental impact of wastewater reuse can be considered practically negligible.

5 CONCLUSIONS

In accordance with the results obtained in the study of the arising air emissions in spray drying and firing, and the data from the mass balance of boron in the various streams, it was verified that most boron compounds were incorporated in the finished product, where they were immobilized.

Thus, in wastewater reuse according to the system proposed in this work, it was found there was a negligible environmental impact if spray dryer dust emissions were kept below 50-75 mg/Nm^3, and if green and fired scrap was also reused in the same process.

From an economic and environmental viewpoint, the implementation of a wastewater reuse system, with reuse of the arising green and fired scrap, therefore wastewater reuse system, with reuse of the arising green and fired scrap, therefore represents the Best Available Technique for treating these waste streams.

6 REFERENCES

[1] Blasco, A.; Ginés, F.; Jarque, J.C.; Monfort, E. Adición de fangos reciclados a composiciones de pavimentos y revestimientos cerámicos (I). *Técnica Cerámica*, 195, 470-483, 1991.

[2] Blasco, A.; Ginés, F.; Jarque, J. C.; Monfort, E. Adición de fangos reciclados a composiciones de pavimentos y revestimientos cerámicos (y II). *Técnica Cerámica*, 196, 578-585, 1991.

[3] Blasco, A. et al. *Tratamiento de emisiones gaseosas, efluentes líquidos y residuos sólidos de la industria cerámica*. Castellón. Ed.. AICE-ITC, 1992.

[4] Enrique, J.E.; Monfort, E.; Ferrando, F.; Negre,. Elimination of boron from wastewater produced in the ceramic industry. *7th Mediterranean Congress of Chemical Engineering*. Barcelona. 1996.

[5] P. Mazzali. *L'inquinamento atmosferico*. Bologna:Ed. Pitagora, 1989.

[6] Busani, G.; Palmonari, C.; Timellini, G. *Piastrelle ceramiche & ambiente*. Modena: Edicer. 1995.

[7] López, F. J.; Giménez, E.; Hernández, F. Analytical study on the determination of boron in environmental water samples. *Fresenius J. Anal. Chem.*, 346, 984-987, 1993.

Nuclear Fuel Development

Kinetics of and Atmospheric Effects on Gallium Removal from a CeO_2 Based Mixed Oxide Surrogate

YoungSoo Park, Thomas N. Taylor, Arlette Atencio, and Darryl P. Butt
Los Alamos National Laboratory, Materials Science and Technology Division
Los Alamos, NM 87545

ABSTRACT

This paper describes preliminary experimental results on the kinetics of Ga removal from Ga_2O_3-doped CeO_{2-x}, a surrogate for weapons grade PuO_{2-x}. Ga is removed from the surrogate feedstock material using thermal techniques. An Ar-6% H_2 gas was used in order to reduce the oxide to gaseous Ga_2O. Experiments were done in the temperature range of 600-1200°C as a function of time, gas flow rate, and sample geometry. Samples were analyzed through measurements of weight change, scanning electron microscopy/energy dispersive spectroscopy, x-ray photoelectron spectroscopy, proton induced x-ray emission, and direct current plasma analysis. Gallium levels were reduced by as much as 98% (130 ppm Ga residue) of that in the starting material. Studies to date have shown CeO_{2-x} is a very good surrogate for PuO_{2-x}.

BACKGROUND

Currently there is interest in fissioning weapons grade plutonium in nuclear reactors, making use of its valuable energy while at the same time reducing certain dangers associated with its potential for nuclear weapons proliferation.[1,2] In the process of dismantling and declassifying nuclear weapons, the U.S. intends to convert much of the Pu metal to oxide using a hydride-dehydride process. This process yields a PuO_{2-x} powder which potentially can be incorporated into a mixed oxide nuclear fuel (MOX), a mixture of PuO_2 and UO_2, or potentially in an advanced non-fertile fuel. However, unlike reactor grade Pu, weapons grade Pu contains minor additions of gallium in order to stabilize the δ-phase and improve the machinability of the alloy. Gallium (melting point = 30°C) is a known embrittling agent and alloys rapidly with most metals.[3-7] Consequently, to assure proper cladding and fuel performance, Ga must be largely removed from weapons grade plutonium before it can be processed and used.

In order to avoid aqueous processing of this material, which produces considerable additional waste, we have proceeded toward the development of a relatively simple thermal process for removing gallium from PuO_{2-x} with the objective of achieving parts-per-million levels. Our proposed removal process involves heating the oxide in an Ar-H_2 environment, probably at temperatures in excess of 1000°C, producing as the primary gaseous product $Ga_2O(g)$ from decomposition of the Ga_2O_3 present in the mixture. As described later, the equilibrium partial pressure of $Ga_2O(g)$ above Ga_2O_3 is relatively high in a reducing atmosphere. Once evolved, the $Ga_2O(g)$ is swept away from the PuO_{2-x} by the gas stream and must be collected and removed from the heat treatment system. In this paper, we briefly communicate some of the important fundamentals of the thermodynamics and kinetics of the processes of vaporization and deposition using CeO_{2-x} as a surrogate for PuO_{2-x}. Due to similarities in thermodynamic properties, d-orbital bonding, Ga solubility, oxygen diffusivity, and nonstoichometry, CeO_{2-x} has been proposed as a reasonably good surrogate for PuO_{2-x}.[3-15] In addition, a well characterized surrogate outside of a glove box environment aids evaluations for pilot or full scale production equipment, thus preventing unnecessary radiation exposure to personnel, excessive levels of radioactive waste, and costly errors that could be associated with setting up a large scale system in a glove box without the advantage of a prior operational checks.

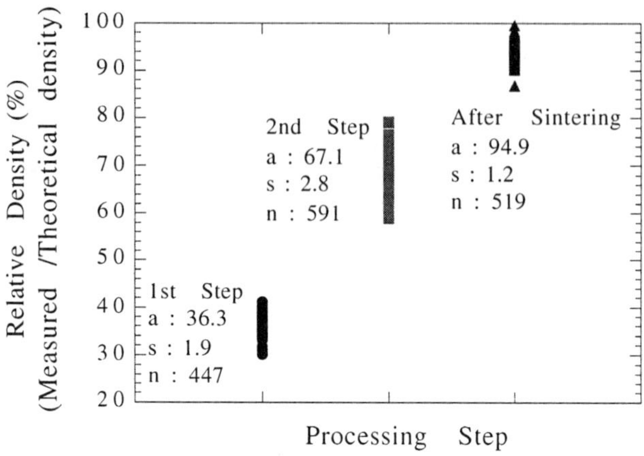

Figure 1. Green and fired densities of CeO_2- 2wt.% Ga_2O_3 pellets after the steps described in the experimental procedure. Note, "n" represents the number of specimens on which measurements were made, "a" denotes average percent theoretical density, and "s" denotes standard deviation.

EXPERIMENTAL PROCEDURES

CeO_2-based MOX surrogate feedstock material was prepared by blending, sintering, and milling CeO_2 plus 2 wt% Ga_2O_3 powders. The overall process were summarized as follows; The desired quantities of CeO_2 and Ga_2O_3 powder were measured and blended, and vibration milled in a small SPEX mill for 20 minutes. The powder was pressed into 1/4" x 1/8" diameter pellets to ≈35% of theoretical density in the green state. The green pellets were milled again and sifted through a -150 μm screen and then pressed into 1.1 cm x 0.65 cm diameter, ≈65% dense pellets. The pellets were then fired at 450°C for 4 h to remove the binder and were immediately sintered at 1650°C for 4 h in air. After sintering, the pellet density was measured by the Archimedes method. The sintered pellet was then re-crushed to desired particle size distributions. The densities of pellets are shown as a function of processing step in Fig. 1. Figure 2 shows the measured Ga content of the surrogate before and after sintering compared with a nominally 1 wt% Ga_2O_3-PuO_2 material calculated. From this figure, it is apparent that Ga loss during sintering provided a surrogate with a Ga content comparable to that in weapons grade PuO_2 feedstock material.

Gallium removal studies were performed by exposing different powder lot sizes of the surrogate feedstock as well as complete pellets to ultra-high purity Ar-6% H_2. In this paper we report the results of experiments performed in the temperature range of 600-1200°C as a function of time, gas flow rate, particle size, and sample size or geometry. The samples consisted of 0.3, 0.9, and 2.5 gram lots, as well as pellets, which rested in small alumina boats. The boats were placed in the hot zone of a controlled atmosphere alumina tube furnace. The oxygen partial pressure was controlled at $\approx 10^{-21}$ ppm during the thermal treatments. Surrogate powder and pellet samples were heated to the desired temperature at 20°C/min and held between 0.5 to 12 hours. The weight change of surrogate was monitored before and after exposure (although some weight gain did occur upon re-exposure to air due to O_2 pick up by substoichiometric CeO_{2-x}). Samples were analyzed by scanning electron microscopy/energy dispersive analysis (SEM/EDS), x-ray photon electron spectroscopy (XPS), Rutherford backscattering spectrometry (RBS), proton induced x-ray emission (PIXE), direct current plasma emission spectroscopy (DCP), and Leco TC-136 oxygen/nitrogen analysis.

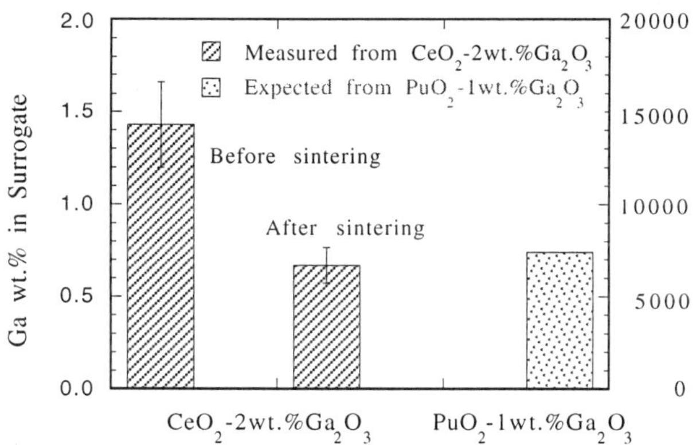

Figure 2. Measured Ga concentration in the ceria based surrogate powders before and after sintering compared with that in a nominal 1 wt% Ga_2O_3-PuO_2 mixture.

RESULTS AND DISCUSSION

Thermodynamic data for the Ga-O-H system were collected[16] and free energies of formation were fit, using stepwise multiple linear regression, to the equation:

$$\Delta G_f = a + bT + cT^{-1} + dT^2 + eT^3 + fT\ln T \quad (1)$$

where a-f are constants, and T is the temperature in degrees Kelvin. Table 1 shows some of the data pertinent to this brief analysis. Note that other gaseous products such as Ga, GaH, GaO, Ga_2O_2, and Ga_2O_3 are not included in these analyses because their equilibrium partial pressures are comparatively low. The constants in Table 1 were used to calculate temperature dependent expressions for the vaporization behavior of gallium oxide from doped PuO_2 in Ar-6% H_2. A temperature dependent expression for the free energy of reaction can be determined from the mass action equation:

$$Ga_2O_3(s) + 2H_2(g) \rightarrow Ga_2O(g) + 2H_2O(g) \quad (2)$$

From this equation the following relationship was derived from which $Ga_2O(g)$ partial pressures can be calculated:

$$p_{Ga_2O} = \frac{a_{Ga_2O_3} p_{H_2}^2}{p_{H_2O}^2} \exp\left[91.3647 + 1.1203 \cdot 10^{-3} T - 7.761944 \cdot 10^{-8} T^2 - \frac{64223}{T} + \frac{157638}{T^2} - 7.8179 \ln T\right] \quad (3)$$

Where a and p represent activity and equilibrium partial pressure, respectively.

Alternatively, the vaporization of $Ga_2O(g)$ may be assessed using the mass action equation:

$$Ga_2O_3(s) \rightarrow Ga_2O(g) + O_2(g) \quad (4)$$

where the partial pressure of $Ga_2O(g)$ may be calculated using the equation:

$$p_{Ga_2O} = \frac{a_{Ga_2O_3}}{p_{O_2}} \exp\left[75.69396 - 3.7071 \cdot 10^{-4} T + 2.50277 \cdot 10^{-8} T^2 - \frac{121066}{T} + \frac{117327}{T^2} - 3.82324 \ln T\right] \quad (5)$$

Equation (5) may be substituted for equation (3) by considering how the partial pressure of oxygen is controlled by the H_2/H_2O ratio according to buffer reaction:

$$H_2(g) + 1/2O_2(g) \rightarrow H_2O(g) \quad (6)$$

It is apparent from equations (5) and (6) that the higher the H_2/H_2O ratio, the higher will be p_{Ga_2O} (i.e., p_{Ga_2O} is inversely related to p_{O_2}). Thus, the vaporization of Ga_2O would be expected to be more rapid in dry versus moist hydrogen. The equations described above can be used to calculate the equilibrium partial pressures of Ga_2O above various PuO_2-Ga_2O_3 solid solutions. In a typical PuO_2 feedstock, it is anticipated that $a_{Ga_2O_3}$ will be on the order of 0.01 prior to any efforts to remove Ga. However, at this date, the activity of Ga_2O_3 in a typical PuO_2

feedstock powder is not precisely known; therefore, for the purpose of illustration we must assume certain values of $a_{Ga_2O_3}$. Figure 3, shows how the vaporization behavior varies with environment as a function of $a_{Ga_2O_3}$. The calculations were done assuming the gas was a dry mixture of 1 atm of 6% H_2 and a balance of inert gas, such as Ar or He. This gas composition represents the experimental gases we are using in our Ga removal studies. It is apparent from Fig. 3, that the vaporization rate under the reducing effect of H_2 is relatively high, for example at 1000°C (1273 K) p_{Ga_2O} varies between 10^{-4} and 10^{-2} for $a_{Ga_2O_3}$ between $1 \cdot 10^{-6}$ and 1, respectively. Thus, in a dry hydrogen environment, at elevated temperatures, Ga_2O_3 will vaporize relatively rapidly according to equation (2).

Table 1. Summary of constants shown in equation 1.

Compound	a	b	c	d	e	f
Ga_2O_3 (s)	-1123572.	574.0020	2267569.	5.3080E-03	-6.8814E-07	-31.7864
Ga_2O(g)	-117031.	-55.3176	1295110.	8.3901E-03	-8.9622E-07	0
H_2O(g)	-236296.	-65.1431	-169071.	-6.2395E-03	4.2670E-07	16.6059

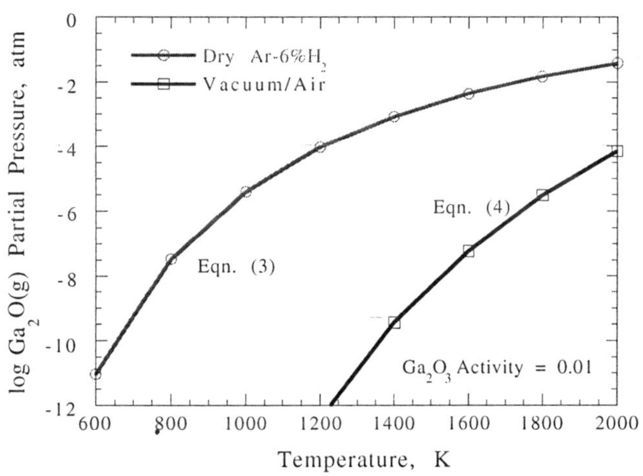

Figure 3. Calculated thermodynamic equilibrium of Ga_2O(g) above Ga_2O_3(s) in 0.06 atm of H_2 as a function of temperature and $a_{Ga_2O_3}$.

The results of studies of the kinetics of Ga removal are presented in Figs. 4 and 5. Figure 4 shows the measured weight change observed as a function of temperature for the surrogate pellets and the three powder lot sizes. The weight changes measured in Ar-6% H_2 are contrasted to weight changes measured in pure Ar. It is noteworthy that following treatments in Ar-6% H_2 the powder appeared grey and metallic, while powder treated in pure Ar remained yellow-white. The data in Fig. 4 demonstrates that the surrogate is reduced in Ar-6% H_2, but is not significantly reduced in pure Ar, which is in agreement with the thermodynamic calculations described above. However, the weight loss observed in Ar-6% H_2 is due to a combination of Ga loss and reduction of CeO_{2-x} to a more substoichiometric compound.[17,18] The data presented in Fig. 5, which shows measured Ga concentrations, are required to fully elucidate the kinetics of Ga removal. The data in this figure clearly show that there is a strong correlation between total weight loss and Ga loss. Chemical analyses of samples heat treated in pure Ar showed no significant loss of Ga. This data serves to illustrate that a significant fraction of Ga can be thermally removed via the reaction between H_2 and the oxide as described above. There is clearly a strong correlation between Ga loss and temperature. It appears temperatures in excess of 1000°C will be required to achieve reasonable low levels of Ga.

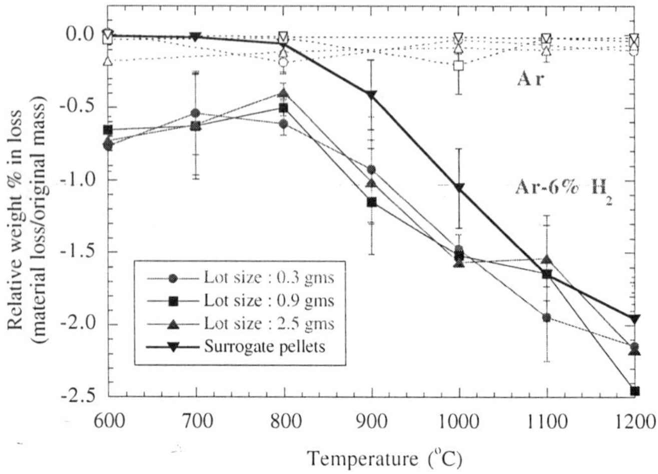

Figure 4. Plot of weight change versus temperature for CeO_2-2wt.% Ga_2O_3 surrogates exposed to Ar-6% H_2 and Ar for 30 minutes.

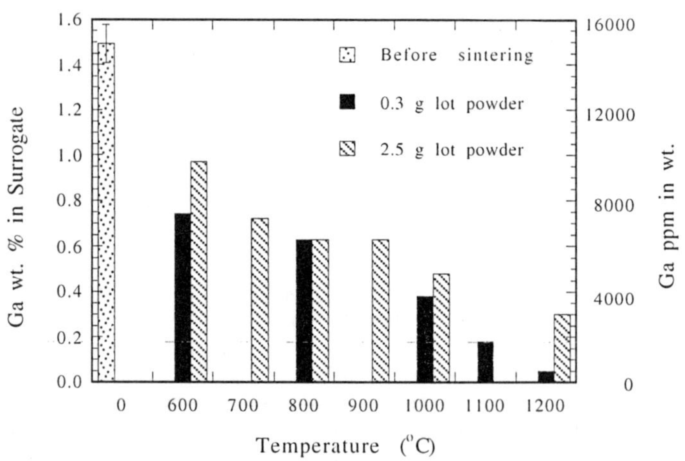

Figure 5. Plot of residual Ga versus temperature for 2.5 gram lots of CeO_2-2wt.% Ga_2O_3 powder exposed to Ar-6% H_2 for 30 minutes.

As illustrated in Fig. 6, the Ga level can be reduced further through longer heat treatments, although the kinetics are nonlinear. As shown in Fig. 6, levels as low as 130 ppm Ga have been achieved. As shown in Fig. 7, there did not appear to be a significant effect of particle size on Ga loss, despite a fair amount of scatter in the data. It is anticipated that Ga transport through PuO_{2-x} is at least as rapid as through CeO_{2-x}.

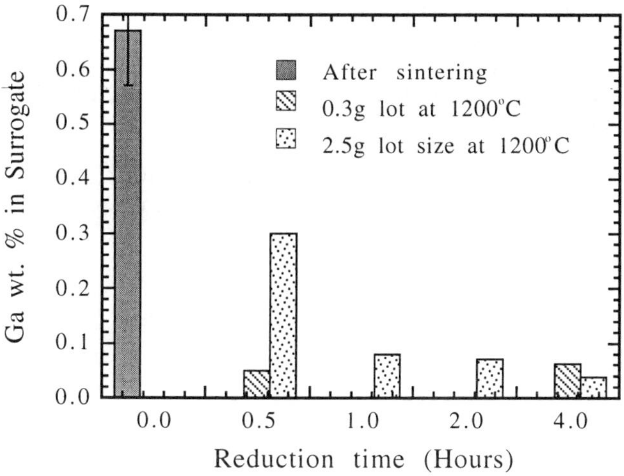

Figure 6. Plot of residual Ga versus time for CeO_2-2wt.% Ga_2O_3 sample exposed to Ar-6% H_2 at 1200°C.

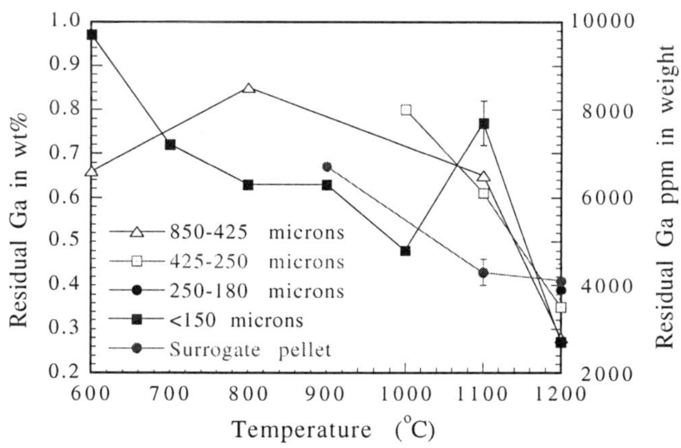

Figure 7. Particle size effect on residual Ga as a function of temperature for CeO_2-2wt.% Ga_2O_3 surrogate exposed to Ar-6% H_2 for 0.5 h at 1.5 cm/sec flow rate.

Studies of lot size have shown that lot size does affect the rate of Ga removal, as shown, for example in Figs. 8 and 9. However, studies of the effects of gas flow rate (between 1.5 and 6.0 cm/sec true velocity) have shown no effect on Ga removal. The combination of these experiments lead us to propose that the rate of reaction is rate limited by transport of $Ga_2O(g)$. Additional experiments are going on for the interpretation. A more extensive publication of these results as well as supporting kinetic modeling will be reported in a future publication.

Gallium deposition (following vaporization) is also an important issue in the design of a Ga removal system. We have conducted a number of studies of the Ga deposition products on the furnace wall, as summarized by the XPS results in Fig. 10. As discussed above, Ga_2O_3 vaporizes in H_2 to $Ga_2O(g)$ at elevated temperatures. During thermal processing the $Ga_2O(g)$ moves to cooler zones of the furnace, back reacts with $H_2(g)$ and $H_2O(g)$ and condenses out as Ga(l) and $Ga_2O_3(s)$. Upon removal from the furnace, the exposed Ga rapidly forms a surface oxide of Ga_2O_3. In agreement with these thermodynamic predictions, the XPS data of Fig. 10 demonstrate that the deposition product contains Ga_2O_3 and metallic Ga. Analysis of the XPS spectra also revealed an intermediate oxidation state for Ga. The precise bonding of this state could not be demonstrated conclusively. Although it is suggested that it may be solid Ga_2O from gas phase deposition, such a suboxide bonding configuration could be due to grading of the atmospherically produced oxide layer.

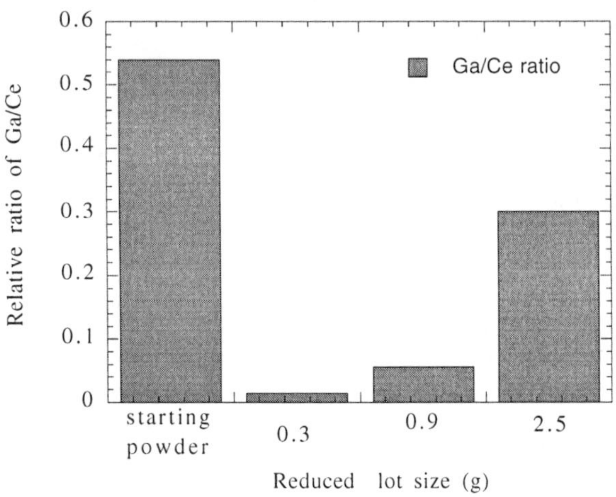

Figure 8. Relative Ga concentration as a function of lot size, measured by PIXE. Samples were exposed to Ar-6% H_2 at 1200°C for 30 minutes.

CONCLUSIONS

Studies of gallium loss were performed using CeO_2 surrogate for PuO_2 system as a function of atmosphere, particle size, sample lot size (configuration), and gas flow rate. It is shown that Ga can be removed thermally in an Ar-6% H_2 environment by the reduction of the oxide to $Ga_2O(g)$, which is subsequently transported away from the solid. Studies have shown that Ga loss depends on sample lot size, exposure time, and temperature but not significantly on flow rate or particle size. Further studies are underway to correlate between results of the kinetics modeling to experimental data and pilot system.

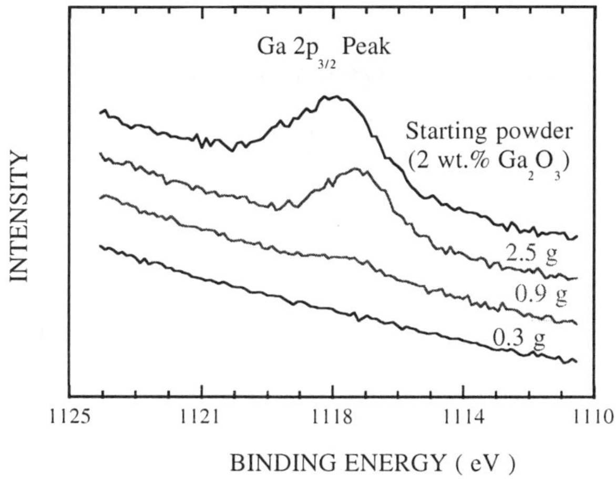

Figure 9. Relative Ga concentration as a function of lot size, measured by XPS. Samples were exposed to Ar-6% H_2 at 1200°C for 30 minutes.

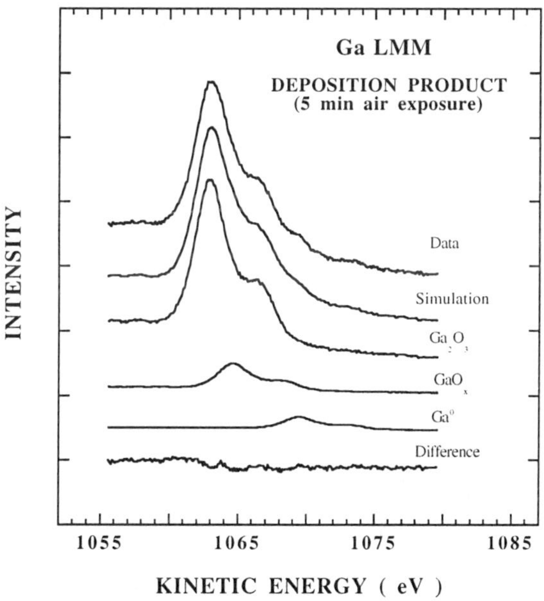

Figure 10. Simulation compared with the measured Ga LMM spectrum from the deposit on the furnace wall (minutes of air exposure following removal from the heat treatment furnace). In addition to the Ga_2O_3 and metallic Ga components used in the simulation, a third contribution (GaO_x) has been introduced that indicates the existence of an intermediate Ga oxidation state (or states). The difference spectrum compares the data with the simulation.

REFERENCES

[1]Department of Energy, "Plutonium Focus Area-Technology Summary," DOE/EM-0297, 1996.

[2]National Academy of Science, "Management and Disposition of Excess Weapons Plutonium," National Academy press, Washington, D.C., 1995.

[3]D. R. Lide, *Handbook of Chemistry and Physics*, 77th Edition, (1996-1997).

[4]D. F. Wilson, J. R. Distefano, J. F. King, E. T. Manneschmidt, and J. P. Strizak, "Interactions of Zircaloy Cladding with Gallium, 1997 Status," ORNL/TM-13505, Oak Ridge National Laboratory, 1997.

[5]P. R. Luebber, W. F. Michayd, and O. K. Chopra, "Compatibility of ITER Candidate Structural Materials with Static Gallium," ANL-93/31, Argonne National Laboratory, 1993.

[6]T. B. Lindemer, "Assessment of Gallium Effects in Processing and Performance of U-Pu Oxide LWR Fuels," ORNL/MD/LTR-38, 1995.

[7]H. R. Trellude, T. Baros, H. T. Blair, J. J. Buska, D. P. Butt, K. Chiester, S. F. Demuth, S. L. Eaton, G. J. Havrilla, R. J. Hanrahan, C. A. James, D. G. Kolman, R. E. Mason, Y. Park, M. Stan, J. H. Steele, Jr., S. S. Voss, T. C. Wallace, Sr., and C. G. Worley, "Nuclear Fuels Technologies Fiscal Year 1997 Research and Development Test Results," Los Alamos National Laboratory Report, LA-UR-97-4423, 1997.

[8]L. R. Morss and J. Fuger, Transuranium Elements, American Chemical Society, Washington, DC, 1992.

[9]A. Nakamura, "A defect-thermodynamic approach to PuO_{2-x} and CeO_{2-x}," *Journal of Nuclear Materials* , **201**, 17-26 (1993).

[10]O. Toffsϕrensen, "Thermodynamic Studies of the Phase Relationships of Nonstoichiometric Cerium Oxides at Higher Temperature," *J. Solid State Chem.*, **18**, 217-233 (1976).

[11]C. Guminski, "Solubility of Metals in Liquid Low-Melting Metals," *Z. Metallkd.*, **81**, 105-110 (1990).

[12]Per Kofstad, *Nonstoichiometry, Diffusion, and Electrical Conductivity in Binary Metal Oxides*, Wiley-Interscience, New York, 1972.

[13]W. D. Kingery, H. K. Bowen, and D. R. Uhlmann, *Introduction to Ceramics*, John Wiley & Sons, New York, 2nd Edition, (1976).

[14]J. L. Smith, Z. Fisk, and S. S. Hecker, "Actinides: from Heavy Fermions to Plutonium Metallurgy," *Physica*, **130B**, 151-158, (1985).

[15]G.V. Samsonov, *The Oxide Handbook*, IFI/Plenum Data Corporation, 1973.

[16] D. P. Butt, Y. S. Park, and T. N. Taylor, "Thermal Vaporization and Condensation of Gallium Oxide in Hydrogen," submitted to *J. Nucl. Mater.*, 1998.

[17]A. Laachir, V. Perrichon, A. Badri, J. Lamontte, E. Catherine, J. C. Lavalley, J. E. Fallah, L. Hilaire, F. I. Normand, E. Quémére, G. N. Sauvion, and O. Touret, "Reduction of CeO_2 by Hydrogen," *J. Chem. Soc. Faraday Trans.,* **87** [10] 1601-09 (1991).

[18]M. Romeo, K. Bak, J.E. Fallah, F.L. Normand, and L. Hilaire, "XPS Study of the Reduction of Cerium Dioxide," *Surface and Interface Analysis*, **20**, 508-12, (1993).

POWDER AND PELLET CHARACTERIZATION IN THE MIMAS-PROCESS

M. De Vleeschhauwer, M. Lippens, F. Eeckhout, E. Pelckmans,
BELGONUCLEAIRE, Belgium.

ABSTRACT

Mixed uranium and plutonium oxide, universally labeled as MOX, is by now an industrial product, fueling a significant number of reactors in Europe. MOX-fuel is currently produced using a process called the 'MIMAS' process (MIcronisation of a MASter blend), which was developed by BELGONUCLEAIRE (BN). In this paper, the main steps in the process, on one hand, and the accompanying characterization of the product, on the other hand, are described.

INTRODUCTION

The fabrication of MOX fuel resembles to a certain degree the fabrication of UO_2. The main steps, from pelletizing, sintering, centerless grinding, fuel rod filling, end plug welding to the fuel assembly, are comparable. But there are also some major differences which have an important impact on the microstructural characteristics of the final pellet. These differences result from the need for a fine incorporation of the PuO_2 in the UO_2 matrix and the lower diffusion kinetics during sintering, related to the presence of Pu in the fuel.

The specifications for MOX-fuel can be divided into two groups. The first group is similar to UO_2 fuel, namely surface quality, geometry, density, thermal stability, stochiometry, chemical purity, grain size and pore distribution. The other group contains those criteria specific for the presence of plutonium; namely homogeneity, solubility, and Pu distribution. Although the specifications of the first group are similar to those of the UO_2-fuel, in the case of MOX, they are harder to reach, due to the influence of the presence of Pu and it's oxidation state.

- The density is more difficult to obtain in comparison with pure UO_2 fuel, due to the lower diffusion kinetics
- The stochiometry has a tendency to decrease, caused by the reduction of PuO_2, which can be compensated by an increased oxygen potential in the furnace, and
- the hydrogen pick up after sintering is greater in MOX-fuel. A reduced open porosity guarantees nevertheless a low content of hydrogen.

The plutonium specific criteria require a particular technology and form the basis for the different fabrication steps for the MOX-production.

- The maximum size of a plutonium rich particle in a MOX-pellet is limited, due to the maximum allowable local heat generation in the reactor, and to ensure a good UO_2 and PuO_2 interdiffusion.
- The solubility must be high enough. This criterion results from the impossibility to dissolve in the standard conditions solid solutions of UO_2 - PuO_2 of which the Pu-content is higher than 40 %.

Now, all the different processes pursue a fine incorporation of the PuO_2-powder into the UO_2 matrix, to meet these two criteria. In the MIMAS-process the mechanical blending of the UO_2 and PuO_2 powder is executed in two steps (see figure 1). Firstly, a master blend with a plutonium enrichment of 25 to 35 % is micronised in a ball mill, and this blend is secondly sieved and mechanically diluted with UO_2-powder in order to obtain the desired plutonium enrichment. BELGONUCLEAIRE has by now produced 350t MOX using the MIMAS process, and the fuel has proven to be very reliable, with an almost zero failure rate.

DESCRIPTION OF THE MIMAS PROCESS

Powder preparation and characterization

The MIMAS process (figure 1) is characterized by the ball milling of a master or a primary blend (UO_2, PuO_2 powders and recycled scraps) with a Pu content of 25-35%, followed by a secondary blending where the master blend is diluted with UO_2 to give a mixed powder for pelletizing at the required Pu content (1-8 %Pu).

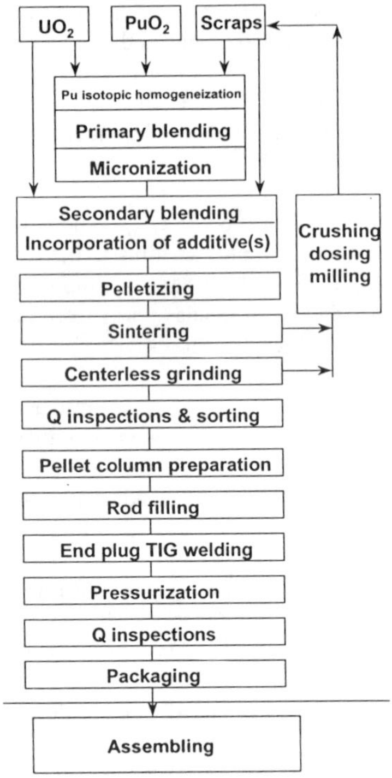

Figure 1: MOX fuel fabrication process (MIMAS)

In figure 2, 3 and 4 SEM-pictures of the different components are shown. The UO_2 powder (ex-AUC) is characterized as rounded grains, with a particle size ranging from a few μm up to 100 μm. The UO_2 powder is not cohesive and free flowing. The initial UO_2 characteristics determine to a great amount the final MOX fuel pellet microstructure.

Unlike PuO_2, UO_2 is delivered in quite large homogenized batches. Besides the specific production characteristics (grain size, densities, pressability, sinterability, etc.), special attention is paid to the impurities:

- boron, cadmium, gadolinium, rare earth content (absorption aspects),
- chlorine, fluorine (corrosion aspects),
- silicon, aluminum (risk of eutectic formation).

The standard isotopic composition corresponds to depleted uranium (0.25% U^{235}).

The ex-oxalate PuO_2 powder consists of plate shaped grains, which are also composed of a relative loose pile of sheets. The cohesive behavior and flowability is minor. The PuO_2 powder is delivered in batches of 50 to 130 kg. As with the UO_2, each lot is delivered with a certificate with the chemical analyses and the determination of the isotopic composition. Particular attention is paid on halide content (corrosion aspects). For the other impurities, the subsequent large dilution of PuO_2 powder by UO_2 allows less severe limits. To avoid an unacceptable dose rate for the personnel, which is a result of gamma and neutron emitters (increasing with age), the ^{241}Am is taken as a convenient expression of the plutonium age and is presently limited to 17000 ppm maximum 12 months after being made available for production.

The third component is scrap, created in the MIMAS-process itself, mainly during the grinding step and the sorting out of the ground pellets. The scrap is homogenized and micronized in a ball mill for several hours. This micronization step leads to submicron particles, which form agglomerates of a few tenths in µm. A fraction consists of irregular particles with a diameter ranging from 1 to 10 µm. This component shows an important cohesive behavior, and has a very poor flowability. Before the homogenized scrap are used in the master blends, a chemical and isotopical analysis is executed.

The master blend which can contain up to 75% scrap is homogenized and micronized in a ball mill for several hours. The final powder structure of the master blend is given in figure 5. The similarity with the structure of the micronized scrap is evident. After the micronization step, the master blend is divided in several cans.

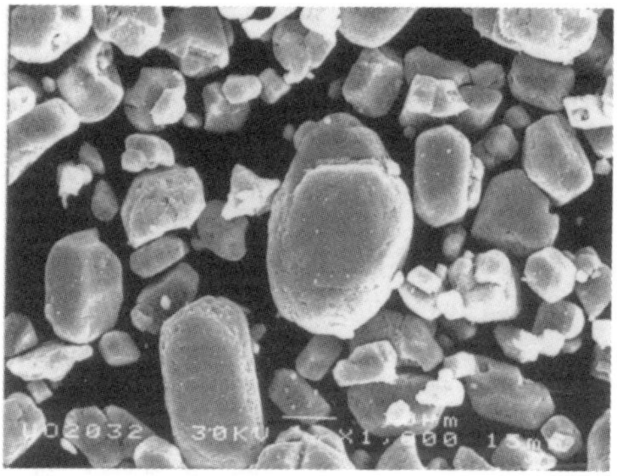

Figure 2: UO_2 (ex-AUC) powder

Figure 3: PuO_2 powder

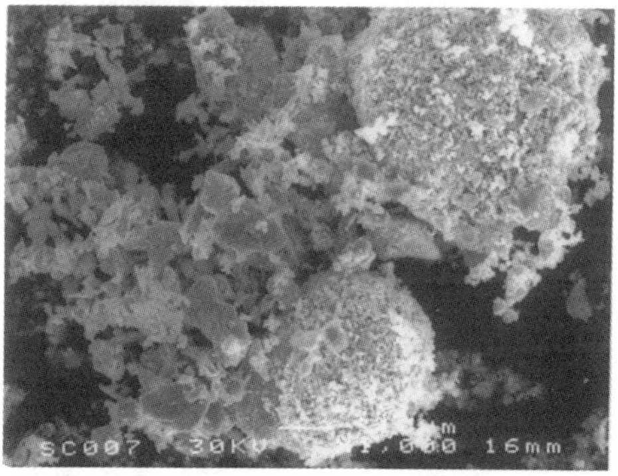

Figure 4: Micronized scrap powder

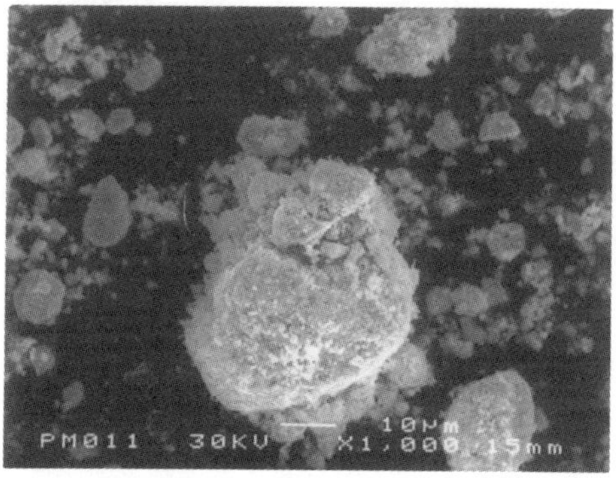

Figure 5: Powder structure of the primary blend

In order to achieve a high level of homogenization in Pu isotopic composition for a large amount of powders, BN has developed and demonstrated a unique computerized system to manage the specific cross blending approach, outlined in figure 6. In this figure, it is shown how the master blends are divided over the secondary blends. Before the several cans of the different master blends are diluted with the free flowing UO_2 powder, they are homogenized and sieved in order to break the larger agglomerates and to reduce the maximum size of the plutonium rich particles. The average effective diameter of the plutonium rich particles is smaller than 50 μm, with 95 % of these particles having a diameter smaller than 100 μm and no particle reaching the 400 μm threshold for safety aspects. It must be emphasized that the maximum plutonium content of these particles is limited from 25 to 35% as given by the master blend composition. Whereas the early fabrications showed important heterogeneities of the plutonium distribution, the present MIMAS fabrication is almost free of heterogeneities and is from this point of view also much closer to U fuel than the MOX fuel prepared by the previous processes. Furthermore during irradiation, the as fabricated heterogeneities disappear progressively by the burn-up effect (production of plutonium in the U-matrix and burn up of plutonium in the Pu rich particles) and by U-Pu interdiffusion.

Finally, the secondary blends are prepared in two steps before they are transported to the presses. First they are homogenized to obtain a fine mixing of the free flowing UO_2 powder with the cohesive powder of the sieved master blend, and secondly the additives like the lubricants and the pore formers are incorporated with a final mixing step. In figure 7 the structure of the powder of the secondary blend is given. It can be seen that besides the bigger agglomerates of the master blend, a large number of the plutonium rich particles are located on the surface of the UO_2 particles.

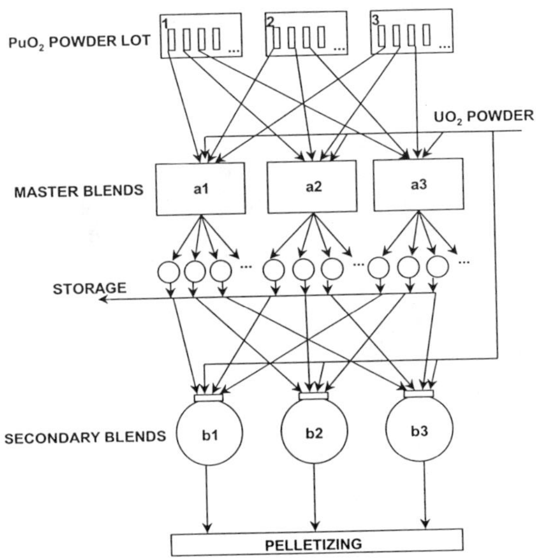

Figure 6: Plutonium isotopic homogenization (MIMAS)

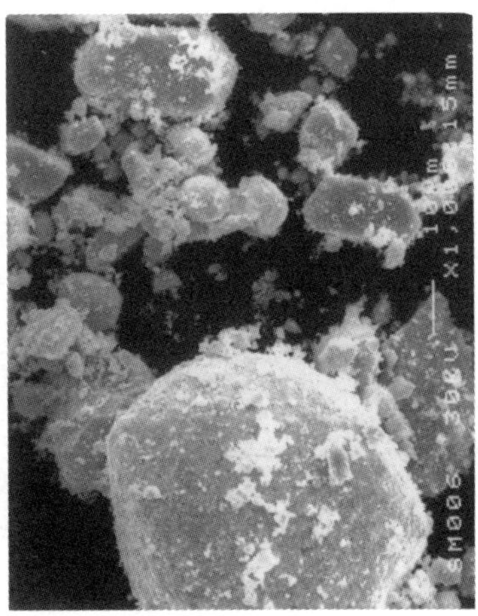

Figure 7: Powder structure of the secondary blend

Pellet fabrication and characterization

It is important that the final homogenized powder doesn't segregate during the transport to or in the filling system of the press itself. The design and concept of the powder transport systems and the filling system of the press are carefully adapted to deal with this topic. Apart from that, the pelletizing of MOX is quite similar to the pelletizing of UO_2. The green MOX pellets are pressed to a density ranging from 6.2 to 6.6 g/cm^3 (± 60 % TD) with an hydraulic, double action press. A pressure in the order of 5 T/cm^2 is applied. On a regular base, the dimensions and density of the green pellets are checked, which allows the fine tuning of the press.

After pelletizing, the green pellets are sintered in a continuous furnace. The furnace is divided into three sections, a pre-sintering zone, a sintering zone and a cooling off zone. In the pre-sintering zone the lubricants and the other additives are removed, and the UO_2 is reduced. This reduction is followed by the reduction of PuO_2, which has the tendency to reduce to Pu_2O_3. The control of the furnace parameters, such as the temperature profiles and the gas chemistry lead nevertheless to stochiometry (2.00). The pellets are sintered at about 1700°C and in a Ar+H_2 atmosphere (with a controlled percentage of moisture). Furthermore, the control of the sintering parameters is one of the key factors to guarantee an appropriate and stable microstructure, a low hydrogen content in the pellets, and a good solubility of the pellet.

The materials, processes and sintering parameters guarantee;

- a stable and constant density of 95%TD
- a maximal insolubility of the BN MOX pellets to an amount of circa 0.3 %(10M HNO_3, 10h)
- a hydrogen content well under control and below 1 ppm.
- a microstructure, characterized with a mean grain size of the UO_2 particles between 8 to 10 µm (figure 8), and a size distribution of the 'plutonium rich' particles, as shown in figure 9.
- a well fixed porosity distribution (see figure 10). The figure shows that typically 70 to 80 % of the pores have a diameter between 1 and 5 µm, which leads to a thermally stable fuel. The addition of a pore former (AZB) creates a bimodal distribution with a supplementary peak around 10 µm. On the other hand, analyses have pointed out that the addition of scrap in the secondary blends augments the fine porosity (5µm). The open porosity is very low, which benefits a low fission gas release during irradiation.
- a moderate homogenization of the plutonium in the MOX fuel. The plutonium diffuses along the grain boundaries of the UO_2, and only for a little amount in the interior of the UO_2 grains, because the volume diffusion kinetics are very poor in comparison with the grain boundary diffusion. Nevertheless during sintering the U-Pu interdiffusion can form a solid solution in the plutonium rich particles. These plutonium rich particles originate from the micronized master blend, and so during sintering the local higher concentrations are smoothed out. In figure 11 the distribution of the local plutonium content is given for three different cases. The first is the reference fuel, obtained by a direct mechanical blending of the PuO_2 and UO_2 powder (not MIMAS). Here peaks up to 100% are seen. The second is the MIMAS prototype fuel, which shows a significant reduction of the plutonium content in the plutonium rich particles. The plutonium content outside the plutonium rich particles is very low, and the content varies very rapidly from point to point, meaning that the plutonium is not diffused inside the UO_2 grains but remains close to the grain boundaries. In the third case the distribution is shown for the MIMAS process where extensive interdiffusion and grain growth has taken place. Here the local plutonium content is significantly smoothed out due to the grain growth (12 µm). The actual MIMAS fuels, characterized with a grain size of 8 µm have a local plutonium distribution typically between the two last MIMAS examples shown here.

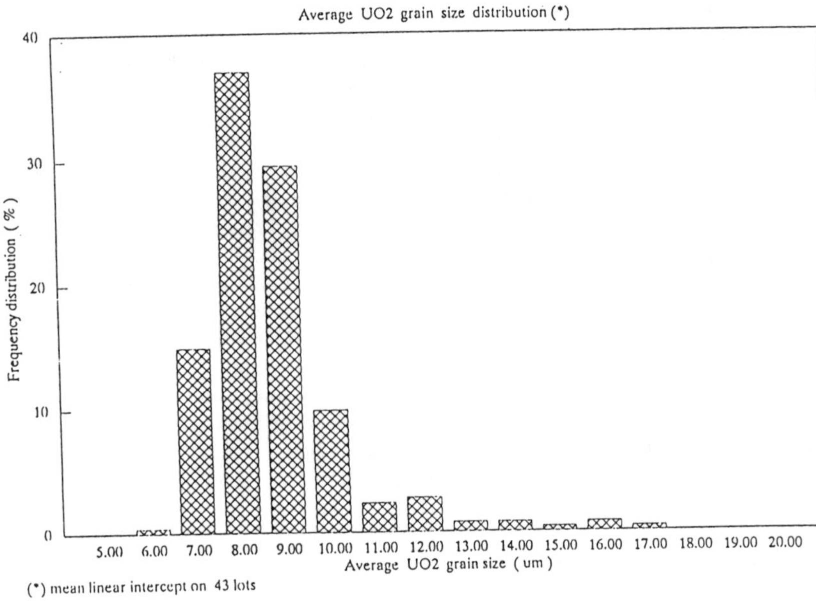

Figure 8: UO_2 grain size distribution of a sintered MOX pellet

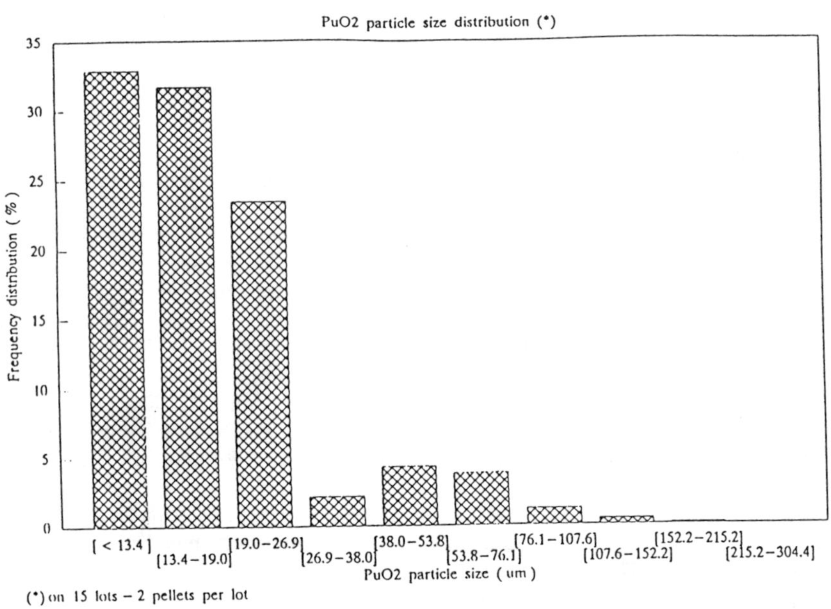

Figure 9: 'Plutonium rich' grain size distribution of a sintered MOX pellet

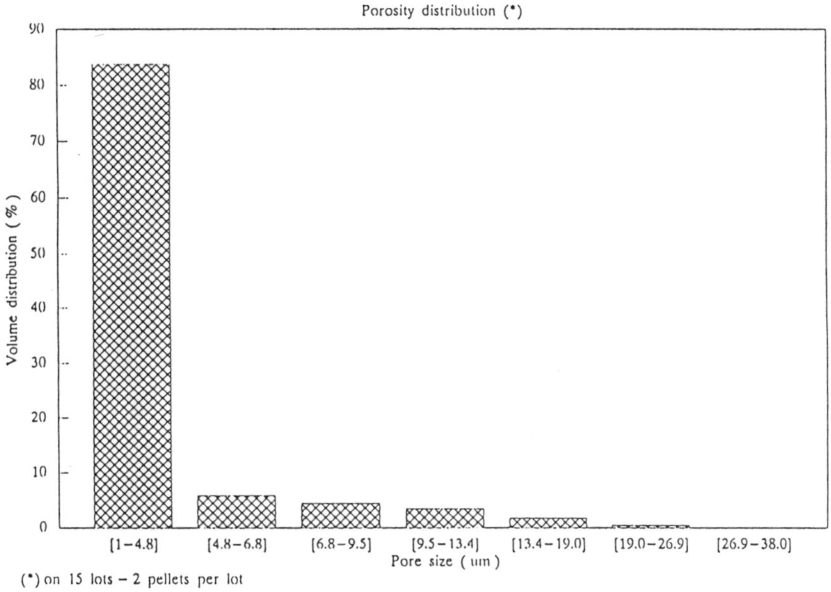

Figure 10: Porosity distribution in a sintered MOX pellet.

CONCLUSION

Due to its microstructural characteristics, the MIMAS fuel is similar to the U fuel prepared from the same UO_2 powder type. These characteristics give MIMAS pellets in-reactor properties approaching very closely those of U pellets. The experimental results accumulated so far on MIMAS fuel demonstrate that the behavior of MOX fuel has always been as good and even superior to U fuel.

However, the presence of plutonium in these pellets means that there are significant differences in the fabrication and control aspects which lead to the delivery of an engineered product fitting the most stringent customer specifications. By coping with the specific Pu-related constraints in the fabrication and control stages, high quality MOX fuel can be readily fabricated with a consistent reliability, meaning that MOX fuel is an industrial solution to the global need to reduce the worldwide plutonium stocks and the production of high level waste.

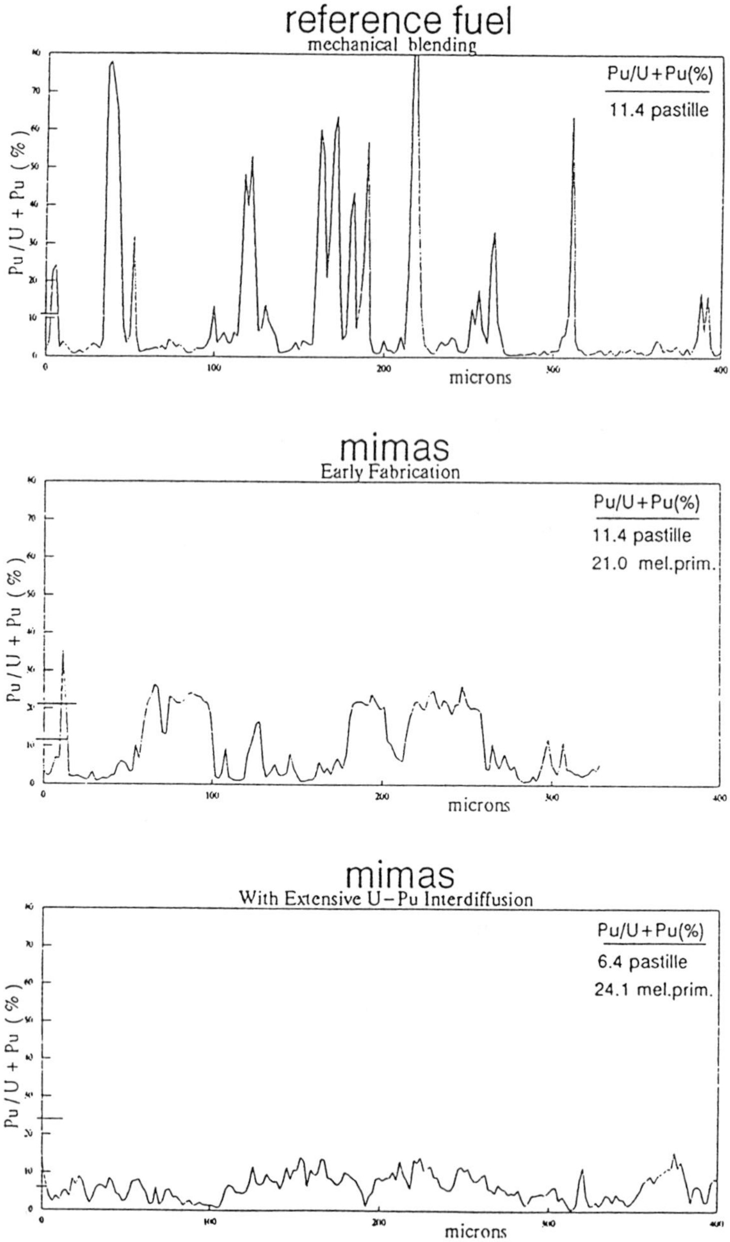

Figure 11: Local plutonium distribution in a sintered MOX pellet

Stabilization of Low-level Radioactive and Hazardous Wastes

MICROWAVE TECHNOLOGY FOR WASTE MANAGEMENT APPLICATIONS INCLUDING DISPOSITION OF ELECTRONIC CIRCUITRY

George G. Wicks
Westinghouse Savannah River Technology Center
Aiken, SC 29802

Rebecca L. Schulz and David E. Clark
University of Florida
Dept. Materials Science & Engineering
Gainesville, FL 32611-6400

ABSTRACT

Advanced microwave technology is being developed nationally and internationally for a variety of waste management and environmental remediation purposes. These efforts include treatment and destruction of a vast array of gaseous, liquid and solid hazardous wastes as well as subsequent immobilization of hazardous components into leach resistant forms. Microwave technology provides an important contribution to an arsenal of existing remediation methods that are designed to protect the public and environment from the undesirable consequences of hazardous materials. One application of special interest is the treatment of discarded electronic circuitry using a new hybrid microwave treatment process and subsequent reclamation of the precious metals within.

INTRODUCTION

Significant quantities of hazardous wastes are being generated from a multitude of processes and products in today's society. Microwave energy represents a unique heating source with the potential of providing a highly flexible means for a) minimizing generation of selected future wastes, b) reducing existing wastes and immobilizing hazardous components and c) reclaiming or recycling reusable and sometimes valuable materials in waste products. Some of the many wastes under study using microwave treatments include radioactive wastes and sludges (high

(HLW), low (LLW) and intermediate level wastes, transuranic (TRU) and mixed wastes), contaminated soils and sediments, incinerator ashes, industrial wastes, sludges, rubber products including tires, medical and infectious wastes, asbestos, groundwaters, volatile organic compounds (VOC's), discarded electronic circuitry and shipboard wastes.

POTENTIAL ADVANTAGES OF MICROWAVE PROCESSING

Microwave energy interacts with matter in a manner different than other thermal treatment processes. As result of this unique feature, the advantages of using microwave energy for treating a vast array of hazardous wastes can include some or all of the following:

- Waste volume reduction
- Rapid heating
- High temperature capabilities
- Selective heating
- Enhanced chemical reactivity
- Ability to treat wastes in-situ
- Treatment or immobilization of hazardous components to meet regulatory requirements for storage, transportation or disposal
- Rapid and flexible process that can also be made remote
- Ease of control
- Process equipment availability, compactness, cost, maintainability
- Portability of equipment and process
- Reduction in personnel radiation exposure for rad wastes (as low as reasonably achievable - ALARA)
- Energy savings
- Cleaner energy source compared to some more conventional systems
- Overall cost effectiveness/savings

The advantages ultimately realized depend both on the type and characteristics of the wastes to be treated.

WSRC/UF MICROWAVE PROGRAMS

A team from the Westinghouse Savannah River Technology Center (DOE Laboratory) and the University of Florida has been active for about a decade in the development of microwave technology for a variety of waste management applications. This interaction has resulted in the development of some unique equipment and uses of microwave energy. The team has now joined forces with industrial partners to tailor this technology for specific industrial needs. Some of these efforts and programs are depicted schematically in Figure 1.

ELECTRONIC CIRCUITRY AND RECLAMATION OF PRECIOUS METALS

Among the existing waste inventory in the United States and world-wide is millions of circuit boards from a vast array of discarded consumer products as well as from the DOE and DOD community. At present, these materials are most often disposed of by discarding them in landfills throughout our nation. This results in a number of concerns. First, because of the large volume of waste material disposed of daily in landfills, many are reaching capacity, thus resulting in the need for new disposal sites causing increased costs both environmentally and economically. Next, there are a variety of hazardous materials that may be contained within electronic components. In landfills, these elements can leach from the waste and migrate into groundwater, which can produce undesirable environmental consequences. Finally, because the products are simply thrown away, the natural resources in the wastes, including precious metals such as gold, are discarded and cannot be reused or recycled. If one can develop a technology to not only treat the large volumes of electronic waste, reduce its volume and immobilize hazardous components into leach resistant forms, but also reclaim precious metals such as gold, an environmentally important contribution can be made with a significant potential return on investment.

Laboratory-scale studies have produced the following important results and have been documented in more detail in previous publications [5-11].

- A wide array of electronic components were able to be treated by a relatively simple and flexible, one-step, hybrid-heated microwave process.

- Geometric waste reductions of more than 50% were observed along with "landfill" volume reductions of many hundreds of percent.

- In this process, circuitry was converted into two waste form products; a glass and a metal form.

- The glass waste form retained hazardous components and met important environmental leaching standards. Furthermore, for most cases, the glass could be produced without the need of any additives.

- The metal form produced was subsequently processed to effectively reclaim precious and valuable metals, including gold and silver.

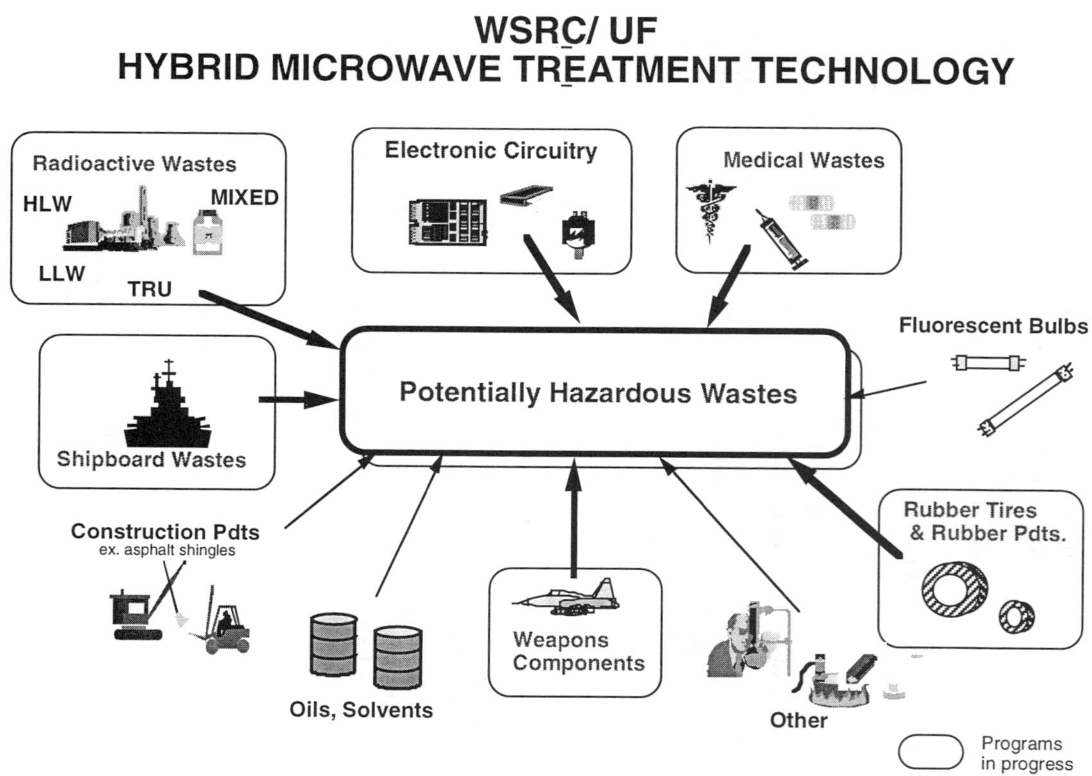

Figure 1. Flowsheet illustrating WSRC/UF's current and potential microwave waste remediation efforts.

• A new "tandem microwave" system was developed which handled both primary as well as secondary wastes in an integrated unit, i.e., the off-gases produced were simultaneous treated along with the electronic circuitry.

• The entire hybrid microwave process and specialized equipment (now being patented) can also be mocked-up to a larger scale and made mobile. The system is particularly suited to treatment of proprietary and/or very hazardous or radioactive materials. One of the many unique features of this treatment system is that since most of the electronics and controls can be made external to the treatment chamber, entire components can be destroyed and/or treated remotely by microwaves, which can be transported via waveguides, to controlled or contained areas.

The microwave studies conducted by SRTC and UF have been applied not only to treatment of primary waste streams, but also to the off-gases produced, using a new tandem microwave treatment concept designed to treat both primary as well as secondary wastes simultaneously.

The tandem microwave unit is shown in Figure 2 and consists of two microwave chambers interfaced together. First, the lower chamber is the combustion chamber that is used to ash or vitrify various solid and liquid waste streams. This unit may be "hybridized" by the use of a susceptor which allows processing of wastes normally not receptive to microwave heating. While the waste is being reduced in volume and transformed in this chamber, the off-gases produced pass into the upper microwave unit which contains a special series of filters. The material is also heated by microwaves which assist in the decomposition of the off-gases produced. Gas chromatography has shown that the amount of hazardous components found in the off gases are significantly reduced by the microwave treatment.

Laboratory scale data of microwave treatment of emissions are shown in Table 1. While only laboratory scale studies have been performed to date, it is believed that the overall concept and process can be mocked-up to a larger scale and even made mobile, if desired.

CONCLUSIONS

Microwave technology provides important advantages for remediation of a wide range of radioactive and non-radioactive hazardous wastes (solids, liquids and

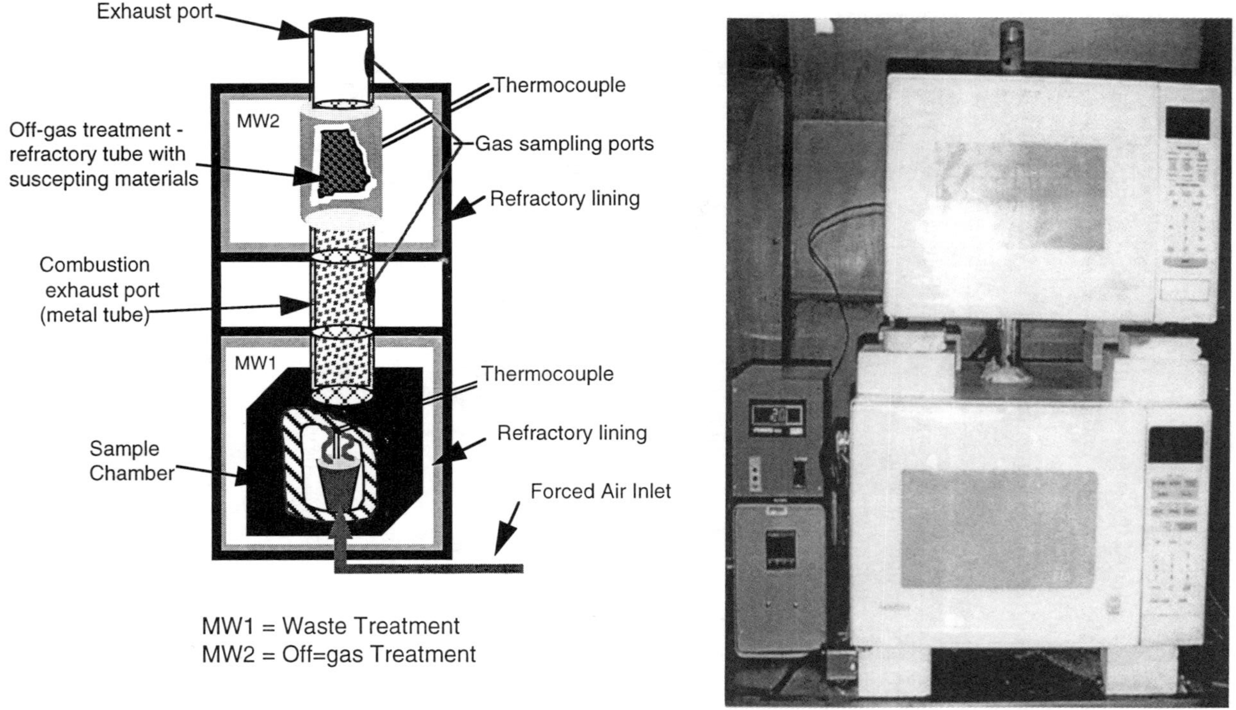

Figure 2. Schematic and digital photograph of microwave waste treatment system.

gases) and provides an important contribution to the growing arsenal of waste remediation tools.

Table 1. Microwave Treatment of Emissions Produced from Combustion of Electronic Circuit Boards.

COMPOUND	SR-8 (ppb)		SR-9 (ppb)	
	A	B	A	B
Benzene*	5838.9	22.2	1415.6	139.5
Toluene*	8146.6	15.7	4215.9	158.7
Ethylbenzene*	1147.4	nd**	4557.0	5.2
Styrene*	1666.9	6.2	20012.0	38.4
Napthalene*	355.5	nd	2403.6	27.9
m/p Xylenes*	2259.0	nd	510.6	nd
1,3,5 Trimethylbenzene	1564.0	nd	378.7	64.3
1.2.4 Trimethylbenzene	904.7	nd	171.8	nd

A = Before microwave off-gas treatment; B= After microwave off-gas treatment
*Listed as hazardous air pollutants in the Clean Air Act, as amended, 1990
** nd= not detected (< 1ppb)

REFERENCES

1. Wicks, G.G., "Nuclear Waste Glasses," Treatise on Materials Science and Technology, Glass IV, M. Tomozawa and R.H. Doremus, eds., Vol. 26, pp. 57-118 Academic Press, Inc. (1985).
2. Oda, S.J., "Microwave Remediation of Hazardous Waste: A Review," Microwave Processing of Materials III, (R.L. Beatty, W.H. Sutton and M.F. Iskander, eds.), Materials Research Society, Vol. 269, pp. 453-464 (1992).
3. Dauerman, L., Windgasse, G., Zhu, N. and He, Y., "Microwave Treatment of Hazardous Wastes: Physical Chemical Mechanisms," Microwave Processing of Materials III, (R.L. Beatty, W.H. Sutton and M.F. Iskander, eds.), Materials Research Society, Vol. 269, pp. 465-469 (1992).
4. Krause, R.T. and Helt, J.E., "Applications of Microwave Radiation in Environmental Remediation Technologies," Microwaves: Theory and Application in Materials Processing II, Ceramic Transactions (D.E. Clark, W.R. Tinga and J.R. Laia, eds.), American Ceramic Society, Vol. 36, pp. 53-59 (1993).

5. Wicks, G.G., Clark, D.E., Schulz, R.L. and Folz, D.C., "Microwave Technology for Waste Management Applications Including Disposition of Electronic Circuitry," Microwaves: Theory & Application in Materials Processing III, Ceramic Transactions, (D.E. Clark, D.C. Folz, S.J. Oda and R. Silberglitt, eds.), Vol. 59, pp. 79-89 (1995).
6. Schulz, R.L., Fathi, Z., Clark, D.E., and Wicks G.G., "Microwave Processing of Simulated Nuclear Waste Glass," presented at the Symposium on Microwaves: Theory and Application in Materials Processing, April 28- May 2, 1991, Cinn. OH, Ceramic Transactions, Nuclear Waste Management IV, (G.G. Wicks, D.F. Bickford and L.R. Bunnell, eds.), Vol. 23, pp. 779- 786 (1991).
7. Schulz, R.L., Clark, D.E., Hutcheon, R.M. and Wicks, G.G., "Microwave Processing of Simulated Nuclear Waste Glass II," Microwaves: Theory and Application in Materials Processing II, Ceramic Transactions (D.E. Clark, W.R. Tinga and J.R. Laia, eds.), American Ceramic Society, Vol. 36, pp. 89-97 (1993).
8. Schulz, R.L., Folz, D.C., Clark, D.E. and Wicks, G.G., "Microwave Destruction/Vitrification of Electronic Components," Microwaves: Theory and Application in Materials Processing II, Ceramic Transactions (D.E. Clark, W.R. Tinga and J.R. Laia, eds.), American Ceramic Society, Vol. 36, pp. 81-88 (1993).
9. Schulz, R.L., Folz, D.C., Clark, D.E., and Wicks, G.G., "Microwave Energy for Waste Remediation Applications," Microwave Processing of Materials IV, (M.F. Iskander, ed.), Materials Research Society Symposium Proceedings, Vol. 347, pp. 401-406 (1994).
10. Schulz, R.L., Folz, D.C., Clark, D.E., Schmidt, C.J. and Wicks, G.G., "Microwave Treatment of Emissions from the Destruction of Electronic Circuitry," Microwaves: Theory & Application in Materials Processing III, Ceramic Transactions, D.E. Clark, D.C. Folz, S.J. Oda and R. Silberglitt, eds., Vol. 59, pp. 107-114 (1995).
11. Schulz, R.L., Folz, D.C., Clark, D.E., Schmidt, C.J. and Wicks, G.G., "Microwave Waste Treatment System," poster presentation at the First World Congress on Microwave Processing, Lake Buena Vista, FL, Jan. 5-9 (1997).

PROCESS CONTROL FOR SIMUTANEOUS VITRIFICATION OF TWO MIXED WASTE STREAMS IN THE TRANSPORTABLE VITRIFICATION SYSTEM (TVS)

A.D. Cozzi
K.G. Brown
C.A. Cicero-Herman
C.M. Jantzen
Westinghouse Savannah River Company, Aiken, SC 29802

ABSTRACT

Two highly variable mixed (radioactive and hazardous) waste sludges were simultaneously vitrified in an EnVitCo Transportable Vitrification System (TVS) deployed at the Oak Ridge Reservation (ORR). The TVS was the result of a cooperative effort between the Westinghouse Savannah River Company (WSRC) and EnVitCo to design and build a transportable melter capable of vitrifying a variety of mixed low level wastes.

The two waste streams for the demonstration were the dried B&C Pond sludges at the K-25 site and waste water sludge produced in the Central Neutralization Facility (CNF) from treatment of incinerator blowdown. Large variations occurred in the sodium, calcium, silicon, phosphorus, fluorine and iron content of the co-blended waste sludges: these elements have a significant effect on the processability and performance of the final glass product. The waste sludges were highly reduced due to organics added during processing (Betz 1138), coal-pile runoff (coal and sulfides), and other organics, including wood chips. A batch-by-batch process control model was developed to control glass viscosity, liquidus, and REDuction/OXidation (REDOX), assuming that the melter behaved as a Continuously Stirred Tank Reactor (CSTR).

INTRODUCTION

Empirical composition/property models were specifically developed for high calcium and iron containing wastes at the Oak Ridge Reservation (ORR). Four unique empirical waste glass composition/property models were used in conjunction with each other to manually control the composition of the feed. The use of the models ensured that the glass compositions would not crystallize in the melter, would be pourable, would not cause excessive refractory wear, would not produce metallic nodules, and would be in the known glass forming region of the soda-lime-silica system.[1] The individual empirical algorithms predict the values of several important processing variables (viscosity, liquidus, and redox) and glass forming ability. These algorithms ensured that the processing variables were within acceptable ranges and that the glasses produced were homogeneous. The control limit processing ranges are derived from constraints that are placed on the Defense Waste Processing Facility (DWPF) to ensure processable, acceptable glasses.[2] A spreadsheet that incorporated these composition/property models was developed to allow the user to tailor the glass properties (viscosity, liquidus, redox, and homogeneity) by balancing the amount of the two waste streams (CNF and B & C Pond sludge), MnO_2 (for redox control) and the glass making additives (silica and sodium carbonate).

With respect to melter performance, careful control of the viscosity, redox and liquidus temperature of the glass to be made was required. The viscosity of a glass directly affects the pourability, melt rate, melt temperature and corrosivity of a glass. For this application, the acceptable processing range for viscosity is 20 - 100 poise.[2,3] Glasses with viscosities in this range can be poured relatively quickly without causing extensive wear on melter refractories. Similar to the viscosity model used for the DWPF,[3] the viscosity model developed for the TVS is three-dimensional, e.g. the viscosity of a glass may be controlled via compositional changes as well as by altering the melter temperature.

Redox is the ratio of the concentration of Fe^{2+} to the total amount of iron present in the glass ($Fe^{2+}/\Sigma Fe$) and is used as an indicator of the oxidation state of all of the other transition metals present in the glass. The redox of a glass can be controlled through additions of either oxidants (e.g. Fe_2O_3, MnO_2, nitrates, peroxides) or reductants (e.g. coal, formates, sulfides, sugar). Each of the oxidants and reductants varies in their strength as oxidizers or reductants. Therefore, when combinations of these are present in a waste, or added during processing, some empirical data must be collected specific to the processing conditions being evaluated. An oxidized melt pool, $Fe^{2+}/\Sigma Fe < 0.09$, can lead to foam formation that insulates the cold cap from the melt, thus reducing the melt rate and slowing production. An oxidized glass is also more corrosive to the molybdenum electrodes and the high chrome refractory bricks used in the throat of the TVS. A glass that is too reducing $Fe^{2+}/\Sigma Fe > 0.5$ has problems associated with it as well. In a highly reduced glass, transition metals can form sulfides or be converted to the metallic state where they can react with the electrodes forming low melting eutectics. There is also the possibility that the metals or sulfides can drop to the bottom and reduce melter efficiency by providing a low resistance path for current. Based upon the materials used to construct the melter (refractories and electrodes) and the composition of the two waste streams, it was determined that the redox of the glass should be targeted to be $0.09<Fe^{2+}/\Sigma Fe<0.5$[4].

The liquidus temperature is defined as the highest temperature at which a glass of specific composition is in equilibrium with the primary crystalline phase associated with that composition.[5] The liquidus temperature determines the minimum melt temperature and can contribute to a reduction in the life of the melter. The melt temperature should be ≥ 100°C above the liquidus temperature of the glass in the melter to prevent conditions favorable for devitrification. Temperature excursions below the liquidus temperature of a melt can lead to unwanted crystal formation in the melter. The crystals then can settle to the bottom of the melter and interfere with melter performance. The formation of a crystalline phase also changes the composition, and thus the properties, of the glass remaining in the melter. The soda-lime silica glass compositions, made necessary by the high calcium content of the ORR waste, are particularly prone to crystallization.[6] When a glass crystallizes, the composition of the residual glass becomes depleted in the species that are enriched in the crystalline phase. This can hinder the acceptance of the waste form since the durability of the crystallized glass is usually not the same as the durability of a homogeneous glass of the same overall composition. The liquidus model used by DWPF is specific to the primary crystalline phase (spinel, nominally $NiFe_2O_4$) specific to the DWPF glass composition. Therefore it was necessary to develop a new model to accommodate another crystalline phase, usually wollastonite ($CaSiO_3$), associated with the proposed TVS glass composition. Due to the variability in the waste compositions, both spinel and wollastonite are capable of forming in the TVS. Table I shows the glass properties and associated control limits for operation of the TVS. Some of the waste constituents, although present in nominal amounts, need to be tracked in the feed batch compositions. They do not appear in any of the models, but, along with more abundant components, can lead to solubility concerns in the melter. Increased amounts of these materials can cause metal/oxide/spinel precipitation, phase separation, and/or formation of a salt layer.

Table I. Monitored Glass Properties and Operational Ranges for the TVS.

Glass Property	Range	Processing Concern
Viscosity	20-100 (poise)	Pourability; corrosion
Redox	$0.09<Fe^{2+}/\Sigma Fe<0.5$	melt rate; corrosion
Liquidus	T_m* - 100°C	crystallization in melter
Solubility	Fe_2O_3	<20% (Fe_2O_3 + FeO + MnO) or spinel can precipitate
	FeO	<20% (Fe_2O_3 + FeO + MnO) or spinel can precipitate
	MnO	<20% (Fe_2O_3 + FeO + MnO) or spinel can precipitate
	NiO	<1.0% or Ni° can precipitate
	P_2O_5	<10% or glass will phase separate
	Cl^-	<0.6% (if greater check melt pool visually for salt layer)
	SO_4^{2-}	<1% (if greater check melt pool visually for salt layer)
	F^-	<2% (if greater check melt pool visually for salt layer)

*T_m is the operational temperature of the melter.

EXPERIMENTAL

Thirty-three glasses from a statistically designed matrix of compositions in the $CaO-Na_2O-B_2O_3-Fe_2O_3-SiO_2$ system were submitted to Corning Engineering Laboratory Services (CELS) for whole element chemistry, viscosity measurement, and liquidus measurement. Of these glasses, sixteen were in the soda-lime-silica system and the data was used to develop the TVS viscosity and liquidus models. Data was generated on an additional sixteen glasses for viscosity but only ten of these participated in the liquidus model. The viscosity and composition data for these additional sixteen glasses included five simulated WETF (West End Treatment Facility at ORR) waste glasses, four simulated SRS Consolidated Incinerator Facility (CIF) glasses, three simulated B & C Pond waste glasses with fluoride and MnO_2, one simulated B & C Pond waste without fluoride and MnO_2, one TVS shakedown glass, and two B & C Pond/CNF actual waste glasses from mixed waste minimelter testing at Clemson University.

Analysis of the two waste streams investigated for treatment indicated that both the CNF and B & C Pond sludge had significant quantities of calcium. The CNF also had large amounts of iron and, in some cases, phosphorus. The primary constituent in the B & C waste was silica. Initial crucible experiments at ORR and SRTC confirmed that the soda-lime-silica glass (SLS) system would provide a suitable base glass for the intended operation. Soda-lime-silica glass was determined to be able to accommodate approximately 45 wt% B & C waste or 15 wt% CNF waste on a dry, calcine basis. Further laboratory tests with actual wastes suggested that a combination of the two waste streams could be used to achieve a combined 45 wt% (30 wt% B & C/15 wt% CNF) waste loading in the glass. It also was determined that there were substantial quantities of organic matter in the CNF waste. To offset the reducing nature of the organic components, experiments were performed to evaluate the feasibility of using manganese dioxide (MnO_2) as a glass former and an oxidant[7]. Results indicated that the MnO_2 was an acceptable oxidant and the preliminary glass composition could accommodate sufficient quantities ($MnO^* + Fe_2O_3 + FeO \leq 20$ wt.%). Exceeding this quantity of MnO_2 promotes devitrification of the glass through manganese spinel formation. Therefore, this summation, $MnO + Fe_2O_3 + FeO \leq 20$ wt.%, was imposed as an additional processing constraint to ensure an acceptable glass product.

* MnO_2 reacts with carbon to form MnO in the glass and CO gas.

RESULTS

Surrogate Operations

The melter was inventoried with a soda-lime silica cullet. Additional lithium carbonate was added to adjust the cullet to a viscosity similar to the waste glass at 1200°C.[8] Calcia as $CaCO_3$ was added as a surrogate for the calcium-rich waste. This formulation provides a "start-up glass" with a composition in the same glass system as the waste glass to be produced. Before initiation of mixed waste operations, the analysis of a glass sample from the melter was evaluated against all of the process control algorithms. It was noted that the glass in the melter had a low predicted viscosity (< 20 poise) and a predicted liquidus temperature near the operating temperature of the melter. Computer simulations were performed to evaluate the effect on glass properties of a typical waste batch containing fluorine and phosphorus. Anticipation of accelerated refractory corrosion due to low viscosities was predicted. To avert any potential problems, the viscosity of the melt pool was increased through the addition of two silica-rich, lithia poor surrogate batches to the melter.

Mixed Waste Operations

B & C Only: Samples taken of the initial B & C and CNF waste drums to be treated were used to set the initial waste compositions. The complexity associated with handling the highly variable CNF waste prompted the use of only the B & C waste for start up of mixed waste operations. A glass composition was developed using a 45 wt% loading of the B & C waste which met the defined control limits. Duplicate analyses of the B & C drum revealed a higher than anticipated amount of organic/sulfide material. The process model algorithms for redox, the property most affected by organic/sulfide content, were based on a limited data set. To complement the predicted redox from the algorithm, experiments were performed with the waste samples and used to determine the initial MnO_2 amounts required to satisfy the redox constraint. The maximum MnO_2 ($MnO + Fe_2O_3 + FeO = 20$ wt.%) was deemed necessary. This corresponded to a 20 wt% addition of MnO_2.[†] Blended feed batches were vitrified in the Analytical Services Organization (ASO) laboratory and redox analysis was performed on the resulting glasses. During initial melter glass inventory turnover, this method provided quicker and more accurate feedback than redox analysis of the glass exiting the melter. The lack of iron in the surrogate glass used to start up the melter precluded redox analysis of pour stream samples until the fourth feed batch. The TVS holds approximately eight feed batches and considering the amount of iron present in the feed, at least three feed batches must be processed before an accurate iron ratio can be obtained from a poured glass sample.

In the second batch, the manganese concentration was again maintained at the maximum level. Continued laboratory analysis of the feed batches suggested that the manganese dioxide additions could be reduced due to a reduction in the organic carbon in the B & C waste. In melter batch three, the amount of MnO_2 added was reduced to 17 wt%. This action was taken to ease concerns regarding the possible formation of manganese containing spinels.

B & C/CNF: Batch five was formulated to incorporate the contents of one barrel (3.5 wt%) of the CNF waste while maintaining a 45 wt% waste loading. In an effort to reduce the amount of manganese dioxide needed to control the redox, laboratory experiments were conducted to ascertain the benefits of using a second (non-transition metal) as an oxidant. It was determined that a partial substitution of $NaNO_3$ would significantly reduce the amount of MnO_2 needed to control the redox of the glass. Calculations indicated that the two oxidizers, when used in conjunction, were a more powerful oxidant than either material acting independently. The reduction of MnO_2 needed becomes critical when CNF waste is considered, due to the high iron content of the CNF, and the $MnO + Fe_2O_3 + FeO = 20$ wt.% constraint. Batch six contained the

---

[†] MnO_2 reacts with carbon to form MnO in the glass and CO gas.

full complement of waste in the intended ratios (30 wt% B & C and 15 wt% CNF). The MnO_2 levels were then reduced to 9.8 wt%. To evaluate the effect of these changes, batch seven was identical to batch six. It was observed that the cold cap was melting much slower than in previous batches and was attributed to the increased levels of CNF in the feed. To mitigate the problem, the waste ratios were adjusted to include only two barrels of CNF waste (~8 wt%) and the balance of waste B & C (37 wt%). This required that the MnO_2 level for batches eight through twelve be increased to 12.2 wt%. Redox values for the poured glass samples corresponding to these batches gradually increased to the point of concern. It was noted that in some feed batches, wood chips had been removed from the strainers of the feed tank. These wood chips may have contributed to the greater than expected rise in glass redox (i.e., more reduced glass). In batch thirteen the level of the $NaNO_3$ was increased and the MnO_2 level was increased to 14.7 wt%. The levels of both oxidizers were maintained for the next three batches (14 - 16). In batch seventeen, the level of the $NaNO_3$ was again increased. In batch eighteen, the level of the second oxidizer was reduced back to the level of batch seventeen and this feed composition was used for the remaining mixed waste operations batches.

DISCUSSION

A standard process control system was developed to accommodate small variations in the waste stream composition and potential process upsets. However, analyses of waste samples taken over a time showed that notable variability occurred in the composition and chemical makeup of the waste streams. These variations were present in the elements sodium, calcium, silicon, phosphorus, fluorine and iron as well as other elements that varied to a somewhat lesser extent. Figure 1 shows the variability of key elements in the melter feed, including silica, which is both a waste component as well as added as a glass former. All of these elements have a significant effect on the processability and performance of the final glass product. This variability necessitated the implementation of a "batch-by-batch" control of the feed compositions modeled to simulate a continuously stirred tank reactor (CSTR). Whereas the TVS melter behaves more like a CSTR with a plug flow reactor (PFR) in series to accommodate the delay,[9] the CSTR was deemed an acceptable model because the feed composition would not be constant and the CSTR-PFR model would unnecessarily complicate the model without affording a significant benefit.

Throughout the mixed waste operations campaign, properties controlled via the developed algorithms were monitored via the spreadsheet for both the feed material entering the melter and for the glass poured from the melter. Figure 2 tracks the predicted viscosities in the melter after adding each feed batch (assuming one-eighth melter turnover) and the glass poured just after the batch had been fed to the melter. The viscosities predicted after the feed additions were less erratic because the CSTR model replaces only one-eighth of the existing melter inventory with a new glass composition. The viscosities predicted from the glass patties are more representative of the glass in the pour chamber of the melter. Figure 3 shows the predicted liquidus temperature of the glass predicted in the same manner as the viscosities. The predicted liquidus from both sources trended towards a lower liquidus value. The reduced liquidus temperature provides benefits during melter shutdown by minimizing the time at which the melter contents will have favorable conditions for devitrification. The oscillations of the predicted liquidus based upon glass samples can be attributed to the compositional changes experienced during the introduction of CNF into the waste stream. To evaluate the influence of the oxidants on the redox of the glass, the amount of the oxidants added as a percentage of the calcined waste in each feed batch was plotted as a function of the batch ID, figure 4. The transformation from a feed that does not influence glass redox to the waste feed laden with reductants gave rise to a reduced glass product. The presence of a reduced glass became evident shortly after there was sufficient iron present to measure the redox of the glass. The redox ratio continued to rise throughout the course of the first melter turnover. The introduction of an second oxidant was necessary to maintain control of the redox ratio of the glass. The reduced state of the glass was exacerbated by the presence of wood chips found in the strainer of the mix tank during several of the later batches. To account for the lag between the feed addition and the participation of the additives in the melter, the measured redox plotted in figure 5 was shifted two feed batches. The shift matches the "cause" (feed batch - MnO_2, $NaNO_3$, or B & C/CNF ratio change) with the "effect" (redox ratio) so the relative influence of feed batch conditions could be correlated. The shift necessary to explain the "cause and effect" is illustrated when comparing the initial addition of $NaNO_3$ and the corresponding reduction in redox. The delayed effect is also depicted later in the campaign when wood chips (a strong reductant) were found in the feed tank strainer coincided with a large increase in the redox. The two batch delay is predicted by the PFR model for a step change in batch composition (in this case, wood chips).

CONCLUSIONS

A demonstration of the immobilization of low-level mixed waste through vitrification was performed in the Transportable Vitrification System (TVS). Using a Statistical Process Control (SPC) strategy, over 7,300 kg of mixed waste was successfully treated at a 45% waste loading (dry, oxide basis). These results were achieved even with wide compositional variations and the presence of wood chips in the feed. The TVS demonstration resulted in the production of 7,600 kg of glass, corresponding to a volume reduction of >48%.

REFERENCES

[1] Jantzen, C.M., Pickett, J.B., and Ramsey, W.G., "Reactive Additive Stabilization Process for Hazardous and Mixed Waste Vitrification," Mixed Waste: Proceedings of the Second International Symposium, eds. A.A. Moghissi, R.K. Blauvelt, G.A. Benda and N.E. Rothermich, pp. 4.2.1-4.2.13 (1993).

[2] Bickford, D.F., Applewhite-Ramsey, A., Jantzen, C.M., and Brown, K.G., "Control of Radioactive Waste Glass Melters: I, Preliminary General Limits at Savannah River," J. Amer. Ceram. Soc., Vol. 73 no. 10 pp. 2896-2902 (1990).

[3] Jantzen, C.M., "First Principles Process-Product Models for Vitrification of Nuclear Waste: Relationship of Glass Composition to Glass Viscosity, Resistivity, Liquidus Temperature, and Durability," Nuclear Waste Management IV, Ceramic Transactions, Vol. 23, eds. G.G. Wicks, D.F. Bickford and L.R. Bunnell, pp. 37-51 (1992).

[4] Schreiber, H.D. and Hockman, A.L., "Redox Chemistry in Candidate Glasses for Nuclear Waste Immobilization," ," J. Amer. Ceram. Soc., Vol. 70 no. 8 pp. 591-594 (1987).

[5] ASTM Procedure C162, Annual Book of ASTM, Vol. 15.02, ASTM, West Conshohocken, PA (1997).

[6] Morey, G.W., Properties of Glass, American Chemical Society, (1954).

[7] Bickford, D.F. and Cicero-Herman, C.A., "Vitrification of Organics Containing Wastes," U.S. Patent Pending

[8] Jantzen, C.M., Pickett, J.B., and Cicero-Herman, C.A., "Lithia Additive Melting Process, LAMP™," U.S. Patent pending

[9] Zamecnik, J.R., Peeler, D.K., Young, S.R. and Smith, M.E., "Processing of Oak Ridge B&C Pond Sludge Surrogate in the Transportable Vitrification System, in Environmental Issues and Waste Management Technologies in the Ceramic and Nuclear Industries III, Ceramic Transactions , American Ceramic Society, Westerville OH (in press).

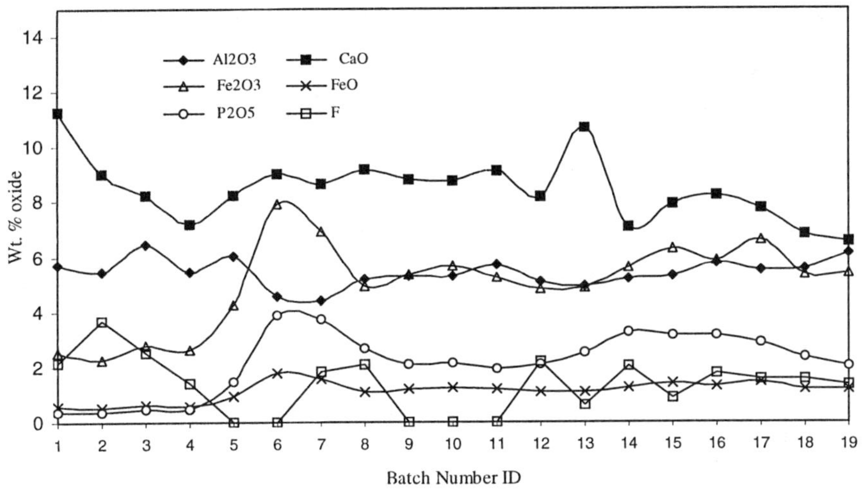

Figure 1. Variability in composition of several components in the melter feed.

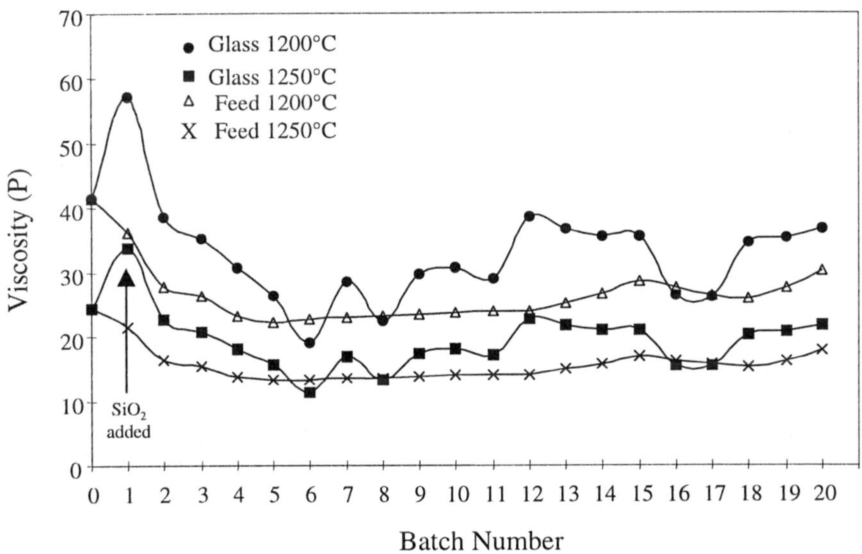

Figure 2. Predicted viscosities of feed batches and glass samples used for process control.

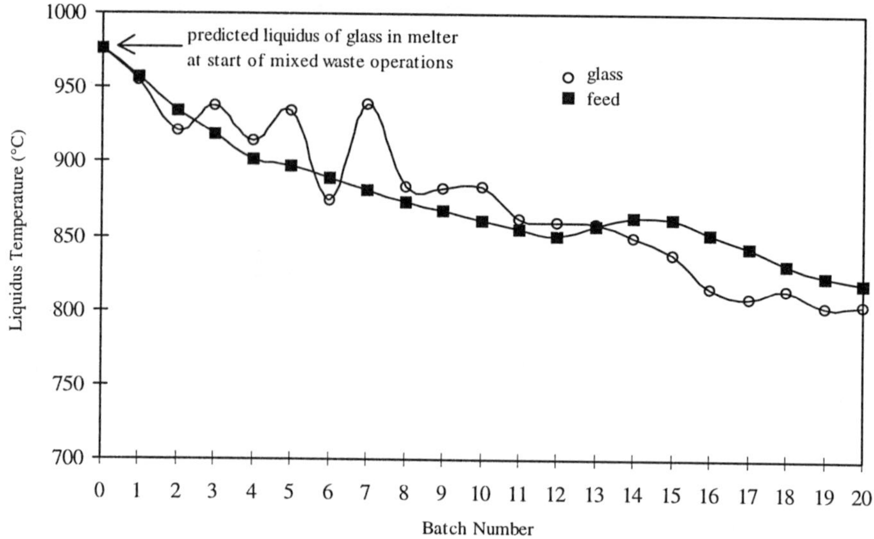

Figure 3. Predicted liquidus temperatures of feed batches and glass samples used for process control.

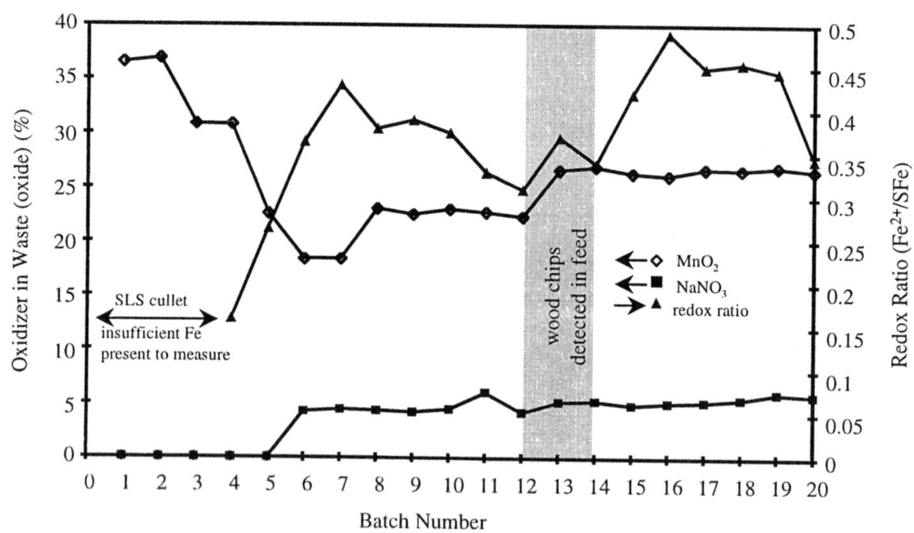

Figure 4. Oxidant added as a wt.% of waste with corresponding redox ratio.

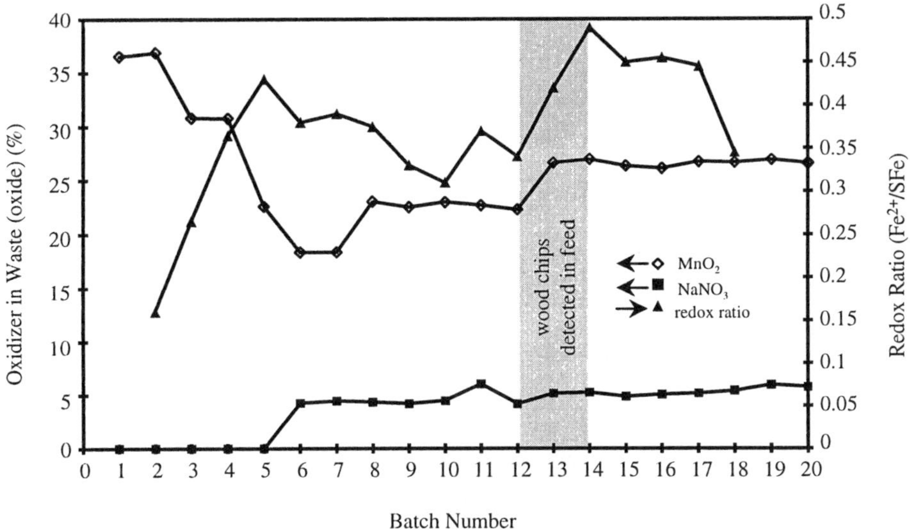

Figure 5. Oxidant added as a wt.% of waste with corresponding redox ratio shifted to account for plugged flow reactor model.

VITRIFICATION OF SIMULATED FERNALD K-65 SILO WASTE AT LOW TEMPERATURE

C. M. Jantzen,† J.B. Pickett,† and R. S. Richards *

† Westinghouse Savannah River Co.
Savannah River Site
Aiken, South Carolina 29808

* Associated Technical Consultants
P.O. Box 352393
Toledo, Ohio 43635-2393

ABSTRACT

Vitrification is one of the technologies that has been chosen to solidify ~18,000 tons of geologic mill tailings, designated as K-65 wastes, at the Fernald Environmental Management Project (FEMP) in Fernald, Ohio. The glass formula developed in this study for the FEMP wastes is a lithia substituted soda-lime-lithia-silica (SLLS) composition which melts at 1050°C[1]. Low melting formulations minimize volatilization of hazardous species such as arsenic, selenium, chromium, and lead during vitrification. Formulation in the SLLS system avoids problematic phase separation known to occur in the MO-B_2O_3-SiO_2 glass forming system (where MO = CaO, MgO, BaO, and PbO which are all constituents of the FEMP wastes). The SLLS glass passed the Environmental Protection Agency (EPA) Toxic Characteristic Leach Procedure (TCLP) for all the hazardous constituents of concern under the previous characteristically waste regulations. The SLLS glass is as durable as high melting soda-lime-silica glasses and is more durable than the borosilicate glasses previously developed for the K-65 wastes. Optimization of glass formulations in the SLLS glass forming system should provide glasses which will pass the newly promulgated Universal Treatment Standards (UTS) which take effect on August 28, 1998.

INTRODUCTION

Vitrification is one of the technologies that has been chosen to solidify geological mill tailings at the FEMP facility in Fernald, Ohio. The geologic mill tailings are residues from the processing of pitchlende ores during 1949-1958. These residues are contained in silos in Operable Unit 4 (OU4) at the FEMP facility. OU4 is one of five operable units at the FEMP. OU4 consists of four concrete storage silos

1 patent pending on SRTC Lithia Additive Melting Process (LAMP™) technology

and their contents. Silos 1 and 2 contain K-65 mill tailing residues, Silo 3 contains non-radioactive metal oxides, and Silo 4 is empty. The 14,262 tons (12.9 million kilograms) of K-65 residues contain radium (Ra^{226}), uranium ($U^{234,235,236,238}$), uranium daughter products (Th^{230}), and heavy metals such as lead and barium plus 1,185 tons of bentonite which was added to the silo headspace in 1991 to lower the radon concentration in the silo domes in 1991 [1].

Vitrification was chosen as the preferred technology for the Silo 1, 2, and 3 wastes [1,2]. Glasses that melted at high temperatures (1289-1595°C) were developed by Pacific Northwest National Laboratory (PNNL) [1,2] and glasses that melted at lower temperatures (1150-1350°C) were developed by the Vitreous State Laboratory (VSL) for the K-65 silo wastes. The PNNL formulated glasses primarily in the soda-lime-silica (SLS) system since the waste is high in BaO which readily substitutes for CaO (lime) in SLS glass. The VSL formulated glasses primarily in the borosilicate glass forming system. The PNNL and the VSL conducted both crucible and pilot scale vitrification studies.

A Vitrification Pilot Plant (VPP) was constructed at FEMP to operate at temperatures up to 1450°C. The VPP began operation at FEMP on June 19, 1996. During the initial startup testing, 12.5 tons of benign startup "frit" were melted. Silo 1, 2, and 3 surrogate melter tests were completed in September 1996 and produced 11 tons of glass. Another campaign using Silo 1 and 2 surrogate with bentonite was subsequently performed. All of the VPP surrogate testing was completed at melt temperatures between 1130°C and 1350°C. A final transition to lower temperature operation at ~1150°C was planned when the VPP failed on December 26, 1996.

This study documents SRTC glass formulation and development for K-65 waste that was completed in April, 1993. The SLS based glass developed in this study takes advantage of the glass forming ability of the waste, e.g. has an 84% waste loading and a volume reduction of ~ 60%, but melts at 1050°C. The low vitrification temperature minimizes volatilization of the RCRA hazardous constituents (As, Pb, Se, Cr) and minimizes corrosion of melter materials of construction.

BACKGROUND

In 1991 PNNL analyzed the composition of K-65 residues from Fernald on a dry wt% calcine (oxide) basis [1]. The waste analyses were comparable to (a) analyses performed by Controls for Environmental Pollution, Inc. (CEP) in 1990 and (b) analyses performed by National Lead of Ohio (NLO) prior to 1989 [1]. In 1993, PNNL published [2] an average composition for the K-65 residues in Silos 1 and 2, an average composition of the Silo 3 waste, and an average composition for the BentoGrout™ added to the tank headspace (Table I). Note that the PNNL analyses are normalized and contain significant contributions of unspecified components designated as "other." The K-65 Silo 1 and 2 residues were reported to contain between 1-3 wt% sulfate and the Silo 3 material is reported to contain up to 18 wt% sulfate on a dry solids basis [2].

Analyses performed by FEMP in 1993 indicate that the sulfate in Silo 1 varies between 0.04 and 0.35 wt% on a wet solids basis while the sulfate in Silo 2 varies between 0.26-1.93 wt% on a wet basis [3]. The total organic carbon (TOC) for Silo 1 wastes varies from 0.52 to 3.48 wt% on a wet basis while the TOC for Silo 2 varies from 0.01 to 2.44 wt% on a wet solids basis [3]. In February, 1998 FEMP developed revised surrogate waste compositions for Silo 1 and 2 wastes and BentoGrout™ [3]. For comparison, the 1993 PNNL average compositions are compared to the 1998 FEMP surrogate waste compositions in Table I.

Due to the high concentrations of silica (SiO_2), lime (CaO), baria (BaO), and PbO in the K-65 Fernald wastes, PNNL chose the use of soda-lime-silica (SLS) glass for solidification [1, 2]. PNNL only formulated glasses in the borosilicate system for Silo 3 wastes which contained no PbO. For the K-65 wastes, the initial PNNL formulations required only one simple additive, Na_2O as NaOH or Na_2CO_3, rather than formulation of a complex glass making frit.

Table I. Chemical Analyses and Oxide Surrogates for K-65 Residues (Silos 1 and 2), Silo 3 FEMP Wastes, and BentoGrout™

	1993 Average K-65 Silos 1 & 2 [2]	**K65 Silo 1 Surrogate with 9 wt% Bento-Grout™*** [4]	**K65 Silo 2 Surrogate with 9 wt% Bento-Grout™*** [4]	**1993 FEMP Silo 3** [2]	**1993 Bento-Grout™** [2]	**1998 Bento-Grout™** [4]
Al_2O_3	3.7	4.77	5.77	8.3	15.8	16.27
BaO	6.3	6.32	3.48	0.1	---	---
CaO	1.5	0.18	3.73	7.3	1.8	1.85
Fe_2O_3	4.8	3.56	6.69	12.6	3.6	3.71
K_2O	0.9	0.81	1.50	2.8	0.8	0.82
MgO	1.7	2.18	2.11	15.8	6.3	6.49
Na_2O	1.7	1.41	1.34	9.3	3.4	3.50
P_2O_5	0.8	0.58	0.68	14.5	1.4	---
PbO	12.3	13.81	6.92	0.3	---	---
SiO_2	62.9	62.08	62.38	22.0	65.4	67.35
SO_4	---	4.30	5.41	---	---	---
Other	3.5	---	---	7.2	1.5	---
SUM	100.00	100.0	100.0	100.0	100.0	100.0

*2-3 grams kerosene added to the surrogate to simulate the Total Organic Carbon

PNNL developed four series of glass compositions for various combinations of actual FEMP wastes [1, 2]. Series A glass was a SLS formulation containing 88-89 wt% K-65 waste and 11-12 wt% Na_2O glass formers. Series B glass was an SLS formulation containing equal amounts of ~45% K-65 and ~45% BentoGrout™. Series C glass was a borosilicate formulation containing ~74% Silo 3 waste. Series D glass was an SLS formulation containing ~37% K-65 and ~37% Silo 3 wastes. Carbon was added to PNNL-A, B and D melts to reduce the sulfate (SO_4) to SO_3 or SO_2 vapor [2]. However, the final glass redox was not measured and several of the glasses exhibited the formation of metallic nodules indicating that too much reduction had occurred in the glass causing sulfides and metallic species to form. For the Series D glasses even significant quantities of carbon (9g of carbon/100g of glass) did not mitigate sulfate formation on the melt surface.

All of the PNNL glasses contained between 74-89 wt% waste on a dry oxide basis which corresponded to volume reductions of 50-68% [1, 2]. All of the PNNL glasses melted at elevated temperatures between 1289° and 1595°C. Glasses PNNL-A, C and D contained unidentified white crystalline inclusions while PNNL-C (a borosilicate glass) and PNNL-D (a high sulfate containing glass) were reported to be potentially phase separated.

The Vitreous State Laboratory (VSL) formulated K-65 waste glasses in the sodium-borosilicate (SBS)[2] system at various boron concentrations (5.88-12.8 wt%). During minimelter tests, urea was added to control foaming attributed to sulfate volatilization. These SBS glasses had waste loadings varying between 60-74 wt% and melted at temperatures between 1000-1050°C.

The K-65 Silo 1 and 2 wastes contain 7-12.3 wt% PbO, 3.5-6.3 wt% BaO, 0.2-3.7 wt% CaO, and 1.7 to 2.2 wt% MgO (Table I). The $PbO\text{-}SiO_2\text{-}B_2O_3$, $BaO\text{-}SiO_2\text{-}B_2O_3$, $CaO\text{-}SiO_2\text{-}B_2O_3$, and $MgO\text{-}SiO_2\text{-}B_2O_3$ glass forming systems all contain large regions of phase separation [5]. At concentrations greater than 40 wt% SiO_2, the phase separation is metastable in the $PbO\text{-}SiO_2\text{-}B_2O_3$ system [6] and causes the separation of an SiO_2 rich glass from a much denser $PbO{\bullet}2B_2O_3$ rich glass [6]. Therefore, formulation of borosilicate waste glasses for the K-65 wastes may readily produce phase separated glasses, as observed in the PNNL-C series glasses, with the PbO being stabilized in a borate rich phase. The borate rich phase, being more soluble than the silica rich phase, would be prone to release more Pb than a homogeneous glass during durability testing.

The low melting borosilicate glasses developed by the VSL and those subsequently developed by FEMP had total MO (where MO = PbO, BaO, CaO and MgO) concentrations within the known regions of phase separation in the $PbO\text{-}SiO_2\text{-}B_2O_3$ and $BaO\text{-}SiO_2\text{-}B_2O_3$, $CaO\text{-}SiO_2\text{-}B_2O_3$, and $MgO\text{-}SiO_2\text{-}B_2O_3$ systems [7].

---

[2]This report distinguishes between borosilicate and soda-lime-silic (SLS) based on the ASTM C-162 definition of borosilicate glass containing >5 wt% B_2O_3.

MAY 1998 ENVIRONMENTAL PROTECTION AGENCY (EPA) REGULATIONS

The K-65 Silo 1 and 2 wastes are a characteristically hazardous waste, as they leach lead at a rate greater than 100 times [1] the allowable Environmental Protection Agency (EPA) Resource, Concentration, and Recovery Act (RCRA) concentration limits [8], (the limits given in Table II, Column A). Therefore, the K-65 Silo 1 and 2 wastes are RCRA hazardous and radioactive (mixed) wastes. The Silo 3 wastes are a dry powdery mixture of metal oxides with a much lower concentration of radionuclides (thorium, uranium and Pb^{210}) than the Silo 1 and 2 wastes. There are approximately 4000 tons of Silo 3 wastes for disposal. The Silo 3 wastes leach arsenic, chromium, cadmium, and selenium at concentrations exceeding characteristically hazardous RCRA limits. Therefore, the K-65 Silo 3 wastes are also RCRA characteristically hazardous and radioactive (mixed) wastes.

Prior to August, 1998, the FEMP wastes would have had to be treated to meet the EPA Characteristically Hazardous Limits given in Table II, Column A. However in August, 1995, the EPA proposed a new rule (Phase IV of the Land Disposal Restrictions) which would require Characteristically Hazardous Waste to be treated to the same Universal Treatment Standards (UTS) as those for listed mixed wastes [9]. The Phase IV Rule was finalized on May 26, 1998, and will take affect August 26, 1998 [10]. Any mixed wastes not stabilized by that date will be required to be treated to the UTS limits, rather than to the Characteristically Hazardous Limits shown in Table II. The final Phase IV UTS limits for the RCRA metals are given in Table II, Column B. This lowers the acceptable Pb TCLP concentration for K-65 waste forms from the previous 5 mg/L to 0.75 mg/L.

The final rule requires that a treated waste form be tested against the UTS standards not only for the constituents which fail the TCLP characteristically hazardous limit, but for all "hazardous constituents which are present in the waste, i.e., underlying hazardous constituents" [10]. The Fernald Silo 1 and 2 wastes exhibit TCLP leaching concentrations for lead which exceed the Toxicity Limit, and leachant levels for nickel (Ni) and selenium (Se) which exceed the UTS in some of the analysis. Therefore, under the final regulations the final waste form will have to meet the UTS for lead, and probably for Ni and Se, as given in Table II, Column B. Similarly, the Silo 3 wastes contain As, Cd, Cr, Pb, and Se as underlying hazardous constituents. The final waste form for the Silo 3 waste would also have to meet the UTS standards for all of these constituents.

Alternatively, a delisting petition could be prepared per 40 CFR 260.22 “Petitions to Exclude a Waste Produced at a Particular Facility,” for the K-65 treated waste glass. The potential delisting limits (Table II, Columns C and D) required for the K-65 waste glass are calculated as follows:

- assume 1.8 million gallons waste x 50% waste loading x 10 yr treatment = 180,000 gallons disposed/yr [7]

This allows a Dilution/Attenuation Factor (DAF) of 100 and generates the Delisting Limits given in Table II, Column C, assuming a 10 year treatment period. If a 5 year treatment time is used, the volume of glass produced per year is increased, and the DAF is reduced to 80 - yielding the delisting limits in Table II, Column D. If the waste form is delisted it will have to meet the delisting limits for all the constituents given in Table II, including fluoride and cyanide. Delisting of the K-65 waste glass product is attractive because it raises the Pb TCLP release limit from 0.75 mg/L to 1.2 or 1.5 mg/L depending on the treatment times.

Table II. Final 1998 EPA TCLP Land Disposal Restrictions, Phase IV [10].

Element	Column A Characteristically Hazardous TCLP Limits (mg/L) [8]	Column B FinaPhase IV TCLP UTS Limits (mg/L) [10]	Column C Delisting Limit TCLP, mg/L 10 Year Treatment DAF =100 [7]	Column D Delisting Limit TCLP, mg/L 5 Year Treatment DAF =80 [7]
Ag	5	0.14	20	16
As	5	5.0	5.0	4.0
Ba	100	21	200	160
Be	---	1.22	0.40	0.32
Cd	1	0.11	0.50	0.40
Cr	5	0.60	10	18.0
Hg	0.2	0.025	0.20	0.16
Ni	---	11	10	8.0
Pb	**5**	**0.75**	**1.5**	**1.2**
Sb	---	1.15	0.60	0.48
Se	1	5.7*	5.0	4.0
Tl	---	0.20	0.20	0.16
V**	---	1.6	20	16
Zn**	---	4.3	---	---
F	---	400	400	320
CN	---	30 (amenable)	590 total mg/kg	590 total mg/kg

* Se must be treated to the characteristic limit (1.0 mg/L) to be non-hazardous, although it may be land disposedas a hazardous waste if <5.7 mg/L TCLP

** V and Zn are not underlying hazardous constituents

This would necessitate less reformulation of the Fernald waste glass product than the alternative of meeting the Phase IV UTS limits.

EXPERIMENTAL

Due to the potential for stable and/or metastable phase separation in the MO-SiO_2-B_2O_3 glass forming systems, this study concentrated on methods to lower the melt temperature of SLS glasses. In the SLS glass forming system PbO, MgO and BaO can all substitute for CaO in the glass structure without separation of a second phase.

Only one SLS glass was formulated by SRTC as an initial scouting test for developing a low melting SLS glass. The glass had a targeted waste loading of ~82% based on an the average Silo 1 and Silo 2 waste composition given in Table I. Lithia (Li_2O) was partially substituted for the Na_2O so that about 8 wt% Li_2O and 8 wt% Na_2O were the only two additives. This surrogate waste glass composition melted at 1050°C. The target and analyzed composition of this glass is given in Table III. The final waste loading in the as analyzed glass was 84%.

The TCLP results for the SRTC glass formulation are given in Table IV and compared with published TCLP results of the raw K-65 waste and the glasses formulated by PNNL. The SRTC formulation melts 300°C lower than the PNNL formulation but is of equivalent durability and waste loading to the high temperature K-65 glass and more durable than the low melting VSL borosilicate glass. The PNNL, VSL and SRTC glasses all pass the EPA TCLP concentration limits before the Phase IV UTS limits went into effect, e.g. pre-August, 1998 (Table IV). The PNNL and SRTC glasses pass at or below the delisting limits while the VSL glasses do not. The SLLS glass formulations require further optimization based on the 1998 waste compositions in reference 4 (Table I) now that the final Phase IV UTS limits (Table II) have been adopted.

The glass attributes of the SRTC SLLS glass are compared to the glasses formulated for FEMP by PNNL and VSL in Table V.

Table III. SRTC Soda-Lithia-Lime-Silica (SLLS) Glass

GLASS COMPONENT	OXIDE WT% Target	OXIDE WT% As Analyzed
Al_2O_3	3.14	4.32
BaO	5.35	4.95
CaO	1.27	1.44
Cr_2O_3	---	0.14
Fe_2O_3	4.08	4.51
K_2O	0.76	0.73
Li_2O	8.50	7.92
MgO	1.44	1.63
Na_2O	10.88	9.85
NiO	-	0.09
P_2O_5	0.68	0.96
PbO	10.45	10.30
SiO_2	53.44	53.10
SUM	99.99	100.03

Table IV. Comparative EPA TCLP Testing for K-65 Waste and for K-65 Waste Glasses Formulated by PNNL, VSL, and SRTC

RCRA Metal	1997 Proposed UTS Limits (mg/L)	Raw K-65 Residue [1] (mg/L)	PNNL-A SLS K-65 Waste Glass [2] (mg/L)	VSL Borosilicate K-65 Waste Glass [2] (mg/L)	SRTC SLLS K-65 Waste Glass Replicate A (mg/L)	SRTC SLLS K-65 Waste Glass Replicate B (mg/L)
Ag	0.11	<0.1	<0.01	---	0.026	0.020
As	5.0	<1.0	0.004-0.005	---	<0.002	<0.002
Ba	21	0.76	0.69-0.87	---	2.62	2.73
Cd	0.20	0.1	<0.005	---	<0.010	<0.010
Cr	0.85	<0.2	<0.01	---	<0.04	<0.04
Hg	0.025	<0.03	<0.0002	---	<0.0096	<0.0096
Ni	13.6	3.2	---	---	0.062	<0.050
Pb	**0.75**	**630**	**0.81-1.2**	**1.9-2.76**	**1.272**	**1.619**
Se	5.7	<0.1	<0.002	---	0.158	0.297

Table V. Comparison of Attributes of Glasses Formulated for FEMP K-65 Silo Wastes.

GLASS ATTRIBUTES	SRTC Formulation	PNNL Formulations	PNNL Formulations	VSL Formulations
Glass Forming System	Soda-Lithia-Lime-Silica	Soda-Lime-Silica and Borosilicate	Borosilicate	Borosilicate
Phase Separation	No Potential	No Potential	Observed	Potential
Melt Temperature (°C)	1050	1289-1595	1150	1000-1050
Waste Loading (%)	84	80-84	<50	60-74
Volume Reduction (%)	~68	55-68	Not Determined	Not Determined
Density (g/cm^3)	2.86	3.1	Not Determined	Not Determined
Inconel®690 Compatible	Demonstrated	Not Determined	Not Determined	Not Determined

CONCLUSIONS

Glasses for the FEMP K-65 silo wastes should be formulated in the Soda-Lime-Lithia-Silica (SLLS) system or the to avoid problematic phase separation known to occur in borosilicate glass forming systems, e.g. the $MO-B_2O_3-SiO_2$ glass forming systems where MO = CaO, MgO, BaO, and PbO. The K-65 wastes contain all of these MO oxides which makes phase separation likely. Phase separation is known to compromise glass durability.

Glasses for the FEMP K-65 silo wastes should be formulated to melt at the lowest temperatures reasonably achievable to minimize volatilization of radioactive radon and hazardous species such as lead, arsenic, selenium, and chromium. The one scouting test performed in this study indicates that acceptable low melting SLS glasses can be formulated which melt at temperatures as low as 1050°C. High waste loading, e.g. 84 wt%, can still be maintained by the addition of only two glass forming additives, Li_2O and Na_2O using the SRTC Lithia Additive Stabilization Process (LAMP™).* Further optimization of SLLS glasses using the LAMP technology needs to be investigated for FEMP.

The SRTC SLLS glass passes the Environmental Protection Agency (EPA) Toxic Characteristic Leach Procedure (TCLP) for all the hazardous constituents of concern under the pre August, 1998 regulations and at the delisting limits. The SLLS glass is as durable as the high melting SLS glasses formulated by Pacific Northwest Laboratory (PNNL) and is more durable than the borosilicate glasses formulated by the Vitreous State Laboratory (VSL) for the K-65 wastes.

SRTC successes in late 1997 with co-vitrification of two highly variable mixed wastes at the Oak Ridge Reservation (ORR) into SLLS glass indicate that high sulfate and high organic containing (including wood chips) wastes can successfully be vitrified [11] using the LAMP™ technology. These vitrification successes were completed in a field scale Transportable Vitrification System (TVS) at the ORR.

ACKNOWLEDGMENTS

This paper was prepared in connection with work done under Contract No. DE-AC09-96SR18500 with the U.S. Department of Energy. The current study was funded by the Department of Energy, EM-50, Mixed Waste Focus Area.

---

* Patent pending.

REFERENCES

1. D.S. Janke, C.C. Chapman, and R.A. Vogel, **"Results of Vitrifying Fernald K-65 Residue,"** Nuclear Waste Management, IV, Ceramic Transactions, V.23, G.G. Wicks, D.F. Bickford and L. Bunnel (Eds.), American Ceramic Society, Westerville, OH, 53-61 (1991).

2. R.A. Merrill, and D.S. Janke, **"Results of Vitrifying Fernald OU-4 Wastes,"** Waste Management '93, 1369-1374 (1993).

3. U.S. DOE Report FEMP-O4RT-6 Final (November 3, 1993).

4. Fluor Daniel Fernald Acquisitions, **"Silos 1 and 2 Proof of Principle Testing,"** World Wide Web.Fernald.gov (February 11, 1998).

5. M.B. Volf, **Chemical Approach to Glass**, Glass Science and Technology, V. 7, Elsevier, New York, 594p. (1984).

6. D.W. Johnson and F.A. Hummel, **"Phase Equilibria and Liquid Immiscibility in the System $PbO-B_2O_3-SiO_2$,"** J. Am. Ceram. Soc., 51[4], 197-201 (1968).

7. C.M. Jantzen and J.B. Pickett, **"Vitrification of Simulated Fernald K-65 Silo Waste at Low Temperature,"** U.S. DOE Report WSRC-TR-0061, Westinghouse Savannah River Company, Aiken, SC (January 14, 1998).

8. Environmental Protection Agency Code of Regulations 40-CFR-261.24(b).

9. **Land Disposal Restrictions, Phase IV - Proposed Rule**, Federal Register, 60 FR 43654, p. 43683-4 (August 22, 1995) and Federal Register, **"Land Disposal Restrictions-Phase IV: Second Supplemental Proposal on Treatment Standards for Metal Wastes , Mineral Processing and Bevill Exclusion Issues, and the Use of Hazardous Wastes as Fill, Supplemental Proposed Rule,"** 62 FR 26041 (May 12, 1997).

10. Environmental Protection Agency, **"Land Disposal Restrictions - Phase IV: Final Rule Promulgating Treatment Standards for Metal Wastes and Mineral Processing Wastes, Mineral Processing and Bevill Exclusion Issues; Treatment Standards for Hazardous Soils, and Exclusion of Recycled Wood Preserving Wastewaters; Final Rule,"** 62 FR 62079 (May 26, 1998).

11. A. D. Cozzi, C.M. Jantzen, K.G. Brown and C. Cicero-Herman, **"Process Control for Simultaneous Vitrification of Two Mixed Waste Streams in the Transportable Vitrification System (TVS),"** this proceedings.

HIGH GRADE ABRASIVE DEVELOPMENT IN THE $MgO-Al_2O_3-SiO_2-TiO_2$ WASTE GLASS SYSTEM UTILIZING THE SEILER HIGH TEMPERATURE VITRIFICATION PROCESS

Russell D. Blume*, Seiler Pollution Control Systems, Dublin OH 43017 and Charles H. Drummond,III, Department of Materials Science, The Ohio State University, Columbus OH 43210

ABSTRACT:

A waste formulation consisting of wastes enriched in MgO, Al_2O_3, SiO_2, and TiO_2 was developed for subsequent vitrification utilizing an oxy-gas fired vitirification system. This formulation shows processing and product attributes suitable for the production of high grade abrasive blast media. The as-quenched glass was amber-brown and possessed moderate hardness making this product ideal for medium grade abrasive and roofing granular applications. Upon heat treatment, the product hardness increased substantially to a value approaching 1100 kgF/mm^2. X-ray diffraction (XRD) established that the principal crystalline phases were $Mg_4Al_2Ti_9O_{25}$, sapphirine ($Mg_{3.5}Al_9Si_{1.5}O_{20}$) and cordierite ($Mg_2Al_4Si_5O_{18}$) with $MgSiO_3$ present as an intermediate phase. Differential thermal analysis coupled with quantitative XRD established that the optimum nucleation and crystallization temperatures for this formulation are 755°C and 1155°C, respectively. The optimal heat treatment conditions necessary to produce a high-grade abrasive product were nucleation at 755°C for 1 h followed by growth at 1155°C for 30 minutes.

INTRODUCTION:

A product based approach to the manufacturing of recycled glass-ceramic products using industrial waste as feedstocks for subsequent vitrification utilizing the Seiler High Temperature Vitrification (HTV) process has been validated[1]. Initial abrasive formulations utilizing Electric Arc Furnace (EAF) dust and foundry sand as the principal waste constituents have provided the basis for commercial production of medium to high grade abrasive products. The resultant products have novel ferrimagnetic properties and superior resistance to particle size deterioration when subjected to high energy impacts. Subsequent findings[2] have shown a linear increase in Vicker's hardness of the base formulation with $MgO-Al_2O_3$ waste additions. However, processing constraints of the Seiler system in terms of liquidus temperature and melt viscosity have established an upper limit of 1000 kgF/mm^2 in this waste system. Additions of additional flux, although effective in reducing melt viscosity are thought to inhibit the crystallization of the hard spinel phase and is not indicated for optimum hardness. Therefore, alternate waste formulations which possess increased processability and superior product hardnesses are desired.

Current research has targeted waste glass systems utilizing waste feedstocks enriched in MgO, Al_2O_3, SiO_2 and TiO_2 as the $MgO-Al_2O_3-SiO_2$ (MAS) system with TiO_2 additions are widely utilized in the production of commercial glass-ceramics. Although TiO_2 is considered to be an acidic oxide constituent, it is known that TiO_2 acts to decrease melt viscosity[3] at high temperatures thereby allowing for increased processibility of highly acidic melts. Additionally, TiO_2 is known to be an effective nucleating agent in the MAS system. Although the exact formulation methodology and the identity of the waste feedstocks are considered to be proprietary, waste formulations produced from a variety of SiO_2 rich waste including contaminated soils and foundry sands, combined with waste enriched in MgO, Al_2O_3 and TiO_2 have exhibited similar product attributes.

EXPERIMENTAL:

Base Formulation:

Waste characterization was performed on the as-received, wet basis, feedstocks in order to ascertain the bulk density, moisture and organic content, and the elemental constituents of the waste feedstocks. Moisture content, organic content and elemental compositions was determined by percent solids, Loss on Ignition (LOI), and Energy Dispersive Spectroscopy (EDS), respectively. The wastes utilized in this study are designated as *Waste M, A, S,* and *T* respectively. *Waste M* was shown to have negligible moisture and organic content and contained MgO in excess of 80% (inorganic, dry basis) with Al_2O_3, SiO_2, and Fe_2O_3 comprising the balance. *Waste A* was shown to have a moisture content of 1% with negligible organic content and contained Al_2O_3 (inorganic, dry basis) in excess of 75 % with the balance consisting primarily of Fe_2O_3 and SiO_2. Foundry sand (*Waste S*) was utilized as a SiO_2 source and was shown to have negligible moisture and organic content and contained SiO_2 in excess of 95%. *Waste T* contained considerable moisture content (65 %) with negligible organic content and was determined to contained TiO_2 (inorganic, dry basis) in excess of 95 % with SiO_2 comprising the balance.

These waste were mixed on a dry weight basis in an approximate ratio of M:A:S:T of 2:3:4:1 and melted in Al_2O_3 crucibles in an electric furnace at 1500°C for 2 hours. The resultant melt was then water quenched and utilized for subsequent product characterization and heat treatment studies. The melt viscosity was observed to be exceedingly low at 1500°C and the liquidus temperature of the MAST formulation is estimated to be between 1300°C and 1400°C. The water quenched glass is amber brown and has a fracture pattern and optical characteristics consistent with a completely amorphous product.

Nucleation Study:

Differential thermal analysis (DTA) was conducted on the water quenched MAST product utilizing a Perkin-Elmer DTA Model 1700 unit. The methodology of Gao et. al[4] was used with a heating rate of 10°C/min to ascertain the optimum nucleation temperature of the crystalline phase which formed in this system. Upon attaining the crystallization peak temperatures of the crystalline phases, samples of the MAST product were heat treated at temperatures ranging from 630 to 815°C for one hour. DTA was performed on the nucleated samples and the shift in the crystallization temperatures was measured. As discussed by Gao et. al, the magnitude of the peak shift is proportional to the nucleation rate at the indicated temperature. A calibration curve using pure gold established that the instrument is accurate to within 0.1°C of the melting temperature of gold (1064.4°C) at a heating rate of 10°C/min.

Crystallization Study:
A crystallization study was conducted on water quenched samples which had been nucleated for 5 hours at the optimal nucleation temperature (T_n). A series of nucleated samples were then individually heated at 10°C/min in an electric furnace to the following temperatures: the first crystallization peak (811°C), the second crystallization peak (898°C), 1000, 1100, 1200 and 1300°C. The heat treated samples were then ground to -100 mesh and mixed with 10 weight percent of reagent grade crystalline Si of the same particle size. X-ray diffraction (XRD) was performed utilizing Cu-Kα radiation at a scan rate of 1°/min over a 2θ range of 20 - 80°. The resultant crystalline phases were identified by comparing the XRD patterns to JCPDS files. The peak intensity corresponding to the strongest reflection for each phase was normalized with respect to the peak intensity corresponding to the strongest reflection of crystalline Si. The normalized peak intensities were then plotted as a function of crystal growth temperature. Additionally, product hardness and fracture toughness were measured using microindentation techniques and plotted as a function of crystal growth temperature.

Heat Treatment Optimization:
As the optimum nucleation (T_n) and crystal growth (T_c) temperatures have been identified by the above methodology, the effects of nucleation and growth time on product hardness were examined. An experimental matrix consisting of 25 heat treatment conditions with nucleation times at T_n of 0, 1, 2, 3, 4 h and crystal growth times at T_c of 0, 0.5, 1, 2, and 4 h. Ten values of Vicker's hardness were measured on each sample in order to insure statistical validity. This data will be utilized to minimize post-processing cost by minimizing heat treatment time and temperature.

RESULTS AND DISSICUSION:

Nucleation Study:
The DTA scan of the as-quenched MAST product manifested two crystallization exotherms at 811°C and 898°C, respectively. These peak temperatures were identified as $T°_{p(1)}$ and $T°_{p(2)}$, respectively. With increasing nucleation temperature, it was observed that the crystallization peak corresponding to $T°_{p(1)}$ was no longer present at temperatures exceeding 755°C while at still higher temperatures above 815°C, the peak corresponding to $T°_{p(2)}$ was also absent. The disappearance of the crystallization peaks above a critical nucleation temperature is thought to represent the minimum temperature at which complete crystallization of the corresponding crystal phase can occur when held for 5 hours.

As shown in Figure 1, the plots of $1/T_p$ - $1/T°_p$ versus T_n exhibited decidedly non-parabolic behavior. This behavior, as well as the presence of critical temperatures above which the crystallization peaks disappear, is not typical of nucleation phenomena for the nucleation of a single crystalline phase. This fact suggest that the observed behavior is indicative of coupled nucleation reactions among multiple crystalline phases. This coupling is also thought to be responsible for the plateau in the nucleation curve for $T_{(p2)}$ in the temperature range corresponding to the critical temperature of $T_{(p1)}$. The optimum nucleation temperature was determined to be 755°C as the nucleation rate of $T_{(p1)}$ is maximized and the crystallization rate of $T_{(p2)}$ is high.

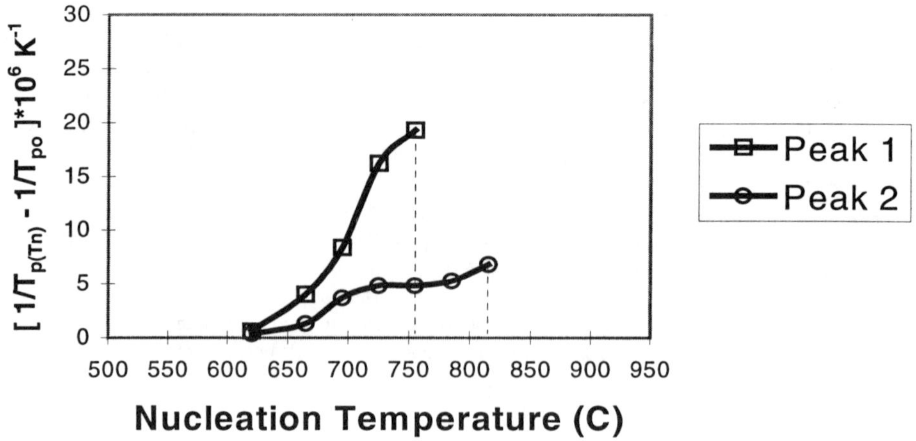

Figure 1: $1/T_p$ - $1/T^{\circ}_p$ versus T_n for the as-quenched MAST product.

Crystallization Study:

XRD established that the as-quenched MAST sample was completely amorphous thereby providing an ideal basis for subsequent crystallization studies utilizing quantitative XRD. The crystal phase in the MAST product, heat treated at the optimal nucleation temperature for 5 hours and then crystallized at 811°C for 1 hour, was determined to be a $Mg_4Al_2Ti_9O_{25}$. At 898°C and 1000°C, the heat-treated product was found to contain both $Mg_4Al_2Ti_9O_{25}$ and $MgSiO_3$ thereby establishing that peak 1 corresponds to the crystallization of $Mg_4Al_2Ti_9O_{25}$ while peak 2 corresponds to the crystallization of $MgSiO_3$. At a crystallization temperature of 1100°C, $Mg_4Al_2Ti_9O_{25}$ and sapphirine ($Mg_{3.5}Al_9Si_{1.5}O_{20}$) were detected indicating a possible peritectic reaction between the previously formed $MgSiO_3$ and the remaining glass. $Mg_4Al_2Ti_9O_{25}$, $Mg_{3.5}Al_9Si_{1.5}O_{20}$, and cordierite ($Mg_2Al_4Si_5O_{18}$) were observed at 1200°C and 1300°C. It should be noted that at 1300°C, the microstructure of the product transformed from non-resolvable crystalline phases to much larger lathe-like crystals Based on this data it is thought that the equilibrium crystalline phases within this system are cordierite, sapphirine, rutile (TiO_2) and cristobalite (SiO_2) with the occurrence of various metastable phases during the transformation of the glass product to the equilibrium state. Future work will establish the crystallization sequences in this system.

As shown in Figure 2, the normalized peak intensity of the $Mg_4Al_2Ti_9O_{25}$ phase increases dramatically at 1100°C. The normalized peak intensity of the sapphirine phase increases linearly with increasing temperature above 1100°C. Hardness (H) and fracture toughness (K_{IC}) data were measured as a function of growth temperature. Vicker's hardness was observed to remain relatively constant at approximately 960 kgF/mm^2 between 900 and 1200°C suggesting that any temperature between these limits will result in acceptable high grade abrasive formulations. Fracture toughness data has many uncertainties primarily associated with the assumption that the elastic modulus of the material does not change as a function of heat treatment and estimated to be 90 GPa for this material. With these uncertainties in mind, the fracture toughness was found to have a maximum (1.8 MPa-$m^{1/2}$) at 1200°C.

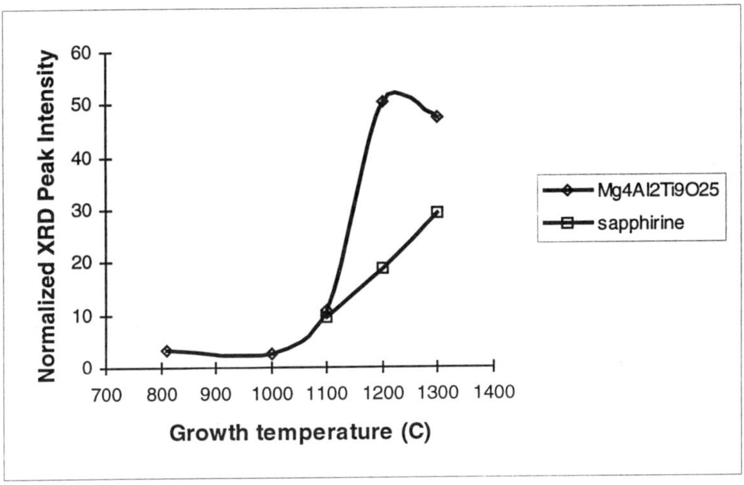

Figure 2: Normalized XRD Peak Intensity Versus Crystal Growth Temperature.

A literature review[5] has established that a material's resistance to abrasive wear is proportional to $H^{1/2}K_{IC}^{3/4}$ thereby establishing the utility of this parameter for developing abrasive media with optimum balance between fracture toughness (related to the granular media's resistance to dusting) and hardness. At crystallization temperatures of 811, 898, 1000, 1100, 1200 and 1300°C, the values of $H^{1/2}K_{IC}^{3/4}$ were 3.1, 3.8, 3.3, 4.3, 4.7 and 3.0 respectively. A parabolic regression was calculated for this parameter between 1000 and 1300°C with an r^2 value of 0.92. Using the obtained relation; $H^{1/2}K_{IC}^{3/4} = -7*10^{-5}\,T_c^2 + 0.1523T_c - 82.587$, the maximum value was calculated to occur at 1155°C. Therefore, this data in conjunction with the XRD data established the optimum crystallization temperature to be 1155°C.

Heat Treatment Optimization:

In order to optimize the economics of the heat treatment process, it is necessary to know the minimum heat-treatment time in which the water quenched sample can be transformed into a high-grade abrasive with a Vicker's hardness in excess of 1000 kgF/mm^2. Table I summarizes the effect of both nucleation and growth time on the resultant product hardness. Analysis of Variance (ANOVA) calculations indicate that the effect of nucleation time on sample hardness is statistical significant at a confidence level in excess of 97% while growth time is statistically significant at a confidence level of 99%. Based on the fact the product hardness remains relatively constant at a value of 980 ± 26 kgF/mm^2 with increasing crystallization time, it was concluded that at a crystal growth temperature of 1155°C, the crystallization occurs quite rapidly and approaches completion within 30 minutes. Based on this observation, it was concluded that further data is needed at times less than 30 minutes or alternately, the study could be repeated utilizing a lower growth temperature. In any case, it appears that the ideal heat treatment conditions for the MAST system consists of nucleating the sample at 755°C for 1 hour followed by growth at 1155°C for 30 minutes.

Table I: Vicker's Hardness of the MAST Sample As A Function Of Nucleation and Growth Time.

	Time at a growth temperature of 1155°C				
H (kgF/mm^2) Time at 755°C↓	0 h	0.5 h	1 h	2 h	4 h
0 h	688 ± 24	980 ± 24	959 ± 34	962 ± 27	875 ± 18
1 h	760 ± 22	1028 ± 24	1015 ± 30	982 ± 23	1044 ± 36
2 h	755 ± 34	986 ± 33	1027 ± 30	945 ± 26	939 ± 48
3 h	727 ± 22	1028 ± 56	1004 ± 28	975 ± 51	994 ± 41
4 h	740 ± 10	1056 ± 34	957 ± 41	935 ± 20	904 ± 38

CONCLUSION:

Abrasive products produced by the vitrification of industrial wastes enriched in MgO, Al_2O_3, SiO_2 and TiO_2 offer an optimum balance between processability and hardness. The MAST system is readily vitrified with both low melt viscosity and low liquidus temperatures. The as-quenched product can be utilized as a brown roofing granule and has product hardness suitable for applications as a medium grade abrasive. Additionally, the moderate hardness of the as-quenched product reduces equipment wear during subsequent grinding for the production of sized granules. The product can then be heat treated to develop high hardness in excess of 1000 kgF/mm^2 thereby increasing the market value of the resultant product.

ACKNOWLEDGMENTS:

This work is sponsored by the Edison Materials Technology Center (EMTEC); an organization comprised of members from academia, industry and government dedicated to the development of new materials and processing systems in order to make Ohio and U.S. business more competitive.

REFERENCES:

1. Blume, R.D. et al.,"High Grade Abrasive Product Development From Vitrified Industrial Waste",Ceramic Transactions, Vol 72, 1996, pp.229-239.
2. Blume, R.D., "The Effect of Partial Pressure of Oxygen and Cooling Rate On The Crystallization and Properties of Vitrified Industrial Waste for Abrasive Applications", masters thesis, The Ohio State University, 1997.
3. Volf, M.B., Chemical Approach to Glass, Elsevier, Volume 7, 1984, p.320.
4. Gao, J. and Drummond, C.H., "Heat Treatment of Vitrified Industrial Waste for Abrasive Applications", Submitted for publication, Journal of the American Ceramic Society, 1997
5. Jahanmir, S., Friction and Wear of Ceramics, Marcel Dekker, Inc., New York, 1994, pp.267-268.

COMMERCIAL ION EXCHANGE RESIN VITRIFICATION IN BOROSILICATE GLASS

Connie A. Cicero-Herman and Phyllis Workman
Westinghouse Savannah River Company
Savannah River Technology Center
Aiken, SC 29801

Kevin Poole, Donald L. Erich and John M. Harden
Clemson Environmental Technologies Laboratory
100 Technology Drive
Anderson, SC 29625

ABSTRACT

Bench-scale studies were performed to determine the feasibility of vitrification treatment of six resins representative of those used in the commercial nuclear industry. Each resin was successfully immobilized using the same proprietary borosilicate glass formulation. Waste loadings varied from 38 to 70 g of resin/100 g of glass produced depending on the particular resin, with volume reductions of 28% to 68%. The bench-scale results were used to perform a melter demonstration with one of the resins at the Clemson Environmental Technologies Laboratory (CETL). The resin used was a weakly acidic methacrylic cation exchange resin. The vitrification process utilized represented a ~64% volume reduction. Glass characterization, radionuclide retention, offgas analyses, and system compatibility results will be discussed in this paper.

INTRODUCTION AND BACKGROUND

The commercial nuclear industry utilizes ion exchange resins to clarify their process and storage waters. The resins are used to remove unwanted impurities, such as radioactive, hazardous, or other contaminants that could potentially harm the equipment or corrode reactor fuel rods. Over time, these resins have to be re-generated or replaced. When this happens, the spent resins have to be disposed. Disposal often becomes an economic problem because of the large volumes of resin produced and the relatively few technologies capable of economically stabilizing this waste.

Resins for removing Cs from Department of Energy (DOE) High Level Waste (HLW) are being investigated by DOE. The DOE also uses divinylbenzene/styrene resins in their reactor facilities to purify fuel rod storage basin water and in other nuclear facilities. Thus, a viable and economic treatment option would benefit both the commercial nuclear industry and DOE.

Vitrification is a preferred technology because it is capable of consistently producing a durable, leach resistant wasteform, while simultaneously minimizing disposal volumes through organic destruction, moisture evaporation, and porosity reduction. Vitrification of organic ion exchange resins presents a challenge because of the high organic content and the volatile Cs-137 that is usually present. High organics tend to induce reducing environments in melters, which can result in metals reduction and melter materials corrosion. The presence of organics can also result in reduced glasses, which has been shown to affect durability.(1) Cs-137 has been shown to be extremely volatile at high temperatures; thus, stabilization of this contaminant in the glass matrix presents a challenge which must be met if the waste is to be successfully stabilized.

Vitrification has been shown to be a feasible treatment method for organic ion exchange resins. Vitrification studies have been performed at the Savannah River Technology Center (SRTC) with both a resorcinol based organic ion exchange resin and divinylbenzene/styrene copolymer resins, but met with only limited success from a volume reduction standpoint.(2,3) Recent vitrification studies with the divinylbenzene/styrene resin matrices have met with more success through the use of a SRTC borosilicate glass formulation. Bench and pilot-scale studies with this formulation have produced homogeneous and durable glass products when tested on ion exchange resins from DOE reactor facilities and Argentina's commercial nuclear industry.

For this study, samples of six organic resins were obtained that are considered representative of commercial nuclear industry resins. The resins will be identified as A-550, IRC-50, 21H, 900-OH, 200-H, and 650-C, and a brief description of the structure and properties is given in Table I.

Table I. Structure and Properties of Commercial Ion Exchange Resins

Resin	Ion Form	Active Group	Matrix	Structure	Type
A-550	Cl	Quaternary Amine Type II Alkyl	Polystyrene	Gellular	Strong Anion
IRC-50	H	Carboxylic Acid	Methacrylic	Beads	Weak Cation
21H	H/OH	N/A	Polystyrene/ Cellulose	Powdered	2 part Cation/ 1 part Anion
900-OH	OH	Quaternary Amine Type I Alkyl	Polystyrene	Macroreticular	Strong Cation
200-H	H	Sulphonic Acid	Polystyrene	Macroreticular	Strong Cation
650-C	H	Sulphonic Acid	Polystyrene	Macroreticular	Strong Cation

Typical commercial nuclear industry radioactive contaminants include Mn, Co, Sr, C, and Cs. Based on data available, typical contaminant concentrations are 5.8×10^{-4} µCi/g of Co-60, 7.9×10^{-4} µCi/g of Cs-137, 6.2×10^{-4} µCi/g of Cs-134, and 5.4×10^{-4} µCi/g of Mn-54.(4) Other known non-radioactive contaminants include iron, borate, and lithium depending on the type of reactor. In this instance, the non-radioactive contaminants don't present a vitrification problem because these elements can easily be incorporated into glass.

BENCH-SCALE STUDIES

These studies were funded by the DOE - Office of Science and Technology Development, and the focus was to determine optimum vitrification processing parameters. Since the commercial nuclear industry ion exchange resins were very similar to other ion exchange resins used in previous SRTC research, attempts were made to vitrify these resins using the same borosilicate glass composition. Previous testing with this base composition and divinylbenzene/styrene ion exchange resins had indicated that a resin loading of approximately 42 g of resin/100 g of glass produced was optimal. Therefore, this loading was used as the starting point.

Experimental

The resins were mixed with the reagent grade chemicals to produce the borosilicate glass composition. All glasses were melted at 1150°C for four hours. The melt temperature was limited to 1150°C to help avoid volatilizing the Cs normally found on spent resins. All glasses were examined for completeness of reaction, visual homogeneity, and presence of metal sulfides. Testing with the resins involved slowly increasing or decreasing the amount of resin until the limit was found for producing acceptable glass. In some cases, the glasses were made by allowing the batch materials to react before melting to help with oxidation of the resin.

The optimum compositions were used to melt surrogate doped resin glasses. Dopants used were based on the representative radioactive contaminant concentrations given earlier. Higher levels of dopants (six to twelve orders of magnitude) had to be used because the actual amounts of the isotopes only equated to very small gram quantities of the elements, which would have been difficult to detect in the glasses when they were analyzed. Non-radioactive compounds were used as the dopants. These compounds were dissolved in water and mixed with the appropriate amounts of resin. The borosilicate glasses were analyzed for Co, Cs, and Mn content so the radioactive retention could be determined.

Results

Most of the resins produced homogeneous glasses. However, in a few instances, small pellets were found in the glass matrix that were determined by Scanning Electron Microscopy coupled with Energy Dispersive Spectroscopy to be metal sulfides. These have been shown to form when glasses become too reduced. Therefore, glasses which produced these pellets were not considered acceptable, and either the waste loading was lowered or the batch reaction time was increased.

After visual examination, several of the resin glasses were analyzed by the SRTC Mobile Laboratory to determine the glass redox ratio by the colorimetric method. Since it is likely that the resins would be treated in a joule-heated vitrification unit, acceptable $Fe(2+)/\Sigma Fe$ was considered to be less than 0.33, which is consistent with the operational limits established for the Defense Waste Processing Facility melter(5). If the redox ratio was found to be acceptable, the glass was further characterized for chemical composition, crystalline content, and durability. The $Fe(2+)/\Sigma Fe$ ratios are given in Table II. The base glass (without resin added) redox ratio was previously determined to be 0.054. The Glass IDs given in Table II denote the resin type and the resin loading per 100 g of glass produced.

Table II. Measured $Fe(2+)/\Sigma Fe$ Ratios for Bench-Scale Studies

Glass ID	Ratio	Glass ID	Ratio
550-50	0.259	900-70	0.124
50-50	0.119	200-38	0.269
21-62	0.054	650-38	0.330

For the 21H and 900-OH resin glasses, the redox ratio was not strongly affected by the resin loading. The redox ratios of the 200-H and 650-C resin glasses, on the other hand, were very dependent on the amount of resin in the glass. These glasses had the most problems with metal sulfides forming and incorporating high waste loadings. For the A-550 and IRC-50 resin glasses, the redox ratio gradually increased and became a limiting factor above 50 g of resin/100 g of glass.

The glasses were analyzed for chemical composition by the Analytical Development Section (ADS) of the SRTC. The glass compositions should be fairly consistent because the resin solids contribute only a minor amount to the glass composition. Overall, the glass compositions were very similar, and the type of resin did not seem to greatly impact the glass composition.

All glasses appeared to be visually homogeneous upon cooling. In order to ensure that no crystalline phases were present, X-Ray Diffraction (XRD) analyses were performed, and all glasses were found to be amorphous.

To determine the final product durability, the Product Consistency Test (PCT), ASTM C1285-94(6), was performed. Glass samples were run in triplicate, and the results were averaged and normalized. The normalized PCT results are given in Table III. The measured leachate pH is also listed, since this provides a secondary indication of durability. The PCT for the base glass composition is given for comparison. Since no acceptance criteria have been established for waste glasses other than HLW glasses, the durabilities of the glasses produced were compared against the Environmental Assessment (EA) glass(7) , which is the benchmark for HLW.

The PCT results were fairly consistent between the different glasses, with the 200-H and 650-C glasses having slightly higher PCT releases. Most of the glasses had similar releases to the base glass composition for B and Si, with Na releases that were slightly higher. These results indicate that the presence of the resin had little affect on the glass durability performance. The pH results also indicate that the glasses had similar durabilities, with the 200-H and 650-C glasses once again having the slightly poorer durabilities. No apparent trend was observed between the amount of

Table III. Normalized PCT Results (g/l)

Glass ID	Base	550-50	50-50	21-62	900-70	200-38	650-38	EA(7)
B	0.16	0.22	0.19	0.17	0.19	0.25	0.23	16.695
Si	0.07	0.07	0.07	0.07	0.06	0.07	0.08	3.922
Na	0.27	0.50	0.41	0.31	0.48	0.99	0.90	13.346
pH	9.66	10.16	10.05	9.87	10.11	10.59	10.64	11.85

resin in the glass and the PCT release; however a slight increase in B and Na release was seen for the glasses with the higher redox ratios. All glasses had excellent durability compared to the EA glass based both on the normalized releases and on the measured pH.

In order to determine the potential volume reduction of using vitrification treatment on each resin, bulk densities were measured and volume reduction calculations were performed. A measured glass density of 2.854 g/mL was used. The calculated volume reductions are shown in Table IV.

Table IV. Calculated Volume Reductions

Resin	Waste Loading	Resin Density (g/mL)	Volume Reduction
A-550	50 g/100 g or 33.3 wt%	0.6365	55.9%
IRC-50	50 g/100 g or 33.3 wt%	0.6235	55.9%
21-H	62 g/100 g or 33.3 wt%	0.6410	63.5%
900-OH	70 g/100 g or 41.2 wt%	0.6305	68.3%
200-H	38 g/100 g or 27.5 wt%	0.7230	34.0%
650-C	38 g/100 g or 27.5 wt%	0.7790	28.0%

No information on the water associated with these resins when they are in use or storage was found. When the resins are disposed of, they usually have an associated volume of water that also has to be disposed. In many cases, the resins also have a large volume of water that is adsorbed in the resin matrix. Volume reductions for the slurried spent resins will be higher because of this extra water that is evaporated during the vitrification process.

For the radioactive dopants studies, no apparent effect on melt behavior or glass formation was seen. The calculated retentions are shown in Table V. Calculated retentions were fairly consistent from glass to glass with the exception of the Cs retention for the glass made from the 900-OH resin. The calculated retentions were lower than anticipated, but were consistent with other vitrification processes. One explanation is that the radioactive surrogates were not processed through a resin column as would be done in practice, so the materials were not adsorbed as well as they would be in a resin column. With the radioactive materials strongly bonded to the resin, the potential for higher retentions exists.

Table V. Calculated Retentions for Glasses

Glass ID	550-50	50-50	21-62	900-70	200-38	650-38
Cs	73.1%	73.5%	76.9%	100%	70.3%	72.2%
Mn	89.1%	81.1%	83.5%	84.8%	83.3%	87.1%
Co	83.3%	76.9%	81.1%	84.2%	86.4%	79.0%

Based on all of the analytical results, the glasses with the optimal waste loading were homogeneous and very durable glasses. The SRTC borosilicate glass formulation seemed well suited for incorporating these wastes.

MELTER DEMONSTRATION

The primary objective of the melter test was to verify the feasibility of stabilizing commercial industry organic ion exchange resins using vitrification technology. From the six commercial resins studied, IRC-50 was selected because its matrix was the most different from the divinylbenzene/styrene resin material already demonstrated. This resin was shown to produce acceptable glasses in a crucible at a loading of 50 g of resin/100 g of glass or 33 wt%. Visual observations of the operating behavior of the melter was very important, including the presence of foam or salt layers. Corrosion of the melter components was also an important consideration.

Melter System Description

The melter used for the demonstration was a 1/4 square foot stirred tank melter. The melter utilizes an Inconel® melt chamber, which serves as one of the electrodes. The impeller is also fabricated from Inconel® 690 and serves as the other electrode. It provides a stirring action that results in much higher mixing and production rates than can be achieved by convective mixing alone. The maximum temperature obtainable in the melter is ~1050°C because of the temperature limitations of the Inconel® components. Pouring of the glass occurs through an overflow drain. Flow through the drain is controlled by controlling the temperature of the spout.

Slurry feeding is accomplished through the use of a slurry feed port that disperses the feed over the melter surface, so it can be rapidly mixed in with the molten glass through the impeller stirring action. The melter can also operate in a cold-top mode by using slower stirring motion to incorporate the new feed into the molten pool. For this demonstration, the slurry was fed from a continuously stirred feed vessel (~19 L capacity). A peristaltic pump in the feed line directed the feed to the melter at a constant rate, which could be adjusted to accommodate the feed melt rate.

For this demonstration, an oxygen sparger was located at the bottom of the melt tank. Oxygen could be metered into the melt at a rate of 0.6 - 6.10 standard liters per minute to help maintain oxidizing conditions in the melt or to help break down organics in the feed.

The melter was connected to an offgas treatment system that included a filter system as a prefilter to the laboratory exhaust system. A sampling manifold was installed between the melter and the laboratory exhaust system, which enabled isokinetic sampling for EPA Method 29.

Feed Fabrication

Three different batches were made for the demonstration, equaling approximately one melter volume of feed. The Stir-Melter can be started with very little glass in the melter tank, so very little turn-over of the melter contents is required before reaching steady-state. The glass in the melter was of the same base composition. Cobalt was not included in feed batches #2 and #3 because it was not available. After initial feeding to the melter, the glass appeared to be viscous, so additional Na_2O was added directly to the melter vessel through the dry batch feed port and the pour riser. The remaining batches were adjusted to this Na_2O level.

Since no information was available on water content of resin slurries, the resin used at the CETL was only mixed with minimal water to help feeding and adsorption of radionuclide surrogates. The radioactive surrogates were dissolved in the water and then the resin was well mixed with the radioactive surrogate containing water. Once again, the actual adsorption onto the resins was not believed to be optimized. Therefore, the radioactive surrogates may have been more volatile during the demonstration because they were not bound to a matrix.

The representative radiolytical characterization given earlier was used as the basis for spiking the resins with surrogate contaminants for the demonstration. The amounts actually used were 9.7×10^7 μCi/g for Co, 5.0×10^5 μCi/g for Cs, and 5.4×10^4 μCi/g for Mn. Higher amounts were used because of the small mass associated with the radiation dose. Cs was only spiked for the Cs-134 portion since the corresponding mass of Cs-134 compared to Cs-137 was much greater for their particular activities.

Melter Operation

During the demonstration, an optimum feed rate was determined based on the slurry properties and feed melt rate. The optimum feed rate was defined to ensure that the melt surface was evenly covered and that no unnecessary build-up of glass or cold-cap material occurred. Slurry was initially added at a rate of 80 mL/min. This rate was maintained for the first day of feeding. However, as more of the feed built up in the melter, the feed rate had to be decreased because of cold-cap build-up and foaming. The foam layer was controlled by adjusting the impeller height, turning off the sidewall resistance heaters, or adjusting spindle speed. The optimal feed rate was determined to be 30 mL/min.

In addition to monitoring the feed rate, system temperatures were also monitored to obtain the desired temperatures and melt rates. The drain temperature was set at 1025 or 1030°C when pouring, and the melt temperature was set at 1050°C. The actual temperature readings were a little lower for the pour temperature at first then hit the set point during pouring. Melt temperature was a few degrees higher than the target. Both sets of resistance heaters were kept on to maintain the melt temperature and glass viscosity. They were at 940 to 970°C during pouring. No problems with drain pluggage or glass exit viscosity were seen during the demonstration.

The offgas system was operational throughout the demonstration. Samples of the offgas were taken near the end of the campaign when the melter was at steady-state operating conditions. An EPA Method 29 Multiple Metals Sampling Train (MMST) was performed to determine the distribution of metals within the offgas system. The train can also be used to determine particulate emissions according to EPA Method 5. For the demonstration, an isokinetic source sampler was used to perform the Method 29 test. Due to past problems experienced with standard EPA organic sampling methods, the sampling and analysis approach used was to primarily identify the constituents and, thus, determine the efficiency of combustion within the melt chamber.

At test completion, the melter was shut-down according to CETL procedures. The melter system was visually inspected and examined for wear or corrosion.

Results

Samples of each feed batch were taken during the demonstration to determine the physical properties and chemical composition. The physical properties of each feed sample are shown in Table VI. Weight percent solids was determined at 105°C, while calcined solids were determined at 1050°C, which was the melting temperature.

Table VI. Physical Properties of Feed Samples

Sample ID	Wt% Solids	Calcined Solids (Wt%)	Density (g/mL)
#1	51.8	46.95	N/A
#2	29.0	28.62	1.66
#3	31.0	29.16	1.35

Sample #1 had a much higher solids content than the rest of the materials. Part of this batch of feed was used to start-up the melter, so less water was added. The density of sample #1 could not be measured because the material was so thick that it was hard to get a representative sample.

The chemical compositions of the material fed to the melter and the glass produced were determined by the Mobile Laboratory of SRTC and ADS. Only feed sample #3 was analyzed for Cs concentration, and the concentrations in the other feed batches were calculated based on the proportions batched to this feed. It was assumed that no batching errors occurred and proportional quantities of Cs would be found in the other batches.

The melter feed composition varied from sample to sample. It is believed that some heterogeneity or settling of glass additives may have occurred. Additional problems may have been experienced with getting a representative feed sample for analyses because of the rheology of the feed. The entire feed sample would likely have to be vitrified to get a truly representative sample because of the feed settling and unmixing that occurs. Compared to the target compositions, the feed samples were lower in B_2O_3, Co_2O_3 and Cs_2O and higher in SiO_2 than targeted.

For the analyzed glass, some deviation from the target composition existed. This deviation may have resulted from incomplete mixing when Na_2O was added directly to the melter. If the melt contents were not completely homogenized, portions of the glass could contain excess levels of Na_2O. It also appears that excess SiO_2 was added, consistent with the batch sheet calculations.

The glass composition was compared to the Sample #3 feed composition. Theoretically, the feed and glass compositions should be roughly the same with minor differences due to volatility during vitrification. Most oxides were comparable except Na_2O and B_2O_3 which were lower than found in the feed, which was expected since they are volatile components.

In order to understand the radioactive retention in the glass matrix, retention calculations were performed. Percent retention was calculated based on the total glass output and total feed input. These results indicated that most of the feed was incorporated in the glass matrix. Approximately 1% Cs_2O and 7% Co_2O_3 were emitted or not accounted for, while all MnO was retained. Two possible explanations exist for the low Co_2O_3 number. One is that the Co_2O_3 was not evenly mixed in the glass because it was only added in one batch, while the other is that the Co_2O_3 was not completely sorbed onto the resin and so it volatilized with the water in the resin slurry.

The $Fe(2+)/\Sigma Fe$ ratios for glass samples taken throughout the demonstration were measured in duplicate by the Mobile Laboratory. Some samples were taken from the melter itself, while other samples were taken from the pour spout. Redox ratios for the glass samples varied from 0.000 to 0.007, while a sample of the foam layer had an average ratio of 0.026. All glasses were very oxidized. The measured ratios were lower than limits established at SRTC for joule-heated melters.(5) For joule-heated melters, very oxidizing conditions can result in foaming in the melter. Foaming was seen during the latter part of the demonstration. It is interesting to note that the foam layer was less oxidized than the melt pool based on the redox results. This may have been the result of the oxygen sparging to the melt pool.

Visual examination of the resulting glass indicated that the glass appeared amorphous. The glass had no visible signs of crystallinity. No devitrification was noticed in the glass containers from the demonstration. XRD analysis confirmed that the glass was amorphous.

The PCT was performed on glass from the end of the demonstration. The standard procedure was performed and the triplicate results were averaged and normalized for the glass elemental content. The normalized PCT results were 0.70 g/L for B, 0.21 g/L for Si, and 0.81 g/L for Na, while the leachate pH was 11.80. The normalized releases for B, Si, and Na were substantially less than the EA glass. The leachate pH was very high for this glass, which sometimes indicates poor durability. However, the elemental releases indicated that the glass was acceptable. The radionuclide surrogates were not detected in the leachates at any appreciable amounts.

Glass viscosity was measured as a function of temperature by CETL. Viscosity at 1150°C was 45.5P, while it was ~135P at 1050°C. The measured viscosities indicate that the glass would be considered acceptable for processing in typical joule-heated melters that are maintained above ~1075°C. For the Stir-Melter, the viscosity was near the upper processing limit.

The samples from the MMST were analyzed according to EPA methods. The front half of the impactor used in the sampling was analyzed separately from the back half. The front half represents the particulate matter captured during the test, while the back half represents the condensable species that were in a volatilized state.

The glass matrix components were the biggest contributors to the offgas pollutants. Cs was found only on the front end of the sampler, indicating it was in the particulate form. The Mn concentration was relatively low and was found in both the particulate and volatile forms, while Cs concentration was a little higher. Co was not measured. The emissions are typical of emissions found during vitrification of waste glasses of similar compositions. The mass of particulate metals captured was orders of magnitudes greater than the mass of volatile metals, which allows for easier capture in an offgas system.

Organic samples were drawn through a condenser and 100 mL of water in the sampling manifold. The samples were analyzed by Gas Chromatography/Mass Spectroscopy (GC/MS). No peaks found by the GC/MS had greater than 90% correspondence. The compounds identified were believed to be heterocyclic ring compounds with nitrile side groups and high molecular weight cyclic and heterocyclic nitriles and cynates. These compounds are expected from pyrolysis of organics. These compounds would present environmental problems (e.g., potential carcinogens) and could cause tarring in an offgas system. Therefore, possible means of breaking down these compounds should be explored.

Total feed to the melter was 32.8 kg. Of this total amount, the resin waste was 7.36 kg with an approximate density of 0.6235 g/mL. The total mass of glass produced was 11.95 kg. Density was determined to be 2.844 g/mL using the buoyancy method. Therefore, the volume reduction associated with using vitrification treatment was 64.4%. Given that resins to be stabilized would have an associated water content, the volume reduction would be higher.

Minimal wear or corrosion was seen to the melt vessel and impeller due to the vitrification of the commercial ion exchange resin.

CONCLUSIONS

Based on the studies performed with the six resins representative of commercial industry resins, it appears that using the SRTC borosilicate glass composition will produce very durable and homogeneous glasses. For the resins, different optimum waste loadings were determined, and they were strongly dependent on the type of resin being vitrified. The waste loadings ranged from 38 to 70 g of resin/100 g of glass produced, with associated volume reductions of 28.0-68.3%. These waste loadings and associated volume reductions are higher than what has been seen in other studies with organic resin, but are similar to volume reductions seen with this glass composition. The resins that were the most difficult to vitrify were the strong cation sulfonated styrene resins.

The melter demonstration showed that vitrification of commercial organic ion exchange material is viable. Processing of this feed did not result in any significant problems. The resulting glass product was homogeneous and durable. The most significant particulate and volatile offgas pollutants were attributed to the glass matrix components. The radionuclide surrogate releases were relatively small compared to the particulate releases of the major glass components. The radioactive material retention (demonstrated with surrogate chemical compounds) was ~93% for Co, ~99% for Cs, and 100% for Mn. For this demonstration, a 64.4% volume reduction was demonstrated based on a 33.3 wt% waste loading.

REFERENCES

(1)X. Feng, I.L. Pegg, E. Saad, S. Cucinell, and A.A. Barkatt, "Redox Effects on the Durability and Viscosity of Nuclear Waste Glasses", *Nuclear Waste Management IV*, 23.

(2)N. E. Bibler, J. P. Bibler, M. K. Andrews, and C. M. Jantzen, "Initial Demonstration of the Vitrification of High-Level Nuclear Waste Sludge Containing an Organic Cs-Loaded Ion Exchange Resin", Presented at the Materials Research Society Meeting 1992, WSRC-MS-91-465.

(3)T.N. Sargent and T.J. Overcamp, "Vitrification of Cesium-Laden Organic Ion Exchange Resin in a Stirred Melter", CU/ESE/VRL-96-003, Clemson University - Environmental Systems Engineering Report, June 1997.

(4)B.A. Weber, "Direct Vitrification of Fermi 2 Bead Resin", Presented at the Electric Power Research Institute - Vitrification of Low-Level Waste Meeting, San Antonio, December 5-6, 1995.

(5)D.F. Bickford, A.A. Ramsey, C.M. Jantzen, and K.G. Brown, "Control of Radioactive Waste Glass Melters: I, Preliminary General Limits at Savannah River", *J. Am. Ceram. Soc.*, **73** [10], 2896-2902 (1190).

(6)ASTM C1285-94, "Standard Test Methods for Determining Chemical Durability of Nuclear Waste Glasses: The Product Consistency Test (PCT)", 1994.

(7)C.M. Jantzen, N.E. Bibler, D.C. Beam, C.L. Crawford, M.A. Pickett, "Characterization of the Defense Waste Processing Facility (DWPF) Environmental Assessment (EA) Glass Standard Reference Material," WSRC-TR-92-346, Rev. 1, Westinghouse Savannah River Company Report, June 1993.

High-level Radioactive Waste Management

NRC'S ACTIVITIES DURING THE DEMONSTRATION PHASE OF THE HANFORD TANK WASTE REMEDIATION SYSTEM PRIVATIZATION PROJECT

Michael Tokar
U.S. Nuclear Regulatory Commission
Washington, D.C. 20555

Gary C. Comfort, Jr.
U.S. Nuclear Regulatory Commission
Washington, D.C. 20555

ABSTRACT

On January 29, 1997, NRC and DOE signed a Memorandum of Understanding (MOU) that established the basis for cooperation and mutual support between the two agencies during the demonstration phase (defined as Phase I) of DOE's Tank Waste Remediation System (TWRS) Privatization project at Hanford, Washington. The objective of the TWRS project is to place the radioactive and toxic materials currently in the Hanford tanks into a solidified waste form. In this paper, the activities conducted by the NRC's TWRS Section during Phase I of the project are described. These activities include gaining familiarity with various likely methods of waste pretreatment and solidification technologies, reviewing and providing input to DOE on TWRS contractor's submittals, and providing DOE with established NRC guidance and position documents for DOE consideration in the development and updating of DOE's regulatory review guidance.

BACKGROUND

At the U.S. Department of Energy's (DOE) Hanford Site in southeast Washington State there are 177 underground storage tanks, with holding capacities ranging from 210 to 4160 m^3 (55 000 to 1.1 million gallons), that contain approximately 208 000 m^3 (55 million gallons) of solid, liquid, and sludge-like radioactive and chemical wastes. The wastes are the residual byproduct of more than 40 years of operation of chemical processes at facilities whose principal mission was the production of plutonium for use in nuclear weapons. The tanks are cylindrical reinforced-concrete structures with inner carbon steel liners, and are of two types: the first 149 tanks constructed at Hanford have a single carbon steel liner (and are thus called single-shell), while 28 tanks have two steel liners separated by an annulus (and are thus known as double-shell). Because the single-shell tanks were expected (designed) to last only 10 to 20 years, all of them have substantially exceeded their design life and several (over 60) are either known or are suspected to be leaking.

To deal with the environmental concerns at Hanford, the Washington State Department of Ecology, U.S. Environmental Protection Agency (EPA) Region 10, and DOE signed an agreement (known as the Tri-Party Agreement[1], or TPA) that legally requires DOE to safely manage and dispose of liquid and solid wastes (including, but not limited to the wastes in the 177 tanks), at the Hanford Site. The TPA contains milestones for tracking cleanup progress. To accomplish the cleanup of the tank wastes on the schedules called for in the TPA, and to limit the associated costs, the DOE plans to privatize the waste treatment and immobilization operations.

As an initial step in accomplishing the privatized cleanup of the Hanford tank wastes, DOE approved a policy[2] in July 1996 for the radiological, nuclear, and process safety regulation of the tank waste remediation contractors. The policy statement was augmented with a Memorandum of Agreement[3] (MOA) that specified roles and responsibilities among DOE officials with regard to the Tank Waste Remediation System (TWRS) privatization. In part, the MOA defines the role and responsibilities of a TWRS Regulatory Official, who acts as the head of a specially created and dedicated organizational group known as the TWRS Regulatory Unit (RU). The DOE RU is responsible for the regulatory oversight of all design, construction, operational, and event-response activities involving the TWRS Privatization contractors during Phase I.

DOE/NRC MEMORANDUM OF UNDERSTANDING

The TWRS Privatization is divided into two phases: (1) a demonstration phase (defined as Phase I); (2) a full-scale operations phase (defined as Phase II). The Phase I program is subdivided into Part A (a pre-design feasibility program) and Part B (demonstration pilot plant study). In September 1996, DOE awarded separate contracts to two groups, headed up by British Nuclear Fuels, Limited (BNFL) and Lockheed Martin Advanced Environmental Systems (LMAES), that called for the purchase by DOE of waste treatment services during Phase I. As described in the contracts, during both phases, DOE would purchase waste treatment services from a DOE contractor-owned, contractor-operated facility under a fixed-price type of contract, where the DOE TWRS Privatization Contractor would finance the project, design the equipment and facility, apply for and receive required permits and licenses, construct the facility and bring it on line, operate the facility to treat waste, and deactivate the facility.

Although DOE has regulatory responsibility for all of the Phase I privatization activities, in January 1997, DOE and the Nuclear Regulatory Commission (NRC) signed a Memorandum of Understanding[4] (MOU) that establishes a basis for cooperation and mutual support between the two agencies. The MOU provides for cooperation and mutual support in an integrated effort between DOE and NRC that provides for DOE to acquire the capability to implement a comprehensive regulatory program that is consistent with NRC's regulatory approach for protecting workers, the general public, and the environment. Also during Phase I, NRC is to acquire sufficient knowledge and understanding of the physical and operational situation at the Hanford waste tanks and the hazards associated with the processes and technology involved in Phase I activities to enable NRC to (a) assist DOE in performing technical reviews in a manner consistent with NRC's technical approach and (b) to be prepared to develop an effective and efficient regulatory program for the licensing of DOE contractor-owned and contractor-operated facilities that will process waste at Hanford during Phase II. These objectives are to be accomplished through interactions with the RU, review of TWRS Privatization Contractor submittals, and site visits.

FAMILIARIZATION WITH TANK WASTE HAZARDS

At the start of fiscal year 1997 (October 1996) the NRC formed a group known as the TWRS Section whose mission is to implement the provisions of the MOU. Following a rigorous search and recruitment effort, the TWRS Section is now comprised of 14 staff with extensive experience and expertise in areas related to chemical process safety, configuration management, environmental protection, fire protection, hazards analysis, maintenance, nuclear criticality safety, plant systems, quality assurance, radiation protection, structural analysis, vitrification technology, and waste management. In keeping with the MOU, the staff has been (a) gaining understanding of the likely processes to be used in the remediation of the Hanford tank wastes and the hazards associated with such processes, (b) developing guidance that can be used by NRC technical reviewers in their reviews of licensing documents, and (c) assisting the DOE RU in their review of contractor submittals during Part A of Phase I.

There is an enormous volume of information available on the Hanford site and current cleanup activities; thousands of documents accessible via the Internet are in the DOE bibliographic database, and many of these documents address potential hazards posted by the wastes contained in the tanks. Given the availability and quantity of that information, hazards posed by the material presently contained in the tanks will be discussed here only briefly and in the context of potential similar hazards that may be encountered in the processing of the wastes.

The safety issues currently associated with the Hanford tanks have been placed into four categories: (1) flammable gas; (2) potentially explosive concentrations of ferrocyanide; (3) potential for flammable organic nitrate reactions; and (4) high-heat generation. DOE developed a set of criteria to identify tanks with potential safety concerns that fell into these categories and placed those tanks on a "Watch List." As of October 31, 1997, there were 32 single-shell and 6 double-shell tanks on the Watch List. The 6 double-shell tanks are all on the Hydrogen Watch List. Eight single-shell tanks are on more than one Watch List. All tanks have been removed from the Ferrocyanide Watch List. Hazards associated with the remaining issues, i.e., flammable gas, presence of organic compounds (which when mixed with nitrites and nitrates can ignite at elevated temperatures) and high-heat, if they are satisfactorily resolved at the affected tanks, may or may not carry over to the remediation activities, depending on the designs of the particular processes that will be utilized for the remediation. For example, radiolytic hydrogen generation could be a problem for a TWRS process where the wastes would be highly concentrated during pretreatment. NRC, with the assistance of its contractor, the Center for Nuclear Waste Regulatory Analyses, is developing information about the potential chemical interactions specific to TWRS operations at Hanford that may need to be addressed during the review of processing activities. This information is being developed in the form of a High-Level Waste Chemistry Manual which will be used the NRC staff in its review of the proposed technologies.

DEVELOPMENT OF REGULATORY GUIDANCE

As part of its responsibilities for the regulation of the TWRS privatization, DOE prepares and issues guidance specific to the regulation of the TWRS privatization contractors. Such guidance is for use by the DOE RU in its execution of regulatory reviews and resulting regulatory actions and is provided as information to the DOE TWRS privatization contractors for their preparation of regulatory submittals. In accordance with the provisions of the DOE/NRC MOU, however, NRC has provided the DOE RU with established NRC guidance and position documents as input for DOE consideration in the development and updating of its guidance for the DOE regulatory review.

A key guidance document that NRC has been developing internally for use with its own staff is a Standard Review Plan (SRP). The SRP is being developed on a schedule that is intended to coincide with the development of proposed revisions to NRC's regulation for the Domestic Licensing of Special Nuclear Material, 10 CFR Part 70. Part 70 would be the regulation that would be used by NRC to regulate a TWRS facility, if NRC were to assume regulatory responsibility for a privatized facility during Phase II.

The SRP will provide guidance to the NRC technical staff on how to review information submitted by a license applicant in areas dealing with facility, process and site description, management organization, radiation safety, nuclear criticality safety, chemical process safety, fire safety, emergency management, environmental protection, and conduct of operations. Under conduct of operations, some additional areas of review include configuration management, quality assurance, and human factors. An integrated safety analysis, which provides a systematic examination of the facility's processes, equipment, structures and personnel activities and that addresses, in an integrated fashion, safety measures that might otherwise conflict, in the areas of fire protection, radiation safety, chemical safety, etc., is expected to be central to the applicant's submittal and will provide a focus for the staff's review.

REVIEW OF DOE CONTRACTOR SUBMITTALS

One of the major regulatory interaction activities called for in the DOE/NRC MOU involves regulatory program implementation. Specifically, NRC has agreed to review and provide comments on DOE TWRS Privatization Contractors' submittals to identify any areas that are not consistent with NRC's regulatory approach. In particular, NRC committed to assisting DOE in evaluating submittals and in verifying the effective implementation of design bases, level of design detail and documentation, design specifications, calculations and drawings, and procurement specifications, as well as quality assurance (for design, procurement, construction, pre-operational testing and operation), operator training and qualification, human factors, and emergency response. The primary submittals that NRC has reviewed from the privatization contractors has been in the form "Standards Approval Packages" (SAPs) and "Initial Safety Assessment Reports" (ISARs). The SAP consists of two documents: (1) a Safety Requirements Document (SRD) and (2) an Integrated Safety Management Plan (ISMP). DOE's regulatory approach for the TWRS Privatization requires the contractors to take the lead in identifying and recommending the standards and requirements that will be used to achieve adequate safety for specific TWRS activities. The contracts require the privatization contractors to recommend (in their SRDs) hazard controls, including radiological, nuclear, and process safety standards and requirements, for their TWRS facilities that are commensurate with the risk expected by the proposed operations and are in conformance with DOE's overall top level safety principles. Likewise, the contractors were required to provide an ISMP for DOE review and approval, where the ISMP, in brief, describes the contractors management system that is intended to ensure control of the agreed-upon work definition. The final documents to be reviewed prior to the close of Part A of Phase I, viz., the ISARs, were expected to (1) define the projected safety basis for the facility (safety envelope) in terms of the physical design, structures, systems, equipment with prescribed safety functions, operating modes, modes and conditions, representative internal and external events, representative safety analyses and results and major uncertainties in data and analyses, (2) describe how the facility should perform , and (3) describe how adequate protection of the public, workers and the environment should be achieved.

NRC staff have also reviewed and provided comments on related documents, including Hazard Analysis Reports (which are part of the Standards Approval Packages), Employee Concerns Management Systems documents, Radiation Exposure Standards for Workers Under Accident Conditions, and Quality Assurance Program documents.

The NRC technical staff, in its reviews of the above documents, focussed on those items that might impact the health and safety of the public, workers and environment, or which could lead to an inconsistency with the NRC's regulatory approach. Thus, the NRC did not review areas of the documents that are related to contractual agreements between the contractors and DOE and that were considered outside the purview of NRC's role as defined by the MOU. The ISMP, which explains how safety standards will be implemented, fell into the latter category.

DIFFERENCES IN APPROACH

Although a few differences in approach have been flagged at the current stage of review, many of these differences in regulatory approach will likely be worked out as the privatization contractors' designs mature and become more detailed. In reviewing the DOE contractors' submittals, the NRC staff has, to the extent practicable, applied applicable parts of the draft SRP. This has presented some difficulties, in that the draft SRP is intended to be utilized in the review of a completed design, while the documents submitted by the DOE contractors during Part A of Phase I of the TWRS Privatization have addressed only a very preliminary conceptual design. Thus, the acceptance criteria in the draft SRP could not be used directly to assess the adequacy of the information presented on the preliminary designs. Rather, the NRC staff used the draft SRP acceptance criteria to identify technical issues and the type and level of detailed information that would be expected to be submitted for a full-blown review of a completed design; i.e., at the submittal of a preliminary safety analysis report (PSAR). The NRC provided this information to the DOE RU in the form of technical review comments on the contractors' SAPs and ISARs. DOE reviewed the NRC comments and incorporated a number of them into their requests for additional information from the privatization contractors. The comments that DOE did not incorporate were either already included in DOE's requests or found to be premature for the level of information expected from the contractors at this stage. All of NRC's comments were eventually provided to each respective privatization contractor to allow them the chance to incorporate the intent of the comments into future submittals as the design matured. NRC agreed that DOE's methods of dispositioning the comments was appropriate for the current level of design.

REVISED 10 CFR PART 70

As mentioned earlier, NRC is currently developing a SRP specific to TWRS privatization-type facilities. Because of the content of special nuclear material in the process streams, NRC would plan to license the facilities under 10 CFR Part 70 which is currently undergoing revision. In the past, the primary large-scale facilities that this regulation was used for were for major fuel cycle facilities. These fuel cycle facilities are generally licensed to process special nuclear material (e.g., enriched uranium or plutonium) without any associated byproduct materials; therefore, the threat from these facilities tends to be from ingestion or inhalation pathways of the radioactive material or associated chemicals. In the case of the TWRS privatization facilities, the most significant threat results from the external exposure pathway from the byproducts in the high-level waste process stream. As such, it will be necessary to focus guidance provided by a SRP toward such a hazard.

NRC's early interactions with DOE are permitting the NRC staff to familiarize themselves with these technologies and develop guidance that is focused toward the potential hazards associated with such operations. Furthermore, NRC staff has been reviewing the revised 10 CFR Part 70 to ensure that the proposed TWRS privatization facilities will be licensable under the revised regulation. In particular, as a result of interactions on this process, the NRC staff is proposing some regulatory changes to address aspects of the design of new special nuclear material processing facilities; these proposed changes are still undergoing Commission review. NRC's aim is to issue the proposed TWRS guidance in a time frame consistent with publication of the revised rule and early enough to be of use to the TWRS privatization contractors during their design of the new facilities.

CONCLUSION

Overall, NRC believes that interactions with DOE at this early stage has been very successful. The primary goals of the NRC's interactions with DOE are (1) to familiarize NRC with the technologies and possible hazards associated with these operations in order to permit NRC to develop a sufficient regulatory policy to efficiently handle the expected licensing of new full-scale facilities during Phase II of the TWRS privatization project, and (2) to inform DOE of any possible inconsistencies between DOE's regulatory approach of the facilities and the regulatory approach that NRC would undertake during Phase II. To date, only a few inconsistencies have been noted, primarily because NRC generally is not involved in such an early design stage of licensed facilities. As the design matures, more inconsistencies in regulatory approach may be noted by NRC and discussed with DOE; however, just because an inconsistency in regulatory approach is detected, it is important to realize that there is not necessarily a reduction of safety under DOE's regulatory approach - in fact, some of the standards adopted by the TWRS privatization contractors may be much more conservative than those required by NRC. DOE's strategy in permitting the TWRS privatization contractor's to select and implement those standards believed to be appropriate to the proposed facilities, may, in fact, result in levels of safety equivalent to those provided by NRC's regulatory approach; however, the codification of certain standards in NRC's regulations would still result in a finding of inconsistency when the standards chosen do not specifically fall within NRC's regulations. Upon notification of such an inconsistency, DOE may (1) require the contractor to adopt a standard that fits within NRC's regulations during Phase I, or (2) notify the contractor to be prepared to adopt such a standard under Phase II operations, or (3) request an exemption or rulemaking change to permit the DOE adopted approach. The early notification allowed by NRC's participation in the project at this time, should permit DOE to make informed decisions early enough to help reduce costs and delays toward the possible NRC licensing of facilities during Phase II of this project.

REFERENCES

[1]*Hanford Federal Facility Agreement and Consent Order (Tri-Party Agreement)*.

[2]*Policy for Radiological, Nuclear, and Process Safety Regulation of TWRS Privatization Contractors*, DOE/RL-96-25, Revision 0 (July 1996).

[3]*Memorandum of Agreement for the Radiological, Nuclear, and Process Safety Regulation for TWRS Privatization Contractors*, DOE/RL-96-26, Revision 0 (July 1996).

[4]*Memorandum of Understanding Between the Nuclear Regulatory Commission and the Department of Energy: Cooperation and Support for Demonstration Phase (Phase I) of DOE Hanford Tank Waste Remediation System Privatization Activities* (January 29, 2997).

RADIOACTIVE WASTE VITRIFICATION AT THE WEST VALLEY DEMONSTRATION PROJECT

R. A. Palmer and S. M. Barnes
West Valley Nuclear Services Co., Inc.,
West Valley, NY.

ABSTRACT

The West Valley Demonstration Project has produced over 190 canisters of glass, high-level waste form suitable for deep geologic disposal (as of early May 1998). This paper will discuss the current status of the Project and how various challenges were met and overcome in the nearly two-year campaign of making glass. The discussion will include the expected and unexpected successes, as well as important lessons learned to improve future vitrification processes.

INTRODUCTION

The Vitrification Facility at the West Valley Demonstration Project (WVDP) was designed and constructed to convert high-level radioactive waste (HLW), stored in underground tanks, into a stable waste form (borosilicate glass) suitable for disposal in a federal repository. After completion of nonradioactive testing with simulated waste, radioactive waste processing began in June 1996. The HLW, consisting of a blend of washed plutonium-uranium extraction (PUREX) sludge, cesium-loaded zeolite, and neutralized thorium extraction (THOREX) waste, is batch-transferred from the Waste Tank Farm to the Vitrification Facility. There the waste is mixed with glass-forming chemicals and fed to a joule-heated melter at 1,150° C. The glass is then poured into stainless steel canisters. The canisters are welded shut, decontaminated, and stored temporarily on site.

Steady-state glass production rates in excess of 50 kilograms per hour have been achieved. As of May 1998, more than 190 canisters have been filled containing over 390,000 kilograms of borosilicate glass that contains nearly nine million curies of cesium-137 and strontium-90. Glass production system availability has surpassed 70%, due in part to design modifications and changes in operational and maintenance strategies developed from operational experience.

OPERATIONAL EXPERIENCES AND LESSONS LEARNED

Processing experience gained during the five years of initial nonradioactive melter operational testing, combined with recipe experimentation in a scaled vitrification (mini-melter) system, has been successfully applied to radioactive operation. During these tests, glass chemistry (largely oxidation-reduction) was studied, and methods of chemistry control were developed and have yielded excellent correlation with radioactive operations.

Initial waste transfers from the Waste Tank Farm had a lower curie content than anticipated (i.e., less than 100,000 curies of cesium-137/strontium-90 per batch). An additional sludge mobilization pump was installed in an existing HLW tank riser above a low-agitation area within the tank where solids were apparently accumulating. Careful management of the HLW tank level also increased batch curie inventory. By periodically decanting the accumulated liquid, the solids concentration in the transfers was optimized while minimizing the transfer pump plugging problems associated with pumping slurries. Many waste transfer batches have exceeded 200,000 curies.

Batch makeup cycle time has been reduced since the start of radioactive operations. Due to the relative consistency of the waste, a chemical premix recipe can be prepared well in advance of the final (post waste analysis) recipe. Independent verification of recipe calculations has virtually eliminated the need for corrective (and time consuming) chemical shims. As the waste tank is emptied, the chemistry of the waste has gradually changed. By using individual raw materials (sand, borax, aluminum hydroxide, etc.), the flexibility of the process allows for effective, efficient tailoring of the glass recipe.

A wider range of glass compositions has been studied to expand the range of acceptable glass recipes. These compositions include glasses with little or no uranium and thorium, which have become more depleted from the waste as the tank empties. With a broader range of acceptable compositions, the necessity for small adjustments in the feed will virtually be eliminated.

The turnaround times for chemical analyses in support of feed batch preparation have shown dramatic improvement relative to initial estimates. Early Analytical Cell mockup training and the experience gained by the lab technicians has reduced the time required for the various feed batch analyses by more than 50%, thereby saving 60 hours on the total batch cycle time. This is significant in that if this were not the case, batch preparation would be the limiting factor in the vitrification process. Instead, the melter has generally been fed continuously from the melter feed hold tank (MFHT).

Slurry transport within the Vitrification Cell is accomplished by either steam jet transfer or air-displacement-slurry (ADS) pumps. Steam jet operation has been as expected with no significant problems associated with these routine transfers. The throughput capability of the WVDP melter requires an average slurry feed flow rate of approximately 100L/h. The intermittent character of the ADS pump has provided this low average feed rate while generating high slurry velocity and high back pressure. These characteristics have resulted in essentially none of the plugging problems often encountered with slurry transport.

The Vitrification Cell in-line slurry sampling units were modified, based upon input from plant operators, to improve the ease of remote operation. Typical sampling time was reduced from several hours to less than one hour.

Shortly following the start of radioactive operations, a significant restriction in the melter off-gas piping (between the melter and the submerged bed scrubber [SBS]) developed. The restriction was suspected to be caused by the collection of dried melter feed adhering to the inside surface of the piping. Remote radiation surveys showed that the buildup occurred at an acute (45 degree) elbow just a few feet from the melter. Design modifications incorporated to mitigate the blockage included the following:

* A water flush line was installed into the melter air injection line. Water is periodically introduced into the off-gas stream to steam clean the piping.

* A flow-restricting orifice was installed in the air line supplying motive force air for the melter ADS pump. Reduction in the airflow rate reduced the atomization of the melter feed as it was pulsed into the melter plenum. This made it less likely for the feed material to be swept into the off-gas piping.

Melter pressure control has been a challenge from the start. Irregular off-gas flow characteristics attributed to the SBS, combined with the pulsing action of the ADS feed pump, create a dynamic environment for pressure control. The control system was modified by installing a high-speed data acquisition system with a quick-acting control valve. An adaptive gain feature was also incorporated into the control scheme. Although, these modifications made melter pressure control more manageable, the melter pressure still fluctuates. Melter pressure, controlled at -12.7 centimeters H_2O, varies as much as 7.6 centimeters H_2O from the nominal set point. Consequently, the melter glass level is maintained at least 2.5 centimeters below the overflow level to preclude excessive dripping of glass into a canister between glass pours, when the melter pressure is at its maximum.

Maintaining a steady pour stream and clear discharge port is essential to continued glass production. Two conditions compromising these were identified and corrected before radioactive operations. During startup testing with nonradioactive glass, the WVDP melter experienced chronic blockages of the glass pour stream. The blockages were the result of a failed barrier between the melt chamber and the discharge chamber, allowing glass migration and subsequent buildup at the bottom of the discharge chamber. The melter was repaired and placed back into service. Also just prior to radioactive operations, an accumulation of very thin glass fibers (angel hair) led to blockages in the glass pour stream. Limiting the airflow through the discharge chamber successfully reduced angel hair production. Angel hair formation, however, has not been,and is likely to never be completely eliminated.

The first set of silicon-carbide discharge heater elements exceeded their expected life by approximately three months (12 months vs 9 months). Changes in operational strategies for the heaters and providing automatic backup power minimizes heater cycling and prolongs heater life. The following modifications were made when the second set of heaters were installed:

* Heaters use maximum length for heating, thereby lowering the watt density/square inch of surface area (original heaters did not heat the full length of the bar in the chamber)

* Heaters are closely matched within each zone; this was accomplished by selecting heaters with closely matched initial resistance

* Control was modified so that the watt density is the same on each heater

* The heaters are now supplied with standby power so that thermal cycling does not occur during power outages.

These changes in installation and day-to-day operation have resulted in lengthening the life of the discharge heaters to about two years.

A significant lesson learned is that a large canister opening is vital to the success of the WVDP vitrification process. The glass pour stream must fall a distance of over 1.5 meters from the end of the trough before entering the canister. Lateral movement in the pour stream can (and does) occur due to melter pressure fluctuations, variations in the airlift flow rate, or effects of air in-leakage to the discharge chamber. The large canister opening (0.42 meters in diameter) has proven to be an effective design feature to accommodate pour stream deflections.

The WVDP HLW canisters are required to be filled at least 80% full. For the first two months, canisters were typically filled to the 85% level. Through the field experience gained with the video imaging capabilities of the infrared level detection system (ILDS), canisters are now normally filled approximately 90% full. While the ILDS-based levels are backed up by mass balance calculations, the accuracy provided by the ILDS system for determining the canister fill level has been outstanding. The final (reported) canister fill verification is performed by direct measurement of the glass level (using a measuring stick device) at the vitrification weld station.

Most of the downtime associated with the WVDP vitrification campaign can be attributed to the repair of remotely operated equipment. For example, mechanical failure necessitated the repair and replacement of the melter feed jumper, resulting in nearly a two-month outage. However, considerable pre-planned vitrification maintenance was rescheduled and accomplished during this unscheduled downtime. In addition, the WVDP was successful in the first attempt to perform hands-on maintenance to a Vitrification Cell remote component. Remote decontamination in the Vitrification Cell allowed hands-on repair of the remotely operated piping connectors.

SUMMARY

The vitrification process systems have been performing essentially as designed since initiating radioactive operations. Exhaustive functional and integrated testing using waste simulant, combined with adherence to design requirements, has limited the unscheduled downtime within the WVDP vitrification process. Strong problem solving skills and efficient use of unscheduled process downtime for scheduled maintenance have contributed greatly to high vitrification system availability at the WVDP.

THE DEFENSE WASTE PROCESSING FACILITY: TWO YEARS OF RADIOACTIVE OPERATION

Sharon L. Marra, James T. Gee, James F. Sproull

Westinghouse Savannah River Company
Savannah River Site
Aiken, SC 29808

ABSTRACT

The Defense Waste Processing Facility (DWPF) at the Savannah River Site in Aiken, SC is currently immobilizing high level radioactive sludge waste in borosilicate glass. The DWPF began vitrification of radioactive waste in May, 1996. Prior to that time, an extensive startup test program was completed with simulated waste. The DWPF is a first of its kind facility. The experience gained and data collected during the startup program and early years of operation can provide valuable information to other similar facilities. This experience involves many areas such as process enhancements, analytical improvements, glass pouring issues, and documentation/data collection and tracking. A summary of this experience and the results of the first two years of operation will be presented.

INTRODUCTION

Approximately 130 million liters of high-level radioactive waste, currently stored in underground carbon steel tanks at the Savannah River Site (SRS), is being immobilized in stable borosilicate glass in the Defense Waste Processing Facility (DWPF). The glass is poured into stainless steel canisters for eventual disposal in a geologic repository. To be acceptable for disposal, the DWPF product (i.e. the canistered waste form) must comply with the Department of Energy Office of Environmental Management's Waste Acceptance Product Specifications (WAPS).[1]

DWPF has been vitrifying radioactive waste for approximately two years. Since the start of radioactive operations, DWPF has operated on a sludge only flowsheet. Prior to the start of Radioactive Operations, DWPF went though an extensive Startup Test Program. This structured program allowed the facility to demonstrate its ability to comply with the WAPS, refine operating procedures, and provide training and experience to personnel. However, even with the Startup Test Program, operational and equipment issues have been encountered during Radioactive Operations. The experiences and lessons learned from a first of its kind vitrification facility like DWPF or the facility at West Valley can provide valuable insight to the designers and operating staff for future radioactive vitrification/immobilization facilities.

DWPF PROCESS/PRODUCT OVERVIEW

The radioactive waste in the SRS Tank Farms has been separated into a water soluble salt solution and saltcake, and an insoluble sludge of metal hydroxides and oxides. The salt solution and saltcake are decontaminated for disposal as low-level radioactive waste by removing the radionuclides by precipitation and sorption. The current baseline process is to add sodium tetraphenyl borate to precipitate soluble salts of primarily potassium and cesium and sodium titanate to adsorb residual strontium and plutonium. The resulting slurry is filtered and the decontaminated filtrate is blended with cement, slag and flyash for disposal at SRS as a low-level radioactive waste. The slurry of the concentrated solids is transferred to DWPF for immobilization. The sludge portion of the waste is washed to remove soluble salts. Thus, the radioactive waste from the SRS Tank Farms is transferred to the DWPF in two forms: precipitate slurry and sludge slurry. The waste is then processed and blended in the DWPF before it is vitrified, poured into canisters, sealed, and placed in interim storage.[2]

The DWPF is currently operating on a sludge only process - no precipitate is being fed to the facility (see Figure 1). The sludge is transferred directly to the Sludge Receipt and Adjustment Tank (SRAT) for processing and blending with the precipitate (when that part of the process becomes operable). The current baseline process includes processing of the precipitate in the DWPF Salt Process Cell to remove most of the organic material. When the alkaline sludge is transferred to the SRAT it is neutralized with nitric acid and formic acid (which is added to take the place of the precipitate material).[3] The excess formic acid reduces the mercuric oxide in the sludge to elemental mercury. The elemental mercury is then steam stripped from the SRAT into a holding tank from which it is later pumped and decontaminated. After processing in the SRAT, the slurry is transferred to the Slurry Mix Evaporator (SME) where a borosilicate glass frit is added and the slurry is concentrated to produce melter feed.

The amount of SRAT material and frit to be blended in the SME is determined by the desired glass composition. Any point within the acceptable glass processing region can be selected as the target for a particular batch. The SME is the hold point in the process where feed acceptability is determined as part of the Glass Product Control Program.[4] The Glass Product Control Program ensures that the DWPF produces an acceptable glass product by controlling the composition of the melter feed material. The analyses of samples from the SME are used by DWPF engineers to determine the acceptability of the melter feed. The acceptability of the melter feed is determined using glass property models and statistical algorithms which take into account analytical uncertainty to ensure the glass product will be acceptable and that processing constraints (such as viscosity and liquidus) are met. In order for the glass to be acceptable it must be more durable than the Environmental Assessment glass[5] as measured by the Product Consistency Test (PCT).[6]

Once the melter feed material in the SME is determined to be acceptable, it is transferred to the Melter Feed Tank (MFT) and then fed to the joule heated melter. The DWPF melter has two pairs of electrodes. The feed slurry is introduced from the top of the melter and forms a crust, or cold cap, on the surface of the melt pool as the water is evaporated and removed via the off-gas system. The cold cap melts from the bottom and forms the borosilicate glass matrix. The nominal glass melt pool temperature is 1150°C. The glass is removed from the melter near the bottom through a riser and pour spout. A vacuum is drawn on the pour spout to pour the glass. A glass pour stream sample is taken occasionally during filling of a canister to confirm that the glass durability (as determined by the Product Consistency Test) is acceptable. The mixing behavior of the melter was evaluated during Waste Qualification Runs.[7] During the first year of radioactive operations several glass pour stream samples were taken and analyzed.[8]

After a canister is filled, a temporary seal is installed to prevent free liquid from entering the canister during the decontamination process. Decontamination of the canister surface consists of blasting an air-injected frit slurry against the canister. The frit slurry from the decontamination process is used in the next SME batch as part of the required frit addition. The canister is then

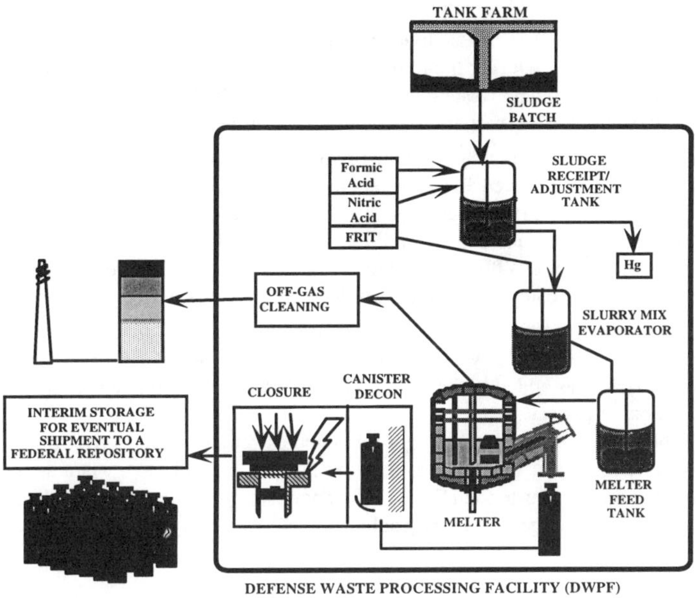

Figure 1. High Level Waste Process.

welded closed and transferred to an interim storage building. The DWPF canistered waste form contains approximately 1800 kg of glass. It is 300 cm in length and 61 cm in diameter.

PRODUCTION STATUS

Prior Operations

Prior to the start of Radioactive Operations DWPF underwent an extensive Startup Test Program. In 1992 DWPF completed integrated water runs testing. In 1993, DWPF began Cold Chemical Runs testing and initiated construction for hydrogen/ammonia upgrades (ammonium nitrate was found in the vent system of the DWPF pilot facility). Melter heatup, the initiation of melter feeding, and the first glass pour were completed in 1994. The Waste Qualification Runs portion of the DWPF Startup Test Program was completed in 1995. During Waste Qualification Runs varying feed compositions were used to demonstrate that the DWPF could control the glass product over a range of materials using the DWPF Glass Product Control Program.[4] The simulated waste was transferred into the DWPF and processed using the same methods as for the radioactive waste. The glass and canistered waste forms produced during Waste Qualification Runs were extensively characterized.[7] The results of this characterization demonstrated the DWPF's ability to comply with the WAPS. The Westinghouse Savannah River Company Operational Readiness Review (ORR) was completed and the DOE ORR was started in 1995. In 1996 the DOE ORR was completed and Radioactive Operations began.

Radioactive Operations

DWPF received the first batch of radioactive waste in March 1996 and began vitrification of radioactive waste in April 1996. During the remainder of the 1996 fiscal year (FY) DWPF produced 69 canisters of borosilicate waste glass. In FY 1997 DWPF produced 169 canisters of borosilicate waste glass and 29 feed preparation cycles. Two of the feed batches required remediation in the SME to correct an unacceptable predicted glass property.[3]

Through April DWPF has produced 146 canisters in FY 1998. The DWPF production goal for FY 1998 is 200 canisters. Since DWPF began Radioactive Operations almost 300,000 gallons of sludge has been processed to produce over 1.5 million pounds of glass.

Throughout this time period, DWPF has continued to operate using a sludge only process with the waste from the first sludge batch (batch 1A). Sludge batch 1A was washed and prepared for transfer to DWPF at the end of 1995.[9] Since there is no precipitate waste stream, slight modifications had to be made to the DWPF flowsheet. A frit composition higher in alkali was developed and formic acid was added to the sludge in the SRAT to drive the reduction of certain metals to the correct oxidation state.

As shown in Table 1 the feed composition has remained extremely uniform over time. DWPF will continue to operate off of this first sludge batch until the summer of 1998 when the next batch of sludge will be ready for transfer to the DWPF. The composition of the second sludge batch is relatively similar to the first sludge batch (see Table 2).

Table 1. SRAT Batch Compositions

Component	Batch 30	Batch 40	Batch 50	Batch 60	Batch 70
Al	6.69	6.08	5.92	5.96	6.62
Fe	24.03	23.67	22.50	22.61	24.60
Mg	2.64	2.38	2.37	2.41	2.60
Mn	6.87	6.60	7.98	7.84	6.97
U	2.81	2.70	2.56	2.53	2.78

Table 2. Sludge Compositions

Component	Sludge Batch 1A	Sludge Batch 1B
Al	6.36	7.04
Fe	25.8	19.87
Mg	2.61	3.17
Mn	9.40	6.14
U	3.21	2.49
Hg	0.16	0.13

EQUIPMENT/PROCESS ISSUES

As with the startup of any new facility, equipment and process issues provide unique challenges. The following sections include discussions of some of the results from the first two years of radioactive operation.

Melter Pouring

To initiate glass pouring, a vacuum is pulled relative to the melter vapor space on the canister and pour spout. The glass travels up the riser and overflows into the pour spout. The glass travels vertically down the side wall approximately where the wall of the pour spout is cut back to form a knife edge to allow the glass to disengage and fall into the canister. The glass falls through an unheated bellows and into the canister. At the end of simulant testing the glass pour stream began to contact the unheated bellows or lower section of the pour spout below the knife edge. This caused the glass to build up and potentially block the flow. Modifications were made to the pour spout to open the lower bore and create a second knife edge. This improved pouring performance, but after the start of Radioactive Operations pouring behavior degraded. In an effort to mitigate the problem, a number of factors were investigated including: melter and pour spout pressure control systems, using a tele-robotic manipulator (TRM) to physically remove glass buildup, and varying glass pour rate.[3] The TRM has a reach of 330 cm and when fully extended is capable of lifting 45 kg. This is a prototype design and while it has been an extremley useful tool, its robustness has been below expectations. All failures have been electrical in nature.

Some improvement was accomplished by optimizing several variables, however, the pouring behavior was still not adequate. It was determined that a physical modification to the pour spout would be needed. A remotely installed, replaceable "insert" (Inconel 690) was installed in the pour spout to lower the glass disengagement point and provide a new disengagement point. Production
rate dramatically increased after installation of the insert. The first insert was installed in May 1997. Since that time the insert has been successfully replaced several times. The first insert was characterized after replacement. It was found that some erosion of the knife edge did occur.

Material Performance

DWPF process equipment can be subjected to harsh environments. Chemical corrosion due to halides, sulfates, mercury, elevated temperatures, etc. and erosion due to glass frit particles can be a concern. Performance of the DWPF process equipment has been evaluated through a materials evaluation program prior to radioactive operations[10] and some experience has been gained during the first two years of radioactive operation.

Removable Process Cooling/Heating Coils - Four of the main process tanks (SRAT, SME, MFT, and SMECT) have removable coil assemblies of similar design. The SRAT and SME have cooling and heating coils while the SMECT and MFT have just cooling coils. SMECT cooling coils are constructed of S31603 stainless steel, while the other assemblies are constructed of alloy N10276.

Initial evaluation of the SME and MFT coils following Waste Qualification Runs, prior to Radioactive Operations showed significant localized erosion on the bottom inner coil run at support bracing, coil supports, and process supply piping that extends below the bottom coil loop directly impacted by frit slurry. In March 1997, the SME cooling coil failed due to erosion (i.e. penetration) in the cooling water supply pipe extending below the coil. The coil was removed from service and repaired using Stellite 6 (W73006) weld metal overlays on curved sections and ULTIMET alloy (R31233) plate wraps on straight pipe sections. The SME cooling coil again failed in March 1998. It was removed from the tank and replaced with a spare unit of similar design, although the bottom supply piping was redesigned to minimize the extent protruding below the bottom coils. The failed coil was pressurized in a remote cell and visual evidence suggests that the lower cooling coil failed near a support. Replacement SME process coil assemblies are being procured which contain protective R31233 wear resistant pipe and fittings in areas sensitive to erosion.

Agitators - The SRAT, SME and MFT have agitator blade assemblies constructed of alloy N10276 which allow mixing of the sludge/slurry/frit mixtures. The agitators consist of two sets of blades - an upper set of three hydrofoil blades and a lower set of four rectangular blades.

Initial evaluation of the agitators following Waste Qualification Runs, prior to Radioactive Operations showed severe erosion on the backside of the SME and to a lesser extent the lower MFT agitator blades. The erosion was again caused by the abrasive frit slurry on the alloy N10276 base metal and may have been enhanced by the boiling cycles (i.e., bubble formation, etc.) in the SME. Prior to radioactive operations spare agitators were installed in both the SME and MFT which had been modified to add W73006 weld metal coatings to high wear areas on the backside of the lower agitator blades.

In March 1997, the SME agitator was removed and inspected. Severe erosive wear patterns were observed on the backside of the lower blades in areas not protected by the W73006 weld metal coatings. In addition, significant wear was also observed on the leading edges of the top hydrofoil blades (alloy N10276 construction). The SME agitator shaft assembly was further modified to add W73006 weld metal coatings to all uncovered areas on the backside of the lower agitator blades. No alterations to the top hydrofoil blades were made.

The SME agitator was again pulled for inspection in March 1998 after a total of two years of radioactive service and approximately 50 batches of slurry. It was replaced by a similar, although, unprotected, spare agitator assembly for short term service. During subsequent inspections of the failed agitator shaft, it was determined that the backside of the lower agitator blades that were protected by W73006 weld metal coatings appeared to be in excellent condition. The upper hydrofoil blades were heavily eroded from the leading edges, with thorough wall penetration on two of the blades. The hydrofoil blades were repaired using alloy N10276 weld metal. It is anticipated that the SME agitator will be placed back in-service during May 1998.

Future permanent design changes may allow fabrication of both the upper and lower SME (and MFT) agitator blades from alloy R31233. Limited bench scale testing suggests that this cobalt based alloy may provide optimum performance in regard to both corrosion and erosion.

Glass Level Detection

DWPF utilizes an infrared camera to determine glass level in a canister. Rectangular stainless steel targets are located in front of the canister. The dimensions and positioning of the targets relative to the canister are known. These targets remain at the melt cell's ambient temperature during canister filling. As the canister is filled its temperature increases from ambient to approximately 800°C. This provides color contrast and, thus, by noting the "line" of highest temperature across the diameter of the canister the glass level can be determined using the targets.

The infrared camera is utilized at the start of pouring and the last 4-5 hours of canister filling. The infrared level is used to report glass level to demonstrate compliance with the WAPS. A load cell that measures canister weight monitors canister filling throughout the pour process. The load cell provides a reliable method for tracking canister weight. However, an absolute measure of canister weight is difficult due to the following:

- A bellows attaches the canister to the pour spout. The average bellows force varies from 174 to 286 pounds.
- The expansion of the canister during filling increases the force on the bellows thus increasing apparent canister weight. The expansion force can vary from 15 to 350 pounds.
- The canister has a throat protector in place during filling which weighs about 11 pounds.

Work is currently underway to improve the reliability of weight measurements. This work involves setting up a program for routine calibration of the load cell and for checking the variations of the factors discussed above that affect weight. In addition, a program is already in place for monitoring canister weight against the glass level determined from the infrared camera.

LESSONS LEARNED

Startup Test Program

The DWPF Startup Test Program was modeled on the testing required for startup of a commercial nuclear reactor. The Startup Test Program:

- Demonstrated the DWPF's ability to reliably produce an acceptable product that meets the requirements of the WAPS. This was accomplished during the Waste Qualification Runs portion of the Startup Test Program.
- Demonstrated the operability, reliability, and integrity of the major process systems.
- Provided operating experience to operations, maintenance, and engineering personnel.
- Baselined equipment and system operating parameters.

The Startup Test Program was a formalized program that had senior management involvement. Senior management from the Operations, Engineering, Startup, and Quality Assurance organizations formed a Joint Test Group. The Joint Test Group approved all Startup Test Plans and Test Procedures. Acceptance criteria were well defined and documented in the test plans and procedures. The test procedures included standard operating procedures to allow operators to gain experience and refine procedures.

Early in the startup testing process a structured process for evaluating schedules and priorities was initiated. This process consisted of plan of the day or plan of the week meetings with upper management and other key personnel. Schedules were statused, roadblocks flagged, and priorities evaluated. These meetings continue today to retain focus on operational issues.

Milestone Preparations

DWPF found that it was important to perform a readiness self assessment (RSA) as close as possible to the operational readiness review (ORR) or have a continuous self assessment to bridge the time span. This ensured that the facility was ready. It was also helpful to enlist peer review of site wide knowledgeable personnel to evaluate the RSA and provide an independent review of the paperwork.

DWPF underwent an ORR prior to the Radioactive Operations milestone. The Westinghouse Savannah River Company (WSRC) ORR was held first. This WSRC internal ORR was very focused and tough questions were asked. The team limited observations and focused on providing detailed findings. The DOE ORR reviewed the internal ORR prior to starting their own review. Thus, the more thorough the internal ORR the more smoothly the DOE ORR should progress.

A contact person was named for each ORR team to provide focus and obtain information as requested. Upper level management was involved in the process and any ORR team issues were resolved as quickly as possible prior to the generation of a finding. At DWPF certain facility modifications weren't complete so DOE couldn't observe process and emergency response actions. Thus, the DOE ORR had to stop their activities and come back again a couple of months later to finish the review. An important lesson to be learned is that it is critical to have the facility ready for ORR.

For each major milestone (e.g. Waste Qualification Runs, Radioactive Operations) DWPF named a milestone manager to focus on the milestone. This individual tracked any open items to closure to ensure readiness for the milestone. Planning and scheduling are critical to meeting milestones and completing programs. DWPF was also fortunate to have continuity of key personnel in the various organizations. Cognizant engineers were designated for each system. These engineers were then responsible for the testing and operation of their systems.

CONCLUSIONS

DWPF has almost completed two years of radioactive waste vitrification. As of April 30, 1998, almost three hundred canisters have been filled with radioactive waste glass. The experience gained during startup testing and radioactive operations can be extremely valuable for future operations at DWPF and other vitrification facilities. DWPF will continue to process sludge waste and is expected to start the second sludge batch in the summer of 1998.

ACKNOWLEDGMENT

This paper was prepared in connection with work done under Contract No. DE-AC09-96SR18500 with the U.S. Department of Energy.

REFERENCES

1. Office of Environmental Management, **Waste Acceptance Product Specifications for Vitrified High-Level Waste Forms,** Rev. 2, USDOE Document EM-WAPS, U.S. Department of Energy, Germantown, MD (1996).
2. E. C. DILLMAN, H. H. ELDER, and D. C. NICHOLS, "Technical and Project Highlights of the Defense Waste Processing Facility," **Waste Management 90**, 2, Roy G. Post (ed.), 737-46 (1990).
3. J.T. CARTER, R.E. EDWARDS, J.E. OCCHIPINTI, R.S. BECK, and D.C. IVERSON, "Defense Waste Processing Facility Radioactive Operations - Year Two," WSRC-MS-98-00108, Westinghouse Savannah River Company (1998). (To be published in proceedings from Waste Management '98.)
4. S.L. MARRA and M.J. PLODINEC, "The DWPF Waste Qualification Program," **Proceedings - Spectrum 94, Nuclear and Hazardous Waste Management International Topical Meeting**, American Nuclear Society, La Grange Park, Il, 1496-1501 (1994).
5. U. S. Department of Energy, **Environmental Assessment - Waste Form Selection for SRP High-Level Waste,** USDOE Document DOE/EA-0179, Washington, DC (1982).
6. C. M. JANTZEN, N. E. BIBLER, D.C. BEAM, W.G. RAMSEY, and B.J. WATERS, "Nuclear Waste Product Consistency Test (PCT) - Version 7.0," USDOE Report WSRC-TR-90-539, Rev. 3, Savannah River Technology Center, Aiken, SC (1994).
7. S.L. MARRA, D.E. SNYDER, H.H. ELDER, and J.E. OCCHIPINTI, "The DWPF: Results of Full Scale Qualification Runs Leading to Radioactive Operations," **Proceedings - Waste Management "96, CD-ROM Session 8-2**.
8. N.E. BIBLER, J.W. RAY, T.L. FELLINGER, O.B. HODOH, R.S. BECK, and O.G. LIEN, "Characterization of the Radioactive Glass Currently Being Produced by the Defense Waste Processing Facility at Savannah River Site," WSRC-MS-97-00617, Westinghouse Savannah River Company (1998). (To be published in proceedings from Waste Management '98.)
9. J.E. MARRA, W.M. BENNETT, H.H ELDER, E.D. LEE, S.L. MARRA, and P.L. RUTLAND, "High-Level Waste Processing at the Savannah River Site: An Update," **Proceedings - International Symposium on Waste Management Technologies in Ceramic and Nuclear Industry**, (1997).
10. D.B. LITTLE, J.T. GEE, W.M. BARNES, "Defense Waste Processing Facility Radioactive Operations - Part I Operating Experience," WSRC-MS-97-0080, Westinghouse Savannah River Company (1997). (Published in proceedings from Waste Management '97.)

THE MULTI-CANISTER OVERPACK — A CONTAINER FOR THE HANFORD SITE'S N REACTOR SPENT NUCLEAR FUEL

Louis H Goldmann
DE&S Hanford, Inc.
P. O. Box 350, M/S R3-86
Richland, Washington 99352-0350

INTRODUCTION

Beginning in 1944, nuclear reactors operated to produce plutonium at the U.S. Department of Energy's Hanford Site; production continued until 1988. Some 2,100 metric tons of the irradiated fuel from these reactors is still unprocessed due to the operational shutdown of the processing facilities in 1990. This fuel inventory now in storage at the Hanford Site's K Basins is mostly N Reactor fuel, a co-extruded, metallic uranium alloy clad in Zircaloy-2. Before irradiation, the fuel's enrichment varied up to 1.25% uranium-235.

The K Basins are water pools that provide fuel cooling, personnel shielding, and storage of the spent nuclear fuel (SNF). These concrete pools, their covering buildings, and support facilities, which are very near the Columbia River, are approaching retirement, and the fuel will be removed from its location close to the river and put into a more modern upland dry-storage facility.

The Hanford Site has developed an integrated process strategy for this irradiated fuel, including sorting and cleaning the SNF in the K Basins, loading the SNF into multi-canister overpacks (MCO), drying the fuel at the Cold Vacuum Drying Facility (CVDF), and transporting the dried fuel to the Canister Storage Building (CSB) for validation testing and, finally, interim storage. This presentation provides a description of the MCO and an overview of the proposed use of the MCO as a container for spent fuel.

BASIC CRITERIA

The most basic criteria for the MCO are to prevent a nuclear criticality and contain the SNF and associated byproducts. The MCO also serves a functional need as (1) the innermost SNF container in the on-site transportation packaging, (2) the process vessel for the SNF during bulk water removal and processing at the CVDF, and (3) the container for the SNF during handling, process validation, and interim storage at the CSB.

The selected design pressure for the MCO is 3.2 MPa (450 lb/in^2 gauge) at 405 K (270 °F). The maximum selected leakage rate for the mechanical seal is 1×10^{-5} standard cm^3/s. The maximum selected leakage rate for the welded cover cap is 1×10^{-7} standard cm^3/s. The design life is 40 years, potentially extendable to 75 years.

THE MULTI-CANISTER OVERPACK DESIGN

Early in the performance specification phase of MCO development, the literature was reviewed and the applicable design code selected. The desire was for the MCO to be a most robust and durable container for the SNF. Transportation packages traditionally have the most stringent requirements, so even though the MCO is not a shipping container, NUREG/CR-3854, *Fabrication Criteria for Shipping Containers* (1), was selected as a guideline. Because of the curie content and activity of the expected payload, the American Society of Mechanical Engineers (ASME) *Boiler and Pressure Vessel Code* (2), Section III, Subsection NB, was selected for the structural boundary and Subsection NG for the criticality control features. As shipping packages do not normally contain pressure relief systems, process ports, process tubes, or high-efficiency particulate air (HEPA) filters, the MCO Project realized there likely would be exceptions to the requirements in Subsection NB in the finished MCO design.

Structural Boundary

In the mature design, (Figure 1) the MCO represents the first boundary for the SNF during shipping, processing, and storage. The structural boundary for the MCO is composed of the lower end cap, the shell wall, collar, and the shield plug assembly. Later in the process and on the time line, the MCO cover cap becomes part of the MCO assembly. Typical attributes of the MCO are as follows.

(1) The lower end of the MCO is a machined cap made of 304L stainless steel, nominally 5.1 cm (2 in.) thick on the end face. A pocket at the bottom of the cap collects bulk water for removal from the MCO. The minimum thickness for the lower end cap is 1.3 cm (0.5 in.) where it is welded to the shell wall.

(2) The shell wall of the MCO is a 61-cm- (24-in.-) outside-diameter pipe section with a 1.3-cm- (0.5-in.-) thick wall section. The shell wall is made of 304L stainless steel.

(3) The collar is made of XM-19 (Nitronic 50™). The shell wall, collar, and lower end of the MCO will be manufactured and pressure and leak tested together with the shield plug assembly before use at the Hanford Site.

(4) The shield plug assembly includes the locking and lifting ring (304N stainless steel), plug body (304L stainless steel), process tubes (304/304L stainless steel), and the penetrations and devices necessary for interface to operating and safety equipment. The devices at the ends of the penetrations from the interior of the MCO form a portion of the MCO pressure boundary. The diameter of the MCO at the shield plug is less than 64.3 cm (25.31 in.). The over all length of the MCO at this configuration is 416.6 cm (160 in.). The process plug valves, port plugs, cover plate bolts, and 3.8-cm (1.5-in.) diameter set screws used to make the MCO main seal are made of Nitronic 60™ austenitic stainless steel. The selection of this material allows the use of these threaded devices without benefit of lubricants.

(5) The MCO cover cap includes a lifting rim similar to the one on the locking and lifting ring. The cover cap is made of 304L stainless steel and is 9 cm (3.5 in.) thick on the end face. The cap is secured to the collar with a circumferential weld. When the cover cap is installed and sealed, the importance of the underlying shield plug assembly is diminished somewhat. The MCO's length after the cover cap is installed is less than 426 cm (167.7 in.).

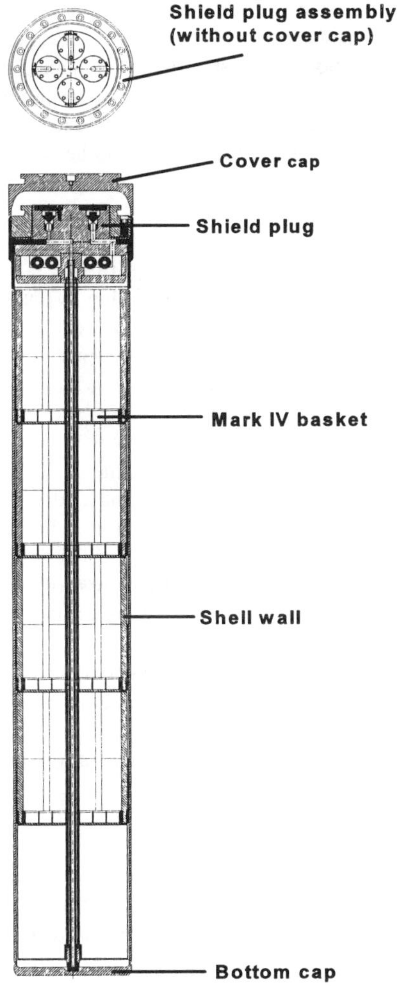

Figure 1. Multi-Canister Overpack Longitudinal Section.

The MCO structural boundary is designed and will be manufactured in accordance with the standards in the ASME *Boiler and Pressure Vessel Code* (2), Subsection NB, with some exceptions. The MCOs will not be "N" stamped. The maximum temperature allowed after fuel is inserted into the MCO is 405 K (270 °F) at the design pressure of 3.2 MPa (450 lb/in^2 gauge). Temperatures as high as 648 K (707 °F) would be allowed with a slight reduction in allowable operating pressure. Fully loaded with fuel, the heaviest MCO weighs less than approximately 8,911 Kg (19,640 lb).

Shield Plug Penetrations

Penetrations are bored through the shield plug (Figure 2) to the interior of the MCO. The penetrations terminate on the outside of the shield plug as ports with devices and fittings that allow the MCO to interface with operating and safety-related devices. Three devices are associated with the penetrations: two process valves (ports 2 and 3, shown in Figure 2) and one pressure relief device (port 4, not shown). Fittings and covers that go over the devices offer additional sealing features for the MCO. A fourth port (port 1, not shown) is available if needed but will be plugged during normal conditions.

The two process valves on the external face of the MCO are connected to process equipment. These process connections are used to introduce fluids into and remove fluids from the MCO. The long axial process tube path (port 3) has a nominal inside diameter of 1.5 cm (0.6 in.); the path connecting to the internal HEPA filters (port 2) has a nominal diameter of 2.5 cm (1 in.). Both port paths 3 and 4 are protected by wire screens (2-mm mesh openings) on the interior of the MCO to prevent larger particles from being moved out of the MCO.

A bank of four HEPA filters is welded to the underside of the shield plug. The HEPA filters, which are not testable after the MCO is assembled, are connected to two penetrations (ports 1 and 2) in the MCO shield plug. During vacuum drying, gases exit through port 2 via the HEPA filters. These HEPA filters serve to minimize the potential for contamination leaving the MCO.

Seals, Welds, and Closures

MCO pressure boundary and structure are established by a combination of seals and welds. The main seal between the shield plug and the collar is a Helicoflex™ seal. This seal includes a high alloy wound spring center that is covered with austenitic stainless steel and silver. The seals for the process valves, plugs, and cover plates are all metal O-rings made of 700-series Inconel™ and covered with gold or silver. These O-rings are constructed of a "C" section and are compliant in use.

The shell wall pipe, collar, and bottom are joined by welding, and the joints are inspected and examined before the MCO is loaded with fuel. This weld is also radiographically examined before it leaves the fabricator.

The shield plug is secured into the MCO using a mechanical closure. A threaded locking ring is placed into the MCO neck extension after the shield plug is inserted. Eighteen set screws move the shield plug into the seal to make a seal with the collar. The seal is maintained during thermal transients and drop accidents by developing and maintaining sufficient preload through the seal. This type of seal system was qualified during the prototypic testing program and by analysis. This closure operation is performed at the K Basin load-out pit.

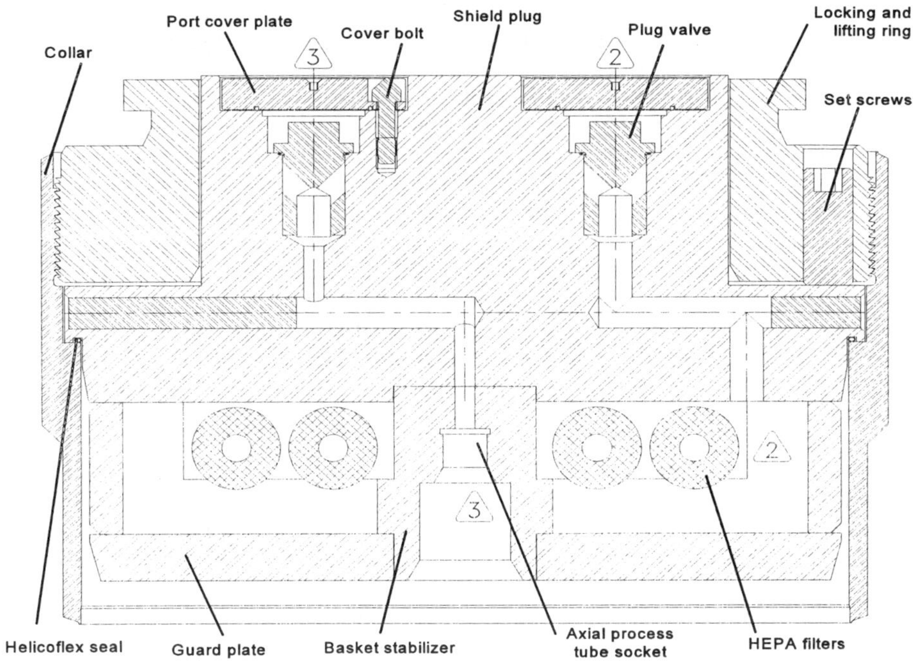

Figure 2. Shield Plug and Collar Section.

The final closure of the MCO is accomplished when the cover cap is installed over the entire shield plug and then welded to the MCO shell wall collar. This field-performed closure weld is then leakage rate tested by using helium, which is introduced through the penetration in the cover cap, as the tell-tale. The closure weld also is designed to be examined by ultrasonic means. After examination of the closure weld, the cap penetration is sealed and similarly leakage rate tested. These operations are performed at the CSB.

Basket Design

Criticality control features are designed to the ASME *Boiler and Pressure Vessel Code* (2), Subsection NG, as a minimum. The fuel element payloads come in two different enrichments. Fuel assemblies enriched to 0.95% uranium-235 or less are considered Mark IV fuel. Fuel assemblies enriched to more than 0.95% uranium-235 are considered Mark IA fuel. Mark IA fuel baskets require application of the Subsection NG design allowables and will not be "N" stamped.

Fuel exists in the basins as essentially whole fuel elements and as scrap. As a result, the basket designs evolved into scrap and whole fuel baskets for each of the fuel enrichments. Type 1 baskets (Figure 3) hold 48 Mark IA (higher enrichment) intact fuel elements and have a criticality control exclusion void built into the basket. Type 2 baskets hold 54 Mark IV intact fuel elements and do not need the exclusion void. Type 3 baskets hold Mark IA (higher enrichment) scrap fuel (fragments) and also have a criticality control exclusion void built into the basket. Type 4 baskets (Figure 4) hold Mark IV scrap fuel (fragments) and do not need the exclusion void.

The zircaloy cladding on some fuel elements was damaged when the fuel was downloaded from the reactor. After years of water pool storage, the uranium metal has slowly corroded and further breached the cladding. Whole fuel sections capable of being loaded into the fuel baskets will be put there. Pieces that will not fit into the fuel baskets and smaller fragments will be put into the scrap baskets. In general, irradiated material fragments less than 2.5 cm (1.0 in.) in size will go into the inner pocket of the scrap baskets. One scrap basket will go into the top position of each MCO. Once the MCO is loaded, the entire payload is subjected to the same process treatment and storage.

The baskets all contain a center tube and a structural plate to support the fuel. The center tube for the Mark IA baskets is 16.8 cm (6.6 in.) in diameter, and the center tube for the Mark IV basket is 9.5 cm (3.7 in) in diameter. This portion of the baskets is 304L stainless steel. In addition, the scrap baskets have a copper shroud, fins, and inner pocket for the scrap fines. This copper assembly aids in the transfer of heat out of the central basket fines area to the MCO shell wall. The copper is needed during the final stages of the MCO drying process.

The maximum length varies for each type of fuel, and as a result the overall heights of the baskets are different. Six Mark IA baskets or five Mark IV baskets will fit into an MCO. Any commingling of Mark IA and Mark IV baskets will be sensed by the MCO loading system at the K Basin load-out pits.

Each basket has a grapple interface that is on the inside surface of the central tube near the top of the basket. This circular grapple groove is used to pick up and move the baskets in the basins and to load the fueled baskets into the MCO at the load-out pits. The grapple interface is large enough to accommodate the axial process tube's outer diameter of 3.3 cm (1.3 in.) and its insertion down through to the bottom of the MCO. This insertion occurs when the completed shield plug assembly is lowered into the MCO at the K Basin load-out pit.

The MCO baskets also have to accommodate such functions as reliable engagement and stacking, water drainage, and gas flows inside the MCO.

PATH SUMMARY

The MCO's path begins at the CSB where an empty MCO, with the locking and lifting ring installed, is placed into the onsite shipping cask. The cask–MCO assembly is put on the conveyance, and the MCO is taken to the K Basin facility and parked near a load-out pit. The cask–MCO assembly is removed from the conveyance, the locking and lifting ring is removed, and the MCO is further prepared for receiving fuel and placed underwater in the load-out pit. A loading funnel is placed in the MCO neck, and the MCO loading machine loads underwater the queued up payload of baskets. When loading is complete, the loading funnel is removed, the Helicoflex™ sealing area of the collar is cleaned, and a fully assembled shield plug with the long axial process tube installed is lowered into the MCO. The cask–MCO assembly's top is raised above the pool water, the shield plug area is dewatered, and the shield plug lifting fixture is removed. The exposed portions of the shield plug and collar of the MCO are decontaminated as necessary, and the locking and lifting collar is installed into the MCO collar's neck. The set screws are tightened and torqued to the design requirement.

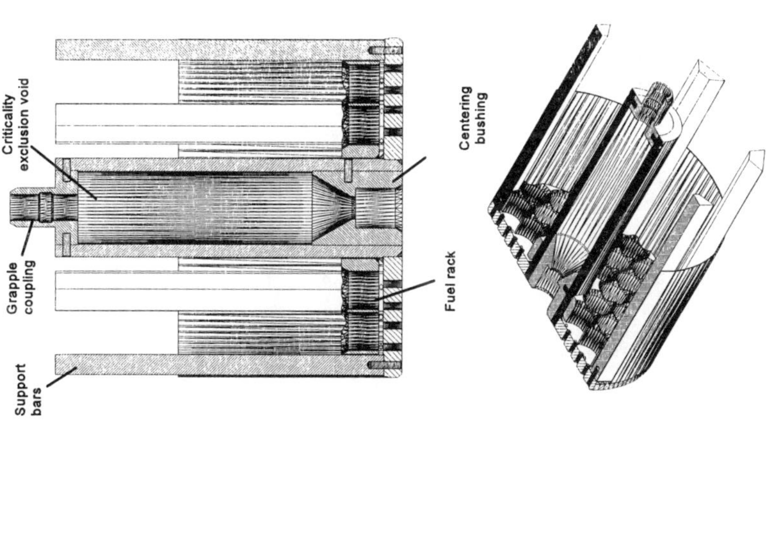

Figure 3. Mark IA Whole Fuel Basket

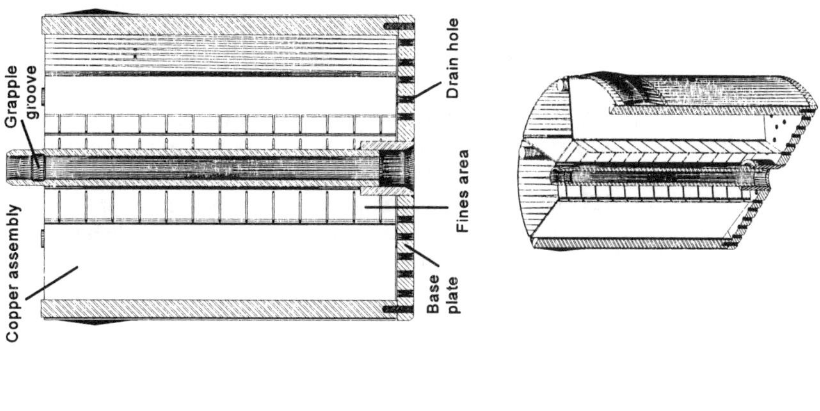

Figure 4. Mark IV Scrap Basket.

The cask is now closed, and the cask–MCO assembly is inerted and placed on the conveyance. The conveyance is taken to the CVDF and brought into one of the processing bays. The cask lid is removed, and a process hood is mounted onto the MCO. Valve operators are mounted on the MCO's ports 2 and 3, the ports are opened, and the cask is heated slightly to assist in the drying of the MCO later in the cycle. The bulk water is drained out through the axial process tube, helium is introduced into the interior of the MCO, and the vacuum drying cycle begins. When the drying cycle has been completed, the MCO is backfilled with helium, the process equipment is disconnected, and the Helicoflex™ main seal is leakage rate tested. The MCO ports are sealed with blind cover plates, and the cask is sealed.

The conveyance with the cask–MCO assembly still on board is now taken to the CSB. While at the CSB, the cask–MCO assembly is moved from the conveyance to the receiving pit. There the cask is opened, and the facility handling machine removes the MCO from the cask. The machine places the MCO in a storage tube or in the MCO welding station. The first several MCOs taken to the CSB will be monitored by gas sampling at the welding station. Eventually all the MCOs will be moved through the welding station, where the cover cap will be installed on the MCO, and then back to the CSB storage tubes for interim storage.

The empty cask is refilled with a new MCO, and the cycle is repeated. This SNF removal campaign is not expected to begin before July of 1999.

ONGOING ACTIVITIES

The MCO is undergoing a review in regards to eventual repository placement. The activity includes a review of Title 10, *Code of Regulations*, Part 71, "Packaging and Transportation of Radioactive Material" (3), for suitability for transport to the repository and an associated nuclear criticality review of the MCO and the basket structure and makeup; an evaluation of the MCO in regard to post-repository receipt; and a long-term performance evaluation of the MCO once in the closed repository.

REFERENCES

1. L. E. Fischer and W. Lai, *Fabrication Criteria for Shipping Containers*, NUREG/CR-3854, U.S. Nuclear Regulatory Commission, Washington, D.C., 1985.

2. *Boiler and Pressure Vessel Code*, American Society of Mechanical Engineers, New York, New York.

3. 10 CFR 71, "Packaging and Transportation of Radioactive Material," *Code of Federal Regulations*, as amended.

PROCESSING SIMULATED OXALATED HIGH-LEVEL WASTE THROUGH A VITRIFICATION FEED PREPARATION FLOWSHEET

G.K. Patello, R.L. Russell, G.R. Golcar, H.D. Smith, G.L. Smith, and M.L. Elliott, Pacific Northwest National Laboratory*, P.O. Box 999, Richland, WA 99352-0999.

ABSTRACT

Following an initial vitrification phase at the West Valley Demonstration Project (WVDP), residual tank waste, consisting of cesium-loaded zeolite in Tank 8D-1 and a mixture of neutralized plutonium-uranium extraction (PUREX) sludge and thorium extraction (THOREX) waste in Tank 8D-2, may require additional processing. A proposed treatment process involves adding oxalic acid to Tanks 8D-1 and 8D-2 to remove the residual activity. The oxalic acid will dissolve solids and increase the mobility of the radioactivity components. One option is to process the oxalated waste directly through the vitrification process. Laboratory-scale tests at the Pacific Northwest National Laboratory (PNNL) with simulated oxalated waste have evaluated the effect of oxalated waste in the melter feed flowsheet, for comparison to the baseline WVDP flowsheet. Comparisons were made on the basis of rheology, pH, and glass redox potential.

INTRODUCTION

Oxalic acid, which has been widely used to decontaminate reactor systems [1], may be applicable to the chemical cleaning of residual waste in high-level waste (HLW) tanks. It has been proposed that oxalic acid be used in the carbon steel tanks at the WVDP. The WVDP has one tank (8D-1) that contains cesium-loaded zeolite and another (8D-2) that contains neutralized PUREX, THOREX HLW, and cesium-loaded zeolite. In previous studies [2][3], oxalic acid was shown to be effective in eluting cesium from zeolite and dissolving simulated HLW. After adding oxalic acid to the HLW tanks, the entire mixture of oxalic acid, eluted cesium, dissolved waste, and corrosion products would be processed though the WVDP vitrification system.

The current HLW processing flowsheet consists of a concentration step; an addition of a glass- former slurry, which includes nitric acid; and a sugar addition. The nitric acid is added to control rheology and the sugar is added to control reduction-oxidation (redox) potential in the melter. The addition of oxalic acid to this system may require changes in the flowsheet in order to prepare feed that is acceptable in its rheology for adequate slurry transport and in redox for adequate melter control. The oxalic acid is added to the tanks based on a ratio of oxalic acid to residual waste required for effective removal of residual activity. Work performed at PNNL investigated the effect of oxalic acid on the behavior of a laboratory-scale version of the WVDP flowsheet. Previous unpublished work had determined that nitrate is

*Pacific Northwest National Laboratory is operated by Battelle for the U.S. Department of Energy.

required as an oxidizing agent to control the redox of the glass melt. Therefore, the slurry rheology and the redox of crucible melts of feeds were evaluated at different nitrate-to-carbon ratios. These laboratory-scale results are intended to provide guidance for a research-scale melter (RSM) test at PNNL and a scale vitrification system (SVS) test at the WVDP. They are not intended to be directly applicable to the WVDP melter.

APPROACH

A nonradioactive simulant of the WVDP HLW was prepared with the components shown in Table I. An oxalic acid solution that had contacted cesium-loaded zeolite to simulate cesium elution in Tank 8D-1 was added to the simulant. More oxalic acid was added until the total mixture contained 5 L of 8 wt% oxalic acid per kg of 12 wt% total solids simulant. The slurry of simulant and oxalic acid was concentrated to 55 wt% total solids and then glass-forming chemicals were added to it.

Table I. FY 1998 Simulant Composition Comparison

Element	Weight Percent (%)	Element	Weight Percent (%)
Al	0.66	Si	1.55
Ca	0.14	Sr	0.13
Fe	3.69	Ti	0.01
K	0.11	Zr	0.02
Mg	0.04	NO_3^-	3.0 mg/g
Mn	0.27	NO_2^-	5.8 mg/g
Na	1.28	pH	12.97
Ni	0.03	Density	1.09 g/mL
P	0.08	wt% solids	13.5%
S	0.03		

This final slurry was divided into six aliquots. Nitric acid was added to each aliquot at a different nitrate level. The inital aliquot pHs were between 2 and 7. The testing was designed so that no aliquot had a pH below 2, which is an operational constraint in the WVDP plant.

Rheological measurements were performed on the six aliquots after nitric acid addition, using a Haake® rotational viscometer with an M-5 concentric cylinder geometry with an MV1 concentric cylinder geometry sensor which is suitable for medium viscosity slurries. This geometry is appropriate for a simple shear flow in which a steady-state shear flow is generated by one cylinder moving at a constant speed relative to the other. In these controlled-rate experiments, shear stress was measured as a function of shear rate. The temperature of the system was kept at 25°C by a water bath. The rheological measurements of each slurry aliquot were repeated after one month.

Samples of the six aliquots were then melted in a quartz crucible with 21% by volume oxygen in helium flowing through the crucible. The glasses produced from these melts were analyzed for the ferrous ion and total iron concentration, as a measure of the glass redox state. In a second test series, an additional five slurries with different nitrate levels were prepared, melted, and analyzed for glass redox state.

RESULTS AND DISCUSSION

Table II shows the initial and one-month yield stresses and pHs of the six aliquots at different nitrate-to-carbon ratios. The source of the carbon is entirely oxalic acid. The same amount of oxalic acid was added to all aliquots. Only the nitrate concentration was adjusted. As expected, the pH of the melter feed decreases as the nitrate-to-carbon ratio increases, because the source of nitrate was nitric acid. Some drift in pH was observed during the testing, as shown in Table II.

The shear stress plots as a function of shear rate are presented in Figure 1. The values B0 and B1 represent the y-intercept and the slope of the best fit line shown in the plot. The rheology of these samples may be mathematically represented as a Bingham or as a yield-pseudoplastic model. Assuming a Bingham model, the yield stress was determined by extrapolating the linear portion of the measured shear stress vs. shear rate function to shear rate equal to 0 s^{-1}. Table II shows the numerical values of yield stress for all six aliquots, initially and after one month, along with the corresponding pH and mass ratio values of nitrate-to-oxalate carbon. Furthermore, the time-dependent nature of the rheology of these samples is evident from the observed thixotropy in the shear-stress plots; i.e., the shear stress versus shear rate plots are not exactly the same when the shear rate is first increased from 0 to 450 s^{-1} and then decreased from 450 to 0 s^{-1}.

Table II. Rheology Results from Slurry at Varying Nitrate-to-Carbon Ratios

Aliquot	Nitrate/ Carbon	Initial pH	pH after 1 hour	pH after 24 hours	pH after aging	Initial Yield Stress [Pa]	Aged Yield Stress [Pa]
1	0.66	7.1	7.0	7.3	7.0	0.29	11.43
2	1.11	6.2	6.3	6.6	6.5	0.35	3.72
3	1.49	5.2	5.3	5.5	5.4	0.62	2.13
4	1.70	4.2	4.5	4.7	4.7	2.17	1.05
5	1.95	3.3	3.6	3.9	3.8	3.62	0.83
6	2.20	2.5	3.0	3.4	3.3	2.67	1.34

From the steady state shear stress - shear rate measurements, values were determined for yield stress (by linear extrapolation) and for the variation of viscosity with the shear rate by dividing shear stress by shear rate. Figure 2 shows the viscosity as a function of shear rate on a logarithmic scale for all six aliquots, immediately after titration. The plots indicate that the viscosity of all six aliquots decreases rapidly over time, as the shear rate increases from about 0 to 100 s^{-1}. This represents a severe shear thinning behavior for these slurries. This behavior may be explained by the presence of agglomerates. As the shear rate increases, the agglomerates break down and produce lower measured viscosities. In addition, viscous forces may reduce the size of the agglomerates, and therefore relatively tightly packed spherical agglomerates are formed. [4]

The viscosities of aliquots 4, 5, and 6 decrease from approximately 800 mPa·s at a shear rate of 0.01 s^{-1} to about 20 mPa·s at a shear rate of 450 s^{-1}; whereas the viscosities of aliquots 1, 2, and 3 decrease from approximately 200 mPa·s at a shear rate of 0.01 s^{-1} to about 2 to 20 mPa·s over the same range of shear rates. The higher viscosities at lower pH values may be

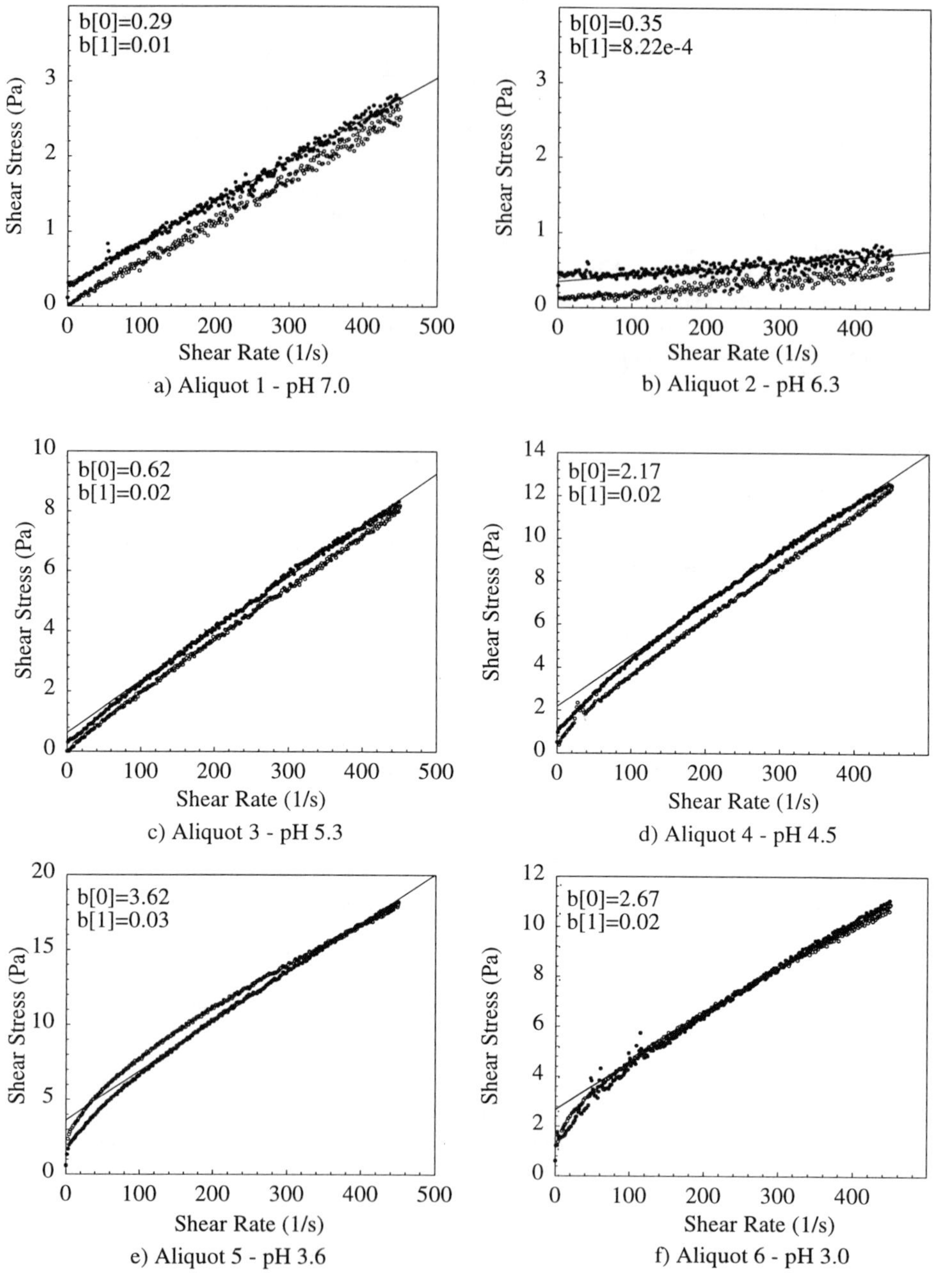

Figure 1. Shear Stress versus Shear Rate for Six Aliquots Measured on Initial Slurries

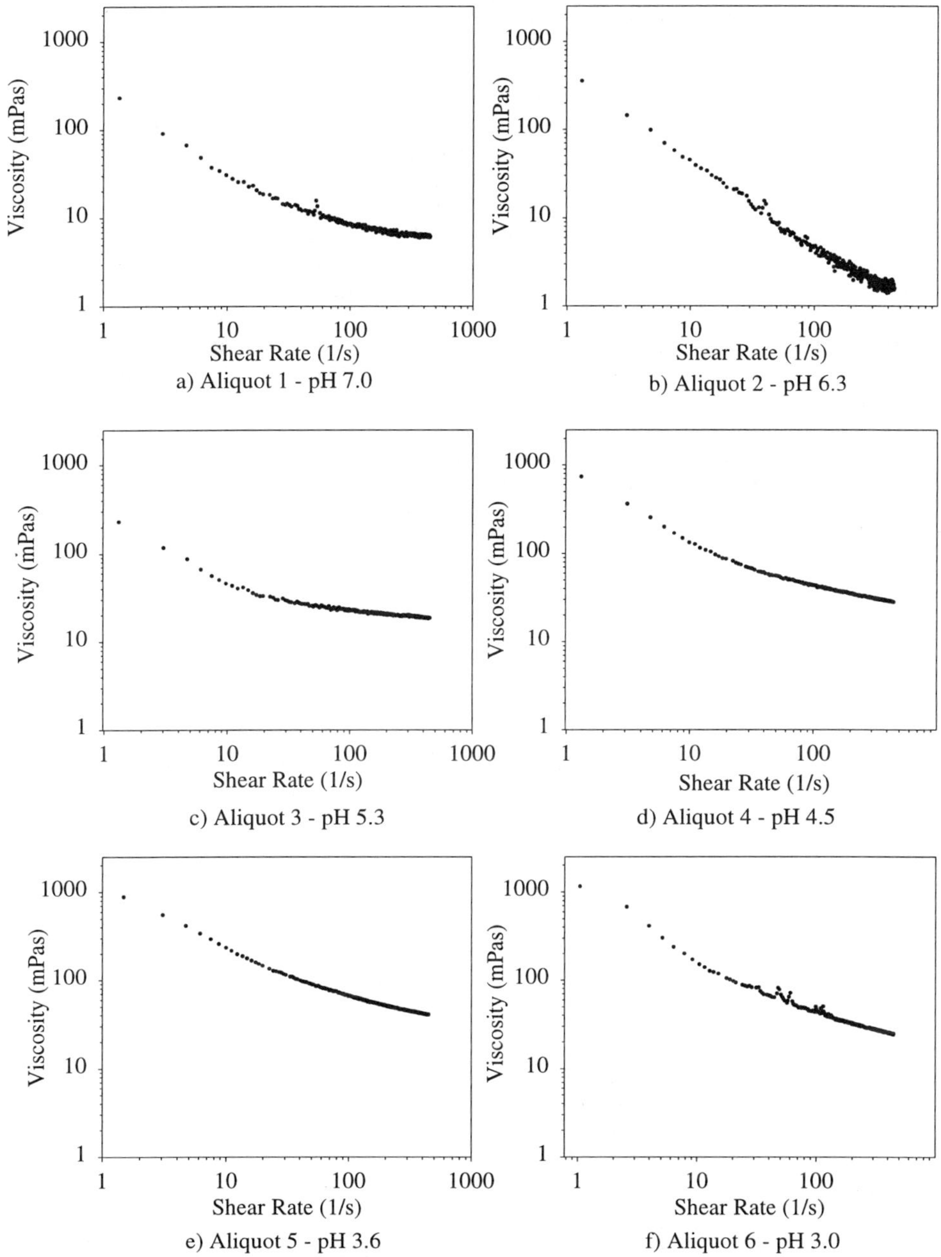

Figure 2. Logarithmic Scale Plots of Viscosity versus Shear Rate for Initial Slurry Samples

attributable to the higher fraction of finer-sized (colloidal) particles present. It is speculated that at a lower pH, a larger fraction of fine particles exists because of dissolution and deflocculation effects.

After aging for one month at room temperature with no stirring, aliquot 1 (pH 7.0) was strongly gelled, and aliquots 2 and 3 (pH 6.5 and 5.4) were weakly gelled. The gels were broken upon mixing. No gelation occurred in the aliquots at lower pH. Figure 3 shows the viscosity as a function of shear rate for all the aliquots, on a logarithmic scale. These results show that aging for one month increases the viscosities at low shear rate (0 to 10 s^{-1}) of aliquots 1, 2, and 3. This increase in viscosity is attributable to gelation during the aging. The yield stress of these samples also increased as the samples gelled as shown in Table II.

In contrast, for aliquots 4, 5, and 6, aging for one month decreases the viscosity and the yield stress. This behavior may be explained by the dissolution of some of solid phases as the samples aged in the acidic environment. As the solid particles dissolve, the total undissolved solids content of the slurry decreases, which results in a lower apparent viscosity and yield stress. Based on the pH behavior, the dissolution may have occurred in the first 24 hours of aging.

The redox measurements are shown in Table III and plotted in Figure 4, where the horizontal line represents a maximum Fe^{+2}/Fe_{tot} limit for the WVDP process. In Table III, both Fe^{+2}/Fe_{tot} and Fe^{+2}/Fe^{+3} ratios are provided along with the approximate target pH of the melter feed. A high Fe^{+2}/Fe_{tot} ratio is an indicator of a reduced glass melt that could potentially precipitate noble metals. The data indicates that a NO_3^-/total organic carbon (TOC) ratio greater than 1.3 will produce acceptable laboratory-scale glass by keeping the redox ratio below the limit. The results provide an approximate NO_3^-/TOC ratio that will be used in future RSM testing. The data provided in this laboratory-scale study will not be directly applied to the WVDP melter.

Table III. Fe^{+2}/Fe_{tot} Results at Varying Nitrate-to-TOC Ratio

g NO_3^-/ g TOC	Fe^{+2}/Fe_{tot}	Fe^{+2}/Fe^{+3}	Approximate Target pH Endpoints
0.66	0.62	1.65	7
1.10	0.79	3.66	6
1.06	0.81	4.32	6
1.23	0.36	0.57	5.5
1.33	0.13	0.15	5
1.45	0.07	0.08	4.5
1.68	0.14	0.16	4
1.52	0.16	0.19	3.5
1.94	0.03	0.03	3
2.19	0.01	0.01	2

CONCLUSIONS

The objective of this work was to determine if oxalic acid could be incorporated into a HLW melter feed preparation flowsheet. The feed must exhibit acceptable rheology during feed preparation for processability, and appropriate redox values for a well-controlled glass melt. The rheology is acceptable for nitrate-to-carbon ratios equal to or greater than 1.3, which

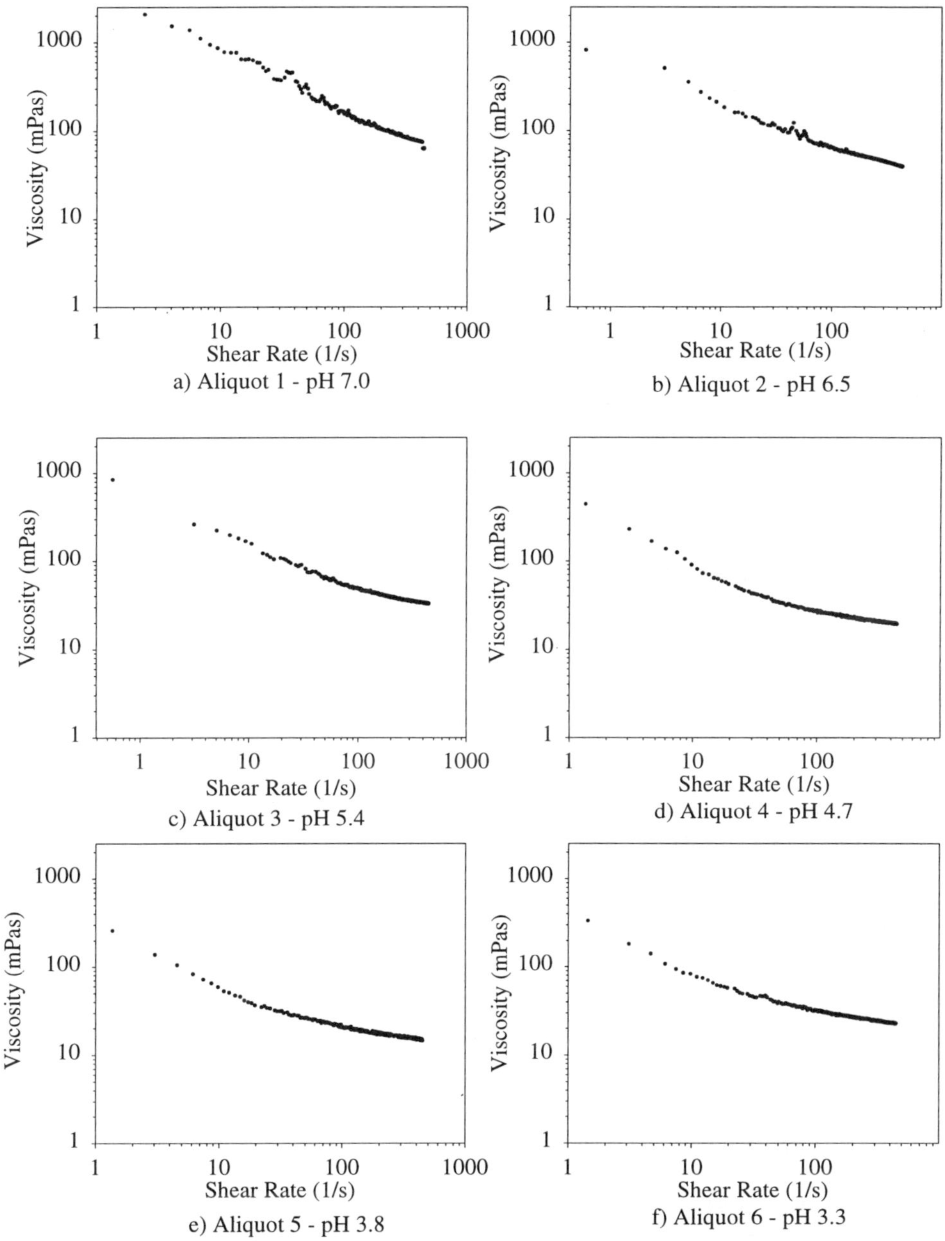

Figure 3. Logarithmic Scale Plots of Viscosity versus Shear Rate for Slurry Samples Aged for One Month

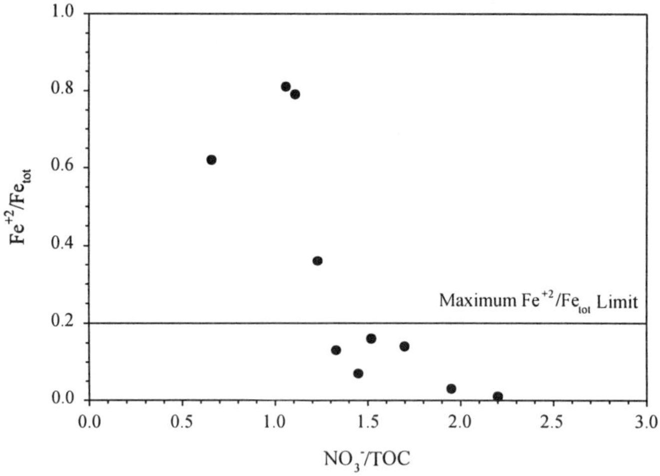

Figure 4. Iron Redox (Fe^{+2}/Fe_{tot}) versus Nitrate-to TOC Ratio

in these tests, corresponded to a pH of less than 5. Under these conditions, the yield stress and viscosity decrease with aging. In the slurries with nitrate-to-carbon ratios less than 1.3, the yield stress and viscosity increase with aging, which is undesirable for a plant in which the slurry may remain in a tank for a long period of time under upset conditions.

The viscosity of all samples, initial and aged, was low at high shear rates (> 150 s^{-1}), which is desirable for processablity, mixing, and pumping. The glass redox is acceptable ($Fe^{+2}/Fe_{tot} <$ 0.2) for nitrate-to-carbon ratios greater than or equal to 1.3 (pH ~5). The redox data is from laboratory-scale crucible melts, and provides a starting point for large-scale melter tests.

REFERENCES

[1] A.B. Meservey, " Decontamination and Film Removal;" pp. 137-159 in *Decontamination of Nuclear Reactors and Equipment*. Edited by J.A. Ayres. The Ronald Press Company, New York, 1970.

[2] J.A. Sills, G.K. Patello, J.S. Roberts, K.D. Wiemers, M.R. Elmore, W.G. Richmond, and R.L Russell, *Fiscal Year 1995 Laboratory-Scale Studies of Cesium Elution in Tank 8D-1 and Sludge Dissolution in Tank 8D-2*. PNNL-10945. Pacific Northwest National Laboratory, Richland, Washington, 1996.

[3] R.L. Russell, G.K. Patello, J.A. Sills, *Fiscal Year 1995 Laboratory Scale Studies of Cesium Elution in Tank 8D-1*. PNNL-11274, Pacific Northwest National Laboratory, Richland, Washington, 1996.

[4] J.E. Funk, D.R. Dinger, *Predictive Process Control of Crowded Particulate Suspensions*. Kluwer Academic Publishers, Boston, Massachusetts, 1994.

ACKNOWLEDGEMENT

This work was funded by the U.S. Department of Energy under Contract DE-AC06-76RLO 1830.

GLASS FORMULATION DEVELOPMENT FOR THE IMMOBILIZATION OF HIGH-LEVEL WASTE COUPLED WITH CESIUM-LOADED CRYSTALLINE SILICOTITANATE

M. K. Andrews and P. J. Workman
Westinghouse Savannah River Company
Aiken, SC, 29808

ABSTRACT

Crystalline silicotitanate (CST) is a highly specific ion exchange material for cesium. In particular, CST has been successfully demonstrated with both simulants and actual supernates from Savannah River and Hanford tank wastes. As a disposal option, vitrification of the cesium-loaded CST coupled with High-Level Waste (HLW) has been proposed. However, the CST sorbent contains significant quantities of titanium which historically have been difficult to incorporate into the glass structure. Therefore, nonradioactive tests using CST coupled with simulated HLW were performed to develop a glass formulation that would be both processable and durable. The results of the simulated crucible melts were verified using radioactive waste and cesium-loaded CST. This paper will detail the results of the formulation efforts.

INTRODUCTION

An alternative to the In-Tank Precipitation (ITP) and sodium titanate processes at the Savannah River Site (SRS) is the removal of cesium, strontium and plutonium from the tank supernate by ion exchange using crystalline silicotitanate (CST). CST was developed jointly by Sandia National Laboratory and Texas A&M. The engineered form of this material was developed by UOP (Universal Oil Products). Both the engineered and powdered form are now commercially available. This inorganic sorbent has been shown to effectively separate cesium (and strontium) from tank supernates containing high amounts of sodium and potassium.[1]

One option for disposal of the loaded CST is immobilization with High-Level Waste (HLW) sludge during vitrification. Because of the composition of the CST, concerns exist that some of the minor component (such as titania) solubility limits in the melt may be exceeded at higher CST loadings. The CST loading required in the glass is a function of the amount of cesium that can be sorbed by the CST. Therefore variable CST loadings, probably between 5 and 10 wt%, may be required.[2] These CST loadings correspond to a titania content which exceeds the current limit of 1 wt% in HLW glass produced at the Defense Waste Processing Facility (DWPF). Therefore, a new glass formulation for HLW coupled with CST had to be developed.

Crucible tests with simulated sludge and unloaded CST focused on glass formulations with approximately 5 wt% CST. Additional tests were also performed to try to maximize the CST loadings. In each case, the sludge loadings were around 28 wt% sludge oxides in the glass. The results of the glass formulation efforts are presented in this paper.

EXPERIMENTAL PROCEDURE

Scoping studies were performed using simulated sludge and unloaded CST. The main difference between unloaded and loaded CST is the cesium concentration, which is not expected to impact the glass formulation. The composition of the Batch 1 sludge that was used during the scoping studies is provided in Table 1. These studies were performed to determine the type and amount of reagent grade chemicals that would be required to produce a durable glass with the appropriate liquidus and viscosity processing properties. The samples were thoroughly mixed and then vitrified in platinum crucibles. After approximately four hours at 1150°C, the glasses were removed from the furnace and either air cooled to room temperature, or slow cooled to room temperature following the DWPF canister centerline cooling regime.[3] (The DWPF canister centerline cooling regime is an extremely slow cooling rate that is essentially equivalent to the slowest cooling rate experienced in a DWPF canister.) A few of the glass samples were removed from the furnace and poured to determine the viscosity range of the glass.

Table 1 - Major Elemental Composition of the Tested Simulated DWPF Sludge (in weight % elemental)

Element	Batch 1	Blend	Purex	HM
Al	6.85	5.81	4.72	10.2
Ca	2.09	1.69	2.47	1.15
Fe	27.6	22.3	27.7	14.6
Mg	1.35	0.20	0.27	0.32
Mn	2.92	5.36	5.76	4.96
Na	6.90	6.34	3.91	7.74
Si	0.94	2.16	1.28	3.48

To determine the effects of sludge composition on the glass formulation, studies were performed with three additional sludge-types: Blend, Purex and HM. The compositions of these sludges is also shown in Table 1. Blend is a hypothetical combination of the sludge in all of the SRS waste tanks. Purex is a high-iron sludge that when vitrified with frit has a lower viscosity. HM is a high-aluminum sludge that has a higher viscosity when vitrified. Crucible studies were performed with each of these sludges.

After the glasses had completely cooled, they were removed from the crucible and analyzed. The density of the glass was measured by buoyancy in water, similar to the method described in ASTM C-693. Samples of the ground glass were dissolved and characterized to determine the chemical composition. The samples were dissolved by both sodium peroxide fusion and HF/HCl/HNO_3 acids. The resulting solutions were analyzed by Inductively Coupled Plasma Excitation Spectroscopy and Atomic Absorption Spectroscopy. The crystalline content of the glass was measured by X-Ray Diffraction (XRD), and the durability of the glass was determined by the Product Consistency Test (PCT).[4] The PCT is a crushed glass leach test that measures the releases of several elements from the glass at 90°C in ASTM Type 1 water over a period of 7 days. The elemental releases are used along with the composition of the glass to calculate the average normalized release for boron, sodium and lithium. The leachate pH is measured as part of the PCT protocol and provides a secondary indication of glass durability.

After a glass formulation was developed using simulants, a confirmation test was performed using radioactive sludge and Cs-loaded CST. This testing occurred in the Shielded Cells facility of the Savannah River Technology Center. Using radioactive sludge from SRS Tank 51, loaded CST from testing performed at Oak Ridge, and reagent-grade glass-forming chemicals, a crucible melt was performed at a waste loading of 28 wt% sludge oxides and 5 wt% CST.

RESULTS AND DISCUSSION

Scoping Studies

Scoping studies were performed to determine the type and amount of chemicals to combine with the nonradioactive sludge and unloaded CST to form a glass. Both the sludge and CST loadings were varied but the sludge loadings were kept close to the target loading of 28 wt% sludge oxides. The sludge used for this testing was Batch 1 sludge, whose composition is given in Table 1. Table 2 shows a sample of the waste loadings that were examined. Each of these samples represent a different glass composition.

Table 2 - Waste Loadings Tested during Scoping Studies

Sample	CST wt%	Sludge wt%	Glass Formers wt%
1	5	25	70
2	7	28	65
3	8	28	64
4	10	30	60
5	15	30	55
6	15	35	50
7	18	28	54
8	20	25	55
9	20	30	50
10	22	32	46

All of the samples listed in Table 2 were analyzed by XRD to determine the type of crystals, if any, that were present in the glass. Two samples (numbers 7 and 10) contained trace amounts of crystals. Since crystallization can affect the durability of the glass, all ten of the glass samples were subjected to the PCT. Each sample was performed in triplicate and each test included the appropriate blanks and standards. The results of the standards and blanks indicated that the tests were acceptable.

The average releases of several elements were used along with the composition of the glass to calculate the average normalized release for boron, sodium and lithium. Table 3 presents the normalized PCT results for the glasses listed in Table 2, along with the accepted values for the Environmental Assessment (EA) glass.[5] The HLW acceptance criteria states that any glass produced must be at least two standard deviations below the results for the EA glass in order to be considered acceptable. Therefore, all of these glasses can be considered durable. The results shown in Table 3 also indicate that the crystals detected in the two samples did not affect the glass durability.

Table 3 - PCT Results (in g/L) during Scoping Studies

Sample	B	Na	Li
1	1.29	1.21	1.27
2	1.30	1.33	1.26
3	0.88	0.85	0.94
4	0.86	0.85	0.97
5	0.82	0.86	0.94
6	0.86	0.91	0.93
7	0.33	0.17	0.61
8	0.50	0.43	0.70
9	0.37	0.15	0.82
10	0.55	0.62	0.67
EA[5]	16.7	13.3	9.6

Sludge Variation Studies

After the successful demonstration of a glass formulation using Batch 1 sludge, it was decided to test the formulation to variations in the sludge composition. To accomplish this, three SRS hypothetical compositions (Blend, Purex and HM) were studied. Purex (high iron) and HM (high aluminum) are extremes in composition, while Blend is an overall average sludge composition.

During the sludge variation studies, CST loadings of 5, 7.5 and 10 wt% were tested with each sludge type using the compositions developed during the scoping studies. Table 4 provides the PCT measured boron release for each of the tested glasses, along with the measured density of each glass sample.

Table 4 - Sludge Variation PCT and Density Results

Sludge Type	CST (wt%)	B release (g/L)	Density (g/cc)
Blend	5	0.53	2.71
Blend	7.5	0.49	2.72
Blend	10	0.49	2.76
Purex	5	0.63	2.73
Purex	7.5	0.58	2.75
Purex	10	0.53	2.80
HM	5	0.42	2.65
HM	7.5	0.44	2.67
HM	10	0.41	2.69

Each of the glasses was significantly more durable than the EA glass, which is shown in Table 3. In addition to this frit composition, several additional frit compositions were tested with the three sludge types to provide comparison information. All of the tested frit compositions had acceptable durabilities but some of the additional compositions did not have acceptable processing properties (viscosity and liquidus).

The current work is focusing on more detailed determinations of the viscosity and liquidus properties. Previously, only quick determinations of the viscosity (by pouring the glass from the crucible) and the liquidus (by limited heat treatments) were made. These studies will measure the viscosity of the glass using a bushing melter. A better estimate of the liquidus will be determined by performing several heat treatments at different temperatures for various times.

Radioactive Demonstration

Based on the results of the nonradioactive glass formulation development work, a waste loading of 28 wt% sludge oxides and 5 wt% CST was chosen for the radioactive demonstration. The radioactive demonstration used actual sludge from SRS Tank 51 (the composition is given in Table 5), along with Cs-loaded CST from a demonstration at the Oak Ridge National Laboratory. The appropriate glass-forming chemicals were thoroughly combined with the radioactive sludge and CST, poured into a platinum crucible, and then vitrified in a muffle furnace.

Table 5 - Major Elemental Composition of Radioactive DWPF Sludge

Element	Weight %
Al	6.62
Ca	2.43
Fe	25.0
Mg	1.19
Mn	2.63
Na	7.17
Si	1.14
U	3.24

After approximately four hours at the melt temperature of 1150°C, the crucible was removed from the muffle furnace and the glass poured into a stainless steel pan. There were no visible differences between the radioactive and nonradioactive glasses. The glass poured easily from the crucible and was in the same viscosity range as the nonradioactive glass sample, indicating that it would be acceptable in the DWPF melter. Durability tests on the glass are in progress, but they are expected to yield results similar to the simulant testing.

CONCLUSIONS

A glass formulation was developed using nonradioactive sludge and unloaded CST. The resulting glass was more durable than the HLW EA glass, thus meeting the criteria of the Waste Acceptance Product Specifications (WAPS). The glass also had viscosity and liquidus properties within the acceptable processing ranges. The radioactive demonstration of this formulation using Tank 51 sludge and Cs-loaded CST successfully incorporated 28 wt% sludge oxides and 5 wt% CST.

REFERENCES

1. R. Braun et al, "Crystalline Silicotitanate - Novel Commercial Cesium Ion Exchangers", Spectrum '96, Proceedings - International Topical Meeting on Nuclear and Hazardous Waste Management, (1996).
2. M. K. Andrews and J. R. Harbour, "Glass Formulation Requirements for DWPF Coupled Operations using Crystalline Silicotitanates (U)", WSRC-TR-97-0004, (1997).
3. R. E. Edwards, "SGM Run 8 - Canister and Glass Temperatures During Filling and Cooldown", DPST-87-801, (1987).
4. ASTM Procedure, C-1286-94, "Determining Chemical Durability of Nuclear Waste Glasses: The Product Consistency Test (PCT)".
5. U. S. Department of Energy, "Environmental Assessment Waste Form Selection for SRP High-Level Waste", USDOE Report DOE-EA-0179, (1982).

OXALATE DESTRUCTION IN A SIMULATED HIGH-LEVEL TANK RESIDUAL WASTE USING NITRIC ACID

Gertrude K. Patello and Renee L. Russell
Pacific Northwest National Laboratory*
P.O. Box 999
Richland, WA 99352-0999

ABSTRACT

Following an initial vitrification phase at the West Valley Demonstration Project (WVDP), residual tank waste consisting of cesium-loaded zeolite in Tank 8D-1 and a mixture of neutralized plutonium-uranium extraction (PUREX) sludge and thorium extraction (THOREX) waste in Tank 8D-2 may require additional processing. A proposed treatment involves adding oxalic acid to reduce residual activity in 8D-1 and to dissolve sludge and increase mobility of the waste in 8D-2. However, it may then become necessary to destroy the oxalic acid to produce a melter feed with acceptable melting properties. One option is to destroy the oxalate ion in the vitrification plant by a manganese-catalyzed chemical reaction with nitric acid. Laboratory-scale tests with simulated waste at Pacific Northwest National Laboratory (PNNL) examined the conditions for successful oxalate destruction. Variables tested include manganese concentration and nitric acid concentration.

INTRODUCTION

Oxalic acid has long been used as a decontamination agent in the nuclear power industry [1]. It can dissolve a large variety of compounds because of its acidity and complexing properties. These properties make oxalic acid a good candidate reagent for dissolving residual high-level waste (HLW) sludge in storage tanks. It has been proposed for use in the HLW storage tanks at the WVDP. The WVDP has one tank (8D-1) in which cesium-loaded zeolite was staged and one tank (8D-2) that contains neutralized PUREX waste, THOREX waste, and zeolite transferred from 8D-1. In previous laboratory-scale studies [2][3], oxalic acid has been shown to be effective at eluting cesium from zeolite and in dissolving the nonradioactive HLW simulant.

---

*Pacific Northwest National Laboratory is operated by Battelle for the U.S. Department of Energy.

After oxalic acid is added to the HLW tanks, the oxalate-containing waste sludge will be processed in the WVDP Vitrification Facility. To minimize changes to the melter feed flowsheet, a manganese-catalyzed oxidation by nitrate is being studied as a possible way to eliminate the oxalic acid within the Vitrification Facility. Work presented in this paper is the result of laboratory-scale testing at PNNL to determine the feasibility of oxalate destruction in simulated WVDP waste containing oxalic acid.

BACKGROUND

Treating residual HLW sludge with oxalic acid was investigated at the Savannah River Laboratory (SRL)[3]. Initial work was performed in 1981, using nitric acid with manganous ion as a catalyst to destroy the oxalate. It was determined that oxidation by nitric acid in the presence of manganous ion as a catalyst is an excellent method for large-scale decomposition of the oxalic acid used to clean waste tanks; with more than 90% of the oxalate destroyed under most test conditions. The value of the reaction rate constant depends on the concentrations of manganous ion, dissolved iron present due to oxalic acid contacting the tank, and nitric acid. The rate constant increases with temperature. Table I shows how the SRL HLW sludge composition compares to the WVDP HLW sludge composition.

Table I. SRL and WVDP HLW Sludge Composition Comparison

	Avg. SRL Sludge Composition (mol%)*	WVDP Sludge Composition (mol%)**
Fe	52	63.7
Al	19	30.6
Ni	6.4	2.1
Mn	14	3.3
Ca	9	0.4

* Based on the ions analyzed.
** Normalized only to these ions.

Reactions leading to oxalate destruction by nitric acid are shown below.

$$2HNO_3 + H_2C_2O_4 \rightarrow 2NO_2\uparrow + 2CO_2\uparrow + 2H_2O \qquad \textbf{(1)}$$

$$2HNO_3 + 3H_2C_2O_4 \rightarrow 2NO\uparrow + 6CO_2\uparrow + 4H_2O \qquad \textbf{(2)}$$

$$2HNO_3 + 4H_2C_2O_4 \rightarrow N_2O\uparrow + 8CO_2\uparrow + 5H_2O \quad (3)$$

The moles of nitrate consumed per mole of carbon reacting are 1, 0.33, and 0.25, respectively, in reactions 1 through 3. Based on work performed at the SRL [4], Reaction 1 is dominant at HNO_3 concentrations greater than 3M. Reaction 2 is dominant at HNO_3 concentrations less than or equal to 2.5M. Reaction 3 is dominant only at HNO_3 concentrations of 1M or less and only if the dissolved sludge does not contain Fe^{3+}. Therefore, Reaction 3 is not expected to occur in the presence of the iron-containing WVDP simulant.

EXPERIMENTAL

Eight tests were performed at varying concentrations of HNO_3 and Mn^{+2} to identify the optimum conditions for oxalate destruction. The experimental conditions for the tests are summarized in Table II. For each test, 200 g of oxalated simulant was used. This simulant was prepared by adding 8 wt% oxalic acid to HLW simulant at a ratio of 5 L of 8 wt% oxalic acid per kg of 12 wt% total solids simulant. Then MnO and 15.6 M HNO_3 were added to the simulant in the quantities shown in Table II, and an initial sample was taken and analyzed for nitrate and organic carbon. There was approximately 0.002M Mn^{+2} present in the oxalated simulant. The mixture was refluxed at 90 to 100°C for 6 hours with continuous stirring. After 6 hours, a final sample was taken and analyzed for nitrate and organic carbon. In Tests #7 and #8, the pH of the starting mixture was adjusted to 5 using a KOH, NaOH, and LiOH solution, before the MnO and HNO_3 were added. This simulates proposed pH adjustment for tank corrosion control in WVDP Tank 8D-2.

A final test (Test #9) was then performed in which the reaction gases were measured using a gas chromatograph (CO_2 and N_2O) and a NO_x analyzer (NO and NO_x). The pH of the WVDP oxalated HLW simulant was adjusted to 5 before the test began. Then, MnO and HNO_3 were added to the simulant to achieve molarities of 0.015M and 2M, respectively. The offgas was measured as the simulant was concentrated over three days to 39 wt% total solids by simmering at 90 to 95°C. This temperature range was maintained only during daytime hours. Overnight, the temperature was decreased until little gas generation was observed; however, the offgas was measured continuously overnight.

RESULTS AND DISCUSSION

The results of the first eight tests are shown in Table III. Nitrate concentrations were measured by ion chromatography, and total organic carbon concentrations were measured by the hot persulfate wet oxidation method. These measurements were used to determine the extent of reaction and the nitrate:carbon reaction ratios.

Table II. Conditions for Oxalate Destruction Tests

Conditions	Test #1	Test #2	Test #3	Test #4	Test #5	Test #6	Test #7	Test #8
Amount of oxalated simulant used	199.8 g	199.9 g	199.9 g	199.9 g	200.2 g	200.2 g	200.4 g	200.4 g
Amount of 15.6 M HNO_3 added	52.4 g (2.5 M)	85.0 g (3.7 M)	85.0 g (3.7 M)	52.4 g (2.5 M)	52.4 g (2.5 M)	19.3 g (1 M)	52.4 g (2.5 M)	19.3 g (1 M)
Amount of MnO added	0.0 g (0 M)	0.0 g (0 M)	0.27 g (.015M)	0.25 g (.015M)	0.12 g (.0075M)	0.23 g (.015M)	0.25 g (.015M)	0.23 g (.015M)
Total amount of Mn present	0.04 g (.002 M)	0.04 g (.002 M)	0.31 g (.017M)	0.29 g (.017M)	0.16 g (.0095M)	0.27 g (.017M)	0.29 g (.017M)	0.27 g (.017M)

Table III. Initial and Final Oxalate, Total Organic Carbon, and Nitrate Measurements

Ion		Test #1	Test #2	Test #3	Test #4	Test #5	Test #6	Test #7	Test #8
Total Organic Carbon	Initial moles	0.33	0.33	0.33	0.33	0.22	0.22	0.23	0.23
	Final moles	0.077	0.0046	0.024	0.0099	0.044	0.039	0.023	0.11
	Moles reacted	0.25	0.33	0.31	0.32	0.18	0.18	0.21	0.12
	Percent reacted	**76**	**99**	**93**	**97**	**80**	**83**	**90**	**52**
Nitrate	Initial moles	0.84	1.35	1.35	0.84	0.83	0.31	0.83	0.31
	Final moles	0.27	0.66	0.85	0.50	0.47	0.14	0.50	0.16
	Moles reacted	0.57	0.69	0.50	0.34	0.36	0.17	0.33	0.15
	Percent reacted	**68**	**51**	**37**	**40**	**43**	**55**	**40**	**48**
Nitrate:Carbon reacted		2.28	2.09	1.61	1.06	2.00	0.94	1.57	1.25

Table III shows that over 90% of the oxalate (organic carbon) was destroyed in Tests #2, #3, #4, and #7. Also, the tests with only 0.017M manganous ion present (Tests #3 through #8) destroyed less nitrate than the tests with only 0.002M total manganese (Tests #1 and #2).

Similar results were achieved in Test #9, as shown in Table IV. In this test, 53% of the nitrate and 99.9% of the carbon were destroyed. Profiles of the gas generated during the concentration of the slurry are shown in Figures 1 through 3. During the first day of concentration, CO_2, NO_x, and a very small amount of N_2O were measured in the offgas. The NO_x was composed entirely of NO. On the second and third days, CO_2 and NO were detected. By the difference between the total NO_x and the NO, a slight amount of NO_2 was also generated. These results indicated that Reaction 2 dominated and agreed with the SRL results that show that the NO reaction dominates when the HNO_3 molarity is less than or equal to 2.5 M. In Test #9, as the slurry concentrated and, therefore, the HNO_3 concentrated, a slight amount of NO_2 was detected, as formed by Reaction 1. Results from the SRL show that as the HNO_3 concentration increases, the amount of NO_2 formed increases until it is the dominant NO_x species at HNO_3 concentrations above 3M.

The quantity of each gas measured was calculated from the areas under curves of the gas generation rate graphs. These values are provided in Table IV for comparison to the analytical data. The carbon destroyed was entirely accounted for in the carbon dioxide measured. But the amount of nitrogen gases measured accounted for only 31% of the nitrate that was destroyed. Also, the offgas results show that 1.6 moles of CO_2 were produced per mole of NO. Reaction 2 indicates that this ratio should be 3. Therefore, additional NO is being produced by a reaction other than Reaction 2. The analytical results for Test #9 indicate that 2 moles of nitrate reacted for every mole of carbon. This indicates that more nitric acid reacted than the 0.33 moles required by the stoichiometry of Reaction 2. This is another indication that other reactions are taking place that consume nitrate and produce NO. In the SRL tests, it was determined that for successful oxalate destruction, an excess of HNO_3 is required; 2.67 moles of HNO_3 for each mole of oxalate destroyed. This is due to the additional acid required for dissolution of the sludge components.

CONCLUSIONS

The results of these laboratory tests demonstrate that it is feasible to destroy oxalate in a WVDP simulant almost completely by a manganese-catalyzed reaction with nitrate. This was shown in the tests with refluxing and those where the slurry was allowed to concentrate. The concentration test better mimics how oxalate destruction would be carried out on a large scale where refluxing is not possible. Nitric oxide was the dominant gas produced. It is desirable to produce NO rather than NO_2 because, as shown by Reactions 1 and 2, the NO-producing Reaction 2 requires less HNO_3.

If oxalate destruction is to be further considered as an option, more testing would be required to determine how the results of this study may be applied to the WVDP melter feed preparation flowsheet. One consideration would be that the additional nitric acid in the system may result in a slurry with an unacceptably low pH.

Table IV. Results of Test #9

	Total Organic Carbon	Nitrate
Initial moles	1.45	5.36
Final moles	0.0021	2.53
Moles reacted	1.45	2.83
Percent reacted	~100	53
Moles of gas measured	1.43 as CO_2	0.87 of N as NO_x and N_2O

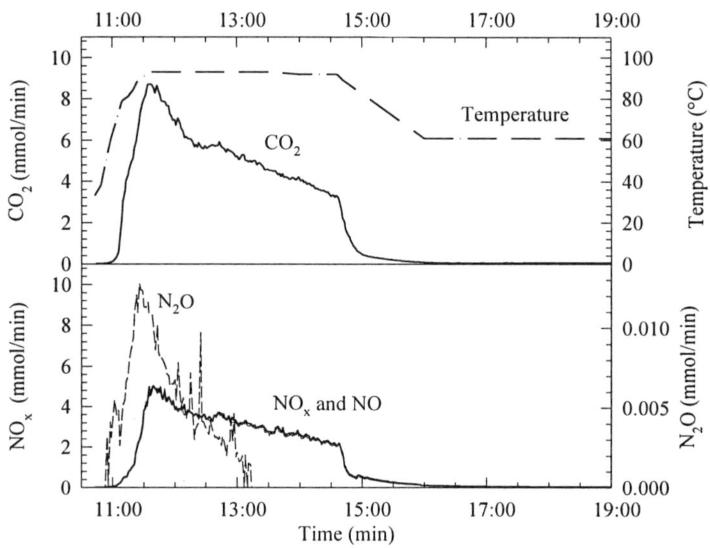

Figure 1. Gas Generation Rates for CO_2, N_2O, and NO_x During the First Day of Concentration/Oxalate Destruction

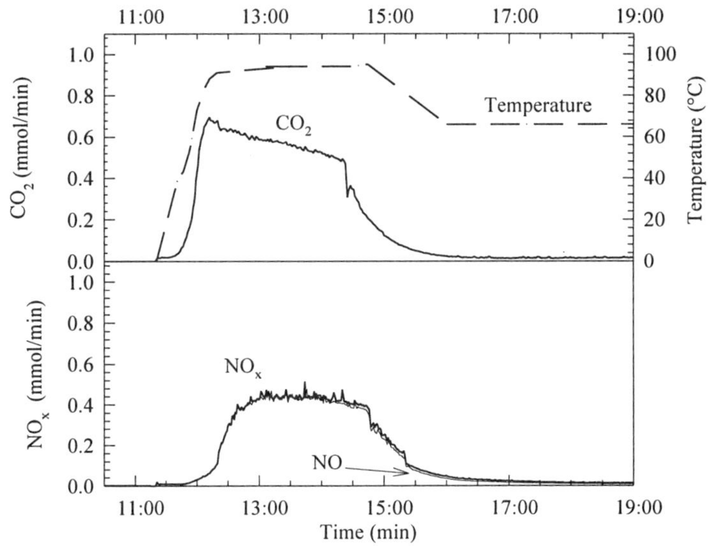

Figure 2. Gas Generation Rates for CO_2 and NO_x During the Second Day of Concentration/Oxalate Destruction

REFERENCES

[1] A.B. Meservey, "Decontamination and Film Removal"; pp. 137-159 in *Decontamination of Nuclear Reactors and Equipment.* Edited by J.A. Ayres. The Ronald Press Company, New York, 1970.

[2] J.A. Sills, G.K. Patello, J.S. Roberts, K.D. Wiemers, M.R. Elmore, W.G. Richmond, and R.L Russell, *Fiscal Year 1995 Laboratory Scale Studies of cesium Elution in Tank 8D-1 and Sludge Dissolution in Tank 8D-2.* PNNL-10945. Pacific Northwest National Laboratory, Richland, Washington, 1996.

[3] R.L. Russell, G.K. Patello, J.A. Sills, *Fiscal Year 1995 Laboratory Scale Studies of Cesium Elution in Tank 8D-1.* PNNL-11274, Pacific Northwest National Laboratory, Richland, Washington, 1996.

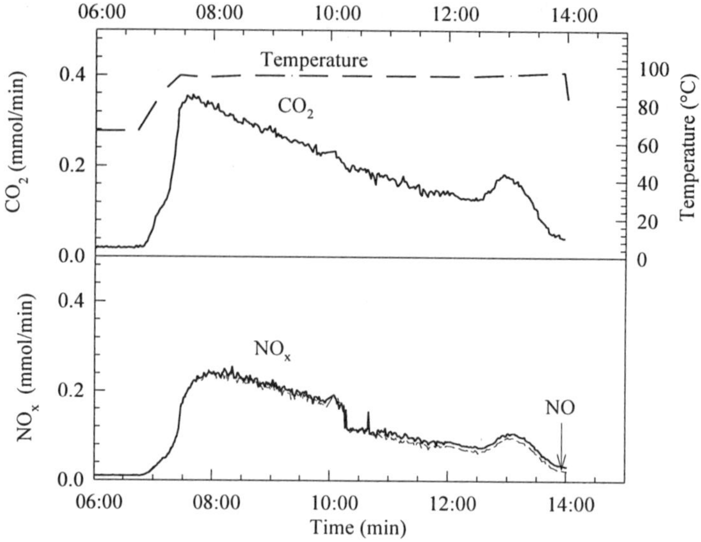

Figure 3. Gas Generation Rates for CO_2 and NO_x During the Third Day of Concentration/Oxalate Destruction

[4] N.E. Bibler, J.E. Hoisington, E.W. Holtzscheiter, *Technical Data Summary Decomposition of Oxalic Acid by the Manganese Catalyzed Nitric Acid Reaction.* DPSTD-80-36. Savannah River Laboratory, Aiken, South Carolina, 1981.

ACKNOWLEDGEMENT

This work was funded by the U.S. Department of Energy under Contract DE-AC06-76RLO 1830.

IRON REDOX AND CRYSTALLIZATION OF IRON PHOSPHATE GLASSES

Chandra S. Ray, Xiangyu Fang, Mevlut Karabulut, Gaya K. Marasinghe and Delbert E. Day
Graduate Center for Materials Research
University of Missouri-Rolla
Rolla, MO 65409.

ABSTRACT

The equilibrium between Fe^{2+} and Fe^{3+} redox couple in iron phosphate glasses prepared by melting a $40Fe_2O_3$-$60P_2O_5$ mol% batch at temperatures between 1150° and 1450° C in air for 1 h and at 1200° C between 1 and 24 h, was investigated as a function of melting temperature and time. Mössbauer spectroscopy was used to measure the concentration of Fe^{2+} (or Fe^{3+}) ions in these as-made glasses. The crystallization of the glasses and phase transformations of the crystals were investigated using DTA, XRD, and TGA. The dissolution rate in deionized water was measured by a weight loss method. Increasing melting temperature increased the concentration of Fe^{2+} ions in these glasses, but melting time longer than 1 h had little effect on the Fe^{2+} or Fe^{3+} concentration. With increasing Fe^{2+} content, the crystallization tendency of these glasses also increased. They crystallized to $Fe_3(P_2O_7)_2$ and $Fe_4(P_2O_7)_3$ when heated in nitrogen between 600° and 820° C, but with additional heating in air at about 820° C the $Fe_3(P_2O_7)_2$ oxidized to $Fe(PO_4)$ and produced a weight gain of the sample. The chemical durability of these iron phosphate glasses was exceptionally high (better than borosilicate or window glasses), independent of the Fe^{2+} or Fe^{3+} concentration, and improved with increasing total iron content or Fe/P ratio in the glass.

INTRODUCTION

Iron phosphate glasses are of interest[1-4] for immobilizing intermediate and high level nuclear wastes because of their exceptionally high chemical durability in water (dissolution rate ~ 10^{-9} g/cm²/min), relatively low melting temperatures (~ 1000° C), and moderately high density (~ 3.0 to 3.2 g/cm³ compared to 2.5 to 2.6 g/cm³ for most silicate glasses). Iron phosphate melts are practically non-corrosive[5] to most commercial refractories such as dense alumina, zircon, or chrome, which not only helps prolong the melter life, but controls the wasteform composition efficiently. Iron phosphate glasses should be particularly useful for vitrifying those wastes that contain significant amounts of phosphates and iron oxides such as some of the wastes at the Hanford site in Washington. By utilizing the P_2O_5 and Fe_2O_3 present in the waste, the waste loading can be increased significantly in iron phosphate wasteforms.

Since iron can exist in two valence states as Fe^{2+} and Fe^{3+}, the iron phosphate glasses are expected to contain both Fe^{2+} and Fe^{3+} ions even if the starting batch composition contains only Fe^{2+} as FeO or Fe^{3+} as Fe_2O_3. The equilibria between the iron redox couple in the melt is controlled by the following reversible reaction,

$$4Fe^{3+} \text{ (melt)} + 2O^{2-} \text{ (melt)} \rightleftharpoons 4Fe^{2+} \text{ (melt)} + O_2 \text{ (gas)} \quad .... \quad .... \quad ....(1)$$

whose equilibrium should depend upon several factors such as the melting temperature, time, and atmosphere, types of raw materials, and the presence of other components. The equilibrium concentration of Fe^{2+} and Fe^{3+} ions in the melt is retained when the melt is quenched to glass, and melting a batch of a binary composition containing, say, Fe_2O_3 and P_2O_5, results in a ternary glass composition containing FeO, Fe_2O_3, and P_2O_5. It has been reported[6,7] that the fraction of Fe^{2+} always reaches a value between 0.20 and 0.34 when iron phosphate compositions with an Fe/P ratio of 0.67 are melted at 1200° C for 2 h in air, nitrogen, or oxygen using various raw materials such as FeO, Fe_2O_3, Fe_3O_4, $FeC_2O_4.2H_2O$, P_2O_5 and $NH_4H_2PO_4$. Apparently, the melting atmosphere or the type of raw materials have little effect on the redox equilibria of iron ions in these iron phosphate melts. However, the effect of melting temperature or time on the iron redox equilibria of the iron phosphate melts is currently unknown.

Since the properties of a glass depend upon its composition, and the composition of an iron phosphate glass may change due to a change in the iron valence state, it is important to know how the iron redox equilibria for iron phosphate melts depends upon the temperature or time of melting. The primary objective of the present work was to obtain data for the iron redox equilibria that would be important to selecting the proper melting temperature and time for large scale production of iron phosphate-based wasteforms that have desirable properties for disposing of specific types of nuclear wastes.

Glasses were prepared by melting a batch composition of $40Fe_2O_3$-$60P_2O_5$ mol% (F40M), which initially contained only Fe^{3+} ions and had an Fe/P ratio of 0.67. The batch was melted between 1150° and 1450° C in air for 1 h and at 1200° C for times between 1 and 24 h in dense alumina (Coors AD998) crucibles. Three other glasses, which had different Fe/P ratios, were also prepared by melting batch compositions of xFe_2O_3 - $(100-x)P_2O_5$ with x = 10, 20, and 30 mol% at 1200° C for 1 h in air. The concentration of Fe^{2+} (or Fe^{3+}) ions was measured by Mössbauer spectroscopy (ASA 600 spectrometer). The dissolution rate in deionized water at 90° C was measured for these glasses for 4, 8, and 16 days using a weight loss method. Differential thermal analysis (DTA-7, Perkin Elmer Corporation) combined with x-ray diffraction analysis (XRD, SCINTAG XDS2000) and thermogravimetric analysis (TGA, Mettler TA-1) were used to investigate the crystallization of these glasses and phase transformations of the crystals. Further details of the experimental procedures are described elsewhere[6-8].

RESULTS AND DISCUSSION

The typical room temperature Mössbauer spectra for the glasses made by melting the F40M batch at 1150, 1350, and 1450° C for 1 h in air are shown in Fig. 1. The Fe^{2+} and Fe^{3+} components in the top spectra (1150° C) are the sum of four Lorentzian doublets used to fit[7] each of the two iron sites. Clearly, the absorption lines for the Fe^{2+} ions become larger with increasing melting temperature indicating a higher concentration of the Fe^{2+} ions in the glasses melted at higher temperatures. On the other hand, there is practically no difference (not shown) between the Mössbauer spectra for the glasses prepared by melting at 1200° C for different times.

The concentration of Fe^{2+} ions, as calculated from the Mössbauer spectra, for the glasses prepared by melting at different temperatures and for different times (at 1200° C) is shown in Fig. 2. Assuming the as-made glasses have the same 0.67 Fe/P ratio as that of the starting F40M batch composition, the final composition of the glasses was calculated using the measured Fe^{2+} concentration and the results

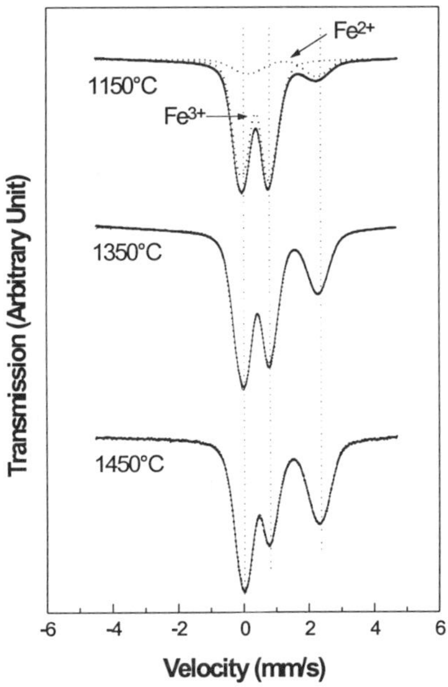

Fig. 1. Room temperature Mössbauer spectra for glasses made by melting a $40Fe_2O_3$-$60P_2O_5$ mol% batch in air at the temperatures shown.

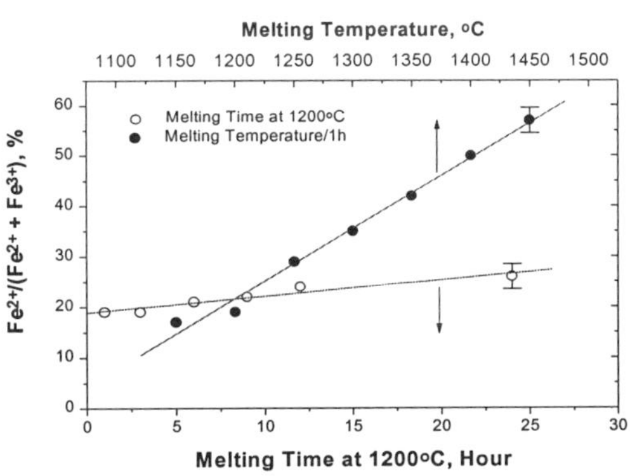

Fig. 2. Concentration of Fe^{2+} ions measured by Mössbauer spectroscopy for glasses made by melting a $40Fe_2O_3$-$60P_2O_5$ mol% batch in air as a function of melting temperature (top axis) or melting time at 1200° C (bottom axis). Typical experimental error is shown by the error bar on one data point.

are shown in Fig. 3. The data point that is further away from the starting binary batch composition represents the composition of a glass melted either at a higher temperature (solid circles) or for a longer period of time (open circles).

As shown in Fig. 2, the Fe^{2+} concentration increased from ~ 17% for the glass melted at 1150° C to ~ 57% for the glass melted at 1450° C. This increase is clearly noticeable in the larger absorption lines for Fe^{2+} ions in Fig. 1 for the glasses melted at higher temperatures. When melted at 1200° C for different times, the Fe^{2+} concentration quickly reached ~ 19% in 1 h and then increased very slowly to ~ 25% in 24 h.

The experimental error involved in the measurements of Fe^{2+} concentration by Mössbauer spectroscopy is ~ ±2.5%. Although it is systematic, the small increase in the Fe^{2+} concentration for the glasses melted for different times falls almost within this experimental error, see open circles in Fig. 2. This means that the concentration of Fe^{2+} ions in the glasses melted for different times may be considered essentially constant at 22±3%. Thus, a value of 22±3% which is reached in about an hour of melting, appears to be the equilibrium value of the Fe^{2+} concentration in this iron phosphate melt at 1200° C in air. However, the melting temperature has a larger effect on the reduction of Fe^{3+} to Fe^{2+} than the melting time. The larger effect of melting temperature on the reduction of Fe^{3+} ions can be seen also in Fig. 3 where the spread in the calculated composition for the glasses melted at different temperatures (solid circles) is much wider than that for the glasses melted for different times (open circles).

Glasses prepared by melting a batch composition of xFe_2O_3 - $(100-x)P_2O_5$ with x = 10, 20, 30 and 40 mol% all have nearly the same Fe^{2+} concentration (25±2.5% as measured by Mösbauer spectroscopy), even though these glasses have different starting Fe/P ratios. Since all of these

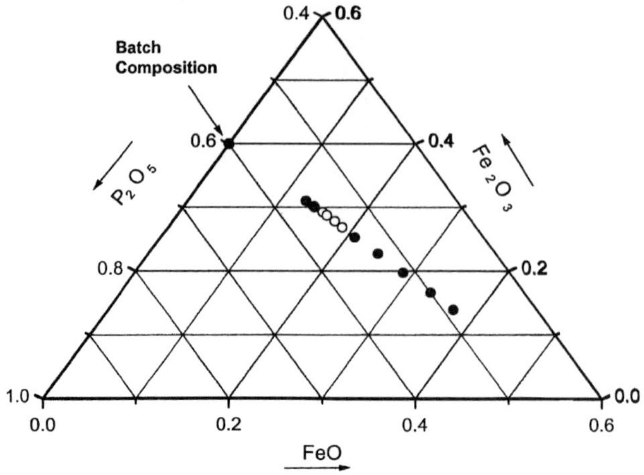

Fig. 3. Calculated compositions of the glasses made by melting a $40Fe_2O_3$-$60P_2O_5$ mol% batch in air at different temperatures (solid circles) and for different times at 1200° C (open circles). The Fe^{2+} concentration measured by Mössbauer spectroscopy (Fig. 2) for the glasses was used to calculate the final composition.

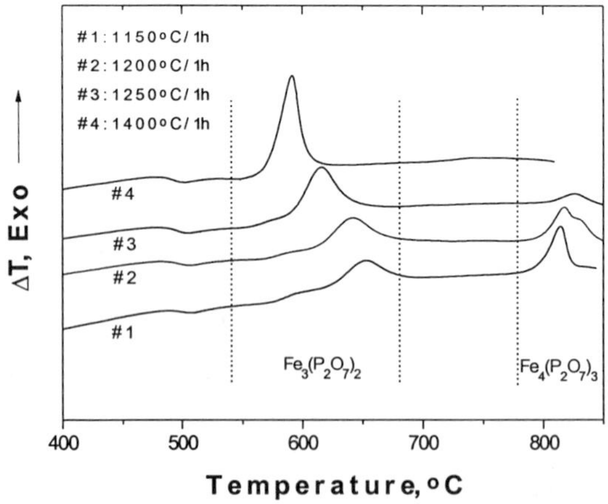

Fig. 4. DTA patterns for glasses made by melting a $40Fe_2O_3$-$60P_2O_5$ mol% batch in air at the temperatures shown and heated at 10° C/min in nitrogen.

glasses were melted at 1200° C, a constant Fe^{2+} concentration for these glasses again demonstrates that the Fe^{2+} and Fe^{3+} redox equilibria depends primarily on the melting temperature.

The weighted average Mössbauer hyperfine parameters, namely, the quadrupole splitting (QS) and isomer shift (IS), for the Fe^{2+} or Fe^{3+} ions did not change for the F40M glasses when melted either at different temperatures or for different times, which suggests that the local coordination environment of either iron ions did not change significantly with melting temperature or time. The values for QS and IS for the Fe^{2+} ions were 2.10±0.05 and 1.23±0.02, respectively, and those for the Fe^{3+} ions were 0.90±0.02 and 0.39±0.01, respectively. These QS and IS values for the Fe^{2+} and Fe^{3+} ions are in excellent agreement with those reported[6,7,9] previously for iron phosphate glasses, which typically represent an octahedral or distorted octahedral coordination for both iron ions. However, the presence of some Fe^{3+} ions in tetrahedral coordination cannot be ruled out.

Fig. 4 shows the DTA curves for the glasses melted at 1150, 1200, 1250 and 1400° C for 1 h and heated at 10° C/min in nitrogen. These curves are typical for iron phosphate glasses. A trend similar to what was observed in the

Mössbauer spectra was also observed in the DTA curves i.e., there is a considerable change in the DTA curves for the glasses melted at different temperatures (Fig. 4), but no significant change is apparent in the DTA curves for the glasses melted at 1200° C for different times (not shown).

In most cases, the DTA curves for these glasses contain two crystallization peaks. The first or lower temperature peak, which ranges from 590 to 655° C, becomes larger and shifts to lower temperature with increasing melting temperature i.e., with increasing Fe^{2+} concentration in the glass. Likewise, the glass transition temperature for these glasses decreases from 512 to 500° C as the Fe^{2+} concentration increases from 17 to 50%. These results suggest that the crystallization tendency for these glasses increases with increasing concentration of the Fe^{2+} ions. With increasing melting temperature or Fe^{2+} concentration, the second or higher-temperature peak becomes smaller and disappears completely for the glasses melted at temperatures ≥ 1300° C. However, the temperature for the second crystallization peak shows little dependence on the melting temperature, and occurs at 820±5° C for all the glasses that showed a detectable second peak.

XRD analysis for the glasses crystallized in nitrogen for 24 h at the temperatures indicated by their DTA exotherms reveal that the low temperature DTA peak is due to the crystallization of ferrous-ferric pyrophosphate, $Fe_3(P_2O_7)_2$, and the high-temperature peak is due to the crystallization of ferric pyrophosphate, $Fe_4(P_2O_7)_3$. Glasses melted at temperatures ≥ 1300° C i.e., which have only one (lower temperature) DTA peak, contain only $Fe_3(P_2O_7)_2$, but no $Fe_4(P_2O_7)_3$ even when heated at an arbitrary high temperature close to 820° C in nitrogen.

Crystalline $Fe_3(P_2O_7)_2$ contains one Fe^{2+} and two Fe^{3+} ions per molecule. With increasing melting temperature the concentration of Fe^{2+} in the glass increases (Fig. 2) and so does the amount of $Fe_3(P_2O_7)_2$ crystallized from glasses melted at higher temperatures. The growing low temperature DTA peak, which is due to the crystallization of $Fe_3(P_2O_7)_2$, for the glasses containing a higher concentration of Fe^{2+} ions is, therefore, an expected result. Since crystalline $Fe_4(P_2O_7)_3$ contains only Fe^{3+} ions, whose concentration decreases with increasing melting temperature, a smaller second DTA peak is also expected for glasses melted at higher temperatures. At an Fe^{2+} concentration of ≥ 33% and assuming, all the Fe^{3+} ions (≤ 67%) are used in the crystallization of $Fe_3(P_2O_7)_2$ (1st DTA peak), no Fe^{3+} will be available for the crystallization of $Fe_4(P_2O_7)_3$ (2nd DTA peak). The complete absence of the second DTA peak in Fig. 4 for the glasses melted ≥ 1300° C is consistent with their Fe^{2+} concentration being ≥ 35%, see Fig. 2. A slight increase (< 5° C) in the 2nd DTA peak temperature for the glasses melted at higher temperatures may be due to the reason that the crystallization of $Fe_4(P_2O_7)_3$ which contains only Fe^{3+} ions, becomes increasingly difficult as the Fe^{3+} concentration in glasses melted at higher temperatures decreases.

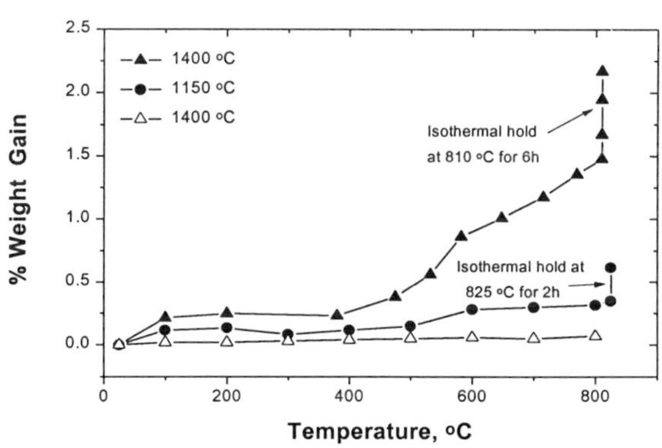

Fig. 5. TGA in air (solid circles and solid triangles) or in nitrogen (open triangles) for glasses made by melting a $40Fe_2O_3$-$60P_2O_5$ mol% batch in air at temperatures shown.

With an additional heat treatment for 24 h in air around 820° C, all these samples including those which initially contained only $Fe_3(P_2O_7)_2$ after the first stage heat treatment in nitrogen at 820° C for 24 h, are found to contain (XRD) $Fe_4(P_2O_7)_3$ and ferric orthophosphate, $Fe(PO_4)$, but no $Fe_3(P_2O_7)_2$. It is believed that the $Fe_3(P_2O_7)_2$ crystals undergo a phase transformation through oxidation in air, which is represented by the following reaction,

$$2Fe_3(P_2O_7)_2 + 1/2O_2 \rightarrow 2Fe(PO_4) + Fe_4(P_2O_7)_3 \quad .... \quad .... \quad ....(2)$$

The TGA analysis for all these iron phosphate glasses showed a weight gain when they were heated in air, but no weight gain occurred when they were heated in an oxygen-free (nitrogen) atmosphere, see Fig. 5. This is consistent with Eq. 2. Fig. 5 shows the typical TGA results for the glasses melted at 1150 and 1400° C and heated in air. These glasses initially contained 17 and 50% Fe^{2+} ions, respectively. The TGA results for the glass melted at 1400° C, but heated in nitrogen is also shown for comparison. To avoid melting, the temperature was not allowed to increase above 810° C for the 1400° C glass and above 825° C for the 1150° C glass during the dynamic TGA ramp. The temperature was then held constant for about 2 and 6 h for the 1150 and 1400° C glasses, respectively.

Oxidation of the Fe^{2+} to Fe^{3+} ions is clearly indicated by the weight gain for the two samples heated in air, solid circles and solid triangles in Fig. 5. The absence of such a weight gain for the 1400° C glass when heated in nitrogen (open triangles) is further evidence of an oxidation process. The larger weight gain for the 1400° C glass compared to that for the 1150° C glass is consistent with its initially higher Fe^{2+} content (50%). Assuming all of the Fe^{2+} ions in these glasses were oxidized to Fe^{3+} during the TGA experiments, the calculated weight gain, according to Eq. 2, for the 1150 and 1400° C glasses would be 0.74% and 2.19%, respectively. These values are in excellent agreement with the measured (Fig. 5) weight gains of 0.62% and 2.17% for these glasses.

The average dissolution rate, DR, of these iron phosphate glasses in deionized water at 90° C for 4, 8, and 16 days varied only from about 1.5 x 10^{-9} for 4 days to 4.5 x 10^{-9} g/cm²/min for 16 days. The DR was independent of the relative concentration of the Fe^{2+} or Fe^{3+} ions in the glass. These DR values are exceptionally low (chemical durability is exceptionally high) and are even lower (better) than for some borosilicate glasses (~ 10^{-8} g/cm²/min) such as CVS-IS and LD6-54-12.

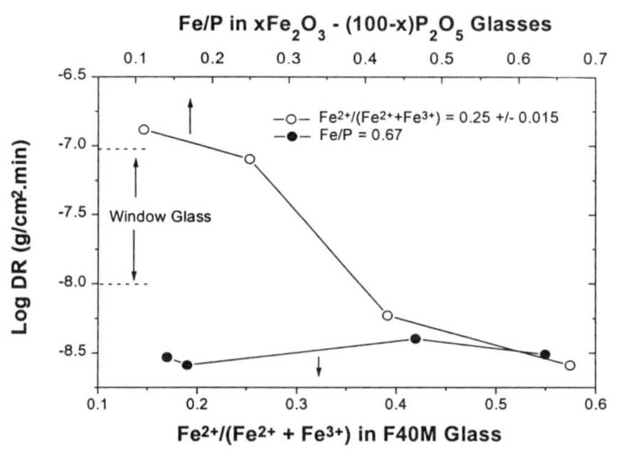

Fig. 6. Log dissolution rate (DR) for the iron phosphate glasses immersed in deionized water at 90° C for 4 days as a function of Fe/P ratio (open circles, top axis) and Fe^{2+} concentration (solid circles, bottom axis). See text for glass compositions and melting schedule.

The measured values of log DR in deionized water at 90° C for 4 days for the glasses prepared by melting a batch composition of xFe_2O_3 - $(100-x)P_2O_5$ with x = 10, 20, 30 and 40 mol%, are shown in Fig. 6 as a function of their Fe/P ratio. The 4-day logDR values for the glasses prepared by melting the F40M batch composition, which had an Fe/P ratio of 0.67, at 1150, 1200, 1300 and 1450° C are also shown as a function of the measured Fe^{2+}

concentration. Fig. 6 clearly demonstrates that the DR is independent of the relative concentration of Fe^{2+} or Fe^{3+} ions in the glass (solid circles), but depends strongly on the Fe/P ratio (open circles). The chemical durability of the iron phosphate glasses improves (DR decreases) with increasing Fe/P ratio or the total iron content in the glass. However, glass formation becomes difficult for binary iron phosphate compositions with a Fe/P ratio > 1.0.

Unlike the structure of conventional (iron-free) phosphate glasses where 100% of the bridging oxygen bonds are of the -P-O-P- type, the structure of iron phosphate glasses contains a significant number of -Fe-O-P- bonds. The high chemical durability of the iron phosphate glasses is attributed to the presence of -Fe-O-P- bonds which are less susceptible to hydration than the -P-O-P- bonds.

As mentioned previously, the iron phosphate glasses crystallize generally to $Fe_3(P_2O_7)_2$ and $Fe_4(P_2O_7)_3$. Although the structure of a glass and its crystalline counterpart will not be identical, it is reasonable to assume that the short range structure of the iron phosphate glasses contains a mixture of structural groups that resembles the structure of crystalline $Fe_3(P_2O_7)_2$ and $Fe_4(P_2O_7)_3$. In the structure of these two compounds only 14% of the oxygen ions are present(7,10,11) as -P-O-P- bonds, the remaining 86% being of the -Fe-O-P- type. Assuming all the Fe^{2+} ions in the glass are grouped in a structural form similar to that of ferrous-ferric pyrophosphate, $Fe_3(P_2O_7)_2$, and the residual Fe^{3+} ions not needed for $Fe_3(P_2O_7)_2$ are grouped in a form similar to the structure of $Fe_4(P_2O_7)_3$, a simple calculation shows that the structure of a F40M glass (Fe/P ratio 0.67) with 20% Fe^{2+} ions contains about 72% -Fe-O-P- bonds. The high concentration of the -Fe-O-P- bonds in these iron phosphate glasses is believed to be the reason for their high chemical durability.

The concentration of $Fe_3(P_2O_7)_2$ and $Fe_4(P_2O_7)_3$ structural groups in the glass should depend upon the concentration of the Fe^{2+} and Fe^{3+} ions and upon the overall Fe/P ratio. For a constant Fe^{2+} concentration, the amount of both structural groups increases in the glass with increasing total iron content or Fe/P ratio. This, in turn, increases the number of -Fe-O-P- bonds compared to that of the -P-O-P- bonds in the glass. Therefore, an increase in chemical durability would be expected with increasing Fe/P ratio in agreement with the open circles in Fig. 6. On the other hand, a change in the relative concentration of Fe^{2+} and Fe^{3+} ions does not change the number of -Fe-O-P- bonds in the glass structure when the Fe/P ratio remains constant. This explains why the DR data represented by the solid circles in Fig. 6 for the F40M glasses (Fe/P ratio of 0.67) is nearly independent of the Fe^{2+} (or Fe^{3+}) concentration.

CONCLUSION

A $40Fe_2O_3$-$60P_2O_5$, mol%, composition which had an Fe/P ratio of 0.67 and which initially contained only Fe^{3+} ions, yielded glasses containing varying amounts Fe^{2+} and Fe^{3+} ions when melted either at different temperatures or for different times in air. Melting an iron phosphate composition reduced a significant fraction of the Fe^{3+} ions to Fe^{2+}. While the concentration of Fe^{2+} ions increased rapidly with increasing melting temperature, from ~ 17% for a glass melted at 1150° C to 55% for a glass melted at 1450° C, melting for times longer than 1 h at 1200° C had little effect on the concentration of Fe^{2+} or Fe^{3+} ions. Reheating the glasses, which contained some Fe^{2+} ions, in air at sub-liquidus temperatures caused all of the Fe^{2+} ions to oxidize to Fe^{3+}. This caused the samples to gain weight.

The crystallization tendency for these iron phosphate glasses generally increased with increasing Fe^{2+} concentration. Glasses containing < 33% Fe^{2+} ions showed two exothermic DTA peaks corresponding to the crystallization of ferrous-ferric pyrophosphate, $Fe_3(P_2O_7)_2$, and ferric pyrophosphate, $Fe_4(P_2O_7)_3$, the latter crystallizing at the higher temperature, ~ 820±5° C. The temperature for the crystallization of $Fe_3(P_2O_7)_2$ ranged from 590 to 650° C for these glasses and decreased with increasing concentration of Fe^{2+} ions. Glasses containing > 33% Fe^{2+} showed only one DTA peak, that being due to the crystallization of $Fe_3(P_2O_7)_2$.

The chemical durability of these glasses in deionized water at 90° C was exceptionally high, the dissolution rate being ~ 10^{-9} g/cm²/min. The dissolution rate was independent of the relative concentration of Fe^{2+} or Fe^{3+} ions, but increased with increasing total iron content or Fe/P ratio in the glass. The exceptionally high chemical durability is attributed to the large concentration of -Fe-O-P- bonds present in the glass network, which are more hydration resistant than the -P-O-P- bonds present in conventional (iron-free) phosphate glasses.

ACKNOWLEDGMENT

This work was supported by the Department of Energy (DoE), contract DE-FG07-96ER45618.

REFERENCES

[1]B. Samuneva, P. Tzvetkova, I. Gugov and V. Dimitrov, "Structural Studies of Phosphate Glasses", *J. Mater. Sci. Letts*, **15** 2180-2183 (1996).

[2]L. Kahl, D. Clark, and W. White, *Adv. In Ceramics*, **20** 141 (1986).

[3]B. C. Sales and L. A. Boatner, *J. Non-Cryst. Solids*, **79** 83 (1986).

[4] M. G. Mesko, D. E. Day, and B. C. Bunker, "Immobilization of High Level Radioactive Sludges in Iron Phosphate Glass", *J. Am. Ceram. Soc.*, (in press).

[5]F. Chen and D. E. Day, "Corrosion of Selected Refractories by Iron Phosphate Melts", *Paper #SXVI-060-98* (this volume).

[6]G. K. Marasinghe, M. Karabulut, C. S. Ray, D. E. Day, C. H. Booth, P. G. Allen and D. K. Shuh, "Redox Characteristics and Structural properties of Iron Phosphate Glasses: A Potential Host Matrix for Vitrifying High Level Nuclear Waste", *Ceramic Transactions*, **87** 261-270 (1998).

[7]G. K. Marasinghe, M. Karabulut, C. S. Ray, D. E. Day, M. G. Shumsky, W. B. Yelon, C. H. Booth, P. G. Allen and D. K. Shuh, "Structural Features of Iron Phosphate Glasses", *J. Non-Cryst. Solids*, **222** 144-152 (1997).

[8]C. S. Ray, X. Fang, M. Karabulut, G. K. Marasinghe and D. E. Day, "Chemical Durability, Crystallization, and Iron Redox Equilibria of Iron Phosphate Glasses", *J. Non-Cryst. Solids*, (in press).

[9]M. Darby Dyar, "A Review of Mössbauer Data on Inorganic Glasses: The Effects of Composition on Iron Valency and Coordination", *American Mineralogist*, **70** 304-316 (1985).

[10]M. Ijjali, G. Venturini, R. Gerardin, B. Malaman and C. Gleitzer, "Synthesis, Structure and Physical Properties of a Mixed-Valence Iron Diphosphate $Fe_3(P_2O_7)_2$: First Example of Trigonal Prismatic Fe^{2+} with O^{2-}Ligands", *Eur. J. Solid State Inorg. Chem.*, **28** 983-998 (1991).

[11]K. K. Palkina, S. I. Maksimova, N. T. Chibiskova, K. Schlesinger and G. Ladwig, "Doppeloktaeder-Cluster [V_2O_9] in der Kristallstruktur von Vanadium (III)-Diphosphat, $V_4(P_2O_7)_2$", *Z. Anorg. Allg. Chem.*, **529** 89-96 (1985).

EFFECTS OF NUCLEAR WASTE COMPONENTS ON REDOX EQUILIBRIA, STRUCTURAL FEATURES, AND CRYSTALLIZATION CHARACTERISTICS OF IRON PHOSPHATE GLASSES

G. K. Marasinghe,[a] M. Karabulut,[a] C. S. Ray,[a] D. E. Day,[a] P. G. Allen,[b] J. J. Bucher,[b] N. M. Edelstein,[b] D. K. Shuh,[b] Y. S. Badyal,[c] M. L. Saboungi,[c] M. Grimsditch,[c] S. D. Shastri,[c] and D. Haeffner[c]

[a] *Graduate Center for Materials Research, Department of Ceramic Engineering and the Department of Physics, University of Missouri-Rolla, Rolla, MO 65409*

[b]*Chemical Sciences Division, Lawrence Berkeley National Laboratory, Berkeley, CA 94720*

[c]*Argonne National Laboratory, Argonne, IL 60439*

ABSTRACT

Glass forming and crystallization characteristics, atomic structure, and redox equilibria in iron phosphate glasses containing varying amounts of a common waste component such as UO_2, Cs_2O, Bi_2O_3, SrO, Na_2O, and MoO_3 have been investigated using differential thermal analysis (DTA), Mössbauer spectroscopy, Fe-*K* edge X-ray absorption spectroscopy (XAS), Raman spectroscopy, and neutron and high energy X-ray scattering. None of the waste components adversely affect the redox equilibria attained during melting in air. Raman spectra, Mössbauer hyperfine parameters, and Fe-*K* edge XAS data indicate that neither the near neighbor environment of the iron ions nor the phosphorus-oxygen network is appreciably altered by the waste components. However, the incorporation of waste elements into the iron phosphate glass matrix is evident in the high energy X-ray and neutron scattering data.

INTRODUCTION

The immobilization of high level nuclear waste (HLW) in a glass host matrix has been studied extensively for several decades. In this process, a suitable glass host is used to dissolve the HLW to form a vitreous, chemically durable, solid. The choice of host material and wasteform composition is a compromise between the waste loading capacity, manageable glass melting conditions, and long-term stability in storage conditions.[1] Even though the chemical compositions of nuclear wastes are complex and diverse, only one glass material, namely borosilicate glasses, is currently approved for vitrifying HLW. Several of the HLWs in the U.S. contain components such as phosphates which make them poorly suited for disposal in borosilicate glasses.[2,3]

However, iron phosphate glasses appear capable of vitrifying a variety of HLWs including those which are not well suited for borosilicate glasses.[4,5] In contrast to the atomic structure of most common phosphate glasses which is dominated by -P-O-P- links and monomer $(PO_4)^{-3}$ units, the atomic structure of above mentioned iron phosphate glasses is dominated by -Fe-O-P- links and dimer $(P_2O_7)^{-4}$ units.[4] The excellent chemical durability of these iron phosphate glasses has been explained in terms of their atomic structure.[4] The present paper describes how the addition of several common waste components affect the glass formation and crystallization characteristics, the atomic structure, and redox equilibria of iron phosphate glasses.

EXPERIMENTAL

Glasses were prepared by melting homogeneous mixtures of reagent grade chemicals in dense alumina crucibles at approximately 1200°C in air for 1-2 hrs. Table I gives the batch composition used for the glasses studied. Each melt was quenched in air by pouring it into 1 x 1 x 5 cm cold steel molds. The rectangular samples were immediately transferred to a furnace and annealed at 450°C for ca. 3 hrs. The Mössbauer spectra were obtained at 23°C on a ASA600 spectrometer which utilized a room temperature 50 mCi rhodium matrix cobalt-57 source. The spectrometer was calibrated at 23°C with an α-iron foil and the linewidth of the α-iron spectrum was 0.27 mm/s. Mössbauer absorbers of approximate thickness $85 mg/cm^2$ were prepared from 200 mesh powders. The Mössbauer spectra were fit with the minimum number of broadened paramagnetic doublets necessary, usually eight, to match the absorption envelope with a misfit of less than 0.003. Details of this fitting procedure and its merits have been discussed elsewhere.[4]

Raman scattering measurements were made on the glasses containing UO_2 and Cs_2O at the Materials Science Division, Argonne National Laboratory (ANL). Raman spectra were excited with 50 to 100 mW of 476 nm radiation from a Kr ion laser. The incident beam impinged on the sample at an angle ~ 45° to the surface. The scattered light was collected along the surface normal with an f/1.4 lens and was analyzed with a triple Jobin-Yvon grating spectrometer equipped with a CCD (charge coupled device).

X-ray absorption spectra were collected on BL 4-1 at the Stanford Synchrotron Radiation Laboratory (SSRL) at the Fe *K*-edge using Si(220) monochromator crystals. The data were reduced and fitted using standard procedures.[6,7] Experimental details have been published elsewhere.[4,5]

The high energy X-ray scattering measurements were carried out on the 1-BM bending magnet beamline operated by the Syncrotron Radiation Instrumentation Collaborative Access Team (SRI-CAT) at the Advanced Photon Source (APS), ANL. A horizontally focusing, single bounce Si (111) monochromator produced the incident 61.332 keV (wavelength, 0.2022 Å) beam. The samples were run in flat-plate symmetrical transmission geometry with the scattering measured by a solid-state Ge detector scanning in the vertical plane to minimize polarization corrections. The raw data were corrected for attenuation, multiple scattering, and Compton scattering using standard methodologies.[8]

The neutron scattering measurements were made on the General Liquids and Amorphous Diffractometer (GLAD) of the Intense Pulsed Neutron Source (IPNS), ANL.[9] In contrast to the X-ray measurements, a pulsed 'white' beam of neutrons with a wavelength range of 0.12 to 3.0 Å was used. The scattered neutrons, as measured by multiple fixed detectors covering the angular range of 8 to 125°, were wavelength-sorted using the time-of-flight technique. The raw data were corrected for dead time, attenuation,

Table I. Batch compositions and Mössbauer parameters for the glasses melted in air.

Sample	Batch Composition (Mol%)	Mössbauer Parameters				
		Fe(II)		Fe(III)		Fe(II)
		δ[a]	Δ[b]	δ[a]	Δ[b]	Frac.[c]
A	$40Fe_2O_3$-$60P_2O_5$	1.22	2.14	0.38	0.94	0.19
UO_2						
B	$36Fe_2O_3$-$54P_2O_5$-$10UO_2$	1.22	2.14	0.37	0.93	0.16
C	$35Fe_2O_3$-$60P_2O_5$-$5UO_2$	1.26	2.08	0.39	0.91	0.19
D	$30Fe_2O_3$-$60P_2O_5$-$10UO_2$	1.25	2.12	0.38	0.88	0.11
E	$25Fe_2O_3$-$60P_2O_5$-$15UO_2$	-	-	0.37	0.90	0.00
F	$28Fe_3O_4$-$62P_2O_5$-$10UO_2$	1.20	2.13	0.37	0.91	0.14
G	$26Fe_3O_4$-$59P_2O_5$-$15UO_2$	1.19	2.26	0.36	0.92	0.08
Bi_2O_3						
H	$39Fe_2O_3$-$58P_2O_5$-$3Bi_2O_3$	1.25	2.14	0.40	0.92	0.33
J	$36Fe_2O_3$-$54P_2O_5$-$10Bi_2O_3$	1.24	2.17	0.40	0.92	0.34
K	$30Fe_2O_3$-$60P_2O_5$-$10Bi_2O_3$	1.24	2.15	0.40	0.89	0.32
L	$20Fe_2O_3$-$60P_2O_5$-$20Bi_2O_3$	1.25	2.14	0.40	0.88	0.34
Na_2O						
M	$36Fe_2O_3$-$54P_2O_5$-$10Na_2O$	1.24	2.13	0.40	0.90	0.28
N	$32Fe_2O_3$-$48P_2O_5$-$20Na_2O$	1.25	2.09	0.41	0.89	0.32
Cs_2O						
O	$38Fe_2O_3$-$57P_2O_5$-$5Cs_2O$	1.22	2.06	0.38	0.92	0.31
P	$36Fe_2O_3$-$54P_2O_5$-$10Cs_2O$	1.26	2.02	0.38	0.89	0.32
Q	$34Fe_2O_3$-$51P_2O_5$-$15Cs_2O$	1.26	1.98	0.37	0.85	0.32
R	$28Fe_2O_3$-$42P_2O_5$-$30Cs_2O$	1.24	1.99	0.39	0.83	0.29
SrO						
S	$38Fe_2O_3$-$57P_2O_5$-5SrO	1.17	2.26	0.42	0.89	0.26
T	$36Fe_2O_3$-$54P_2O_5$-10SrO	1.18	2.26	0.42	0.89	0.27
U	$32Fe_2O_3$-$48P_2O_5$-20SrO	1.18	2.24	0.42	0.90	0.25
MoO_3						
V	$30Fe_2O_3$-$60P_2O_5$-$10MoO_3$	1.25	2.00	0.37	0.93	0.03
W	$20Fe_2O_3$-$60P_2O_5$-$20MoO_3$	1.23	2.00	0.36	0.91	0.05
X	$10Fe_2O_3$-$60P_2O_5$-$30MoO_3$	0.00	0.00	0.37	0.91	0.00

[a] Isomer shift in mm/sec, [b] Quadrupole splitting in mm/sec, [c] Fe(II)/[Fe(II)+Fe(III)]

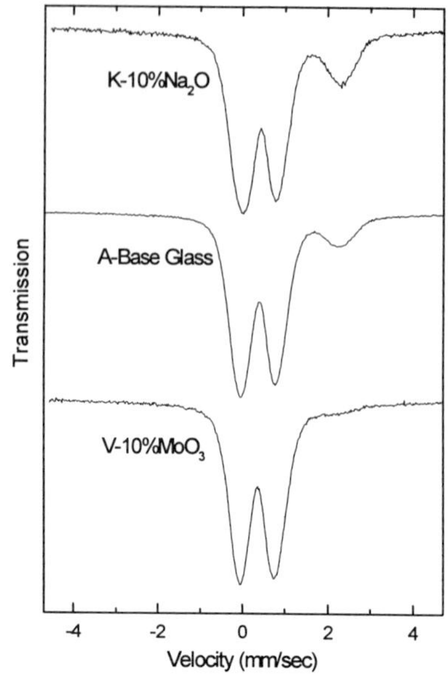

Figure 1. Room temperature Mössbauer spectra of samples K, A, and V. The areas of the absorption line at approximately 2.1 mm/s is indicative of the Fe(II) fraction.

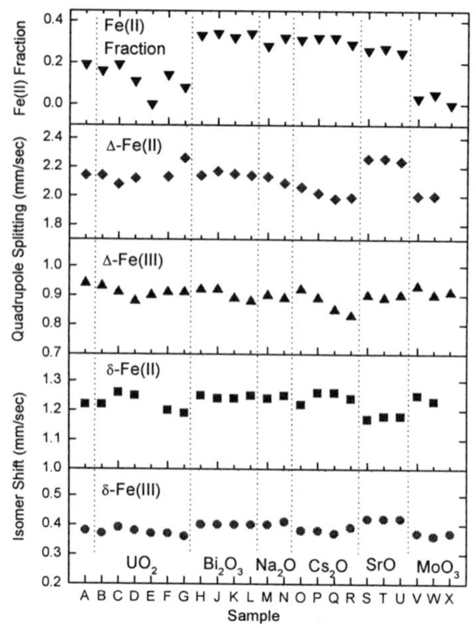

Figure 2. Dependence of the Fe(II) fraction, quadrupole splitting (Δ), and isomer shift (δ), on the sample composition.

multiple scattering, and inelasticity effects.

Differential thermal analysis (DTA) data was collected on a Perkin Elmer DTA-7 instrument. About 300 mg of 425 to 500μm glass particles were used for DTA measurements, with high purity alumina powder as the reference material. The thermocouples of the instrument were calibrated at the melting points of aluminum (660.4°C) and gold (1064.4°C). Experimental details can be found elsewhere.[10]

RESULTS AND DISCUSSION

The Mössbauer spectra shown in Figure 1 are typical of those measured for other glasses. Table I summarizes the Mössbauer parameters obtained. Figure 2 shows the dependence of the Fe(II) fraction, i. e. Fe(II)/[Fe(II)+Fe(III)], and hyperfine parameters on the type and concentration of waste components. Addition of UO_2 or MoO_3 reduces the Fe(II) fraction in the glass, as compared to the base glass (sample A),whereas Na_2O, Cs_2O, SrO, and Bi_2O_3 cause the Fe(II) to increase. However, none of the waste components cause the Fe(II) fraction to exceed 0.4, the level above which the glass forming ability of iron phosphate melts decreases.[4]

The isomer shift is sensitive to changes in the coordination environment, up to the second shell in certain cases, of iron ions. As seen in Figure 2, the isomer shifts, whose approximate uncertainty is ±0.05mm/s, do not depend appreciably on the type or the concentration of the waste elements. Thus we can conclude that, at the concentrations studied herein, none of the waste elements alter the local environment around iron ions to an extent which can be detected by Mössbauer spectroscopy. The spread of quadrupole splittings, see Figure 2, is also within the experimental uncertainty ($\sim \pm 0.1$mm/s), indicating that the coordination symmetry around the iron ions is not altered appreciably by the waste components.

The Raman spectra in Figure 3 are very similar to those measured for other glasses containing UO_2 and Cs_2O. It has been shown[11] that the Raman spectrum of the base glass (sample A) is characteristic of a structure dominated by dimer $(P_2O_7)^{-4}$ groups. The bands at approximately 650, 570, and 1070 cm^{-1} are assigned to the dimer units and the band at approximately 920 cm^{-1} is assigned to monomer $(PO_4)^{-3}$ units. Other than a small increase in the band at 920 cm^{-1}, the Raman spectra do not change appreciably as UO_2 or Cs_2O is added suggesting that the oxygen-phosphorus network is not altered by these waste elements.

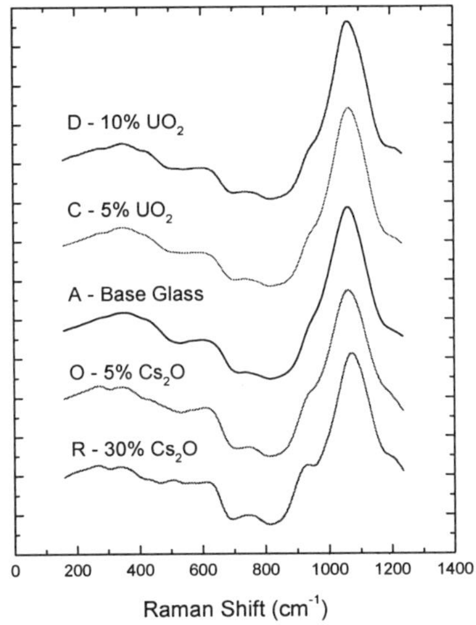

Figure 3. Raman spectra of iron phosphate glasses containing UO_2 and Cs_2O.

Recent XAS studies[4] of glass sample A and its crystallized counterpart have suggested that the iron ions are approximately six-fold coordinated with near neighbor oxygens in the first shell. Figure 4 shows the Fourier transform (FT) of $k^3\chi(k)$ for four of the glasses listed in Table I. The dominant peak centered at $R+\Delta \simeq 1.4$Å of the Fourier transform is due to a shell of near neighbor oxygens centered at approximately 1.8 Å from the center of the iron ions.[4] As seen in Figure 4, the shape and position of this peak do not change with the addition of waste components, consistent with an atomic structure where the waste elements are located far from the iron ions. A comprehensive analysis of the XAS data will be published elsewhere.[12]

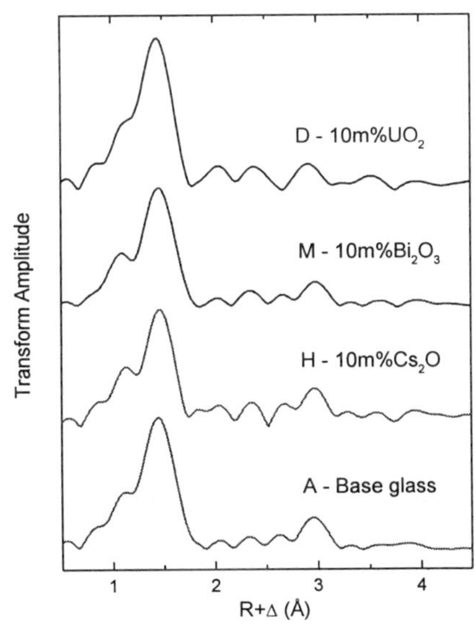

Figure 4. Fourier transform (FT) of $k^3\chi(k)$ for samples A, D, H, and M.

Total correlation functions, T(r),[13]

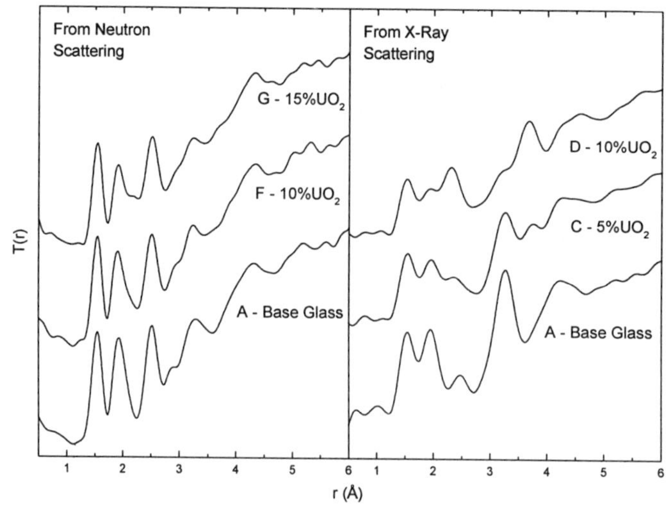

Figure 5. Total correlation functions obtained from neutron scattering data (left) and high energy X-ray scattering data (right) for several iron phosphate glasses containing UO_2.

obtained from neutron and high energy X-ray scattering data for some of the UO_2 containing glasses are shown in Figure 5. Prominent peaks at approximately 1.5, 1.9, 2.4, and 3.3 Å are assigned to P-O, Fe-O, O-O, and Fe-P pairs, respectively. The positions of the above mentioned peaks do not depend appreciably on the sample composition indicating that the structural features of the base glass are not altered by the addition of UO_2. However, the addition of UO_2 results in two new peaks, most appearent in the T(r) plot for X-ray scattering data, at 2.3 and 3.7 Å. Based on the interatomic distances observed in uranium phosphates such as $U_2P_2O_9$,[14] the peak at 2.3 Å can be attributed to U-O pairs. The peak at 3.7 Å is most likely due to U-P and/or U-Fe pairs.

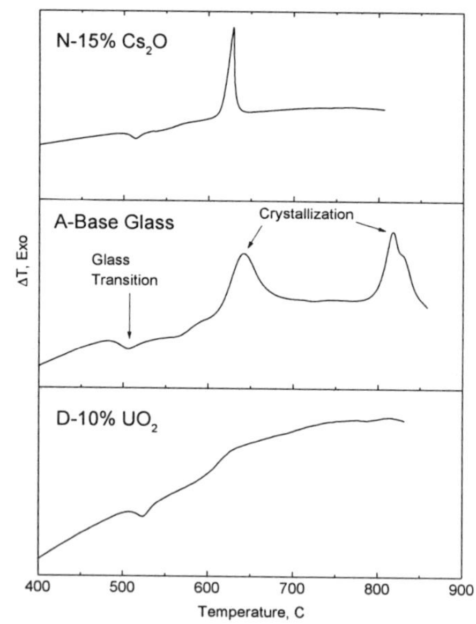

Figure 6. DTA data for three of the iron phosphate glasses listed in Table I.

The endothermic peak at approximately 500°C in the DTA

thermogram is for the glass transition and the exothermic peaks at approximately 580 and 775°C are due to crystallization of the glass. Figure 6 shows typical DTA patterns and how the waste components affect the glass forming and crystallization characteristics. Table II summarizes the DTA results. UO_2, SrO, and MoO_3 suppress crystallization, but Na_2O and Cs_2O promote the crystallization process. However, none of the waste components listed in Table II lowers the glass transition temperature below 400°C, which is the minimum glass transition temperature for an acceptable host matrix.

Table II. Summary of differential thermal analysis data.

Sample		T_g(°C)[a]	T_x(°C)[b]	Comments
Base	A	505	642	
UO_2	B	515	609	
	C	512	661	UO_2 enhances glass formation. No distinct crystallization peak for samples D and E.
	D	512	-	
	E	515	-	
Na_2O	M	510	588	Crystallization peaks are very sharp. Na_2O promotes crystallization.
	N	511	590	
Cs_2O	P	512	634	Crystallization suppressed at low Cs_2O concentrations but enhanced at high concentrations.
	Q	513	629	
	R	529	641	
SrO	S	518	670	Increases glass transition temperature and crystallization temperature. Suppresses crystallization.
	T	531	679	
	U	552	775	
MoO_3	V	503	-	No crystallization peaks. Glass formation enhanced.
	W	487	-	
	X	503	-	

[a] Glass transition temperature.
[b] Temperature corresponding to the first crystallization temperature.

CONCLUSION

Considerable amounts of UO_2, Cs_2O, Bi_2O_3, SrO, Na_2O, or MoO_3 can be added to iron phosphate glasses without adversely affecting the redox equilibria or the glass forming characteristics. Furthermore, the addition of the above waste components does not appreciably alter the near neighborhood of iron ions or the phosphorus-oxygen network. Results presented herein suggest that iron phosphate glasses may be used to successfully vitrify a variety of high level waste components.

ACKNOWLEDGMENTS

The authors acknowledge the financial support of the U. S. Department of Energy through contracts DOE96ER45617 (University of Missouri-Rolla) and DE-AC03-76SF00098 (LBNL). Part of this work was performed at the Stanford Synchrotron Radiation Laboratory and the Argonne National Laboratory which are operated by the Director, Office of Basic Energy Sciences, U. S. Department of Energy.

REFERENCES

[1] I. W. Donald, B. L. Metcalfe, and R. N. J. Taylor, "The Immobilization of High Level Radioactive Wastes Using Ceramics and Glasses," *J. Mat. Sci.*, **32**, (1997) 5851.

[2] S. L. Lambert and D. S. Kim, "Tank Waste Remediation System High-Level Waste Feed Processability Assessment Report", PNNL #WHC-SP-1143, PNNL, Hanford, WA, 1994.

[3] M. J. Kipfer et al. "Best Basis for Total Inventories of Chemicals and Radionucides in Hanford Site Waste", interim report, PNNL #WHC-SD-WM-TI-740, PNNL, Hanford, WA, 1996.

[4] G. K. Marasinghe, M. Karabulut, C. S. Ray, D. E. Day, M. G. Shumsky, W. B. Yelon, C. H. Booth, P. G. Allen, and D. K. Shuh, "Structural Features of Iron Phosphate Glasses," *J. Non-Cryst. Solids*, **222** (1997) 144.

[5] G. K. Marasinghe, M. Karabulut, C. S. Ray, D. E. Day, C. H. Booth, P. G. Allen, and D. K. Shuh, "Redox Characteristics and Structural Properties of Iron Phosphate Glasses: A Potential Host Matrix for Vitrifying High Level Nuclear Waste," Ceramic Transactions, **87** (1998) 261.

[6] T. M. Hayes and J. B. Boyce, in *Solid State Physics*, edited by H. Ehrenreich, F. Seitz, and D. Turnbull (Academic, New York, 1982), Vol. 37, p. 173.

[7] G. G. Li, F. Bridges and C. H. Booth, Phys. Rev. B, 52, (1995) 6332.

[8] Y. S. Badyal, M-L. Saboungi, D. L. Price, D. R. Heaffner, and S. D. Shastri, *Europhys. Lett.*, **39** (1997) 19.

[9] A.J.G. Ellison, R.K. Crawford, D.G. Montague, K.J. Volin & D.L. Price, *J. Neut. Research*, 1993, **1**(4), 61

[10] C. S. Ray, X. Fang, M. Karabulut, G. K. Marasinghe, and D. E. Day, "Iron Redox Equilibria and Crystallization of Iron Phosphate," *These Proceedings*.

[11] A. Mogus-Milankovic, B. Pivac, K. Furic, and D. E. Day, "Structural Study of Iron Phosphate Glasses," *Phys. Chem. Glasses*, **38**, (1997) 63.

[12] C. H. Booth et al., In preparation.

[13] D. L. Price, A. J. G. Ellison, M-L. Saboungi, R.-Z. Hu, T. Egami, and W. S. Howells, "Short-, Intermediate-, and Extended-Range Order in Rubidium Germanate Glasses," *Phys. Rev. B*, **55**, (1997) 11249.

[14] J. H. Albering and W. Jeitschko, "Crystal Strucutre of Diuranium Monoxide Diorthophosphate, $(U_2O)(PO_4)_2$," *Zeitschrift fuer Kristallographie*, **210**, (1995) 878.

HIGH Cr_2O_3 REFRACTORY CORROSION IN OXIDIZING MELTER FEEDS: RELEVANCE TO NUCLEAR AND HAZARDOUS WASTE VITRIFICATION

C. M. Jantzen, K.G. Brown, K.J. Imrich, and J.B. Pickett
Westinghouse Savannah River Co.
Savannah River Site
Aiken, South Carolina 29808

ABSTRACT

Monofrax K-3 refractory is a robust refractory that has been selected to line both High Level Waste (HLW) and Low Level Mixed Waste (LLMW) melters throughout the DOE complex. Monofrax K-3 is tolerant of transition metal oxides but is highly reduced in nature. The corrosion of the reduced high Cr_2O_3 containing Monofrax K-3 refractory in highly oxidizing waste melter feeds was, therefore, studied. Testing of the corrosion resistance of the K-3 refractory was performed in sealed crucibles starting with slurried melter feed. The sealed crucible testing where the K-3 coupon was immersed in melter feed was used preferentially to the ASTM C621 refractory corrosion test which immerses the coupon in pre-reacted glass. The sealed crucible melter feed testing was performed so that the refractory coupon would be exposed to the oxidizing or reducing species being released during the feed to glass conversion, e.g. NO_3^-, NO_2^- ,CO_2, CO, O_2 which may impact refractory performance. The corrosion rates measured in highly oxidizing (high nitrate) feeds were approximately 1.2-1.6 times greater than those determined using the ASTM C621 test protocol. Confirmatory corrosion rates were measured on K-3 coupons immersed in oxidizing feed in a 1/100th scale Defense Waste Processing Facility (DWPF) pilot scale melter. Calculations show that high nitrate oxidizing feeds, which produce highly oxidized melts, act as the driving force for attack of the highly reduced Monofrax K-3. Monofrax K-3 corrosion products are insoluble in the borosilicate waste glass tested. Hence, the corrosion products have the potential to settle to the melter floor and accumulate if they are not eliminated from the melt pool during pouring. For the DWPF melter, such accumulation would still allow the DWPF to operate well beyond (>5 years) its design basis of 2 years.

INTRODUCTION

Monofrax K-3 refractory has been used to line the High Level Waste (HLW) melters at the Savannah River Site (SRS), e.g. DWPF, and at West Valley Nuclear Services (WVNS). In addition, Monofrax K-3 is the refractory lining Duratek's Duramelter 5000 at the SRS M-Area facility where high nitrate LLMW waste is being vitrified. Monofrax K-3 is also being considered for Hanford's HLW and Low Activity Waste (LAW) melter design. Previous scale melter testing of the Monofrax K-3 refractory at the Savannah River Technology Center

(SRTC) had been performed using only reducing feeds containing minimal concentration of oxidizers (nitrate and nitrite). Currently, highly oxidized feeds are being vitrified in the DWPF (Table I). In addition, oxidizing feeds are being vitrified in the SRS M-Area facility and in the WVNS HLW melter. Highly oxidized HLW and LAW wastes will also be vitrified at Hanford. The relative oxidizing nature of the wastes at the SRS, WVNS and Hanford are given in Table I in terms of their nitrate and nitrite concentrations. Vitrification of highly oxidized feeds initiated the re-evaluation of the K-3 refractory corrosion reported in this study.

Table I Comparison of the Concentrations of Sodium and Oxidizers in Sludge and Salt Supernate Wastes at SRS and Hanford

Description	Molar NO_3^- Concentrations	Molar NO_2^- Concentrations
SRS M-Area LLMW Sludge[a]	0.68	---
SRS HLW DWPF	~0.40-0.69	---
Hanford LAW Salt Supernate[b]	3.2	1.7
Hanford HLW[c]	0.18(max)	calculated as total NO_3^-
West Valley HLW	2.4	---

[a] J.B. Pickett, Vendor Treatment Specification X-SPP-M00001, [b] WHC-SD-ER-498, Rev. O, [c] TWRS Privatization Contract No. DE-RP06-96RL13308

BACKGROUND

High Cr_2O_3 refractories are commonly used in steel manufacturing, in the commercial glass industry, and in the waste glass industry [1-5]. The corrosion mechanisms by which high Cr_2O_3 refractories degrade are outlined by Muan and Osborn [3]. The causes include conditions of fluctuating temperature, fluctuating oxygen partial pressure (pO_2), and poor resistance to attack of the refractory by molten iron oxide in the steel, slag, or glass being melted.

The dominant phase in high Cr_2O_3 refractories is a complex solid solution of various spinels. The spinel phases are mixtures of the four end member spinels $MgO{\bullet}Al_2O_3$, $MgO{\bullet}Cr_2O_3$, $FeO{\bullet}Cr_2O_3$, and $FeO{\bullet}Al_2O_3$ and some SiO_2.[53] One mechanism by which high chrome refractories can degrade is by some component in the refractory combining with some component in the material being melted and forming a low melting eutectic liquid [3]. These low melting eutectics can allow the molten slag or glass and refractory to react at temperatures as low as 1083°C [3]. A second mechanism by which chrome refractories can degrade, such as on melter roofs and walls, involves spalling or peeling off of layers on the order of an inch in thickness [3]. The peeling off of layers is believed to be due to a "bursting expansion" accompanying iron oxide absorption by the spinel grains of the chrome spinel in the refractory [3]. When iron oxide goes into solid solution in the spinel, swelling occurs as the porosity is increased.

Testing of fuse-cast chrome-alumina refractory has been carried out in both reducing and oxidizing glasses ($Fe^{+2}/\Sigma Fe$ of 0.24 to 0.71) in the fiberglass industry [6]. The refractory was tested under static and rotating conditions [6] in a borosilicate melt at a temperature of 1316°C. Under static conditions, the glass adjacent to the refractory became saturated with the refractory components and further dissolution was retarded, i.e. the saturated layer was protective of the refractory. During melt rotation or stirring, the saturated protective layer was removed from the refractory wall and replaced by fresh glass melt. The fresh glass melt in contact with the refractory caused increased corrosion. The results indicated that the addition of a reducing agent to fiberglass compositions could decrease the wear rate of fuse-cast refractories.

Miller and Steggs [6] determined that the Al_2O_3 phase of the Monofrax K-3 brick was much more vulnerable to solution in glass than the chrome (Cr_2O_3) phase. This is in agreement with the findings of Muan and Obsorn [3] for the degradation of Cr_2O_3 refractories in molten calcium silicate slags. Miller and Steggs also identified the Monofrax K-3 corrosion mechanism in DWPF type glasses as being a strong function of the nickel oxide and iron oxide in the glass where Ni and Fe in the glass exchange for Mg and Al in the refractory. A Ni-Cr-Fe spinel is formed as a reaction product [1].

Corrosion deposits containing ~30 volume percent spinel formed on the walls of the Project 1941 melter (2/5 DWPF scale) during testing at SRTC with reducing feeds . The formation of this layer was attributed to reaction of the refractory with the glass. This layer was viscous and appeared to remain on the wall of the melter refractory protecting it from further corrosion. Decreased melt viscosity (which caused increased melt convection) was found to remove the protective layer. The reaction zone was studied and considered typical of zones found on Monofrax K-3 in the glass industry and in waste glass melters at the Pacific Northwest National Laboratory (PNNL) [7-8].

Additional Monofrax K-3 corrosion studies were recently conducted at the PNNL [5]. Waste glasses with Fe_2O_3 concentrations varying from 12.6 to 15.5 wt% were tested in the presence of an oxidizer (HNO_3) and a reductant (glycol as a substitute for formic acid/formate salt reductants). Accelerated K-3 corrosion was observed in the highly oxidized feeds. Elevated concentrations of Al_2O_3, Cr_2O_3, MgO, and MoO_3 were observed in the bottom drain and melt surface samples. The concentrations of these constituents increased with increased processing time.

Research at SRTC using the ASTM C621 test protocol [9] measured an overall corrosion rate and defined this corrosion rate as "the average loss of material". Subsequently, SRTC researchers [10] reported two corrosion rates. Rankin [10] delineated an "average loss of material" and a "total penetration depth." Rankin observed that there were two reaction mechanisms which occurred between the specimen and the molten glass: (1) a "corrosion" reaction mechanism which resulted in a change in the dimensions of the specimen and (2) "selective penetration" of a corrosion front into the material forming a new reaction layer.

EXPERIMENTAL

A piece of Monofrax K-3 refractory from Carborundum was analyzed by X-Ray Diffraction (XRD) and Scanning Electron Microscopy (SEM) to determine its phase chemistry. The K-3 redox was analyzed by dissolution in H_2SO_4/HF in the presence of NH_4VO_3 [11] followed by colorimetric analysis for Fe^{2+} and total Fe expressed as ΣFe.

A K-3 brick was sectioned into three replicate rectangular coupons about 1" by 1" by 0.5." The K-3 coupons were immersed in high nitrate oxidized feeds (0.694M NO_3^-, 0.0086M NO_2^- , 0.4M $COOH^-$) [1]in high purity Al_2O_3 crucibles. that were sealed with a gel. The refractory coupons stayed immersed in the feed/glass during the fusion of the feed so that general corrosion rates could be measured. The crucibles were heated at 5°C per minute until a melt temperature of 1150°C was reached. Crucibles were held at 1150°C for various times, e.g. 4, 24, and 120 hours. At the end of the designated time the crucibles were removed from the furnace and air cooled.

The cooled Al_2O_3 crucibles were sectioned and a segment of the K-3 embedded in the glass was removed and mounted in epoxy for examination by SEM. In this manner the corrosion rate of the refractory in the glass and any corrosion products formed at the K-3/glass interface could be examined. K-3 was also tested in an identical fashion in black frit 165 as a control standard for comparison to the ASTM C621 crucible studies performed by SRTC researchers in 1978-1985 [9-10]. The sealed crucible type of corrosion testing provides more information about the mechanisms of refractory corrosion than the ASTM C621 procedure which merely measures a dimensional loss of material.

Refractory corrosion coupons made from K-3 were also immersed in the SRTC 1/100th DWPF scale mini-melter. Corrosion coupons were immersed during operation with the DWPF nitric acid flow sheet (0.616M NO_3^-, <0.002M NO_2^- , 0.853M $COOH^-$). During these campaigns there was virtually no cold cap (<10 %) based on visual observation and video films. Samples were hung half way into the glass melt pool on platinum wires so that the corrosion depth could be measured at the melt line and in the plenum (vapor space).

RESULTS

Chemical and Phase Analysis of Monofrax K-3 Refractory

To better understand the corrosion potential of the Monofrax K-3 refractory, especially the susceptibility to oxidizing or reducing feeds, the chemical and phase composition of the K-3 refractory was determined. The measured redox ratio, expressed as $Fe^{+2}/\Sigma Fe$, of the K-3 was measured in duplicate and found to

---

1 Note that NO_3^- is a three times stronger oxidizer than formate is a reductant (see Brown, et. al., this proceedings

be 0.939 (94% reduced) and 0.917 (92% reduced). This indicates that the Monofrax K-3 refractory is highly reduced and that 5.4 to 5.5 wt% of the 5.9 wt% Fe_2O_3 reported by the Carborundum Co. is actually FeO.

Analysis by XRD and SEM confirmed that the Monofrax K-3 brick was composed of a solid solution of Al_2O_3-Cr_2O_3 (major phase), $MgAlCrO_4$ (major phase) $FeCr_2O_4$ (minor phase), $FeAlCrO_4$,(minor phase), SiO_2 (minor phase), TiO_2 (minor phase), FeO (minor phase), and metallic Fe° (trace). The Fe° is used as a solid state buffer to maintain the chromium in the K-3 as Cr_2O_3 per conversations with Carborundum Co. Coupling the chemical analysis of Monofrax K-3 with the phase analysis allowed the direct comparison of the K-3 refractory composition to the composition and the known degradation mechanisms of other high Cr_2O_3 refractories [3].

Corrosion of Monofrax K-3 Refractory in DWPF Oxidized Feeds

The dual reaction layers observed by Rankin [10] were observed in all of the Monofrax K-3 coupons tested in this study. Where applicable, both measurements are given. It should be noted that, based on the refractory corrosion study of Cooper and Nicholson, [6] glass flow can cause the outer reaction layer to spall off exposing the penetration layer as the next reaction surface. Therefore, "average loss of material" depths may underestimate refractory corrosion in highly convective, highly stirred, or bubbled melts. "Total penetration" depths may be overly conservative in non-agitated, low-flow environments.

The corrosion of Monofrax K-3 in 165 black frit was used as a control standard since the corrosion rate of K-3 had been measured by Corning (Table II) using the ASTM C621 test protocol in this same glass. The as received Monofrax K-3 refractory was exposed to DWPF 165 black frit in sealed crucibles for 5 days at 1150°C. The control gave a measured corrosion rate corresponding to an "average loss of material" of 2.0×10^{-5} m/day (0.79 mils/day). The value of 2.0×10^{-5} m/day (0.79 mils/day) is in good agreement with the 1.7×10^{-5} m/day (0.68 mils/day) and the 2.0×10^{-5} m/day (0.79 mils/day) reported by Corning and is in agreement with the tests of 131 average waste glass which have a measured corrosion rate of 1.6×10^{-5} m/day (0.64 mils/day) [10] (see Table II). Likewise, the total penetration depth was measured at 5.0×10^{-5} m/day (1.97 mils/day). Total penetration data was not reported for the 165 black frit testing done by Corning (Table II). The total penetration depth of 5.0×10^{-5} m/day (1.97 mils/day) measured for K-3 penetration depth in black frit 165 in this study corresponds well with the 4.9×10^{-5} m/day (1.92 mils/day) total penetration depth measured for 131 glass (Table II) [10] which is of similar composition. The comparative corrosion rates for the control standards are shaded in Table II.

The corrosion mechanism is shown in this study to be a depletion in the Al_2O_3 and MgO components of the K-3 brick, which leaves a corroded layer enriched in Cr_2O_3. X-ray mapping indicated that the Cr_2O_3 component in the corroded layer reacts with Fe_2O_3 and NiO in the waste glass as noted previously [1, 3, 6] to form a Ni-Fe-Cr rich K-3 corrosion product.

Figure 1 shows the growth of crystalline K-3 corrosion products, which volume expand (a mechanism suggested by Muan and Osborn [3] but never visualized). These crystalline masses delineate the interface between the corrosion front and the penetration front (SEM in Figure 1). EDAX analyses indicate that the bulk Monofrax K-3 (Region 1) is rich in Al and Cr. Region 2 lies between the penetration front and the corrosion front (defined by the row of expanded protrusions of crystallized material) and is enriched in Al and Cr but depleted in Mg and Fe. The region between the corrosion front and the glass interface becomes progressively more enriched in Cr and Fe over Mg and Al (Regions 3 and 4) until there is no matrix left to hold the Cr and Fe rich deposits remaining onto the refractory. At this point the Cr rich spinel corrosion products, which appear to be insoluble in the melt, can spall off into the glass melt (Region 6).

The depletion of the refractory in Mg is responsible for the formation of the refractory corrosion product, krinovite ($NaMg_2CrSi_3O_{10}$) found by XRD analysis of deposits of the 1/10th DWPF scale melter floor and in the 1/100th DWPF scale mini-melter deposits. The decomposition of high chrome-alumina refractories into magnesium silicates and $MgO \cdot Cr_2O_3$ is predicted by the known phase equilibria in the MgO-Cr_2O_3-SiO_2 system [3]. In addition, analysis conducted at SRTC on samples of the glass/K-3 interface from the sidewall of a 1/100th DWPF scale mini-melter indicated that the spinel reaction product was $Ni(Cr_{0.8}Fe_{0.2})_2O_4$.

By knowing the approximate composition of the K-3 corrosion products, it can be shown from thermodynamic calculations that the $Ni(Cr_{0.8}Fe_{0.2})_2O_4$ spinel K-3 corrosion product that forms is thermodynamically more stable at a melt temperature of ~1150°C than the $MgAlCrO_4$ spinel composing the bulk of the K-3 refractory. Using data from Barin [12] at 1400°K (~1130°C) and writing a partial reaction for Fe_2O_3, NiO, and SiO_2 in a glass reacting with $MgAlCrO_4$, the major spinel phase in the K-3 refractory, to form a $NiCrFeO_4$ spinel reaction product, plus $MgO \cdot SiO_2$ (proto-enstatite),[2] SiO_2 (cristobalite)[2] or $Al_6Si_2O_{13}$ (mullite)[2] [3] yields Equation 1. Writing similar partial reaction for the reaction of the reduced iron containing spinels (minor components) and FeO (minor component) in the K-3 refractory, reacting with NiO and SiO_2 and O_2 in the melt to form the same $NiCrFeO_4$ spinel and mullite yields Equation 2. These reactions indicate that the decomposition of the K-3 into $NiCrFeO_4$ spinel and silicates is favored at the melt temperature of ~1150°C. Reaction 2 is extremely favored under oxidizing melt conditions.

$$\underset{\text{(major K-3 phase)}}{6MgAlCrO_4} + \underset{\text{(melt)}}{3Fe_2O_3} + \underset{\text{(melt)}}{6NiO} + \underset{\text{(melt)}}{8SiO_2} \rightarrow \underset{\text{(observed+predicted chrome refractory corrosion products)}}{6NiCrFeO_4 + 6MgO \cdot SiO_2 + Al_6Si_2O_{13}} \quad (1)$$
$$\Delta G_{fm}(\text{products}) - \Delta G_{fm}(\text{reactants}) = -1759 \text{ kJ / mole}$$

---

2 $MgO \cdot SiO_2$ and SiO_2 or mullite reaction products and $MgO \cdot Cr_2O_3$ from the refractory further react with Na_2O in the glass to form the silicates like krinovite, ($NaMg_2CrSi_3O_{10}$)

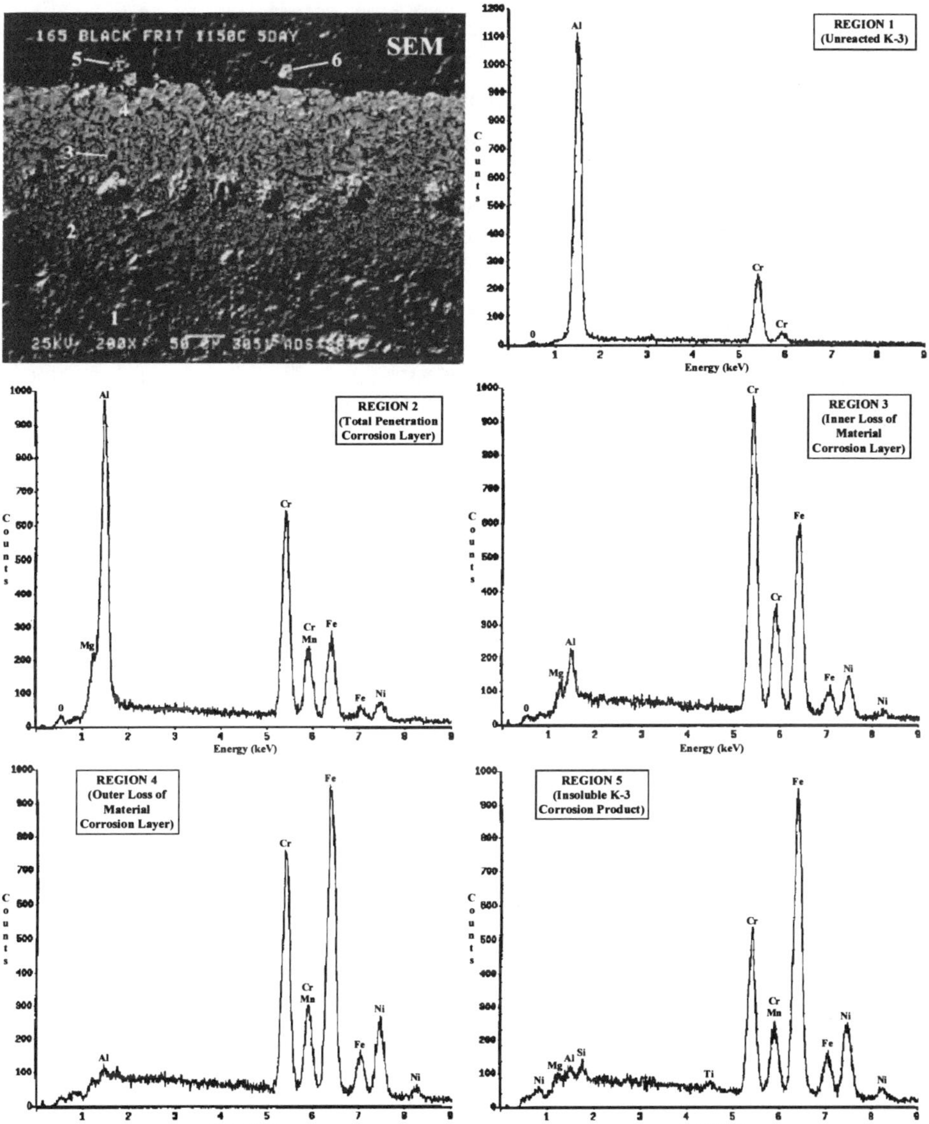

Figure 1. Scanning Electron Micrograph of the K-3 refractory corrosion coupon immersed in glass composed of DWPF black frit 165 for 5 days at 1150°C. Note the thick corrosion layer which is composed of two layers separated by a row of circular precipitates. The outer layer is more porous and is designated the "average loss of material" corrosion layer (Regions 3 and 4). The inner layer is termed the penetration layer (Region 2). Region 1 represents unreacted K-3. Note the corrosion reaction product (Region 5 and 6) that has broken off from the corrosion layer and appears to be insoluble in the glass.

$$6FeAlCrO_4 + 2FeCr_2O_4 + 2FeO + 10NiO + 2SiO_2 + 2\tfrac{1}{2}O_2 \rightarrow 10NiCrFeO_4 + Al_6Si_2O_{13} \quad (2)$$

(K-3 phase) (K-3 phase) (K-3 phase) (melt) (melt) (melt) (observed+predicted chrome refractory corrosion products)

$$\Delta G_{fm}(\text{products}) - \Delta G_{fm}(\text{reactants}) = -3128 \text{ kJ / mole}$$

The dual corrosion zones were observed in all of the Monofrax K-3 crucible studies and in the 1/100th scale mini-melter refractory coupon tests in this study. It was difficult to tell exactly where the melt line was on the refractory coupon immersed in the mini-melter due to splatter from the melt pool (there was <10% cold cap coverage during all of the campaigns). However, measurements on this coupon in the vicinity of the melt line indicated that the ratio of the two corrosion depths, e.g. the "average loss of material"/"total penetration," was the highest of any of the K-3 coupons tested in this study (Table II), e.g. the outer corrosion layer was the largest fraction of the total penetration depth. The top of this same Monofrax K-3 coupon had been exposed to the mini-melter vapor space and exhibited a lower "average loss of material"/"total penetration" ratio (Table II).

Table II. Comparison of Monofrax K-3 Corrosion Rates in This Study with Previous ASTM C621 Measurements (Temperature = 1150°C)

Physical Location of Measurement	Time (Days)	Frit	Simulated Waste Type or Glass	Avg. Loss Mat'l. (mil/day)	Total Pene-tration (mil/day)	Ratio Loss/ Pene-tration	Ref.
Crucible Melt Line	7	131	TDS Average Glass	0.64	1.92	0.333	10
Crucible Melt Line*	7	165	Black Frit Glass	0.73	ND	ND	CELS
Crucible Melt Line*	7	165	Black Frit Glass	0.23	ND	ND	CELS
Crucible Melt Line**	7	165	Black Frit Glass	0.68	ND	ND	CELS
Crucible Glass Contact†	5	165	Black Frit Glass	0.79	1.97	0.40	This study
Crucible Glass Contact-1†	5	200	Tank 51A, Batch 1A, Simulated High Nitrate Feed	1.57	2.36	0.66	This study
Crucible Glass Contact-2†	5	200	Tank 51A, Batch 1A, Simulated High Nitrate Feed	1.57	3.15	0.50	This study
Crucible Glass Contact-3†	5	200	Tank 51A, Batch 1A, Simulated High Nitrate Feed	1.57	1.97	0.80	This study
1/100th DWPF Mini-melter, Melt Line†	3	200	Tank 51A, Simulated High Nitrate Feed	2.29	2.62	0.87	This study
1/100th DWPF Mini-melter in Vapor Space†	3	200	Tank 51A, Simulated High Nitrate Feed	0.52	0.87	0.60	This study

* As cast K-3 surface; ** Ground K-3 surface; † Diamond cut K-3 surface, CELS = Corning Engineering Laboratory Services, Corning, NY; the comparative corrosion rates for the control standards are shaded as discussed in the text.

Measuring the total corrosion as a function of the thickness of both of the outer and inner reaction zones indicates that the corrosion rates measured in oxidized feeds were 1.2-1.6 times greater than those measured previously (Table II). However, the measured corrosion rates were less than the 3, 5.4 and 7.5 mils/day allowable DWPF Design Basis for the melter bottom, sidewall, and melt line. The corrosion of the Monofrax K-3 replicates were measured as a function of time for three different time periods, e.g. 4, 24, and 120 hours. Plotting of the data provides linear rate equations expressed as $\partial d/\partial t$, where d is depth in mils and t is time in days (see Figure 2). The linearity of the K-3 corrosion as a function of time and the observation that the corrosion products spall off into the surrounding glass in semi-static crucible studies indicates that a protective or passivation layer may not be forming on the K-3 refractory when the melter is processing highly oxidizing feeds.

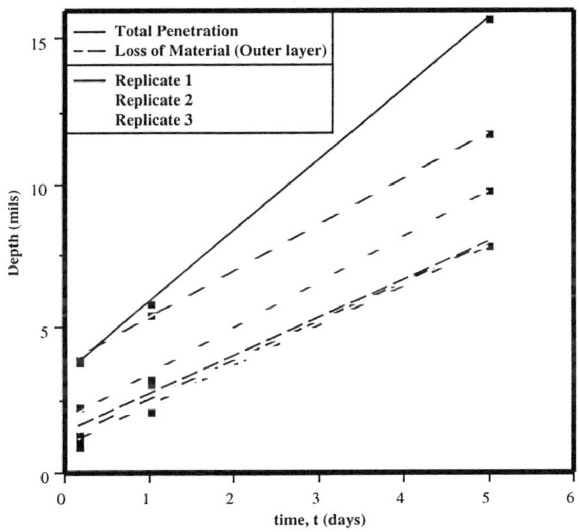

Figure 2. Linear Corrosion Rates of K-3 Refractory measured as depth of attack (mils) versus time in hours for replicates 1, 2, and 3.

Proposed Accelerated Corrosion Mechanisms for Monofrax K-3 Refractory

The data presented above on the corrosion of the Monofrax K-3 in highly oxidized DWPF feeds suggest that the presence of feed additives changes the melt redox which, in turn, impacts the relative corrosion rates of the reduced K-3 refractory. Under the DWPF high nitrate oxidizing feed/melt conditions, the highly reduced Monofrax K-3 is thermodynamically unstable (Equation 2). The oxidizing conditions destabilize both the reduced Fe^{+2} and possibly Cr^{+3} species[3] in the Monofrax K-3 brick which accelerates the Monofrax K-3 corrosion.

---

3 The chemical reactions controlling the redox of melter feed will be discussed in another paper.

CONCLUSIONS

Monofrax K-3 corrosion was determined to be incongruent or heterogeneous. Mechanistically, two different corrosion layers were observed to form on the K-3 surface in contact with molten glass: (1) an outer layer of poorly attached reaction products designated as the region of "average loss of material" and (2) an inner reaction layer designated as the region of "total penetration." Measuring the total corrosion as a function of both reaction layers indicates that the corrosion in oxidized DWPF feeds is ~1.2-1.6 times greater than those measured previously using the ASTM C621 protocol. However, the measured corrosion rates are less than the 3, 5.4 and 7.5 mil/day allowable DWPF Design Basis for the melter bottom, sidewall, and melt line, respectively. Reaction of Monofrax K-3 refractory with molten glass produces $Ni(Cr_{0.8}Fe_{0.2})_2O_4$ spinels which are insoluble in the glass. Corrosion rate data suggests that oxidizing feeds inhibit the accumulation of a spinel rich protective reaction layer at the K-3/glass interface. Therefore, the $Ni(Cr_{0.8}Fe_{0.2})_2O_4$ spinels can spall off, agglomerate in the melt pool, and may ultimately accumulate on the melter floor.

ACKNOWLEDGMENTS
This paper was prepared in connection with work done under Contract No. DE-AC09-96SR18500 with the U.S. Department of Energy.

REFERENCES

1. D.C. Iverson and D.F. Bickford, "Evaluation of Materials Performance in a Large-Scale Glass Melter After Two Years of Vitrifying Simulated SRP Defense Waste," Sci. Basis for Nucl. Waste Mgt., VIII, Mat. Res. Soc., Pittsburgh, PA, 839-845 (1985).
2. S.A. Degterov and A.D. Pelton, "Thermodynamic Calculation of Glass/Slag/Refractory Equilibria in Coal Gassification," Corrosion of Materials by Molten Glass, G.A. Pecoraro, et.al. (Eds), Ceram. Trans., V. 78, Am. Ceram. Soc., Westerville, OH, 91-103 (1996).
3. A. Muan and E.F. Osborn, "Phase Equilibria Among Oxides in Steelmaking," Addison-Wesley Publishing Company, Inc., Reading, MA, 236pp. (1965).
4. M.J. Plodinec, "Vitrification Chemistry and Nuclear Waste," J. Non-Cryst. Solids, 84, 206-214 (1986).
5. G.L. Smith et. al., "Effect of Feed Chemistry on Waste Melter Vitrification Kinetics," U.S. DOE Report, PVTD-T3C-95-129, Pacific Northwest Laboratory, Richland, WA (Sept. 1995).
6. S.A. Cooper and P.S. Nicholson, "Influence of Glass Redox Conditions on the Corrosion of Fusion-Cast Chrome-Alumina Refractories," Ceramic Bulletin, 59[7], 715-717 (1980).
7. R.D. Dierks, "The Design and Performance of a 100-kg/hr, Direct Calcine-Fed Electric Melter System for Nuclear Waste Vitrification," U.S. DOE Report PNL-3387, Richland, WA (1980).
8. S.M. Barnes and D.E. Larson, "Materials and Design Experience in a Slurry-Fed Electric Glass Melter," U.S. DOE Report PNL-3959, Richland, WA (1981).
9. G.G. Wicks, "Compatibility Tests of Materials for A Prototype Ceramic Melter for Defense Waste Products," Ceramics in Nuclear Waste Management, U.S. DOE Report CONF-790420, U.S. Dept. of Energy, Natil. Tech. Info. Serv., Springfield, VA 82-85 (1979).
10. G.G. Wicks, "Corrosion of Melter Materials," U.S. DOE Report DPST-79-580, E.I. duPont deNemours & Co., Savanannah River Laboratory, Aiken, SC (May, 1980).
11. E.W. Baumann, "Colorimetric Determination of Iron (II) and Iron (III) in Glass," Analyst, v. 117, 913-916 (1992).
12. I. Barin, "Thermochemical Data of Pure Substances, Parts 1 and 2," VCH Publishers, New York, NY, 1739pp (1989).

CORROSION OF SELECTED REFRACTORIES BY IRON PHOSPHATE MELTS

Fuyi Chen and Delbert E. Day*
Department of Ceramic Engineering
and the Graduate Center for Materials Research
University of Missouri-Rolla
Rolla, MO 65409-1170

ABSTRACT

The dynamic corrosion rate of six commercial refractories was measured in four iron phosphate melts and one borosilicate melt at temperatures ranging from 1000 to 1300°C. A dense alumina and a chrome refractory had the lowest melt-line corrosion rate in iron phosphate melts, whereas, the corrosion rate for silica, zircon, and alumina-zirconia-silica (AZS) refractories was somewhat higher. In general, the corrosion rate for the alumina and chrome refractory in the iron phosphate melts was no higher than their corrosion rate in a borosilicate melt now used at the Savannah River Site (DWPF). For the chrome refractory, the corrosion rate in three iron phosphate melts containing simulated waste was under 0.1 mm/day at the melt line. It is concluded that commercially manufactured, dense alumina or chrome refractories should be acceptable for melting many iron phosphate waste compositions, even in wastes containing up to 16 weight percent soda.

INTRODUCTION

Vitrification is being used to immobilize high level nuclear waste (HLW) in a stable, chemically durable borosilicate glass.[1-5] In the waste vitrification process, the glass melt is contained in a refractory-lined furnace. The high-temperature melt dissolves the HLW but also corrodes the refractory. Knowledge of the corrosion resistance of refractories to melts containing HLW is of considerable importance to the vitrification technology.

A borosilicate glass is being used to vitrify HLW at the Savannah River Site in Aiken, SC. and by West Valley Nuclear Services at West Valley, NY.[6,7] Borosilicate glasses have a good chemical durability, but may not be suitable for all HLW compositions, especially those wastes containing significant phosphates, halides and heavy metals (Bi, U, Pu).

Iron phosphate glasses are being studied[8-10] as an candidate for vitrifying selected HLW, since iron phosphate glasses can have a chemical durability equal to borosilicate glass, a low melting temperature (950-1200°C), low melt viscosity (1-5 poise), and short melting time (1-4 h). However, at the present time, little is known about the corrosion of refractory oxides by iron phosphate melts

There are few published reports on the corrosion of refractories by phosphate glasses. Rouse[11] compared the corrosion of high alumina crucibles by soda-lime, phosphate and aluminosilicate melts and found that phosphate melts severely corroded the crucible between 1000 and 1300°C. Mel'nikova[12] studied the refractories used in pot furnaces and found that a zircon refractory was best suited for melting phosphate-based optical glasses. Golovin[13] measured the corrosion of zircon, quartz and aluminosilicate refractories in lead phosphate glass and also found that zircon had the lowest corrosion rate between 800 and 1000°C.

In the present study the chemical corrosion of various refractories by iron phosphate melts was determined as a function of melt composition and temperature. The objective was to identify refractories that could be used for melting iron phosphate glasses in a Joule melter.

EXPERIMENTAL PROCEDURES

Preparation of Glass and Refractory Samples.

Table I gives the batch composition and melting temperature of four iron phosphate simulated wasteforms and the borosilicate glass wasteform melted in the Defense Waste Processing Facility (DWPF) at Savannah River. Batches weighing 300 grams were prepared of each composition from reagent grade chemicals and melted in a 250 ml alumina crucible in an electric furnace. All components, except the P_2O_5, were mixed thoroughly and then heated to 160°C for 2h and then to 550°C for 5h in a furnace, for the purpose of excluding water. Just prior to melting, the P_2O_5 was mixed with the calcined batch. Each batch was charged 3 to 4 times between 900°C and 1020°C and melted for about 4 h at the temperature listed in Table I. Each melt was quenched in a steel mold and cooled to room temperature in air. The quenched material was analyzed by X-ray diffraction (XRD) to ascertain that it was vitreous. It was then stored in a sealed container until used.

Table I. Batch Composition* (wt%) of the Iron Phosphate and Borosilicate (DWPF) Glasses

Sample	F43	C112[a]	T111[b]	TFB[c]	DWPF[d]
Fe_2O_3	43.0	10.0	30.0	24.5	14.2
P_2O_5	57.0	54.5	45.5	45.4	3.3 Li_2O
Al_2O_3	--	2.0	1.8 La_2O_3	2.0 Nd_2O_3	4.6
Bi_2O_3	--	--	10.4	2.0	2.7 K_2O
CaO	--	8.0	1.2	1.3 CeO_2	1.2
Mn_2O_3	--	4.8 NiO	3.6	--	1.9
Na_2O	--	4.5	2.4	16.4	11.5
SiO_2	--	1.2	4.0	1.4	47.9
UO_2	--	15.0	1.2	7.0 CaF_2	8.5 B_2O_3
Tm	1200°C	1200°C	1100°C	1000°C	1200°C

(a) iron phosphate wasteform containing 50 wt% of simulated waste from tank C112 at Hanford, WA, (b) iron phosphate wasteform containing 35 wt% of simulated waste from tank T111 at Hanford, WA, (c) iron phosphate wasteform containing 30 wt% of average simulated waste from tank farm B (TFB) at Hanford, WA, (d) borosilicate glass wasteform containing 28 wt% of simulated waste, melted in DWPF at Savannah River Site, SC. *Compositions may not total 100% since they contain small percentages of oxides too numerous to list. See ref. 10 for exact composition.

Cylindrical rods, about 10 to 15 mm in diameter and 50 mm in length, of each of the commercially manufactured refractories listed in Table II were tested in the iron phosphate melts listed in Table I. The cylindrical rods were either purchased directly from the manufacturer or were core-drilled from commercially prepared blocks.

Table II. Refractories Investigated

Refractory Type	Trade Name	Major Composition (wt%)	Bulk Density (g/cm^3)	Porosity (%)
Silica	Fusillac-C[a]	SiO_2(68.7), Al_2O_3(29.5)	2.52	20.5
Zircon	ZS1300[b]	ZrO_2(65.0), SiO_2(32.1), TiO_2(1.2)	4.33	0.5
AZS-1	333[a]	Al_2O_3(73), ZrO_2(11), SiO_2(15)	2.79	20.5
AZS-2	Z-Alac[a]	Al_2O_3(69), ZrO_2(20), SiO_2(10.7)	2.95	21.0
Alumina	AD998[c]	Al_2O_3(99.8)	3.92	0
Chrome	C1215z[b]	Cr_2O_3(91.2), ZrO_2(3.0), TiO_2(3.0)	4.33	13.0

a) Product of Emhart Refractories, Laclede Christy Clay Products; b) Product of Corhart Refractories; c) Product of Coors Porcelain

Refractory Corrosion Measurements

A dynamic corrosion procedure, see Figure 1, was used for the corrosion tests. A cylindrical refractory rod was connected to one end of an alumina rod using Pt/Rh wire. The other end of the alumina rod was fastened in a chunk attached to the drive shaft of a variable speed motor. Once the melt was at the desired temperature, the motor and refractory sample were lowered until the lower portion of the refractory rod was immersed in the melt and the top of the rod extended about 10 mm above the melt. The refractory rod was rotated in the melt at 9.2 rpm. At the end of the corrosion test period, the refractory rod was raised clear of the melt, held above the melt inside the furnace for 2h to allow any adhering melt to drip back into the crucible, and then transferred to an another furnace for slow cooling. After cutting the corroded rod along its longitudinal cross section, the diameter of the corroded rod was measured at the melt line (ML) and at a distance halfway between the melt line and the bottom of the rod. The corrosion rate(CR) in units of mm/day, was calculated from the following equation:

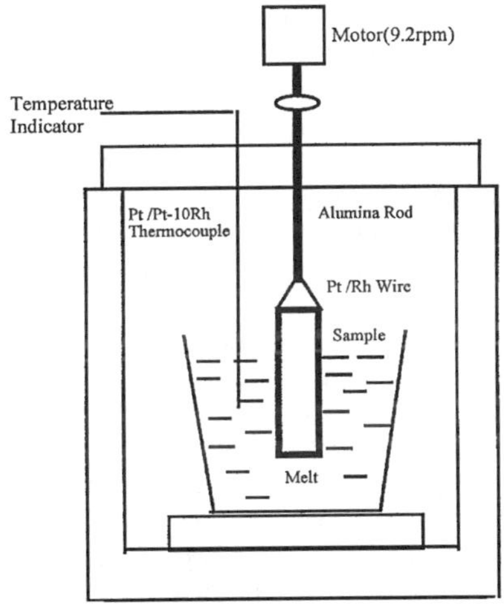

Figure 1. Schematic of the apparatus for measuring dynamic corrosion of a refractory in a high temperature melt.

$$CR(mm/day) = [D - \tfrac{1}{2}(d_1 + d_2)] / 2 \cdot t \qquad (1)$$

where D [mm] is the diameter of the cylindrical rod at the melt line at the beginning of the test, d_1 and d_2 [mm] are the diameter of the two halves of the sectioned rod at the melt line or in the melt at the end of the test, and t is the duration of the test (days). The diameter of the rod was measured several times and the average was used for d_1 and d_2 in Eq(1). The error in the corrosion rate was estimated at less than 5%, based on measurements made on four duplicate rods.

The commercially manufactured refractories listed in Table 2 were measured because of their potential application as melt-contact refractories in a high temperature melter.

RESULTS AND DISCUSSION

Corrosion Rate for Various Refractories.

In a preliminary scouting experiment, all six refractories were tested in the F43 iron phosphate melt at 1270°C for 24h. A Pt/20%Rh crucible was used instead of an alumina crucible. The corrosion rate for all of the refractories are compared in Figure 2. The horizontal line, labeled DWPF, denotes a corrosion rate of 0.33 mm/day which was used in designing the refractory lining in the DWPF melter at the Savannah River Site.[14]

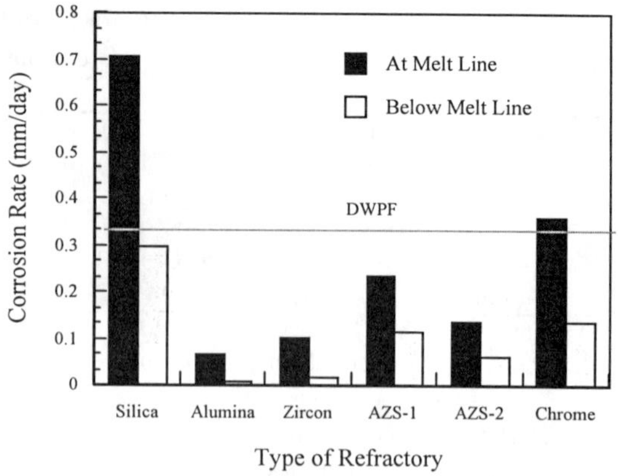

Figure 2. Corrosion Rate of Refractories at 9.2 rpm in F43 Melt for 24h at 1270°C. The horizontal line labeled DWPF is the corrosion rate used in the design of the DWPF melter.

It should be noted from Table I that the F43 melt is normally melted at 1200°C rather than at 1270°C as used for the measurements given in Figure 2. Even at this higher than needed temperature, four of the refractories have a corrosion rate below 0.33 mm/day. The most corrosion resistant materials include the alumina, zircon and AZS-2 refractory while the moderately resistant materials were the chrome and AZS-1 refractory. The silica refractory had the lowest corrosion resistance at the melt line.

The four more resistant refractories, i.e., the alumina, zircon, AZS-2 and chrome refractory, were selected for additional testing in the more realistic C112 waste melt in Table I, which contains many of the metal oxides and other elements typically found in HLW. Figure 3 shows the corrosion rate at the melt line in the C112 melt and F43 melt at 1270°C. These wastes would normally be melted at ~1200°C, but the higher temperature of 1270°C was used to accelerate the corrosion. The higher corrosion rate of the AZS-2 refractory in the C112 melt indicates that AZS bonded refractories may not be suitable for direct contact with iron phosphate melts.

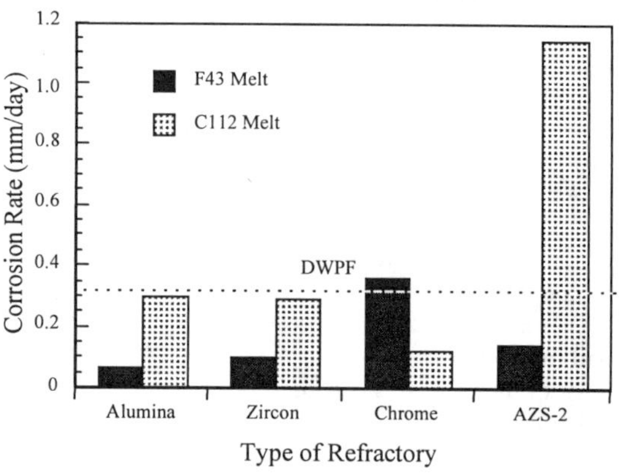

Figure 3. Melt-line Corrosion Rate of Refractory in F43 and C112 melts at 1270°C for 24h using a rotation speed of 9.2 rpm.

It is understandable from mass transport and microstructural principles that refractories with large amounts of internal interfaces between different phases and grain boundaries will normally dissolve faster than single phase refractories. An interface or grain boundary provides a path for ions to diffuse from the melt into the refractory and vice versa. The single phase refractories, namely, zircon, alumina and chrome which had the lower corrosion rates in the C112 melt were chosen for further study.

Effect of Melt Composition on Corrosion Rate.

The melt-line corrosion rate of the alumina, zircon and chrome refractory in the iron phosphate and in the borosilicate melt, now being used in the DWPF melter, are compared in Figure 4. The composition of each melt is given in Table II. In Figure 4, the corrosion rate for each refractory has been measured at that temperature that would be used normally to melt each waste, see Table II.

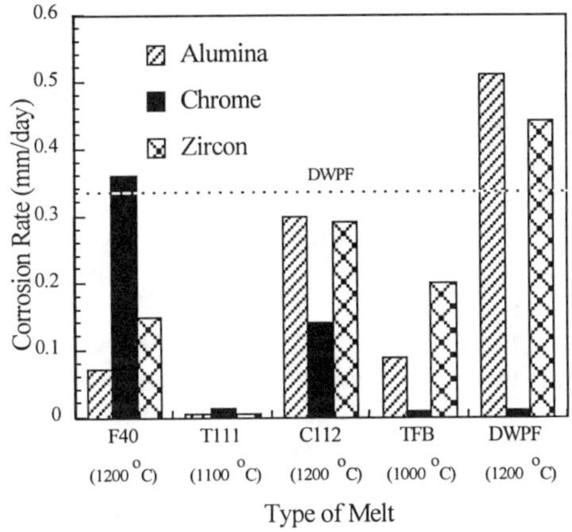

Figure 4. Corrosion rate of refractories in different iron phosphate and a borosilicate melt (DWPF) at the temperature where waste would be melted.

The data in Figure 4 illustrate several important factors about the corrosion of these refractory oxides by iron phosphate melts. The first noteworthy result is that the corrosion rate of all three refractories in these iron phosphate melts is generally low, and in the case of the T111 melt, barely measurable. While the corrosion rates of the alumina and zircon refractories in the TFB melt are higher than in the T111 melt, the values are much lower than anticipated in view of the high soda content (16.4%wt) of the TFB melt. Because of its higher soda content, the TFB melt was expected to be more corrosive than it was found to be. A second point is that the chrome refractory had the lowest corrosion rate in all but the F43 melt which contained only Fe_2O_3 and P_2O_5. Finally, the corrosion rates of all three refractories in the iron phosphate melts were lower than the 0.33 mm/day value used for the DWPF.

Phosphate melts are generally assumed to be more chemically aggressive toward refractory oxides than silicate or borosilicate glasses. However, the data in Figure 4 show that the DWPF borosilicate melt corroded the alumina and zircon refractories more than these iron phosphate melts.

Effect of Melt Temperature on Corrosion Rate

The iron phosphate compositions which have been investigated to date typically melt at temperatures from as low as 950°C up to about 1200°C. In some cases, the corrosion rate at the normal melting temperature is too low to be measured in 24h or 48h of testing, see Figure 4. Consequently, the corrosion rate for the alumina, chrome and zircon refractory was measured from the normal melting temperature to about 200 °C higher.

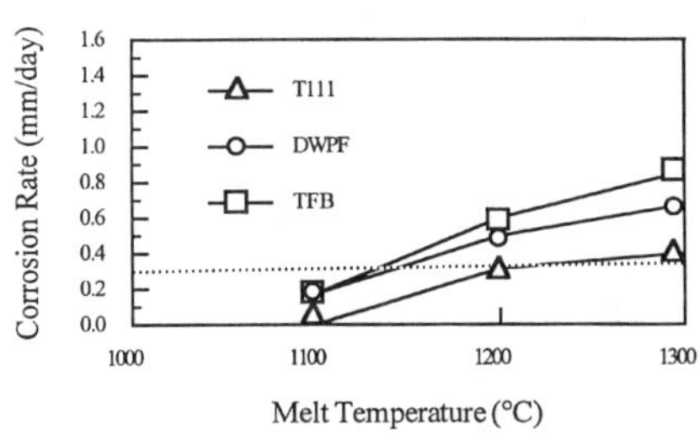

Figure 5. Temperature dependence of melt-line corrosion rate of alumina refractory in iron phosphate and a borosilicate (DWPF) melt. Dotted horizontal line in this figure and figures 6 and 7 denote corrosion rate used in design of DWPF melter.

Figures 5 through 7 show how the corrosion rate varies with temperature for the alumina, zircon and chrome refractories, respectively. As expected, the corrosion rate increased with temperature. In most cases, the corrosion rate increased by a factor of about 4 per 100°C. However, the corrosion rate of the chrome refractory in the T111 melt increased by a factor of about 10-15 (see Figure 7), but in the TFB and borosilicate melts there was hardly any increase with temperature.

Effect of Time upon Corrosion Rate

Two experiments were performed to determine how the corrosion rate varied with time. The zircon and chrome refractories were corroded in the T111 melts for 24h and 48h with a new refractory rod and fresh melts used for each test. The results in Figure 8 indicate that the corrosion rate decreased slightly with the time.

In another test, the zircon refractory was corroded in the TFB melt for 24, 72, and 240 hours with the melt contained in a dense alumina crucible. A separate refractory rod was used for each test period, but the same melt was used throughout. The results in Figure 9 show that the corrosion rate decreased significantly from 24 to 240h. The composition of the melt after the corrosion test was analyzed by EDS and the concentration of silica increased from 1.4 wt% in the starting melt to 3.2 wt% in the 240h melt. The decrease in corrosion rate could be explained by the dissolution of the some portion of the zircon ($ZrSiO_4$) refractory rod which then lowered the corrosion rate of subsequent zircon rods in the TFB melt.

CONCLUSION

The present results, although limited, indicate that commercially manufactured alumina and chrome refractories should be acceptable for melting iron phosphate wasteforms of variable composition. This conclusion is based on the observation that the corrosion rate of a chrome refractory in the iron phosphate melts was essentially the same as the corrosion rate of the same refractory in the borosilicate glass being produced in the DWPF.

The corrosion rate of each refractory varied with the composition and temperature of

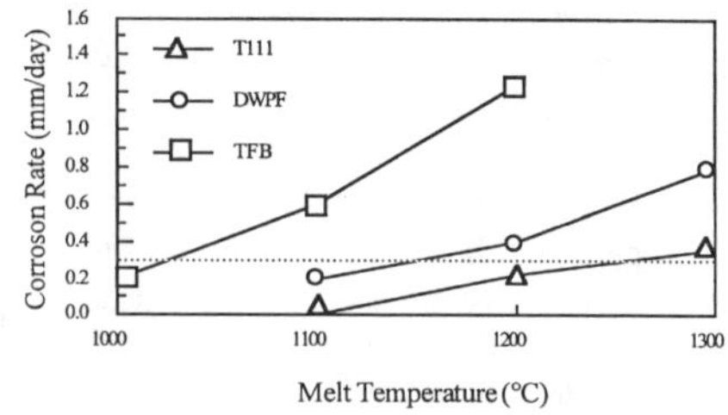

Figure 6. Temperature dependence of melt-line corrosion rate of zircon refractory in iron phosphate and a borosilicate (DWPF) melt.

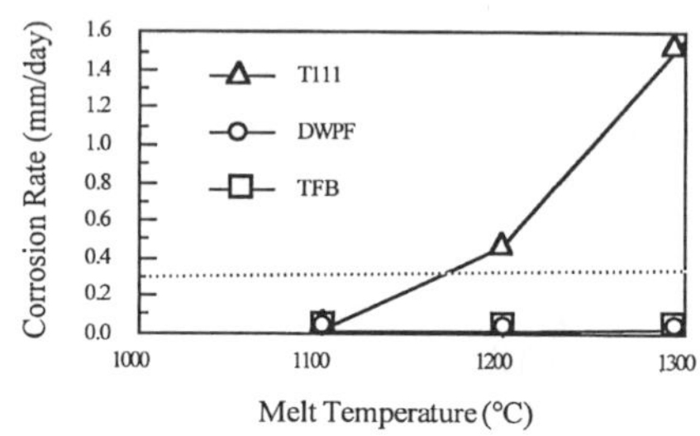

Figure 7. Temperature dependence of melt-line corrosion rate of chrome refractory in iron phosphate and a borosilicate (DWPF) melt.

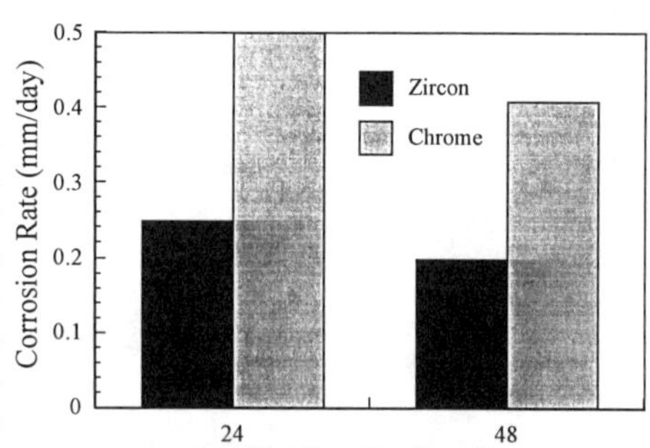

Figure 8. Melt-line corrosion rate of zircon and chrome refractories in the T111 melt at 1200°C for 24h and 48h.

each iron phosphate melt. However, the compositional dependence was not as large as might be expected for the wide difference in soda content, from zero in the F43 melt to 16.4 wt% in the TFB waste. At the temperature where each iron phosphate wasteform composition would normally be melted, the corrosion rate for the alumina, chrome and zircon refractory was below 0.33 mm/day.

Previous work has shown that glassy iron phosphate wasteforms have an unusually high chemical durability in comparison to most other phosphate glasses. The corrosion rates from the present study indicate that another distinguishing feature of the iron phosphate glasses is their relatively low tendency to corrode common refractory oxides for reasons that are not completely clear at this time. The present results confirm earlier qualitative observations that alumina and zircon refractories are not easily wet by iron phosphate melts.

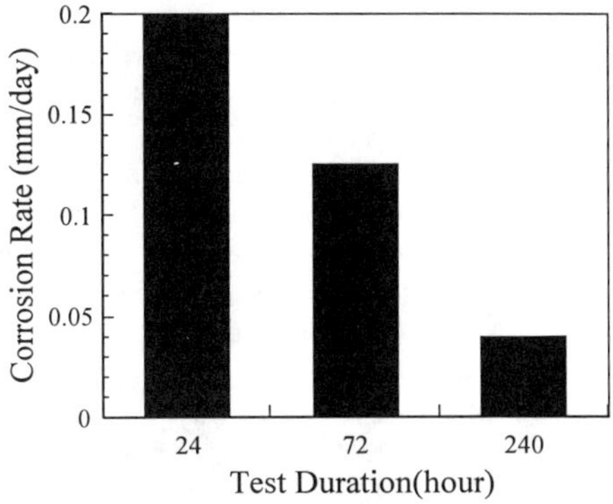

Figure 9. Melt-line corrosion rate of zircon refractory in the TFB melt at 1000°C.

ACKNOWLEDGMENT

The authors acknowledge the partial financial support of this work by the Department of Energy, contract DE-FG07-96ER45618, and the China Scholarship Council, People's Republic of China.

REFERENCES

[1] C. Jantzen, N. Bibber, D. Beam, C. Crawford and M. Picketed,"Development and Characterization of the Defense Waste Processing Facility (DWPF) Environmental Assessment Glass Standard Reference Material," p 313-22 in Ceramic Transactions, Vol 39, Environmental and Waste Management Issues in the Ceramics Industry, American Ceramic Society, Westerville, OH, 1994.

[2] D. Kim, P. Hama, D. Lamar and M. Elliott, "Development of High-Waste Loaded High-Level Nuclear Waste Glasses for High-Temperature Melter," p 39-48 in Ceramic Transactions, Vol. 45, Environmental and Waste Management Issues in the Ceramics Industry, American Ceramic Society, Westerville, OH, 1995.

[3] W. Rankin, "Advances in Ceramics," p 559-566 in Nuclear Waste Management I, Vol. 8, American Ceramic Society, Wesleyville, OH, 1980.

[4] F. Lifanov, S. Stefanovsky, A. Kobelev, A. Savkin, O. Knyazev, T. Lashchenova, S. Merlin, P. Roux," Vitrification of Simulated Intermediate-Level French and Russian Wastes in Cold Crucible Based Plant," p 163-171 in Mat. Res. Soc. Symp. Proc., Vol. 412, Materials Research Society, 1996.

[5] D. Bennert, T. Overcamp, D. Bickford, C. Jantzen and C. Cicero, "Demonstration of Vitrification of Surrogate F006 Wastewater Treatment Sludges," p 73-90 in Ceramic Transactions, Vol 45, Environmental and Waste Management Issues in the Ceramics Industry, American Ceramic Society, Westerville, OH, 1995.

[6] J. C. Marra, T. J. Overcamp and D. L. Erich, "Melter Component Performance During Surrogate Low-level Mixed Waste Vitrification Campaigns," p 241-251 in Ceramics Transactions, Vol 41, Corrosion of Refractory by Molten Glass, American Ceramic Society, Westerville, OH, 1996.

[7] H. Matzke and E. Vernaz, "Thermal and Physicochemical Properties Important for the Long-Term Behavior of Nuclear Waste Glass," J. Nuc. Mat., **201,** 295-309 (1993).

[8] D. E. Day, Z. Wu, and C. Ray, "Chemically Durable Glasses for Vitrifying Phosphate Containing Simulated Nuclear Wastes," Final Technical Report, Contract No 276822-A-F1, Pacific Northwest National Laboratories (1995).

[9] X Yu, D. E. Day, and G.J. Long, "Properties and Structure of Sodium-Iron Phosphate Glasses," J. Non-Cryst. Solids, **215,** 21-31 (1997).

[10] M.G. Mesko, D.E. Day, and B. C. Bunker "Immobilization of High-Level Radioactive Sludges in Iron Phosphate Glass," J. Non-Cryst. Solids (in press).

[11] Cilon G. Rouse, Samuel M. Toffoli, Alvaro B. Lopes, "Glass Corrosion Studies of a High-Alumina Refractory," Adv. Fusion Glass, 561-568 (1988).

[12] I. G. Mel'nikova, I.V. Razdol'skaya, E. M. Dolivo-Dobrovol'skaya, T. A. Nesterova, "Physical and Technological Properties of an Zircon Ceramics," Glass and Ceramics (English Translation of Steklo i Keramika) , **44** [1-2] 26-29 (1987).

[13] A.I. Golovin, V.K. Pavlovskij, O.S.Shchavelev, "Corrosion Resistance of Ceramics Refractories and Platinum to Lead-Phosphate Glasses," Steklo i Keramika. **49** [11-12] 21-22 (1992).

[14] R. Schumacher, Personal communication, Savannah River Labs.

EVALUATION OF HYDROGEN GAS GENERATION FROM OXALIC ACID IN CONTACT WITH THE CARBON STEEL OF A HIGH-LEVEL WASTE STORAGE TANK

H.D. Smith*, R.L. Russell, G.K. Patello, Pacific Northwest National Laboratory, PO Box 999, Richland, WA 99352-0999.

ABSTRACT

Following an initial vitrification phase at the West Valley Demonstration Project (WVDP), residual tank waste consisting of cesium-loaded zeolite and a mixture of neutralized plutonium-uranium extraction (PUREX) sludge, thorium extraction (THOREX) waste, and cesium-loaded zeolite in Tanks 8D-1 and 8D-2 may require additional processing to facilitate removal from the tanks. A proposed treatment involves using oxalic acid to dissolve solids and mobilize the radioactivity. Hydrogen may be generated when oxalic acid comes in contact with the carbon steel storage tanks. Laboratory-scale testing at the Pacific Northwest National Laboratory (PNNL) evaluated the potential for hydrogen generation under simulated tank conditions using carbon steel coupons to mimic the tank wall. Results are presented that show carbon steel (A516) corrodes in 8 wt% oxalic acid solution producing a ferric oxalate complex and hydrogen. Future research will investigate the influence of waste components, zeolite, and temperature.

INTRODUCTION

The residual waste in Tanks 8D-1 and 8D-2 may require treatment to dissolve the waste and to transfer the radionuclides to the vitrification plant. Oxalic acid has been proposed as a reagent for the elution of cesium from zeolite in Tank 8D-1 and for the dissolution of sludge and mobilization of radionuclides in Tank 8D-2. The effects of oxalic acid on hydrogen generation during removal of residual Tank 8D-1 and 8D-2 wastes and during the vitrification process need to be quantified to support a safety analysis that must be conducted prior to implementation of the proposed treatment.

Bryan, et al., 1997[*] have reviewed the literature for hydrogen generation during the corrosion of carbon steels. Additional studies [2][3][4] have investigated the potential for corrosion of the WVDP carbon steel tanks. These studies, using carbon steel coupons submerged in oxalic acid,

---

[*] Bryan, et al., 1997, *Evaluation of Hydrogen Generation from the Addition of Oxalic Acid to the West Valley Demonstration Project Vitrification Process.* WVSP 98-03, West Valley Demonstration Project.

---

have shown a substantial corrosion rate (3.8 mm per year at pH$\leq$ 1). Enhanced hydrogen generation due to iron corrosion of tanks used to store high-level nuclear waste has also been studied under conditions existing at the Hanford Site [5][6]. Devine [5] estimated that the corrosion rate of tank carbon steel under highly alkaline conditions (i.e., pH $\geq$13) is about 0.013 mm per year. This would produce about 170 L per day of hydrogen gas in a Hanford tank such as 241-SY-101. Bryan and Pederson [6] suggested that further hydrogen could be produced by the catalytic action of iron oxide causing a thermolysis of hydrogen-containing species. Smith, et al., 1997* showed that significant hydrogen gas is produced by the reaction between carbon steel and oxalic acid solution.

Based on the fact that the only gas given off in significant amounts indicating an oxidation-reduction reaction was hydrogen (some carbon monoxide was also observed, but that would not involve oxidation of iron), the most likely reactions involving oxalic acid and carbon steel are as follows:

1) $$Fe^0 + 2H^+ \rightarrow Fe^{+2} + H_2\uparrow$$

2) $$2\,Fe^0 + 6\,H^+ \rightarrow 2\,Fe^{+3} + 3\,H_2\uparrow$$

2a) $$Fe^{+3} + 3\,C_2O_4^{-2} \leftrightarrow Fe(C_2O_4)_3^{-3}$$

2b) $$2\,Fe(C_2O_4)_3^{-3} \rightarrow 2\,FeC_2O_4\downarrow + 3\,C_2O_4^{-2} + 2\,CO_2\uparrow$$

Equation 2a indicates the formation of a soluble ferric oxalate complex. This complex is recognizable by its emerald green color in solution. The ferric oxalate complex photocatalytically decomposes to ferrous oxalate (a yellow precipitate), oxalic acid, and carbon dioxide (Equation 2b) over a period of hours or days, depending on the radiation intensity in the visible range [7]. The reactions denoted by Equations 2a and 2b can effectively increase the corrosion rate because the iron first goes into a soluble (ferric) state which carries it away from the corrosion interface (preventing passivation) before reduction to the ferrous state which forms an insoluble precipitate. In addition, the decomposition of oxalic acid might occur at higher temperature, yielding carbon dioxide and hydrogen or water, carbon dioxide, and carbon monoxide as follows:

3) $$2\,H^+ + C_2O_4^{-2} \rightarrow H_2\uparrow + 2\,CO_2\uparrow \ \underline{\text{or}}\ H_2O\uparrow + CO_2\uparrow + CO\uparrow$$

Since hydrogen is an observed product of oxalic acid corrosion of carbon steel, the objective of this study was to determine the time dependence of hydrogen production from carbon steel in contact with an oxalic acid solution. The results (i.e., hydrogen generation rates) are quantified for use in the safety analysis required for the proposed treatment. Also, the effects due to corrosion products on the surface of the steel are evaluated, since the tank steel is known to be covered with a corrosion scale believed to be a hydrated iron oxide. Hence, tests were run with oxalic acid solution in contact with clean and with prefilmed (corroded) carbon steel samples made of steel typical of the WVDP waste tanks.

---

*Smith, HD, GL Smith, RL Russell, and GK Patello., 1997, *Fiscal Year 1997 Laboratory Scale Studies of the WVNS Flowsheet for Remediation of Tank 8D-1 and Tank 8D-2,* Pacific Northwest National Laboratory, Richland, WA.

EXPERIMENTAL PROCEDURES

The hydrogen gas generation experiments were conducted under the following nominal test conditions: ~ 200 mL of 8 wt% oxalic acid (deaerated, continuous stirring, and a vessel volume of 400 mL. Deaeration consisted of sparging the oxalic acid solution with argon until low, constant levels of air components were observed in the argon sweep gas (i.e., ≤ 50 ppm O_2) while the coupon was in a raised position. The argon sweep gas delivery tube was then raised above the level of the solution and the carbon steel coupon was submerged into the oxalic acid solution which was at 50°C. Five tests were run using rectangular coupons with a 0.25 inch hole, three tests on clean coupons (visibly free from corrosion products and contaminants) and two on coupons prefilmed (yellow-brown, semi-transparent film) with iron oxide. The prefilmed coupons of A516-carbon steel had been produced by exposure to tank vapor space conditions (air above a solution of sodium salts of nitrite, nitrate, and chloride) at 50°C for a period of six months. Each test involved continuous (two hour) offgas monitoring of an argon sweep gas for hydrogen, carbon monoxide, carbon dioxide, and nitrous oxide, using a gas chromatograph (GC), and an NO_x analyzer. Estimated detection limits in these tests were:

hydrogen	15 ppm
carbon monoxide	~150 ppm
carbon dioxide	50 ppm
nitrous oxide	50 ppm
nitric oxide and NO_x	5 ppm

The head space gas was sampled and tested after one week of being held at 50°C with constant stirring.

It is believed that the three tests using clean carbon steel coupons represent worst-case conditions for producing maximum hydrogen because there is no chemical barrier between the iron metal and the acid. After it was demonstrated that hydrogen was observed and could be related to the observed coupon corrosion rate, two tests were performed using prefilmed carbon steel coupons in the same environment.

RESULTS AND DISCUSSION

The results of the three baseline tests of hydrogen generation from clean steel, and the two tests on oxide film-coated steel are presented below. Hydrogen generation is calculated from the observed rates of hydrogen production and compared with the values calculated stoichiometrically from the observed iron losses in the steel test coupons. The results are also compared with values reported in the literature.

Hydrogen Generation from Clean Steel

Three baseline tests (Tests 1, 2, and 3) were run using cleaned carbon steel coupons in 8 wt% oxalic acid. The hydrogen generation profiles are shown in Figure 1. Hydrogen was observed immediately in the first of the three tests when condensate from the 50°C oxalic acid solution was observed to form on the coupon while it was still in the raised position. When the coupon was lowered into the oxalic acid solution, accelerated hydrogen generation was observed. The hydrogen generation rate peaked and then declined over a two-hour period to about half the peak value. Over the two-hour period, the total amount of hydrogen produced in Test 1 was 1.46 mmol as determined by integrating the hydrogen concentration profile.

In the next two tests, the coupons were suspended higher in the vessel during the deaeration of the acid solution to reduced formation of condensate on the coupon. This was observed to reduced the hydrogen produced prior to immersion in the oxalic acid solution to below the hydrogen detection limits. When these coupons were submerged in the oxalic acid solution, hydrogen generation commenced immediately and peaked at a value about twice that observed during the first test. However, the generation rate declined to about one-eighth of this peak value over the remainder of the two hours. The hydrogen generation profiles for the second and third tests are nearly identical, as shown in Figure 1. The total amount of hydrogen produced in Tests 2 and 3 were 1.24 and 1.28 mmol respectively. The results of these tests are summarized in Table I.

After each sealed reaction vessel was held for one week in a water bath at 50°C, the gas composition of its plenum was measured. It was expected that hydrogen would accumulate if the reaction had continued during the week. As shown in Table I, after one week all tests except Test 1 were still producing detectable hydrogen, and considerable hydrogen had accumulated in the plenum of each vessel. It was assumed, based on the results of Tests 2 and 3, that the hydrogen in the Test 1 vessel had escaped through inadequate seals since it was known that hydrogen was still being produced after two hours when it was removed from the offgas system. The amount of hydrogen involved is not known.

The hydrogen concentration in the vessel plenum was difficult to measure because of the sweep gas flow rate and the length of the analytical cycle time of the GC. Based on the shape of the hydrogen profiles observed, the bulk of the plenum gases were purged quickly from the plenum (~ 4 to 5 minutes) and since the GC samples the gas about every 97 seconds, it is likely that the peak hydrogen concentration was missed. It was also observed, from the amounts of nitrogen and oxygen measured for each vessel, that some air had leaked back into the vessels. It is believed, therefore, that the amount of hydrogen measured is less than the amount actually produced (i.e., either the peak concentration was not sampled by the GC or some hydrogen had leaked out or both). A combination of these factors could explain the difference in the plenum hydrogen concentration measured for Test 2 and Test 3, 7450 ppm compared to 106,000 ppm, even though the tests appeared to be almost identical during their first two hours. Plenum hydrogen concentration was essentially zero for Test 1, indicating that after a week, any hydrogen that may have accumulated in the plenum had leaked away, and that hydrogen generation was now too low to be measured by the offgas system.

The total oxalic acid contact time for each coupon at 50°C ranged from 168 to 190 hours. (The specific length of time was not the important factor, only that the length of time be known.) The metal coupons were cleaned after test completion using an inhibited HCl solution and the weight loss of each coupon was measured. These weight changes were converted to estimated moles of hydrogen generated, based on the stoichiometry of Equation 2 above. Equation 2 was used because the bright green solution color indicated that ferric oxalate had formed. Other reactions were not considered for this calculation. The amount of hydrogen generated during the tests was also estimated from an average hydrogen

Table I. Initial and Final Hydrogen Generation Rates in ppm from Carbon Steel in 8 wt% Oxalic Acid Solution

Test #	H_2 Generation Rate ppm by Volume			Est. Average H_2 Generation Rate During One Week	Measured H_2 Concentration in Plenum After One Week
	Max.	After 2 hours	After One Week		
1	240	100	< DL	50	< DL
2	380	60	18	39	7450
3	360	37	75	56	106000
4	210	140	40	90	3400
5	200	135	28	82	2160

< DL = less than detection limit

generation rate computed from the measured hydrogen generation rate after two hours and that measured after one week. The results shown in Table II indicate good agreement between the amount of hydrogen produced and the amount of hydrogen that would be predicted based on the amount of iron consumed and assuming that the reaction indicated by Equation 2 occurred.

Table II. Carbon Steel Coupon Weight Loss

Test #	Coupon Weight Loss (g)	Moles of Iron Lost	Moles of H_2 Based on Iron Loss	Moles of H_2 Based on Generation Rates
1	0.73	0.013	0.02	0.022
2	0.8687	0.0155	0.023	0.019
3	0.9218	0.0165	0.025	0.026
4	0.8417	0.0151	0.023	0.046*
5	0.704	0.0126	0.019	0.041*

* See the discussion of these values at the end of this section.

Hydrogen Generation from Prefilmed Carbon Steel

Once a baseline with clean carbon steel coupons was established, two tests were conducted on prefilmed carbon steel coupons. The initial hydrogen generation rates from the prefilmed carbon steel coupons in 8 wt% oxalic acid are shown in Figure 2. The spikes in this figure are due to the release of hydrogen bubbles that accumulated in the stirring vortex under the flat horizontally oriented coupons. This apparently did not occur with the clean coupons because they were smaller and were positioned more to the side of the stirring vortex. The hydrogen generation profiles from these two tests appear almost identical. Unlike the clean coupon tests that showed a hydrogen generation peak with the rate tailing off, the hydrogen generation rate in these tests quickly rose to between 0.010 and 0.015 mmoles/minute and then remained fairly constant throughout the tests. Test 4 produced 0.93 mmol of hydrogen, while Test 5 produced 0.89 mmol when both are integrated over the first 151 minutes of these tests. (Note that ppm is converted to mmoles or mmoles/min. by dividing by 22.4 L/mol (STP) and multiplying by 1000 since the gas flow rate is 1 L/min.). These values indicate that the test gives reproducible results.

It appears that the oxide film (rust) blunts the attack of the oxalic acid on the steel. This could happen for two reasons. One, the oxide film does form a physical barrier that must be removed by ferric oxide reacting with the oxalic acid, forming the complex already mentioned, and not producing hydrogen. Second, the complexation of the ferric iron will momentarily reduce the effective concentration of oxalic acid in the locality of the steel surface. Both of these factors will initially retard the corrosion reaction as is observed.

Tests 4 and 5 showed less hydrogen accumulation after one week than the clean coupon tests, as indicated in Table I. However, the gas chromatogram showed a peak in nitrogen and oxygen levels coincident with the hydrogen peak, indicating that the vessel seal was not complete.

Tests 4 and 5 were still generating hydrogen at the end of the week, as were Tests 2 and 3. The hydrogen production estimated from the measured generation rates for Tests 4 and 5 (the rate after ~2 hrs and the rate after ~ one week averaged) was approximately twice the amount estimated from the measured iron loss. The most likely explanation for this apparent disagreement (but not based on direct evidence from these tests) is that the method of approximating the hydrogen generation rate is inappropriate for these two tests (i.e., a linear decrease of the hydrogen generation rate over the week). For instance, if the generation rate actually decreased to the final rate quickly, the estimated amount of hydrogen generated would be reduced by as much as 50%. However, a definitive reason is not known at this time.

CONCLUSIONS

Carbon steel corrodes in 8 wt % oxalic acid solution by Equation 2, producing 1.5 moles of hydrogen for each mole of iron consumed. Clean carbon steel starts corroding more rapidly than carbon steel with a light oxide film, and the rate of corrosion of clean steel decreases more rapidly from its peak rate than steel with a light oxide film, asymptotically approaching a rate that is lower than that for steel with a light oxide film. For steel with a light oxide film, the initial maximum corrosion rate peak is much more subdued than that for the clean steel.

The ferric iron complex formed in the oxalic acid solution displays a photosensitivity that results in its conversion to ferrous oxalate consistent with the textbook description of this phenomenon.

The corrosion rate indicated in the 8 wt % oxalic acid solution, calculated from iron weight loss rates over a one-week period, ranged from 1.5 to 1.8 mm/year for oxide-filmed steel to 2.2 to 2.7 mm/year for clean steel. These values appear to be consistent within a factor of 2 or 3 with the rate of 3.8 mm/year previously reported for the oxalic acid environment and are certainly much more rapid than observed for high pH conditions.

ACKNOWLEDGEMENT

This work was funded by the U.S. Department of Energy under Contract DE-AC06-76RLO 1830.

REFERENCES

[1]L.M. Cadoff, "Waste Compliance Plan for the West Valley Demonstration Project," WVNS-WCP-001, REV 3. West Valley Nuclear Services Co., West Valley, New York, 1991.

[2]Burger LL. 1988. *Potential for Radiation-Induced Corrosion in the 8D-1 Tank.* WVNS 88-069, prepared by Pacific Northwest National Laboratory, Richland, Washington.

[3]Elmore MR. 1995. *Mild Steel Corrosion in Oxalic Acid Solutions: (Simulated Cesium Removal Process Solutions).* Pacific Northwest Laboratory, Richland, Washington.

[4]Elmore MR. 1996. *Corrosion of Mild Steel in Simulated Cesium Removal Process Solutions.* PNNL-11284, Pacific Northwest National Laboratory, Richland, Washington.

[5]Divine JR et al. 1985. *Prediction Equations for Corrosion Rates of A-537 and A-516 Steels in Double-Shell Slurry, Future PUREX, and Hanford Facilities Wastes.* PNL-5488, Pacific Northwest Laboratory, Richland, Washington.

[6]Bryan SA and LR Pederson. 1993. *Hydrogen Safety Project Simulated Waste Studies: FY 1992 Annual Report.* PNL-8465, Pacific Northwest Laboratory, Richland, Washington

[7]Cowan, Dwaine O. and Ronald L. Drisko. *Elements of Organic Photochemistry*, Plenum, New York and London, 1976.

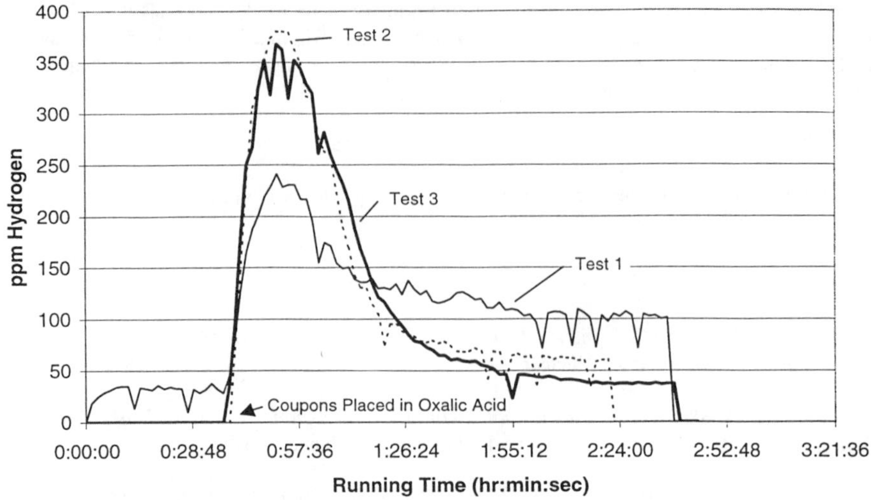

Figure 1 - Hydrogen Generation Profiles for Three Tests using Clean Carbon Steel Coupons

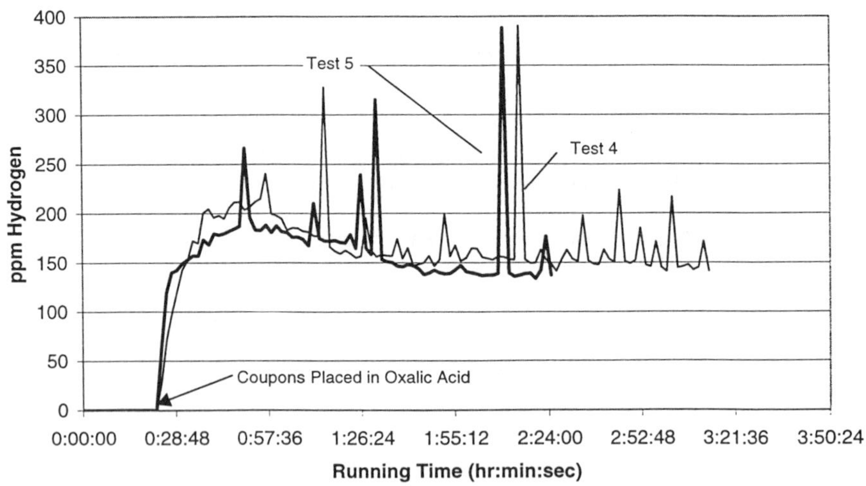

Figure 2 - Hydrogen Generation Profiles for Two Tests Using Oxide Coated Coupons

THE EVALUATION OF SINTERED SODALITE-GLASS COMPOSITES

M. C. Hash, J. P. Thalacker, D. R. Simon, G. L. Burns, and R. H. Woodman
Argonne National Laboratory
Chemical Technology Division
Nuclear Technology Program
9700 South Cass Avenue
Argonne, IL 60439

ABSTRACT

Ceramic waste forms are being developed for immobilizing the chloride waste salt generated from the electrometallurgical treatment of spent fuel. Sodalite-glass composites with high density were made using liquid-phase sintering techniques. At present, these ceramics are fabricated by hot isostatic pressing (HIPing). Sintering technology is being developed to improve productivity as compared with HIPing. Sintering of sodalite-glass composites may also be more amenable to production-scale processing by remote operations. Processing parameters for sintering were controlled to yield desirable structural ceramic properties, such as chemical, mechanical, and thermal stability. Chemical compositions and physical characteristics of raw materials were studied to determine their effect on the ceramic properties, as well as the intermediate powder characteristics. Powder processing techniques were refined to provide enhanced leach resistance of the consolidated composites.

INTRODUCTION

The electrometallurgical treatment of spent nuclear fuel developed at Argonne National Laboratory (ANL) produces two radioactive waste streams[1]. These waste streams must be disposed of appropriately. This implies that the final form or forms of the waste must be intrinsically corrosion resistant and non-dispersible. The first waste stream is a stainless-steel-rich, metal waste form[2]. The second waste stream is a chloride salt that contains salt-borne fission products. This radioactive salt is inherently difficult to dispose of since it is readily dissolved and transported in water.

The sodalite-glass ceramic waste form was developed at ANL-East (Illinois) and is fabricated by hot isostatic pressing (HIPing)[3-5]. This batch processing technique yields high-density (>99% of the theoretical density) composites with appropriate corrosion resistance and physical integrity.

The developed technology has been transferred to ANL-West (Idaho), where large-scale ceramic waste forms are being produced as part of the Electrometallurgical Treatment of Spent Nuclear Fuel Demonstration, scheduled to be completed by the summer of 1999. After this demonstration, or pilot-plant activity, ANL intends to continue treating its entire inventory of spent nuclear fuel from the experimental breeder reactor (EBR-II).

Inventory reduction operations will require higher throughput than is now possible. The current technology will require the design, purchase, and construction of a large hot isostatic press. Unfortunately, HIPing is a batch process (limiting throughput), which requires significant specimen preparation and very high gas pressures (~25,000 psi). These processing constraints are even more troublesome when working with radioactive materials. Specimen preparations require remote handling, including the transfer of powder into HIP canisters, evacuation of these canisters, and final sealing (via welding). Improperly prepared canisters yield final waste forms unsuitable for disposal and difficult to re-process.

Recent developmental activities at ANL-East have focused on consolidation of the sodalite-glass composites by liquid-phase sintering techniques. Liquid-phase sintering, which implies a multi-phase material, requires sufficient heating to cause one or more but not all, phases to exhibit viscous flow. In this case, the liquid phase is a low-viscosity glass. The viscosity of the glass is controlled by the processing temperatures; this allows for maximum densification (mass transfer) without loss of initial form[6].

The sintering technology being developed employs moderate processing temperatures, which are likely to lend well to continuous processing (tunnel kilns). This ultimate goal is to improve throughput over 1000% as compared to batch processing (HIPing or sintering).

EXPERIMENTAL

The crystalline precursor used to form sodalite is salt-loaded zeolite. Both sodalite and zeolite are sodium aluminosilicates with "cage-like" structures[7]. However, the sodalite structure is not suited for sorbing chlorides. The zeolite structure has three times the theoretical capacity and larger cage openings, making sorption possible. By conversion of salt-loaded zeolite to sodalite during consolidation, the chlorides are trapped within the resulting smaller cage structure.

The raw materials for these ceramic waste forms are commercially available. Raw zeolite and glass powders, both fine- and coarse-grained, are provided by UOP (DesPlaines, IL) and Pemco (Baltimore, MD), respectively. Fine-grained materials were nominally -325 mesh (or < 44 μm). Various sieve cuts of coarse-grained raw materials were used, ranging from -60 to +325 mesh (or 250 to 44 μm). The simulant waste salt was made by mixing reagent grade (or better) ingredients to a predicted nominal composition. The mixture, mostly LiCl-KCl eutectic, was melted and homogenized at 500°C overnight prior to quenching.

Both fine- and coarse-grained zeolite powders were first dehydrated to remove moisture, which would otherwise interfere with sorbing of the waste chloride salt.

Dehydration without loss of crystallinity was achieved by a thermal schedule which brought the powders to a maximum temperature of 525°C[8]. The loosely packed powders were heated under a flowing nitrogen purge and were immediately placed in a purified-argon atmosphere glovebox to minimize re-hydration.

The simulant waste salt used was contacted with the dehydrated (activated) zeolite in a processing step referred to as "blending". The salt and zeolite powders were heated under vacuum to a maximum temperature of 500°C. The stainless steel blending vessel, or blender, was slowly rotated during heating to provide a moderate degree of dry mixing during heat-up. Once the maximum temperature was reached, the vacuum was re-established to remove gases evolved during heating. The blender continued to be rotated while the furnace was held at 500°C for 20 hours. Once cooled, the blender was transferred into a purified argon atmosphere glovebox for handling. Premature exposure of blended zeolite to moisture can cause the salt to sweat out of the zeolite, especially when in the form of a fine powder[8].

The blended zeolite powder was then dry mixed with a desired amount of glass. Fine powders were combined in a high-velocity rotating-blade mill. Coarse powders, >44 μm, were combined by means of a conventional V-blender[8]. Glass loading was varied from 25 to 50 wt%, with a balance of blended zeolite.

Forming Methods

Two forming methods were investigated: uniaxial cold pressing and mimicked vibratory packing. Green densities were varied from 35 to 60% of theoretical density. Green densities of large-scale HIPed specimens (made at ANL-West), which employ vibratory packing, average ~35% of theoretical density when fine powders are employed. We mimicked this average initial packing and slightly enhanced packing levels by partially compacting the powders at various pressures less than 1,000 psi; uniaxial.

Specimens were either sintered as cold-pressed pellets (fully compressed at 12,000 psi; uniaxial) on stainless steel foils or as partially pressed compacts within either stainless steel, graphite, or graphite-coated stainless steel canisters (1-in. x 1-in. right cylinders).

Thermal Schedule

Specimens were heated at a variety of maximum processing temperatures, ranging from 750 to 900°C. Heating and cooling rates were held constant at 5°C/min to limit thermally induced stresses. Specimens were heated in either stagnant air or in a purified argon atmosphere glovebox.

RESULTS AND DISCUSSION

Powder Preparation

Dehydration and blending operations were enhanced by the use of coarse-grained raw materials. Our approach, in terms of raw materials, has been to start with particle sizes most appropriate for the applied consolidation techniques. This eliminates the need for particle size tailoring during the processing of radioactive components. Undesired particles of raw zeolite and glass can merely be discarded or recycled to the target particle-size range. This is not possible once the materials have been contacted with radioactive constituents.

Fine-grained raw materials were used initially in laboratory-scale specimens to enhance sintering kinetics and to limit impurities introduced when coarse-grained zeolites are used (from the proprietary binders used to granulate zeolite crystallites). Even at the laboratory-scale, it was obvious that particle sizes would need to be increased to provide free-flowing powders. The fine powder mixtures are prone to agglomerate, clump, and bridge.

Fine-grained zeolite required the use of grinding media within the blender to continuously break up agglomerates. Coarse-grained zeolite does not require this grinding, since it is free-flowing. Mixing of blended zeolite and glass is also greatly simplified using free-flowing blended zeolite and glass powders. Use of V-blending was ineffective for fine-grained materials. The resulting agglomerates required the high-velocity mixing provided by the rotating-blade mill to simultaneously break apart and evenly re-distribute the two types of particles. The coarse-grained raw materials are readily mixed in a V-blender, as established for large-scale processing.

Our powder processing includes two quality-assurance measures that can be used to qualify large-scale products: level of dehydrated zeolite moisture and amount of blended zeolite free salt.

We have recently completed a parametric study in which we made blended zeolites of various moisture contents. We concluded that moisture levels of 1 wt% or less do not adversely affect sorption of salt during blending. Higher amounts of moisture may be tolerable in terms of producing blended zeolite but will most likely lead to excessive corrosion of the blending vessel.

Similarly, various qualities of blended zeolite (or salt-loaded zeolite) were investigated. By rinsing samples of blended zeolite with deionized water for one minute at room temperature, we can quantify the amount of chlorides that are readily dissolved within the zeolite cage structure (referred to as "free salt"). Corrosion resistance data from subsequently produced waste forms showed that free salt levels of <0 to 0.05 wt% will provide appropriate precursors for our ceramic waste forms.

Glass loadings ranging from 25 to 50 wt% glass were investigated. The resulting composites were subjected to static immersion tests in deionized water at 90°C for 3 days. Release of cesium, because it is considered to be the hardest radionuclide to retain, was chosen to provide a conservative quantification of the corrosion resistance for a given waste form. Leach test data are given in Fig. 1. We found no discernible difference in corrosion resistance for sodalite-glass composites made with glass loadings ranging from 30 to 50 wt%. Specimens made at the next lowest glass loading in our study, 25 wt%, showed greatly degraded physical integrity. To provide a safety margin, we selected glass loadings at or above 35 wt% to produce composites of sufficient physical integrity. Corresponding data for HIPed composites are also shown in Fig. 1 illustrating comparable corrosion resistance resulting from the two different consolidation methods.

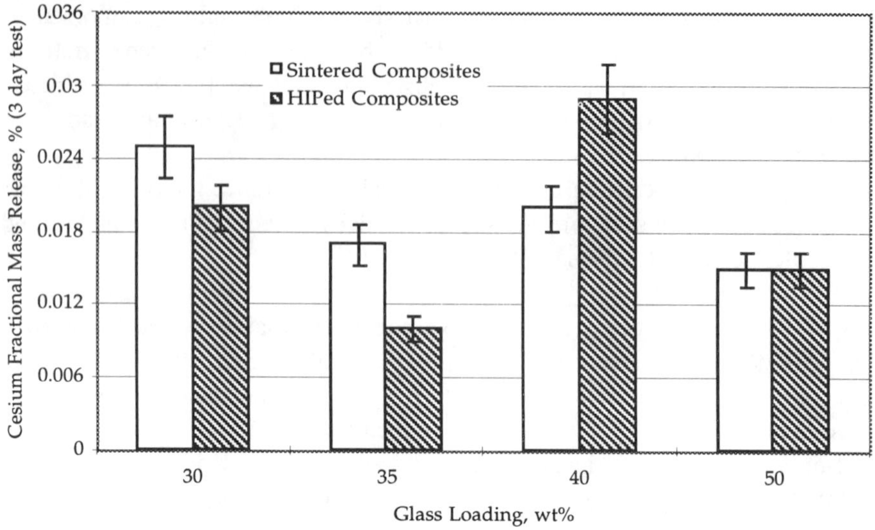

Fig. 1. Cesium Release as Function of Glass Loading in Sodalite-Glass Composites

Forming Methods

As with powder preparation, our goal in developing a forming method was to use existing capabilities derived from HIP processing. This prompted investigation of low-green-density compacts, such as those expected with vibratory packing of fine powder mixtures (~35% of theoretical density). The use of coarse-grained raw materials enhanced packing such that we could not prepare specimens at a green density of 35% of theoretical due to enhanced flow characteristics. Mere loading of the coarse-grained powder mixtures led to green densities exceeding this level without the need for compression or tapping.

As shown in Fig. 2, properly sintered sodalite-glass composites starting from green densities of 35 to 60% of theoretical resulted in comparable final bulk densities. These composites also have comparable apparent porosity (less than 1 vol%) and phase assemblage. These data are encouraging since green densities approaching 60% theoretical require uniaxial pressing. Fabrication of large-scale specimens with relatively large loading areas demands very large pressing equipment capable of providing >10,000 psi onto specimens with diameters >12 in. (this example would require more than a 500 ton press). Uniaxial pressing is also complicated by remote loading, handling of pressed pellets, and die wear.

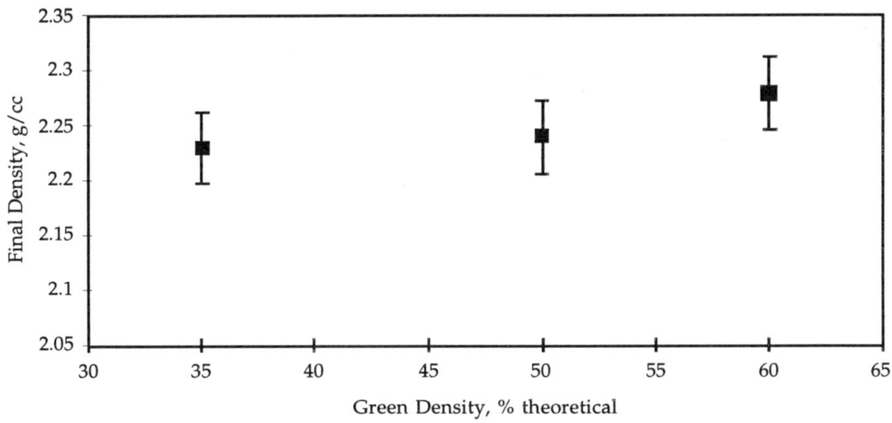

Fig. 2. Effect of Green Density on Final Density of Sodalite-Glass Composites

Thermal Schedule

Bulk density and apparent porosity data were used to compare specimens heated at a variety of maximum temperatures for various soak durations under either stagnant air or purified argon atmosphere. We concluded that high-density (99% of theoretical) corrosion-resistant (comparable to HIPed specimens) sodalite-glass composites can be consistently sintered at temperatures ranging from 825 to 875°C in either air or argon using moderate, heating and cooling rates (5°C/min). This large processing window lends itself well to continuous sinter, or tunnel kiln, technology.

CONCLUSIONS

We have developed a potential alternative for HIPing in the consolidation of sodalite-glass ceramic waste forms. The sintering technology being considered can employ all of the established capabilities of large-scale processing, including vibratory packing. Our test results have shown that this technology is able to uniformly consolidate low-density compacts (as low as 35% of theoretical) into high-density, corrosion-resistant waste forms and avoids many potential difficulties associated with cold-forming techniques, such as uniaxial pressing. The added benefit of sintering at moderate temperatures and rates makes continuous processing possible, which would greatly enhance throughput.

ACKNOWLEDGMENTS

The authors gratefully acknowledge the guidance and support of J. P. Ackerman, M. A. Lewis, C. Pereira, D. Compton, P. Johnson, B. Tani, and K. Parish (ANL).

REFERENCES

1. J. P. Ackerman, L. S. H. Chow, S. M. McDeavitt, C. Pereira, and R. H. Woodman, "Isolating Wastes in the Electrometallurgical Treatment of Spent Nuclear Fuel," JOM, Vol. 49 (7), pp. 26-28 (1997).

2. S. M. McDeavitt, D. P. Abraham, J. Y. Park, and D. D. Keiser, Jr., "Stainless Steel-Zirconium Waste Forms from the Treatment of Spent Nuclear Fuel," JOM Vol. 49 (7), pp. 29-32 (1997).

3. M. A. Lewis, L. J. Smith, and D. F. Fischer, "Salt-Occluded Zeolites as an Immobilization Matrix for Chloride Salt Waste," Am. Ceram. Soc. Vol. 76 (11), 2826-2832 (1993)

4. M. A. Lewis, D. F. Fischer, and C. D. Murphy, "Densification of Salt-Occluded Zeolite A Powders to a Leach Resistant Solid Monolith," Scientific Basis for Nuclear Waste Management, Eds., A. Barkatt and R. Van Konynenburg, Vol. 333, pp. 277-284, Materials Research Society, Pittsburgh, PA (1994)

5. M. C. Hash, C. Pereira, M. A. Lewis, R. J. Blaskovitz, and J. P. Ackerman, "Hot Isostatic Pressing of Glass-Zeolite Composites," Environmental and Waste Management Issues in the Ceramic Industry II, pp. 135-143 (1996)

6. W. D. Kingery, H. K. Bowen, and D. R. Uhlman, Introduction to Ceramics, 2nd Edition, New York, John Wiley & Sons, Inc. (1976).

7. Donald W. Breck, Zeolite Molecular Sieves, New York, John Wiley & Sons, Inc. (1974).

8. M. C. Hash, C. Pereira, V. N. Zyryanov, M. A. Lewis, G. L. Burns, J. J. Smith, J. P. Thalacker, and J. P. Ackerman, "Preparation Techniques for Ceramic Waste Form Powder", Environmental Issues and Waste Management Technologies in the Ceramic and Nuclear Industries III, Eds., David K. Peeler and James C. Marra, pp. 399-407 (1998).

Waste Form Leaching and Property Measurement

COMPOSITIONAL EFFECTS ON CRYSTALLINITY AND DURABILITY OF HIGH-LEVEL RADIOACTIVE WASTE GLASSES

H. Li, D.E. Smith, P. Hrma, G.K Patello, G.L. Smith, and M.L. Elliott
Pacific Northwest National Laboratory,* Richland, WA 99352, U.S.A.

ABSTRACT

To support ongoing vitrification activities and site transition activities at the West Valley Demonstration Project (WVDP), glass compositional effects on crystallinity and durability of high-level waste borosilicate glasses were measured. The glass compositions are near or within the WVDP target glass process region. Spinel formation did not affect glass durability as evaluated by the 7-day Product Consistency Test (PCT). These glasses were more durable than the Environmental Assessment (EA) glass by an order of magnitude. The PCT normalized B and Na releases were found to be well correlated with the molar concentration $\{[R_2O]+[B_2O_3]\}$ or the normalized, molar concentration $\{[R_2O]+[B_2O_3]\}/\{[SiO_2]+[Al_2O_3]\}$ in the glass, where $[R_2O]$ is the total molar concentration of alkalis.

INTRODUCTION

The West Valley Demonstration Project (WVDP) is currently vitrifying a high-level waste (HLW) stream to produce a final, durable borosilicate waste form. As the contents of Tank 8D-2 are processed, several constituents in the WVDP HLW composition could possibly be outside either the expected process region or potential target glass composition region, or be close enough to the boundaries that analytical uncertainty could place some components outside the boundaries. The purpose of this study is to provide documentation that the glasses produced from the vitrification of Tank 8D-2 waste will continue to meet the WVDP Waste Form Qualification Report [1] specification. A test matrix was provided by WVDP personnel. It consisted of two series of glass compositions, denoted by EXP (ten glasses) and TRANS (twenty glasses). Target compositions of these WVDP glasses and the chemicals used to batch them are given in Table 1 and Table 2. Crystallinity and short-term durability, measured by the 7-day Product Consistency Test (PCT), for the isothermally heat-treated EXP glasses showed that spinel was a primary phase within the composition region. Short-term durability of these glasses was not affected by the precipitation of spinel at low concentrations. Hence, in the second stage of the study, short-term durability of the twenty TRANS glasses without the heat-treatment was investigated.

---

*Pacific Northwest National Laboratory is operated by Battelle for the U.S. Department of Energy under Contract DE-AC06-76RLO 1830.

EXPERIMENTAL

Glass Fabrication Glasses, 500 g each, were prepared following a two-step process. The nonradioactive glass components were first mixed together and melted at 1150 °C in a Pt-10%Rh crucible covered with a Pt lid for one hour using a resistance-heated furnace, then quenched by pouring the melt onto a stainless steel plate. The quenched, nonradioactive glass was ground in a tungsten carbide disc mill and then transferred to a radiologically controlled area. Then UO_3 and ThO_2 were mixed with the nonradioactive glass powder, which was re-melted at 1150 °C for two additional hours in a Pt-10%Rh crucible while covered with a Pt lid. The glass melt was quenched by pouring it onto a stainless steel plate. The final glasses were found to be free from crystals using optical microscopy.

Table 1. Compositions of EXP Glass Series (wt%)

EXP	1	2	3	4	5	6	7	8	9	10	Batch
Al_2O_3	6.16	5.85	5.84	5.28	5.50	5.55	5.86	6.18	5.81	6.24	Al_2O_3
B_2O_3	13.24	12.59	12.89	12.89	12.89	12.89	15.34	10.61	15.87	10.33	H_3BO_3[a]
CaO	0.12	0.70	0.12	0.70	0.00	1.00	0.60	0.28	0.58	0.32	$CaCO_3$[a]
Fe_2O_3	12.35	11.74	12.02	12.02	12.02	12.02	11.74	12.38	11.41	12.24	Fe_2O_3
K_2O	5.13	4.89	5.63	5.63	5.63	5.63	4.88	5.15	4.99	5.35	K_2CO_3[a]
Li_2O	3.81	3.63	4.17	4.17	4.17	4.17	3.62	3.82	3.71	3.98	Li_2CO_3[a]
MgO	0.49	1.29	0.49	1.29	0.62	0.76	0.69	1.09	1.11	1.07	MgO
MnO	0.45	1.19	0.45	1.19	1.07	0.60	0.63	1.02	0.98	0.66	MnO
Na_2O	8.21	7.82	9.00	9.00	9.00	9.00	7.82	8.24	7.99	8.58	Na_2CO_3[a]
P_2O_5	0.66	1.74	0.66	1.74	1.56	1.38	0.93	0.91	0.95	1.43	$NaPO_3$[a]
SiO_2	42.08	40.02	41.67	37.67	39.23	39.61	40.03	42.21	39.57	42.47	SiO_2
TiO_2	0.44	1.16	0.44	1.16	0.62	0.98	0.60	1.00	0.96	0.64	TiO_2
ThO_2	3.65	3.48	3.56	3.56	3.56	3.56	3.48	3.67	2.68	2.87	ThO_2
UO_3	0.16	0.91	0.16	0.91	1.30	0.00	0.79	0.38	0.41	0.75	UO_3
ZrO_2	1.35	1.29	1.20	1.09	1.13	1.15	1.29	1.36	1.28	1.37	ZrO_2
Others											
BaO	0.26	0.26	0.26	0.26	0.26	0.26	0.26	0.26	0.26	0.26	$BaCO_3$[a]
Ce_2O_3	0.31	0.31	0.31	0.31	0.31	0.31	0.31	0.31	0.31	0.31	CeO_2[a]
Cr_2O_3	0.14	0.14	0.14	0.14	0.14	0.14	0.14	0.14	0.14	0.14	Cr_2O_3
NiO	0.25	0.25	0.25	0.25	0.25	0.25	0.25	0.25	0.25	0.25	NiO
SO_3	0.23	0.23	0.23	0.23	0.23	0.23	0.23	0.23	0.23	0.23	Na_2SO_4[a]
SrO	0.25	0.25	0.25	0.25	0.25	0.25	0.25	0.25	0.25	0.25	$SrCO_3$[a]
ZnO	0.26	0.26	0.26	0.26	0.26	0.26	0.26	0.26	0.26	0.26	ZnO

(a) Equivalent quantity of chemicals was used in the glass preparation to achieve their corresponding oxide concentrations as specified in the table.

Glass Composition Analysis To determine compositions for each glass, powder samples in triplicate were used. Each sample was divided into one half and separately digested by fusion with Na_2O_2 in a Zr crucible and KOH in a Ni crucible at about 800 °C. Sample aliquots were analyzed by inductively coupled plasma atomic emission spectroscopy (ICP-AES) for all specified components except for U. Uranium was measured using inductively coupled plasma-mass spectroscopy (ICP-MS) because the detection limit for U is 1000× in ICP-MS lower than ICP-AES. A single ICP-MS measurement for U was performed for each composition. Except for U, the average compositions of each glass were determined from three (for Na, K, Ni, and Zr) or six (for the rest of cations) independent ICP-AES measurements.

Heat Treatment A heat-treatment was performed on the ten EXP glasses. Pieces of the broken glass in closed Pt boxes (25×25×25 mm^3) were first fused at 1150°C for 20 minutes and then transferred to a furnace preset at 600°C for a 96-hour heat-treatment. The temperature in the heat-treatment furnace stabilized at 600°C within 30 minutes after the fused samples were placed into the furnace.

Table 2. Compositions of TRANS Glass Series (wt%)

TRANS	1	2	3	4	5	6	7	8	9	10	Batch
Al_2O_3	6.30	5.50	6.50	6.80	6.90	6.00	5.70	7.20	6.10	5.90	Al_2O_3
B_2O_3	13.90	11.89	13.00	12.37	11.95	10.10	14.30	10.20	13.20	12.20	H_3BO_3[(a)]
CaO	0.55	0.20	0.90	0.35	0.95	0.15	0.85	0.40	0.80	0.05	$CaCO_3$[(a)]
Fe_2O_3	12.20	11.50	13.17	10.50	12.10	13.67	9.97	13.51	12.47	11.20	Fe_2O_3
K_2O	4.20	5.40	4.60	5.60	4.60	5.00	4.80	5.60	4.00	5.70	K_2CO_3[(a)]
Li_2O	3.30	4.10	3.30	3.90	3.00	4.00	3.50	4.10	3.50	4.20	Li_2CO_3[(a)]
MgO	1.25	1.15	0.60	0.80	0.55	0.85	1.35	1.00	0.95	1.20	MgO
MnO	0.55	0.30	0.70	1.15	0.75	1.10	0.50	0.50	0.90	0.20	MnO
Na_2O	7.60	9.10	7.00	8.50	7.70	8.80	7.20	8.70	7.30	8.30	Na_2CO_3[(a)]
P_2O_5	0.70	2.10	0.80	1.00	1.10	1.60	1.90	2.00	1.70	1.80	$NaPO_3$[(a)]
SiO_2	44.40	43.00	43.70	42.80	43.90	41.90	44.00	42.00	42.90	41.80	SiO_2
TiO_2	1.25	0.56	1.20	0.80	1.10	0.53	0.50	0.66	0.88	1.05	TiO_2
ThO_2	0.00	0.20	0.30	0.00	0.80	2.00	0.70	0.40	0.10	0.00	ThO_2
UO_3	0.00	0.00	0.03	0.03	0.00	0.00	0.03	0.03	1.50	1.70	UO_3
ZrO_2	2.10	3.30	2.50	3.70	2.90	2.60	3.00	2.00	2.00	3.00	ZrO_2
Others[(b)]	1.70	1.70	1.70	1.70	1.70	1.70	1.70	1.70	1.70	1.70	

TRANS	11	12	13	14	15	16	17	18	19	20	Batch
Al_2O_3	5.00	6.60	6.70	5.00	5.30	5.10	5.70	5.50	5.40	6.60	Al_2O_3
B_2O_3	14.50	14.80	11.00	14.80	11.20	9.70	15.00	11.50	11.30	14.40	H_3BO_3[(a)]
CaO	0.20	0.05	0.70	0.05	0.39	0.65	0.15	0.55	0.65	0.25	$CaCO_3$[(a)]
Fe_2O_3	12.90	9.75	13.80	13.60	14.25	10.20	10.40	10.00	12.70	11.60	Fe_2O_3
K_2O	4.00	4.50	4.36	4.20	4.40	6.10	3.50	6.10	4.70	5.30	K_2CO_3[(a)]
Li_2O	3.00	2.90	3.20	2.60	3.31	4.90	2.50	4.30	3.30	3.20	Li_2CO_3[(a)]
MgO	0.40	0.45	0.60	0.20	1.10	1.40	0.30	1.30	1.10	0.25	MgO
MnO	0.35	0.60	1.20	0.20	1.10	1.30	0.35	0.45	0.50	0.55	MnO
Na_2O	8.90	9.50	6.25	9.00	6.50	6.75	9.65	7.00	8.21	7.30	Na_2CO_3[(a)]
P_2O_5	0.70	0.80	0.59	0.40	1.00	1.80	0.50	1.70	0.90	0.75	$NaPO_3$[(a)]
SiO_2	44.50	45.20	45.50	45.00	46.20	46.20	46.20	46.20	44.60	44.10	SiO_2
TiO_2	0.30	0.50	1.10	0.15	0.60	1.30	0.30	0.50	1.04	0.20	TiO_2
ThO_2	1.60	0.30	0.60	0.30	0.80	0.10	1.20	0.10	0.60	0.50	ThO_2
UO_3	0.15	0.25	0.20	0.10	0.15	0.70	0.15	0.50	1.60	1.80	UO_3
ZrO_2	1.80	2.10	2.50	2.70	2.00	2.10	2.40	2.60	1.70	1.50	ZrO_2
Others[(b)]	1.70	1.70	1.70	1.70	1.70	1.70	1.70	1.70	1.70	1.70	

(a) Equivalent quantity of chemicals was used in the glass preparation to achieve their corresponding oxide concentrations as specified in the table.

(b) Others contains the same group of oxides listed in Table 1.

Characterization of Heat-Treated Samples Transmission optical microscopy was performed on all heat-treated EXP glass samples. To prepare samples that are transparent to light, sample slides were prepared using a cleavage method. Specifically, a heat-treated sample block was notched half way through using a diamond blade. Then, the block was cleaved into two halves by forcing a wedge into the notch, creating a smooth fracture surface. Finally, a thin sample was made by cutting one half of the block parallel to the fracture surface (~0.5 mm behind the fracture surface) using the diamond blade. Optical transmission micrographs were taken on all heat-treated samples. Crystalline phases in the heat-treated samples were identified by a powder x-ray diffraction (XRD) method using a Philips® x-ray diffractometer. The XRD scans covered a two-theta angle (2θ) range of 5 to 75° at a scan rate of 0.02° per 25 sec. The crystallinity in the heat-treated sample was estimated using semi-quantitative XRD analysis.

Glass Durability Testing The short-term durability of each quenched glass was determined by the 7-day PCT method following a standard ASTM C1285-94 procedure [2]. The tests were conducted at 90°C using deionized water. The test matrix was divided into three subsets. In each subset, ten quenched glasses were tested simultaneously with the Environmental Assessment (EA) [3] and the Reference 6 standard glasses [4]. Three replicate samples of each glass were tested using 304L stainless steel vessels. The concentrations of elements in the acidified leachate solutions were analyzed using ICP-AES. The normalized elemental mass loss, $[NL]_i$, was calculated using the formula

$$[NL]_i = C_i / f_i (SA/V), (g/m^2) \qquad (1)$$

where C_i is the concentration of the i-th element in the 0.45-μm filtered leachate solution (g/m^3), f_i is the mass fraction of the i-th element in the glass, and (SA/V) is the ratio of sample surface area to volume of water (m^{-1}), which is approximately 2000 m^{-1}. The initial and final solution pH values were measured at room temperature using an Orion® 601A Digital Ionalyzer that was calibrated using Fisher® standard buffer solutions (pH 4 and pH 7 for deionized water and pH 7 and pH 10 for PCT solutions).

RESULTS AND DISCUSSION

Glass Composition Analysis The average oxide concentrations, determined from the measured cation concentrations, are generally within ±10% of the target concentrations for Al_2O_3, B_2O_3, BaO, Fe_2O_3, Li_2O, Na_2O, SiO_2, and TiO_2. The average concentrations for CaO, MgO, P_2O_5, ThO_2, UO_3, and ZrO_2 are within ±30% of the target concentrations and the average concentrations of the majority of components listed under the "Others" are within ±40% of the target concentrations. Some of the measured CaO concentrations are higher than 30% of the target concentrations; the higher values mainly came from Na_2O_2-fused sample solutions if the concentration of CaO in the glasses was less than 0.5 wt%. The discrepancy between the measured and the target concentrations for the glass constituents are systematic, which are believed to be caused by instrumental limitations, such as interferences among element emission lines during the ICP-AES measurements or the proximity of the cation concentrations to the detection limits. For these reasons, the target glass compositions were used in the 7-day PCT normalized elemental mass loss calculations.

Crystallinity Figure 1 shows transmission optical micrographs of the heat-treated EXP 1, 2, 8, and 9 glasses. The morphology of crystals in all the heat-treated EXP glasses was found to be the same. A spinel phase, trevorite [NiFe(III)$_2$O$_4$], was identified by XRD. The mass fractions of crystals in the heat-treated EXP 2 and EXP 8 were 0.0089 and 0.0081, respectively. Other heat-treated EXP glasses had much smaller volume fractions of spinel crystals. A previous study of the spinel crystals in a similar WVDP glass showed that the spinel phase was close to $Ni_{0.68}Fe(II)_{0.32}Fe(III)_2O_4$ as determined using XRD analysis with an assumption that the spinel phase is a solid solution of $NiFe_2O_4$ and $Fe(II)Fe(III)_2O_4$ or $Ni_{0.61}Fe(II)_{0.39}Fe(III)_2O_4$ as determined using scanning electron microscopy-energy dispersive spectrometry [the proportions of Fe(II) and Fe(III) were calculated based on a charge balance]. Hence, it is likely that the spinel phase precipitated in the heat-treated EXP glasses is made up of a $Ni_{1-x}Fe(II)_xFe(III)_2O_4$ solid solution.

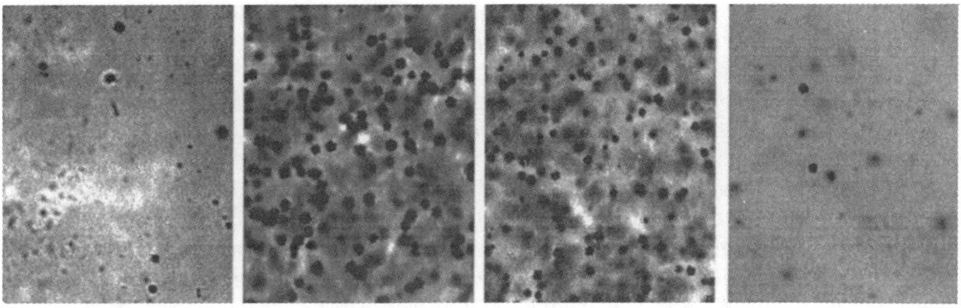

(a) EXP 1 (b) EXP 2 (c) EXP 8 (d) EXP 9

Figure 1. Transmission Optical Micrographs of the Heat-Treated EXP Glasses Showing Spinel Crystals (the dark areas are spinel crystals out of a focal plane).

Liquidus temperature of glass with a primary phase of spinel is expected to affect spinel crystallization. The measurement of the T_L values was not performed under this study for the EXP glasses. Fortunately, a first-order mixture liquidus temperature (T_L) model for HLW glasses precipitating spinel as a primary phase has been developed at PNNL [5]. The model was, therefore, utilized to predict T_L values for the EXP glasses. The model predicted T_L values are 1154 and 1136 °C for EXP 2 and EXP 8, respectively, whereas the T_L values for other EXP glasses are between 1044 and 1101 °C. Comparing the crystallinity data with the predicted T_L values, it follows that the glasses with higher T_L values close to their melting temperature (1150 °C in this case) have a higher tendency to precipitate spinel crystals.

Short-Term Durability In this study, the average, normalized elemental mass loss ($[NL]_i$) values of B, Na, and Si for the EA reference glass are 7.97, 6.10, and 1.72 g/m^2, respectively. The average solution pH after the test is 11.68. For the WVDP Reference 6 glass, the average $[NL]_i$ values (in the same order) are 0.32, 0.30, and 0.12 g/m^2 for B, Na, and Si, respectively; the average solution pH is 9.97. These values are in very good agreement with previous studies for the EA and Reference 6 glasses [2,3]. The average $[NL]_i$ values for EXP and TRANS glasses are 0.19 to 0.88 g/m^2 for B, 0.17 to 0.67 g/m^2 for Na, and 0.10 to 0.18 g/m^2 for Si. The solution pH ranges from 9.36 to 10.40. The durability of the EXP glasses is much more durable than the EA glass by at least one order of magnitude.

EXP Series (Heat-Treated Glasses): The short-term durability of the heat-treated EXP glasses varied depending on composition rather than the extent of spinel crystals precipitated in the glasses. The heat-treated EXP 1 and EXP 2 glasses, which contained widely different concentrations of spinel crystals (cf. Figure 1), had nearly the same values for $[NL]_B$ and $[NL]_{Na}$. The quantity of spinel crystals is reported also to have little effect on glass durability [6]. The PCT results obtained from the heat-treated glasses are consistent with this observation. Among the ten heat-treated glasses, EXP 4, 5, 6, 7, and 9 glasses had $[NL]_B$ values that are 100% higher than those of EXP 1, 2, 8, 10, and the Reference 6 glass. A similar comparison of $[NL]_{Na}$ values can be found between these two sets of glasses.

TRANS Series (Quenched Glasses): In terms of $[NL]_B$ or $[NL]_{Na}$, TRANS glasses are comparable to the EXP glasses. Among the twenty TRANS glasses, TRANS 11, 12, 14, 17, and 20 had $[NL]_B$ values that were about two times higher than those of other TRANS glasses and the Reference 6 glass. A similar comparison of the $[NL]_{Na}$ values showed similar results.

Various statistical models have been developed [1,7,8] and successfully applied to predict 7-day PCT $[NL]_{Na}$ or $[NL]_B$ of HLW glasses. The effects of individual glass constituents may be qualitatively compared among them using the model coefficients. However, the interpretations on those component effects are subject to glass composition regions that are used to fit the models and the model form selection. Although HLW glasses are complex systems, in principle glass resistance to water attack is expected to be influenced by the glass network structure. A primary network building block is SiO_2. B_2O_3 is a glass former; however, it depolymerizes the silicate network. In a silicate network with the presence of alkalis, the addition of B_2O_3 may increase or decrease glass durability depending on composition [9]. Using nuclear magnetic resonance (NMR) spectroscopy, Na_2O was shown to preferentially react with Al_2O_3 and then with B_2O_3 in alkali-alumino-borosilicate glasses [10]. The coordination of boron with oxygen (or a proportion of BO_3 and BO_4) in the glass was also shown to correlate with the net Na_2O concentration (the Na_2O concentration minus the Al_2O_3 concentration) [10]. Additionally, the preferential reaction of Na_2O with B_2O_3 rather than with SiO_2 was demonstrated in a previous study of sodium-borosilicate glasses using NMR [11]. Al_2O_3 in the network, which is charge balanced by alkalis, functions as a glass former polymerizing the silicate network with alkalis. Based on the same study [11], fourfold boron (BO_4) is considered as a glass former and threefold boron (BO_3) as a glass modifier. By considering these published NMR data, it may be expected that some simplified parameters exist, which can be used to describe short-term glass durability. The following data analysis shows that durability of the WVDP glasses is indeed well correlated with the total molar concentration of $[R_2O]+[B_2O_3]$ or the normalized molar concentration of $\{[R_2O]+[B_2O_3]\}/\{[SiO_2]+[Al_2O_3]\}$ in the glass (R_2O is the total concentration of Li_2O, Na_2O, and K_2O).

Figure 2 (a-f) shows the 7-day PCT $[NL]_{Na}$ and $[NL]_B$ values plotted versus the molar concentration of some of major glass constituents expressed in oxide forms. The trend shown in Figure 2(a) suggests a possible minimum in $[NL]_{Na}$ and $[NL]_B$ between 10 and 14 mol% B_2O_3, which has been previously reported for other waste glasses [9]. Figure 2(b) indicates that $[NL]_{Na}$ and $[NL]_B$ generally increase with the increase of total alkali concentrations although a single correlation cannot be found. When $[NL]_{Na}$ and $[NL]_B$ are evaluated against the total concentration of B_2O_3 and R_2O, clear trends exist with a few exceptions shown in Figure 2(c). With respect to Al_2O_3, no systematic trend for $[NL]_{Na}$ and $[NL]_B$ is observed, as illustrated in Figure 2(d). A lack of correlation between $[NL]_{Na}$ or $[NL]_B$ and the SiO_2 concentration in the glass is illustrated in Figure 2(e). $[NL]_{Na}$ and $[NL]_B$ are also evaluated against the total concentration of Al_2O_3 and SiO_2; no clear trend is evident. Other oxides, such as CaO, MgO, Fe_2O_3, ThO_2, TiO_2, UO_3, and ZrO_2, were also evaluated in a similar manner, but no apparent correlations were found with respect to the $[NL]_{Na}$ and $[NL]_B$ values.

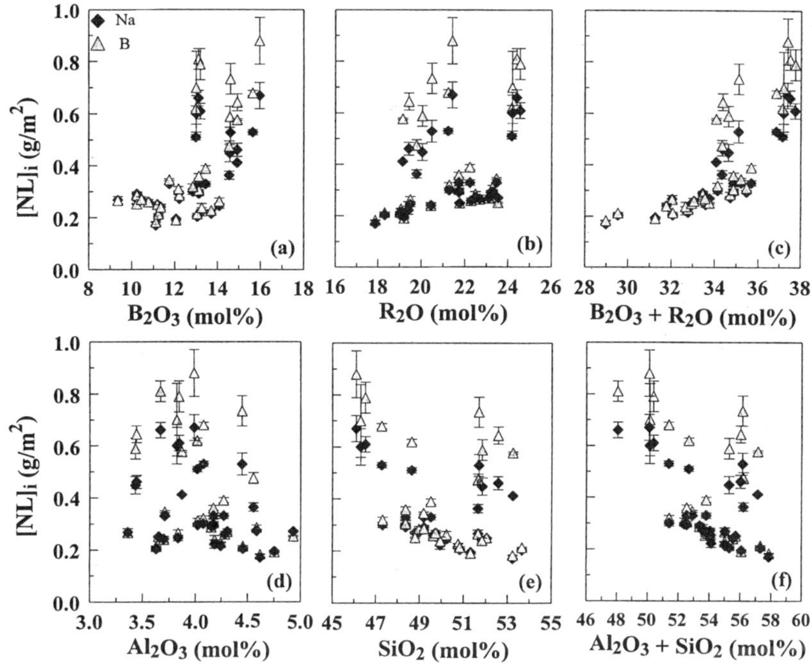

Figure 2. Glass Dissolution as Indicated by Normalized B and Na Mass Loss Plotted as a Function of a Molar Concentration of Oxides.

Figure 3 plots the $[NL]_B$ and $[NL]_{Na}$ values versus the ratio of $([R_2O]+[B_2O_3])$ over $([SiO_2]+[Al_2O_3])$. By normalizing the concentration, the trend for $[NL]_{Na}$ is slightly more separated from the trend for $[NL]_B$ as compared to those shown in Figure 2(c), especially for those five glasses having $[NL]_{Na}$ and $[NL]_B$ that do not fit well with the rest of the data. These five glasses were found to have lower Li_2O/B_2O_3 molar ratios (<0.52) than the rest of the glasses (no other consistent composition correlations could be found for these glasses). Regardless of the use of $[R_2O]+[B_2O_3]$ or $\{[R_2O]+[B_2O_3]\}/\{[SiO_2]+[Al_2O_3]\}$, the observed trends for $[NL]_B$ and $[NL]_{Na}$ suggest that B_2O_3 functions as a glass modifier, decreasing glass durability for the composition region studied. To further examine the reliability of the trends as found by this study, additional 7-day PCT data from testing of 58 WVDP glass compositions previously reported in the literature [8] have been evaluated using $\{[R_2O]+[B_2O_3]\}/\{[SiO_2]+[Al_2O_3]\}$. Similar trends for both $[NL]_B$ and $[NL]_{Na}$ are found. Further evaluation will be performed on more than 200 simulated Hanford HLW glasses tested at Pacific Northwest National Laboratory [12] and the results will be published elsewhere.

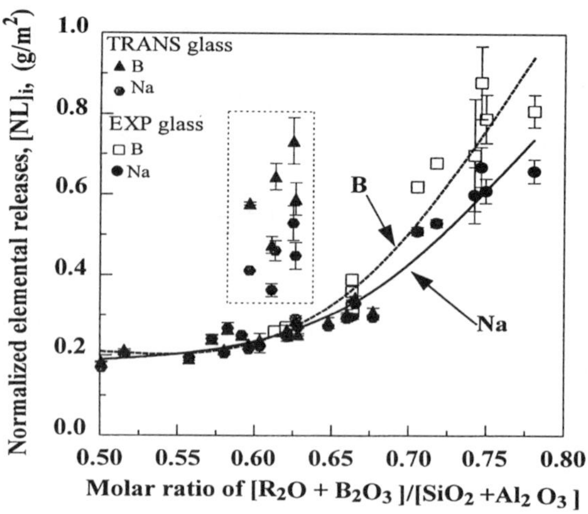

Figure 3. Glass Dissolution as a Function of $[R_2O+B_2O_3]/[SiO_2+Al_2O_3]$ (error bars represent one standard deviation and the curves result from nonlinear regression fitting excluding the data points surrounded by the dashed box).

CONCLUSION

Spinel formation did not affect short-term durability as measured by the 7-day PCT method. The 7-day PCT normalized B and Na releases were found to be well correlated with the concentration $\{[R_2O]+[B_2O_3]\}$ or the normalized concentration $\{[R_2O]+[B_2O_3]\}/\{[SiO_2]+[Al_2O_3]\}$ in the glass with only minor exception for glasses with molar ratios of Li_2O to B_2O_3 less than 0.52.

ACKNOWLEDGMENT

The authors wish to acknowledge and thank M. Gong, R.J. Barbera, and E.J. Wyse for performing glass composition and PCT solution analysis, and E.D. Jenson for performing XRD analysis. Laboratory assistance from M.J. Schweiger, W.C. Buchmiller, and J.V. Crum is also appreciated. Several discussions with J.G. Darab and J.D. Vienna have been stimulating. The authors also thank C.S. Fuez and A. Drobot for their technical advice through a course of this study and beneficial comments on the manuscript.

REFERENCES

[1] S.M. Barnes, "*Waste Form Qualification Report (WQR).*" WVDP-186 Rev. 1, WQR1.1, West Valley Nuclear Services Co., Inc., West Valley, New York, 1996.

[2] ASTM C1285-94, "Standard Test Methods for Determining Chemical Durability of Nuclear Waste Glasses -The Product Consistency Test (PCT)," *Annual Book of ASTM Standards* American Society for Testing and Materials, West Conshohoken, PA, 1995.

[3] C.M. Jantzen, N.E. Bibler, D.C. Beam, C.L. Crawford, and M.A. Pickett, *"Characterization of the Defense Waste Processing Facility (DWPF) Environmental Assessment (EA) Glass Standard Reference Material (U),"* WSRC-TR-92-346, Rev. 1. Westinghouse Savannah River Company, Aiken, South Carolina, 1993.

[4] K.M. Olson, G.L. Smith, and S.C. Marschman, *"Product Consistency Testing of Three Reference Glasses in Stainless Steel and Perfluoroalkoxy Resin Vessels,"* PNL-10497, Pacific Northwest National Laboratory, Richland, Washington, 1995.

[5] P. Hrma, J.D. Vienna, M. Mika, and J.V. Crum, *"Liquidus Temperature Data for DWPF Glass,"* PNNL-11790, Pacific Northwest National Laboratory, Richland, Washington, 1998.

[6] D.F. Bickford and C.M. Jantzen, "Devitrification Behavior of SRL Defense Waste Glass," *Scientific Basis for Nuclear Waste Management VII, Mater. Res. Soc.*, Vol. 26, ed. by G.L. McVay, North-Holland, New York, 1984, p.557-565.

[7] P. Hrma, G. F. Piepel, M. J. Schweiger, D. E. Smith, D.-S. Kim, P. E. Redgate, J. D. Vienna, C. A. LoPresti, D. B. Simpson, D. K. Peeler, and M. H. Langowski, *"Property/Composition Relationships for Hanford High-Level Waste Glasses Melting at 1150 °C,"* PNL-10359, Vol. 1 and 2, Pacific Northwest Laboratory, Richland, Washington, 1994.

[8] G. Piepel and T. Redgate, "Mixture Experiment Techniques for Reducing the Number of Components Applied for Modeling Waste Glass Sodium Release," *J. Am. Ceram. Soc.*, 80. 3038-44 (1997).

[9] D.S. Kim, P. Hrma, S.E. Palmer, D.E. Smith, and M.J. Schweiger, "Effect of B_2O_3, CaO, and Al_2O_3 on the Chemical Durability of Silicate Glasses for Hanford Low-Level Waste Immobilization," *Ceramic Transactions*, Vol. 61, ed. by V. Jain and R. Palmer (Am. Ceram. Soc., Westerville, Ohio, 1995) p. 531-538.

[10] J.G. Darab, X. Feng, J.C. Linehan, P.A. Smith, and I. Roth, "Composition-Structure Relationships in Model Hanford Low-Level Waste Glasses," *Ceramic Transactions*, Vol. 72, ed. by V. Jain and D. Peeler (Am. Ceram. Soc., Westerville, Ohio, 1996) p. 103-110.

[11] Y.H. Yun and P.J. Bray, "Nuclear Magnetic Resonance Studies of the Glasses in the System Na_2O-B_2O_3-SiO_2," *J. Non-Cryst. Solids.*, 27, 363-380 (1978).

[12] P.R. Hrma, G.F. Piepel, M.J. Schweiger, D.E. Smith, D.-S. Kim, P.E. Redgate, J.D. Vienna, C.A. LoPresti, D.B. Simpson, D.K. Peeler, and M.H. Langowski, *Property/Composition Relationships for Hanford High-Level Waste Glasses Melting at 1150 °C*, Pacific Northwest National Laboratory, Richland, Washington, 1994.

HEAVY METALS BEHAVIOR OF ULTIMATE GLASSY WASTES

P. LeCoustumer, A. Gauthier, J. H. Thomassin
Laboratoire des Matériaux et Géologie Environnementale, ESIP,
40 Av. Du recteur Pineau, 86022 Poitiers cedex, France

M. Motelica-Heino
Laboratoire de Chimie Bio-Inorganique et Environnement, EP CNRS 132,
Helioparc, 64000 Pau, France

ABSTRACT

Management of domestic wastes often relies on incineration, a process that eliminates large amounts of wastes but also produces toxic residues that concentrate heavy metals. Vitrification is seen as a powerful method to stabilize those hazardous secondary wastes. However, there is a concern about the long-term behavior of these ultimate glassy wastes and the potential release of toxic species into the environment. This paper investigates the behavior of vitrified domestic fly ashes in simulated aggressive environments using leaching tests. The alteration of the waste matrix and the release of heavy metals such as Pb, Cu, Cr, Sn, and Zn from the matrix were studied. Physical characterization techniques such as electron microscopy were used to assess the evolution of the matrix and mineralogical transformations. Trace analytical techniques were used to monitor the leached matrix elements and heavy metals.

Key words: ultimate vitrified wastes, heavy metals, leaching tests, matrix alteration

INTRODUCTION

Waste management is an important issue for communities and industries in light of increasing and ever-changing regulations requiring wastes to be treated and recycled. Specific treatment procedures and stabilization strategies have been developed for domestic and industrial solid wastes. One common elimination technique of domestic wastes is based on incineration. Incineration plants generate secondary toxic wastes such as ashes and residues of filtered fumes (REFIOM in French) that concentrate metals such as Zn, Cu, Pb, Sn, Cr, and Ni. Vitrification of hazardous wastes in a plasma torch at temperatures over 6000° has been investigated.[1] This original process produces glasses from the wastes themselves. Vitrified glasses from ashes collected after the thermal incineration of domestic wastes are ceramic-like glasses.[2] This matrix seems to be an ideal host for toxic elements. However, it is necessary to understand the long-term behavior of their matrix and evaluate the potential release of toxic species in the environment.

The aim of this work is to present an evaluation methodology, based on experimental predictive models and analytical tools, to monitor both the evolution of vitrified waste matrix and the fate of hazardous elements trapped in the matrix. The alteration of vitrified domestic fly ashes and the behavior of heavy metals such as Pb, Cu, Cr, Sn, and Zn were investigated in simulated aggressive environments. Leach tests were performed on two representative samples using three different alteration reagents simulating rain, acid rain, and soils. Analyses of leaching solutions and characterization of the solid waste matrix at the microscopic scale were performed to assess the evolution of the mineralogical phases of the wastes and to understand the alteration process. In-situ characterization of the heavy metals at trace levels in the solid matrix and analysis of the heavy metals present in the leaching solutions were used to investigate the potential mobility of these hazardous elements.

EXPERIMENTAL

Glassy Wastes

Fly ashes from incinerated domestic wastes were vitrified in a plasma torch by Electricité de France (EDF) at 1616°C for 30 minutes. The material from the melting bath underwent partial crystallization during the cooling process and two phases (vitreous and crystallized) could be distinguished at a macroscopic level. Two typical samples 1V (vitreous) and 1C (crystallized) representative of vitrified domestic wastes generated by plasma techniques were used in the present set of experiments.

Leach Tests and Analysis of Leaching Solutions

Three alteration agents were selected to simulate different aggressive environments for ultimate waste disposal. Permuted water (pH 5.5) and hydrochloric acid (pH 3.5) were used as analogues for rain and acid rain, respectively. Oxalic acid (pH 2.2; concentration = 0.01 mol/L) was used to simulate soil storage conditions. Solid samples were prepared as 2-3 mm thick rectangular sheets. Samples were place in Teflon vessels and the ratio of sample surface/volume of solution was approximately 0.5 cm^{-1}. Static leaching experiments were performed at 25°C from 0.5 h up to 8 days for samples 1V and 1C with mechanical stirring. Another set of experiments were performed with the same samples at 90°C (+/- 1°C) using a thermostatic oven from 0.5 up to 24 h solution. Solution pH was measured by a Metrohm 632 pH meter. Each leaching solution was regularly sampled, acidified with nitric acid (0.1 mol/L) and filtered prior to analysis. Analysis was performed by a Perkin Elmer 2380 atomic absorption spectrometer (AAS) for Ca, Mg, Na, K, Cu, Zn, and by a Jobin Yvon 38+ inductively coupled plasma atomic emission spectrometer (ICP-AES) for Si, Fe, Pb, Cr, and Sn.

Characterization of Waste Matrix

Both original and altered materials were characterized for their physio-chemical properties including determination of chemical composition by AAS according to a standard protocol[3] and by scanning electron microscopy – x-ray energy dispersive spectroscopy (SEM/X-EDS) (Jeol 6400C). Structure and mineralogy were established by x-ray diffraction (XRD) (Philips PW1729) and texture was evaluated by SEM. Laser ablation inductively coupled plasma mass spectrometry (LA-ICPMS) was used to study trace elements at the microscale level in the original samples. The CETAC LSX 100 laser-sampling device was coupled to the standard Perkin Elmer ELAN 5000 ICP-MS. The sampling pattern was a 50 μm width and 2 mm long line. The relative limits of detection were estimated to 0.1 weight %. Si was used as an internal standard from X-EDS data to compensate for differences in ablation between the mineralogical phases.

RESULTS AND DISCUSSION

Analysis of Original Samples

Results of the physio-chemical characterization of the original samples are summarized in Table I. The average chemical composition of the matrix determined by AAS was similar for the two samples. Vitrified wastes are calco-silico-alumino glasses with Fe, Mg, Ti, P, K, Zn occurring as minor elements around a few 10 wt%. XRD analysis revealed typical patterns of amorphous materials (1V) and crystallized sample (1C) with some amorphous parts. SEM analysis confirmed that sample 1V was homogeneous and 1C was heterogeneous. Automorphous crystals observed are spinels and sorosilicates. LA-ICP-MS was used to assess the spatial distribution of heavy metals between the different mineralogical phases. Figure 1 illustrates the laser impact in the automorphous crystals for sample 1C.

Table I. Physio-Chemical Characterization of the Vitrified Wastes Matrix

Sample	1V	1C
Structure (DRX)	Mainly amorphous Minor crystallized phase	Mainly crystallized Sorosilicates: Melilite ($NaCaAlSi_2O_7$) Gehlinite ($Ca_2Al_2SiO_7$) Spinels: Gahnite ($ZnAl_2O4$) Franklinite ($ZnFe_2O_4$) Minor amorphous phase
Texture (OM & SEM)	Vitreous matrix with micrometric Automorphous crystals (<1 μm) Homogeneous texture	Automorphous crystals 10-50 μm (sorosilicates and spinels) Microlitic crystals (sorosilicates) included in vitreous phase Heterogeneous texture Cavities (100 μm) containing spinels and sorosilicates
Bulk composition (AAS) wt%	Si = 31 Na = 1 Al = 10.5 P = 1 Ca = 30 Cr = 1 Fe = 1.5 Zn = 1 Mg = 4 O = 16.5 Ti = 1.5	Si = 30 Na = 1 Al = 12.5 P = 1 Ca = 30 Cr = 0 Fe = 1.5 Zn = 1 Mg = 3 O = 17 Ti = 3

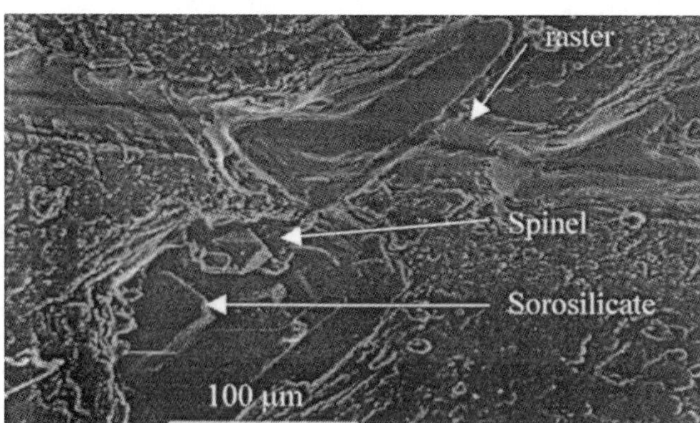

Figure 1. SEM Micrograph of Laser Impact in Sample 1C

The LA-ICP-MS data are presented in Figure 2. It appears that crystallized phases are enriched in several and nearly all heavy metals compared to the amorphous vitreous phase. Major elements such as Ti, Fe, and Zn; minor elements such as Cr, Sn, Mn, Ba; and trace elements such as Ni, Cu, and Pb have an affinity to the spinel or sorosilicate crystals. The microchemistry of the trace metals is consistent with the distribution of heavy metals observed by Clozel by SEM-XEDS in vitrified domestic wastes except for Pb.[4] The micro-chemistry of the trace metals in the vitrified wastes shows that Cr, Sn, Ti, Mn, and Zn have an affinity to the spinel crystals whereas Zn can also be incorporated in soroscilicates and Pb is trapped into the vitreous phase. In the vitrified wastes, crystals concentrate toxic metals. Those hazardous elements can be part of the crystals as Al and Fe for spinels and Al for sorosilicates. The following elements can also be trapped in crystals (spinels and sorosilicates): Sn, Ti, Mn, Fe, Cr, Zn, Ni, Ba, Sb, Cu, and W.

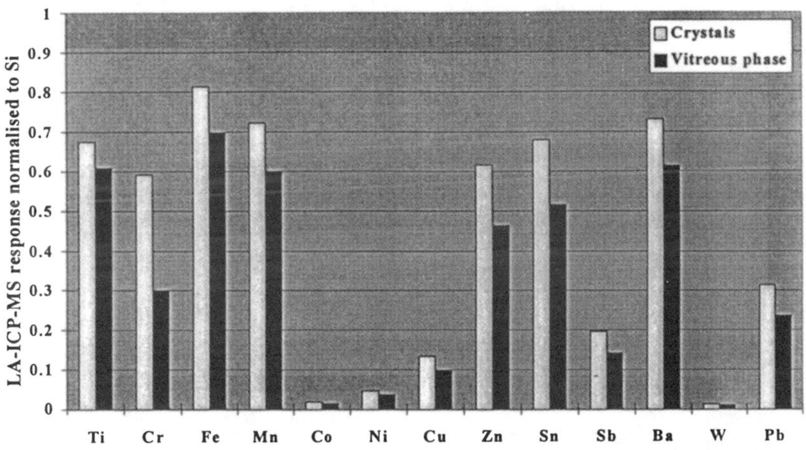

Figure 2. Microanalysis of Heavy Metals by LA-ICP-MS

Analysis of Leachants

Table II (25°C test) and Table III (90°C test) present the evolution of pH and major element concentrations in the leaching solution. The solution pH is governed by the release of Na, Ca, A1, and Si for both samples (1C and 1V). The pH starts to increase at the earliest stage of the test, thus, the structure of the ultimate glass waste does not influence the leaching process. Main differences in

Table II
pH and ICP-AES Analysis of Leaching Solutions at 25°C for Samples 1V/1C
(concentrations are in mg/L)

H2O	Hours	pH	Na	Ca	K	Mg	Al	Si
	0.5	6.33/6.55	0.23/0.31	1.38/1.89	0.1/0.14	0.08/0.11	0.05/0.07	0.04/0
	24	8.31/8.43	0.75/0.63	7.11/6.62	0.53/0.34	0.24/0.2	1.54/1.1	1.57/1.17
	48	6.83/6.97	0.13/0.18	2.57/3.04	0.20/0.0.09	0.14/0.13	0.59/0.67	0.73/0.79
	72	6.82/6.92	0.37/0.44	3.56/3.50	0.44/0.35	0.21/0.17	0.69/0.64	1.28/1.08
	192	7.14/6.87	0.36/0.42	4.12/2.94	0.09/0.11	0.27/0.17	0.29/0.08	1.1/0.33

HCl	Hours	pH	Na	Ca	K	Mg	Al	Si
	0.5	5.7/6.24	0.36/0.49	1.45/2.15	0.11/0.15	0.09/0.12	0.01/0.02	0/.01
	24	8.01/8.15	0.71/0.96	8.29/8.43	0.34/0.55	0.24/0.22	1.54/0.88	1.68/0.94
	48	6.48/6.72	0.18/0.29	2.73/3.26	0.2/0.26	0.13/0.15	0.36/0.43	0.32/0.63
	72	6.63/6.82	0.7/0.44	4.4/3.92	0.66/0.31	0.13/0.18	0.64/0.57	0.65/1.02
	192	6.82/6.80	0.44/0.42	3.64/4.25	0.1/0.12	0.21/0.22	0.12/.08	0.72/0.63

(COOH)2	Hours	pH	Na	Ca	K	Mg	Al	Si
	0.5	5.27/5.28	5.36/6.26	28.44/27.39	0.93/1.11	5/5.43	45.79/45.14	45.08/47.54
	24	5.59/5.55	4.87/5.21	34.25/35.68	1.51/1.74	5.15/5.41	38.94/39.60	49.7/50.98
	48	5.47/5.56	4.71/5.35	25.89/26.11	1.1/1.45	5.3/5.81	32.45/32.78	52.03/49.06
	72	5.24/5.29	4.4/5.26	32.65/32.76	1.03/1.47	6.03/5.74	32.45/32.12	50.2/51.24
	192	5.41/5.42	5.34/6.45	21.40/20.03	0.74/1.03	5.58/5.75	*40.42/37.33*	43.58/46.86

Table III
pH and ICP-AES Analysis of Leaching Solutions at 90ºC of Samples 1V/1C
(concentrations are in mg/L)

H2O	T(hours)	pH	Na	Ca	K	Mg	Al	Si
	0.5	6.85/6.33	0.58/0.61	3.56/2.71	0.49/0.59	0.22/0.15	0.88/0.22	1.48/1.01
	1	7.57/7.68	0.69/0.89	7.66/6.82	0.49/0.77	0.35/0.26	2.42/1.32	2.82/1.84
	4	9.8/9.25	1.34/1.23	5.61/3.94	0.6/0.69	0.33/0.26	3.3/1.76	5.68/4.41
	24	10.22/9.94	1.48/1.29	11.66/6.95	0.7/0.82	0.2/0.26	6.38/4.62	6.81/7.15
HCl	0.5	7.13/6.73	0.65/0.9	4.05/3.41	0.41/0.54	0.2/0.18	0.66/0.22	0.92/1
	1	7.867.9	1.08/0.72	3.48/7.11	0.84/0.81	0.2/0.26	0.88/1.32	1.72/2.02
	4	8.78/8.50	1.18/1.15	7.11/5.26	0.7/0.72	0.29/0.31	3.52/2.42	5.31/4.36
	24	7.61/8.13	1.58/1.6	9.79/6.51	1.47/2.28	0.24/0.48	4.18/3.30	7.52/9.15
(COOH)2	0.5	2.67/2.69	3.83/3.98	6.53/8.23	1.32/1.59	3.78/3.65	27.06/28.60	32.22/34.18
	1	3.09/3.12	4.72/4.78	14.83/15.20	1.83/1.56	4.67/4.51	32.78/34.32	43.25/41.81
	4	5/4.79	5.625.56	19.56/20.83	1.89/1.64	5.54/4.64	34.76/36.96	49.34/46.88
	24	4.56/4.61	6.19/6.18	16.37/19.01	1.96/1.85	5.76/4.82	34.54/36.96	55.13/52.19

the leaching tests are related to the nature of the altering agents. Na, Ca, Al, and Si exhibit the same leaching behavior for both 1V and 1C in H_2O and HCl solutions, however, the oxalic solution strongly attacks the ultimate glassy wastes by a factor of ten more than the other two agents. Therefore, the dissolution mechanisms are likely to be different. Several authors[5, 6, 7] using congruency ratio (relative to Si on calco-alumino-silicate glass) explained that three different dissolution modes exist, namely: congruent ($[Ci/Si]_{liquid} = [Ci/Si]_{solid}$), selective and not congruent ($[Ci/Si]_{liquid} \neq [Ci/Si]_{solid}$). Temperature accelerates the dissolution process of the matrix. For the same experimental time (typically 24 hours), the concentration of the major elements is approximately ten times more at 90°C than at 25°C compared to previous studies.[5]

Behavior of heavy metals (see Table IV) such as Fe, Cu, Zn, Sn, Cr, and Pb is quite different from the major elements. The influence of the structure can be established by comparing the results obtained at 25°C and 90°C between 1V and 1C (cf. Table IV). No significant difference is noticed between the heavy metals concentration in the leaching solutions from crystallized (1C) and amorphous (1V) samples except for Pb and Zn for the water leaching test at 25°C. Enrichment in Pb and Zn in water leaching solutions from the crystallized sample compared to the amorphous sample (1V) can be related to the relative enrichment of large spinel crystals in Pb and Zn. Influence of the temperature is quite different from what has been previously observed with major elements.

Comparing the results at 25ºC and 90ºC for the same time (24h), fine differences are noticed. ICP-MS reveals that the matrix concentrates selectively some heavy metals such as Pb, Sn, and Cr. For longer test times, the heavy metal concentrations remain low in water and hydrochloric, regardless of the structure. With oxalic solution, this trend is not followed by Fe, Zn, and Sn, which are present in higher concentrations, even for long periods of time.

The pH of the solution is an important parameter. While water or hydrochloric acid exhibit the same trend, oxalic acid is more aggressive. For water and HCl, Pb, Sn, and, to some extent, Zn and Fe are easily extracted. For the oxalic solution, heavy elements (except for Cr and Pb) are strongly released (10 times more for Sn, Zn, Cu, and Fe), independent of structure and the temperature. The dissolution of the glassy matrix by the leaching agent is also pH dependent. For low pH, the glassy matrix is dissolved at a higher rate than for medium pH or solution at equilibrium.

Table IV. ICP-AES Analysis of the Leaching Solutions for Heavy Metals (mg/L)

25°C-1C

H2O	Hours	Fe	Cu	Zn	Sn	Cr	Pb
	0.5	0.00	0.00	0.00	0.02	0.00	0.00
	24	0.02	0,00	1.23	0,00	0,00	0.33
	48	0.03	0,00	1.52	0.05	0.01	0.07
	72	0.03	0.02	1.24	0.05	0.01	0.72
	192	0.00	0.01	0.05	0.03	0.00	0.00

HCl	Hours	Fe	Cu	Zn	Sn	Cr	Pb
	0.5	0.00	0.00	0.00	0.00	0.00	0.00
	24	0.02	0.00	0.02	0.00	0.00	0.00
	48	0.04	0.01	0.01	0.05	0.01	0.09
	72	0.02	0.03	0.02	0.08	0.02	0.08
	192	0.00	0.02	0.01	0.01	0.00	0.01

(COOH)2	Hours	Fe	Cu	Zn	Sn	Cr	Pb
	0.5	5.78	0.16	1.41	1.29	0,00	0.03
	24	4.95	0.18	1.23	1.03	0.01	0.08
	48	4.92	0.19	1.52	0.8	0.03	0.11
	72	4.91	0.22	1.24	0.91	0.02	0.16
	192	3.87	0.22	1.52	0.85	0,00	0.03

25°C-1V

H2O	Hours	Fe	Cu	Zn	Sn	Cr	Pb
	0.5	0.00	0.00	0.00	0.01	0.00	0.00
	24	0.19	0.00	0.00	0.00	0.00	0.07
	48	0.02	0.00	0.00	0.05	0.01	0.08
	74	0.03	0.01	0.00	0.07	0.02	0.27
	192	0.00	0.01	0.00	0.03	0,00	0.02

HCl	Hours	Fe	Cu	Zn	Sn	Cr	Pb
	0.5	0.00	0.00	0.01	0.02	0.00	0.00
	24	0.02	0.00	0,00	0,00	0.00	0.02
	48	0.08	0.01	0.04	0.06	0.02	0.34
	72	0.05	0.03	0.04	0.09	0.02	0.06
	192	0.00	0.01	0.02	0.03	0.00	0.02

(COOH)2	Hours	Fe	Cu	Zn	Sn	Cr	Pb
	0.5	5.6	0.14	1.6	1.2	0,00	0.05
	24	5.23	0.11	1.34	1.05	0.01	0.53
	48	5.45	0.18	0.132	0.8	0.01	0.11
	72	5.07	0.13	1.78	0.98	0.09	0.62
	192	4.46	0.18	1.8	0.97	0.03	0.04

90°C-1C

H2O	Hours	Fe	Cu	Zn	Sn	Cr	Pb
	0.5	0.00	0.00	0.02	0.01	0.00	0.02
	1	0.00	0.00	0.02	0.03	0.01	0.05
	4	0.05	0.00	0.02	0.08	0.02	0.09
	24	0.07	0.00	0.02	0.49	0.02	0.24

HCl	Hours	Fe	Cu	Zn	Sn	Cr	Pb
	0.5	0.00	0.00	0.02	0.01	0.00	0.02
	1	0.00	0.00	0.02	0.04	0.01	0.05
	4	0.00	0.00	0.04	0,00	0,00	0.02
	24	0.01	0.00	0.04	0.38	0.02	0.47

(COOH)2	Hours	Fe	Cu	Zn	Sn	Cr	Pb
	0.5	4.16	0.09	0.97	0.65	0.02	0.08
	1	4.41	0.11	1.32	0.82	0.03	0.09
	4	3.78	0.13	1.36	0.74	0.01	0.06
	24	4.02	0.24	1.23	1.02	0.01	0.14

90°C-1V

H2O	Hours	Fe	Cu	Zn	Sn	Cr	Pb
	0.5	0.00	0.00	0.00	0.00	0.00	0.01
	1	0.00	0.00	0.02	0.03	0.03	0.45
	4	0.05	0.00	0.04	0.09	0.09	0.11
	24	0.15	0.00	0.09	0.44	0.44	0.27

HCl	Hours	Fe	Cu	Zn	Sn	Cr	Pb
	0.5	0.00	0.00	0.02	0.00	0.00	0.02
	1	0.00	0.00	0.04	0.01	0.01	0.01
	4	0.00	0.00	0.04	0.01	0.01	0.01
	24	0.00	0.00	0.04	0.46	0.03	0.03

(COOH)2	Hours	Fe	Cu	Zn	Sn	Cr	Pb
	0.5	3.58	0.09	1.1	1,00	0.04	0.04
	1	4.44	0.11	1.36	0.92	0.05	0.05
	4	3.59	0.11	1.65	0.71	0.06	0.06
	24	3.74	0.24	1.87	1.28	0.08	0.08

Characterization of the Altered Samples

Pitting attack was observed on samples 1C and 1V exposed to H_2O and HCl. Pit size distribution varied from less than one micrometer up to several μm for long term tests. The corrosion concept can be applied to the alteration of the solid waste which is typical of a selective mechanism (i.e. the vitreous matrix is strongly attacked while the crystals are not, which is consistent with studies in the literature).[8-9] This feature is more important when temperature of the leaching solution is increased. Materials 1V and 1C altered by oxalic acid, develop large pits of several micrometers in the beginning of the experiment due to corrosion of the amorphous matrix. Meanwhile, large plates with rosace texture enriched in Ca are developed on the surface. These layers are crystallized and the XRD patterns are typical of those of calcium oxalate (Whewellite). This is representative of a dissolution-reprecipitation process under leaching test.

CONCLUSION

Knowledge of the partition of toxic metals between the mineralogical phases and particularly the partition between the vitreous and the crystallized phase is a fundamental requisite to assess the potential mobility and release of hazardous elements in the environment. The temporal dissolution of an amorphous and a crystallized ultimate glassy waste into three different leaching agents (water, HCl, and oxalic acid) at different temperatures (25°C and 90°C) has been determined. Dissolution modes are selective for sample 1C and 1V in water and HCl solutions at 25°C and 90°C, while it is congruent in oxalic solution at 25°C and 90°C. The behavior of the heavy metals released shows that for water and HCl: Pb, Sn, and to some extent Zn and Fe are more easily extracted. For oxalic solution, heavy elements are strongly released regardless of the structure and the temperature.

All of these observations confirm that the pH is the most important parameter concerning the heavy metal extraction from solid to solution rather than microstructure, microtexture, and temperature. In fact, the pH controls the dissolution of the network of the ultimate glassy waste. As soon as the pH does not evolve toward solutions at equilibrium, the dissolution of the glassy network can progress as soon as no interphase layer is formed. Thus, the oxalic leaching agent strongly dissolves the matrix (dissolution reprecipitation process) while the water and HCl are more selective. This implies that crystals that concentrate the heavy metals are in contact with the leaching agent during longer times for low pH values rather than for medium pH. The decrease of the concentration in heavy metals with time, according to the recent TEM[10] data, is controlled by the formation of a fine layer on the surface of the crystals. This phenomenon is analogues to those observed in passive oxidation conditions on ceramic glass. In this study, the interphase developed at the altered samples is a sol-gel type. According to the work of Bowel[11] and Emsley[12], the concentrations of heavy metals within the leaching solution of the two altered samples are under the limits of toxicity. Vaporization of the ultimate glassy wastes elaborated by arc plasma torch is therefore possible.

REFERENCES

[1]R. Meunier, "L'arc Electrique Pour le Traitement de REFIOM Toxiques," Congrés Procedes de Solidification et de Stabilization des Dechets, Nancy, 28 Nov-1 Dec., pp. 138-142 (1995)

[2]A. Gauthier, J. H. Thomassin and P. Le Coustumer, "A New Recyclable Ceramic Glass for Aggressive Environments," First International Conference on Composite Materials and Energy, Enercomp 95, May 8 to 10, Montreal, Canada (1995)

[3]E. Jeanroy, "Analyse Totale par Spectrometrie d'Absorption Atomique des Roches, Sols, Minerais, Ciments apres Fusion au Metaborate de Strontium, "Analusis, 2, pp. 703-712 (1974)

[4]B. Clozel, "Caracterisation Mineralogique et Cristallochimique de Vitrifiats," Congres Procedes de Solidification et de Stabilization des Dechets, Nancy, 28 Nov-1 dec., pp. 53-57 (1995)

[5]J. C. Touray, "La Dissolution des Mineraux: Aspects Cinetiques," p. 109, edited by Masson, Paris (1980)

[6]H. Atassi, "Evaluation de la Resistance a la Corrosion en Solution Acqueuse de Quelques Verres Silicates," p. 111, PhD thesis, Louis Pasteur University, Strasbourg, France (1989)

[7]T. Advocat, "Lew Mecanismes de Corrosion en Phase Acqueuse du Verre Nucleaire R7T7," p. 213, PhD thesis, Louis Pasteur University, Strasbourg, France (1991

[8]H. Scholze, "Glass, Nature, Structure and Properties," pp. 341-353, edited by Springer-Verlag, New York (1991)

[9]E. Yatsu, "The Nature of Weathering," pp. 238-283, edited by Sozosha, Tokyo (1988)

[10]P. Colombel, "Etude du Comportement a Long Terme de Vitrifiats de REFIOM," p.142 PhD thesis, University of Poiters, France (1996)

[11]H.J.M. Bowen, "Environmental Chemistry of the Elements," p. 124, edited by Academic Press, London (1979)

[12]J. Emsley, "The Elements," pp. 64-78, edited by Oxford University Press, Oxford (1992)

STRUCTURAL PARAMETERS AND LEACHABILITY OF A GLASS-BONDED CERAMIC WASTE FORM

Tanya L. Moschetti*
Purdue University
West Lafayette, IN 47907

Steven M. Frank and Stephen G. Johnson
Argonne National Laboratory
PO Box 2528
Idaho Falls, ID 83403-2528

ABSTRACT

Lattice parameters for a crystalline material can be obtained by several methods, notably by analyzing x-ray powder diffraction patterns. By utilizing a computer program to fit a pattern, one can follow the evolution or subtle changes in the structure of a crystalline species in different environments. This work involves such a study for an essential component of the ceramic waste form that is under development at Argonne National Laboratory. Zeolite 4A and zeolite 5A are used to produce two different types of waste forms: a glass-bonded sodalite and a glass-bonded zeolite, respectively. Changes in structure during production of the waste forms are discussed. Specific salt-loadings in the sodalite waste form are related to relative peak intensities of certain reflections in the XRD patterns. Structural parameters for the final waste forms will be given as well as their leachability under standard conditions.

INTRODUCTION

Argonne National Laboratory is developing an electrometallurgical process for the treatment of metallic spent nuclear fuel to condition the material for disposal in a geologic repository. In this process, the spent fuel elements are lowered in a steel basket into a molten salt electrolyte and anodically dissolved from the cladding. Pure uranium is deposited on steel cathodes while the transuranic elements and fission products remain in the salt or with the cladding hulls. The transuranic elements and active fission products are left in the salt to be incorporated into a ceramic waste form. A metal waste form incorporates the fission products that are "noble" to the process and remain with the cladding hulls.

PRODUCTION OF THE CERAMIC WASTE FORM

The ceramic waste forms being studied incorporate a eutectic salt that contains surrogate fission products. Table I gives the general composition of the salt. The surrogate fission products are in these forms: KBr, RbCl, $SrCl_2$, YCl_3, KI, CsCl, $BaCl_2$, $LaCl_3$, $CeCl_3$, $PrCl_3$, $NdCl_3$, $SmCl_3$, and $EuCl_3$. The surrogate fission products were not included in the structural model for lattice parameter determinations and computer simulations of XRD patterns. This assumption is justified because the elements Li, K, Na, and Cl account for 94.8 wt. % or 98.8 at. % of the salt.

Table I. Composition of salt

Compound	Wt. %
(Li,K)Cl	85.4
NaCl	6.1
Surrogate fission products	8.5

Results are presented for two candidate waste forms: a glass-bonded zeolite and a glass-bonded sodalite. To produce the zeolite waste form, zeolite 5A, $Na_{48}Ca_{24}Al_{96}Si_{96}O_{384}$, is dried until the water content is less than 0.5 wt%. The zeolite powder is next hot mixed with the salt to obtain an average salt loading of 60 Cl^- per unit cell (uc), or 21 wt%. The blended zeolite is then mixed with glass in a 50:50 weight ratio and processed in a hot isostatic press (HIP) at about 823K and 155 MPa.[1] In the sodalite waste form, dried zeolite 4A is blended with salt to obtain three different, average salt-loadings: 25.6, 28.0, and 30.4 Cl^-/uc (approximately 10.5 wt%). The zeolite and glass were then mixed so that the weight ratio of zeolite to glass is 75:25. This mixture was hot isostatic pressed (HIPed) at 1123K and 170 MPa.[1] Due to more aggressive processing conditions, the zeolite 4A transforms in-situ to sodalite. Each zeolite unit cell forms 16 sodalite unit cells so that the salt loadings for the sodalite unit cells are 1.6, 1.75, and 1.9 Cl^-/uc.

EXPERIMENTAL

X-ray diffraction patterns were obtained with a Scintag X1 diffractometer (Cupertino, CA), which utilizes Cu-K_α radiation and has a theta-theta goniometer configuration. Samples of the final waste forms were run with LaB_6 (NIST SRM 660) as an internal standard to check peak positions and profiles during structural refinement. Initial pattern matching routines were run using the Diffraction Management System Software by Scintag, Inc. Structure determination and refinement of lattice parameters were done using the General Structure Analysis

System[2]. Computer simulations of XRD patterns were performed with *Crystallographica* from Oxford Cryosystems (Oxford, UK).

Product consistency tests (PCT-B) were also performed according to ASTM C1285-94.[3] The samples were first crushed and sieved to obtain particle sizes between 75-150 μm. Then they were placed in deionized water at 90°C. The surface area-to-volume ratios were 2340 m^{-1} for the zeolite waste form and 2240 m^{-1} for the sodalite waste form. For the sodalite waste form with 1.6 Cl^-/uc and the zeolite waste form, tests were performed for a duration of 7, 14, and 28 days.

Upon completion of the PCT, several analyses were performed. Chloride concentrations and pH were measured with ion specific electrodes. Other elemental concentrations in the leachate were determined with inductively coupled plasma-atomic emission spectroscopy and atomic absorption spectroscopy. Water content of samples before and after the PCT was determined using the coulometric Karl-Fischer method. The normalized release rates (NRR) for elements were calculated using

$$NRR = \frac{C_i}{f_i \cdot SA/V \cdot t}$$

where C_i is the concentration of the ith element in the leachate, f_i is the mass fraction of the ith element in the unleached sample, and t is the test duration.

STRUCTURE OF A CERAMIC WASTE FORM

The starting materials for the ceramic waste form are zeolite 4A and zeolite 5A. The structure for zeolite A has been well documented with numerous crystallographic studies.[4-7] Sodalite has also been studied.[8]

X-ray diffraction patterns illustrating the evolution of a zeolite waste form are shown in Figure 1. Dried zeolite 5A, in which Ca is partially substituted for Na, has a cubic structure described by the space group Fm3c. During salt occlusion, the unit cell contracts, as evident by the shift in peak positions toward higher two-theta. In addition to reflections of the Fm3c space group, superlattice reflections for the space group F432 may be present and some examples are given in Figure 1. Richardson found a space group of F432 for a zeolite 5A blended with various salts.[9] The superlattice reflections become more distinct in the HIPed zeolite waste form, which may be described by space groups F432 and Fm3c with a lattice parameter of 24.32(5) Å.

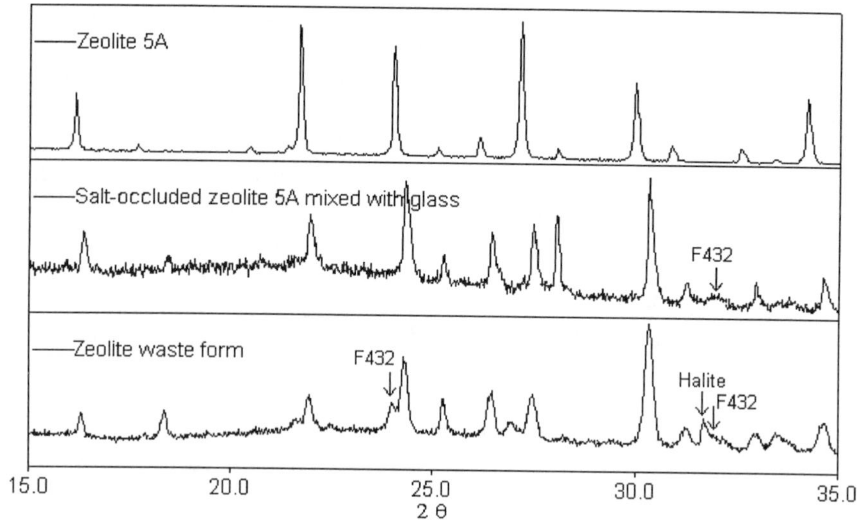

Figure 1. Evolution of a zeolite waste form.

In the HIPed product, minor phases of halite and microsommite are present. The halite phase may result from the ejection of some salt from the zeolite cages as a result of the HIP process itself or partial phase transformation of the zeolite to microsommite. Upon closer inspection, the peak positions of HIPed zeolite waste form appear slightly shifted from those of the salt-occluded zeolite 5A toward lower two-theta, which may be due to ejection of salt. Removal of salt would cause an expansion of the unit cell.

An illustration of the transformation to sodalite is given in Figure 2. Dried zeolite 4A, which contains only Na counter-ions, may be described with the space group Fm3c. The salt-loaded zeolite 4A also appears to show superlattice reflections of F432; an example of such a reflection is indicated in Figure 2. Richardson also determined that zeolite 4A may be described by the space groups of F432 and Fm3c.[9,10] A halite phase is observed after the HIP process for the sodalite waste form with a salt loading of 1.9 Cl^-/uc. In the sodalite waste forms with 1.6 Cl^-/uc of 1.75 Cl^-/uc, there may be a trace of halite but the evidence is inconclusive. The XRD patterns demonstrating this are shown in Figure 3. Nepheline, $NaAlSiO_4$, is also observed as a minor phase. The sodalite phase is described by the space group P43n, and the sodalite waste form with a salt-loading of 1.6 Cl^-/uc has a lattice parameter of 8.838 Å.

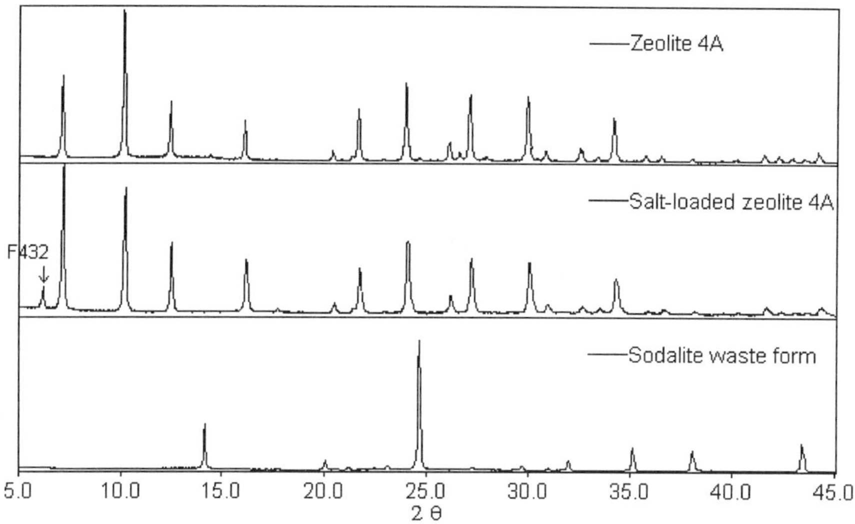

Figure 2. Transformation to a sodalite waste form.

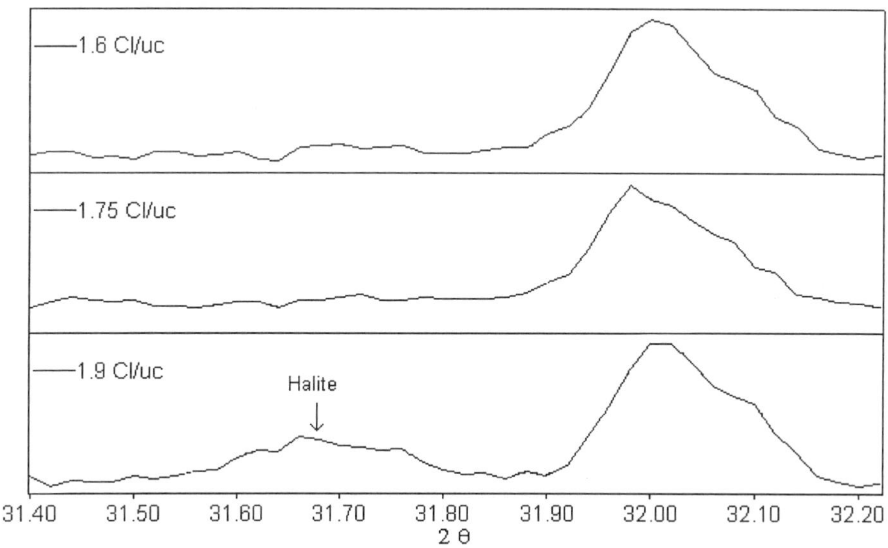

Figure 3. Halite presence in sodalite waste form with different salt loadings.

Different salt-loadings in the sodalite also affect the relative peak intensities in the XRD patterns. The values of relative peak intensities depend on atomic positioning and occupancy. Increased salt content in the sodalite unit cell reduces the intensity of the 100 reflection (I_{100}) relative to the intensity of the 211 reflection (I_{211}). Table II gives ratios for the peak intensities of these reflections for observed and computer-simulated patterns. The ratios for the computer-simulated pattern correspond very closely with those of the observed XRD patterns.

Table II. I_{100} / I_{211} in the sodalite waste form.

XRD pattern	Salt-loading (Cl^- /uc)		
	1.6	1.75	1.9
Observed	0.55	0.49	0.44
Computer simulation	0.55	0.49	0.43

RESULTS OF LEACH TESTS

The waste forms were studied to determine leach resitance of the waste forms under standard conditions. For comparison, the ARM-1 reference waste glass was used. The concentrations of B, Li, Na, and Si in the leachate from the 7-day PCT corresponded well with those found in a reference material study.[11]

Table III lists normalized release rates of elements for the 28-day PCT. Rates are normalized for mass fraction. Elements are assigned to the principal phases: glass (G), zeolite (Z), or sodalite (S). The zeolite waste form preferentially leached alkali cations and chloride ions. The normalized release rates of the sodalite waste form (1.6 Cl^-/uc) give evidence for congruent dissolution of the sodalite phase. The sodalite waste form appears to have better retention of the salt.

Table III. Normalized release rates from 7-day PCT given in $g/m^2 \cdot d$

Element	Phases	ZWF	SWF	ARM1 glass
Li	Z/S	1.4E+00	4.0E-02	3.5E-02
Na	G/Z/S	1.7E+00	1.2E-02	3.0E-02
K	G/Z/S	1.3E+00	9.7E-03	N/A
Rb	Z/S	1.3E+00	N/A	N/A
Cs	G/Z/S	2.4E-01	5.1E-03	N/A
Ca	G/Z/S	3.1E-02	N/A	N/A
B	G	4.1E-02	1.8E-02	3.1E-02
Al	G/Z/S	4.9E-03	1.2E-02	N/A
Si	G/Z/S	2.0E-03	9.8E-03	1.8E-02
Cl	Z/S	3.1E+00	1.7E-02	N/A

A summary of the water content measurements for the zeolite waste form and the sodalite waste form (1.6 Cl^-/uc) is given in Table IV. The waste forms were prepared with material that had a water content of less than 0.5 wt. %. The zeolite waste form, upon sitting out in air has increased to 2.4 wt. %, and the leach test, not surprisingly, increases the water content even more. However, the water content of the sodalite waste form remains below 0.5 wt. %, even after a 28-day PCT.

Table IV. Water content in ceramic waste forms

Sample	Exposure	Water content (wt. %)
Zeolite waste form	Air	2.4
Zeolite waste form	28-day PCT	5.1
Sodalite waste form	Air	0.01
Sodalite waste form	28-day PCT	0.40

CONCLUSION

Salt-occlusion contracts the zeolite and sodalite unit cells. In the sodalite waste form, the change in salt-loading for a range of 1.6 Cl^-/uc to 1.9 Cl^-/uc is more evident in changes in relative peak intensities than in the lattice parameter. This was shown specifically for the 100 and 211 reflections of the sodalite waste form. Computer simulations of the salt-loadings were in good agreement with the observed data.

The performance of the waste forms has been given in terms of normalized release rates of important constituents. The degrees of hydration of the waste forms differ dramatically and will be the subject of further research. The sodalite waste form has a better retention of salt. The lower hydration in the sodalite waste form also suggests that the migration of water into the sodalite waste form is slower. This may be partly due to the smaller apertures in the framework of the sodalite, which are dependent upon lattice parameters, space group, and atomic positions.

REFERENCES

[1]Goff, K. M., R. W. Benedict, K. Bateman, M. A. Lewis, C. Periera, C. A. Musick, "Spent Fuel Treatment and Mineral Waste Form Development at Argonne National Laboratory-West", *Spectrum '96*, ANS, Seattle, WA, 2436-2443 (1996).

[2]Larson, A. C. and R. B. Von Dreele, GSAS, General Structure Analysis System, Los Alamos National Laboratory Report LAUR86-748 (1986).

[3]ASTM C1285-94, "Standard Test Methods for Determining Chemical Durability of Nuclear Waste Glasses: The Product Consistency Test (PCT)", *Annual Book of ASTM Standards*; Philadelphia, PA (1995).

[4]Reed, T. B. and D. W. Breck, "Crystalline Zeolites. II. Crystal Structure of Synthetic Zeolite, Type A", *J. Am. Chem. Soc.* 78:5972-5977 (1956).

[5]Gramlich, V. and W. M. Meier, "The Crystal Structure of Hydrated NaA: A Detailed Refinement of a Pseudosymmetric Zeolite Structure", *Z. Kristallogr.* 133:134-149 (1971).

[6]Adams, J. M., D. A. Haselden, and A. W. Hewat, "The Structure of Dehydrated Na Zeolite A (Si/Al = 1.09) by Neutron Profile Refinement", *J. Solid State Chem.* 44:245-253 (1982).

[7]Faux, D. A., W. Smith, and T. R. Forester, "Molecular Dynamics Studies of Hydrated and Dehydrated Na^+-Zeolite-4A", *J. Phys. Chem. B*, 101:1762-1768 (1997).

[8]Hassan, I. and H. D. Grundy, "The Crystal Structures of Sodalite-Group Minerals", *Acta Cryst.* B40:6-13 (1984).

[9]Richardson, Jr., J. W., M. A. Lewis, and B. R. McCart, "Cation Segregation in Simulated Radioactive-Waste Zeolite-A Mixtures", *Zeolites and Related Microporous Materials: State of the Art 1994*, Studies in Surface Science and Catalysis, Vol. 84., Elsevier Science B.V(1994).

[10]Richardson, J.W., Jr., "Salt-Occluded Zeolite Waste Forms: Crystal Structures and Transformability," *Mat. Res. Soc. Symp. Proc.*, 465:395-400 (1997).

[11]Jantzen,, C. M., N. E. Bibler, D. C. Beam, C. L. Crawford, and M. A. Pickett, "Characterization of the Defense Waste Processing Facility (DWPF) Environmental Assessment (EA) Glass Standard Reference Material (U)", Westinghouse Savannah River Co. report WSRC-TR-92-346 (1993).

ACKNOWLEDGEMENTS

Argonne National Laboratory is operated for the US Department of Energy by the University of Chicago. This work was supported by the Department of Energy, Nuclear Energy Research and Development Program, under contract no. W-31-109-ENG-38. Support was also received from DOE through the Oak Ridge Institute for Science and Education, under contract no. DE-AC05-76OR00033. Special thanks to D.G. Cummings and S.A. Barker for leachate analyses, M. Noy and T. DiSanto for assistance in performing leach tests, and M. Simpson and K. Bateman for preparing the waste forms.

EXAMINATION OF SUBAERIALLY ALTERED BASALTIC GLASS WITH TEM AND EELS

J.S. Luo and W.L. Ebert
Argonne National Laboratory, Argonne, IL 60439

ABSTRACT

We have examined the weathered surfaces of 720 year old Hawaiian basalt glasses that were recovered from a subaerial environment with high-resolution transmission electron microscopy (TEM) and energy filtered imaging and electron energy loss spectroscopy (EELS) techniques. Whereas the alteration products (palagonite) were physically detached from the underlying glass in most samples, a gel-like amorphous layer was observed adjacent to the glass in a few samples. To our knowledge, this is the first time a gel layer has been observed on weathered basalt. This is significant because analogous gel layers have been observed on nuclear waste glasses reacted in laboratory tests, and this demonstrates an important similarity in the mechanisms of the weathering of basalt and the corrosion of waste glasses.

INTRODUCTION

Because of the similarity in composition, basalt glasses have been utilized as a natural analogue for nuclear waste glasses to predict long-term performance in a geologic repository.[1] Basalt and other naturally occurring glasses serve as natural analogues for both thermodynamic and kinetic aspects of waste glass corrosion. The alteration phases formed during the corrosion of basalt under natural conditions can be compared to the phases formed during reaction of samples of the same basalt in laboratory tests designed to accelerate corrosion. This comparison provides confidence that the test method accelerates corrosion to the same end-state that occurs naturally, and an indication of the degree of acceleration provided by specific test conditions. The test method can then be used with confidence to accelerate the corrosion of nuclear waste glasses to a similar degree.

Another use of basalt as a natural analogue is through comparison of the corrosion products that are formed during reaction of nuclear waste glasses in laboratory tests with those formed by the weathering of basaltic glasses in nature. These comparisons provide insight into the effects of glass composition on the phases that form as the glass corrodes and whether radionuclides become incorporated into common mineral phases. Detailed analysis of the altered surface can also be used to gain information regarding the reaction path and the rate-controlling process.

In this work, we have characterized the glass/palagonite interface of subaerially altered Hawaiian basalt glass by using a combination of high-resolution transmission electron microscopy (TEM) with associated energy dispersive x-ray spectroscopy (EDS) and electron energy loss spectroscopy (EELS). In our previous investigation of the natural alteration of Hawaiian basalt glasses, we identified the palagonite layer to contain mainly iron oxide hydroxide (goethite), aluminum hydroxide hydrate (scarbroite), and an amorphous phase that was rich in Fe and Al. The purpose of the present study was to examine the microstructure and chemical nature of the palagonite/glass interface to compare the alteration processes that dominated in a 700-year-old Hawaiian basalt glass with those that dominate the corrosion of waste glasses to characterize the reaction pathway of glass corrosion.

EXPERIMENTAL

The samples of volcanic basaltic glass were provided by curators of Bishop museum from material collected from an archaeological site at Oahu, Hawaii. The site is located about 85 cm below the ground level and provided a wet and cool environment. Although its exposure history is not known, the basalt had likely been altered through contact with meteoric water and the humid atmosphere. The age of the sample had been determined to be about 720 ± 80 years by radiocarbon dating of the occupation layer of charcoal. The palagonite layer formed on the surface had a yellowish color. The palagonite was removed from one sample and the underlying glass was chemically dissolved and analyzed with inductively coupled plasma atomic emission spectroscopy (ICP). The analyzed composition of the glass is given in Table 1 on an oxide basis.

Samples were prepared for TEM analysis of the palagonite/glass interface by fixing small chips of the basalt that had been carefully removed onto an epoxy resin block. The chip was carefully oriented so that samples for analysis of the palagonite/glass interface could be prepared. After a small amount of resin was placed on the surface of the specimens, they were placed in a vacuum to facilitate penetration of the resin into the surface layer and elimination of air bubbles. After the resin had cured, TEM samples were cut with a Reichert-Jung Microtome to a thickness of about 50 nm. Samples were examined with a JEOL 2000FX unit operating at 200 kV. The microscope is equipped with an ultra-thin window energy dispersive X-ray spectrometer (EDS) and a Gatan Filtering Imaging (GIF) system. The dispersion of the spectrometer was set to 0.5 eV/channel and calibrated using the ls/π transitions on the K edges of carbon at ΔE = 285 eV and oxygen at ΔE = 532 eV. The energy resolution was estimated to be about 1.9 eV based on the measured full width at half maximum of the zero-loss peak. Chemical analysis by EDS was performed with a electron beam focused to about 15 nm; this corresponds roughly with the spatial resolution of the EDS analysis. Unreacted glass beneath the interface was used as a standard for the quantitative determination of major components in the alteration products by means of the well-established Cliff-Lorimer method.[7]

RESULTS AND DISCUSSION

Figure 1 shows two types of interfaces that were observed in this investigation. The glass had chattered in all TEM samples during microtome sample preparation parallel to the knife edge. The alteration phases did not chatter because they are less brittle than the glass and because they were impregnated with the epoxy. The interface in Fig. 1a is that typically observe with TEM. The alteration layer was usually seen to be detached from the glass surface. The presence of epoxy between the layer and the glass indicates the layer had separated from the glass prior to microtomy, which suggests that the layer was only loosely fixed to the glass. The high-magnification image of the glass surface shows no evidence of alteration (see inset in Fig. 1a). Characterization of the edge of the glass nearest the palagonite with EDS line profiling indicated a slight depletion of alkali metals at the outer 0.1 μm. No variation in silicon or aluminum was observed. Similar palagonite/glass interfaces have been observed in basalt glasses reacted in sea water,[2] glacial water,[5] and samples recovered from Deep Sea Drilling Project (DSDP).[8]

The majority of the palagonite consisted of hemispherical masses that were identified as goethite [FeO(OH)] by selected area electron diffraction (SAED) and EDS. The details of identification of the phases that were present in palagonite will be published elsewhere.[6] The formation of a layer comprised predominantly of goethite is consistent with the very low solubilities of Fe, Ti, and Mn. Other major elements of basalt (e.g., Na and Ca) are more soluble and remain in solution. Other crystalline precipitates that were identified generally had a layered or fibrous texture characteristic of smectite-type clays.[6]

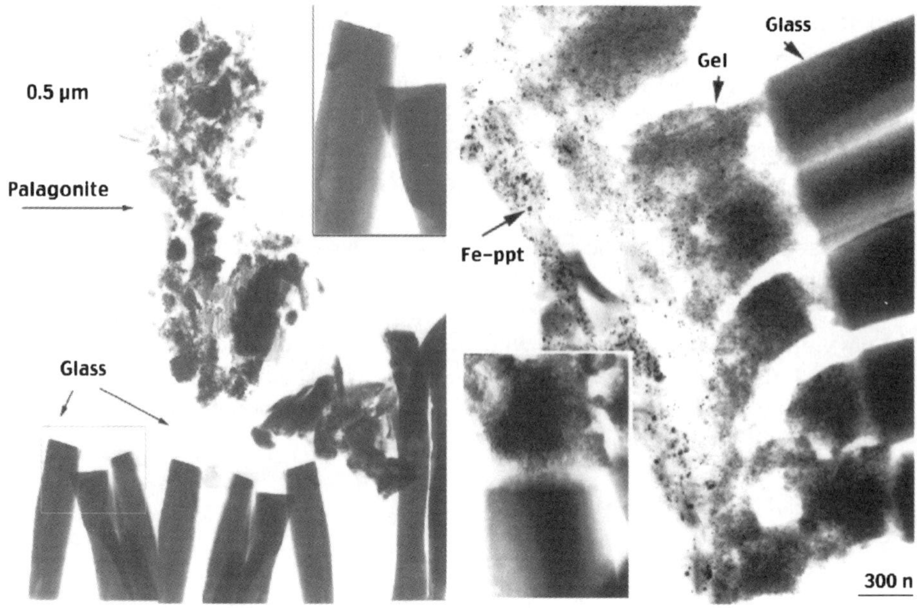

(a) (b)

Fig. 1. Transmission electron micrograph of (a) typically observed glass/palagonite interface and (b) interface with gel layer.

Figure 1b is a TEM image of the glass/palagonite interface on another sample recovered from same location. This sample exhibits a very different microstructure than the sample shown in Fig. 1a. In the sample in Fig. 1b, a layer as thick as 0.5 µm in some areas was observed to remain fixed at the outer surface of the glass. The image of this layer had a uniform amorphous contrast and the layer is referred to as a gel layer. The gel layer remained attached to the underlying glass (see inset in Fig. 1b). Electron diffraction confirmed that the gel layer was mostly amorphous, although small crystalline domains reminiscent of clays were seen within the layer. Many small dark spots were observed to be distributed throughout the layer. These spots were typically about 10 nm in size, and were determined to be iron-rich by EDS analysis. The results of EDS analyses of the glass and the gel layer are summarized in Table 1. The gel layer was significantly depleted in Na, Ca, and Mg, and Si, and enriched in Fe, Ti, and Al relative to the glass. That the sum of oxides for the gel layer is appreciably less than 100% is probably due to the presence of water, which cannot be quantified with EDS.

Although many basalt/palagonite interfaces have been examined for samples recovered from various environments,[2,5,8] this is the first time, to the best of our knowledge, that a gel layer has been observed on a naturally altered basalt glass. On the other hand, gel layers have been frequently observed in laboratory reacted nuclear waste glasses, usually seen at early stages of reaction progress.[9] At more advanced stages of reaction, the gel layers formed on nuclear waste glasses transform into a clay. Elements in the gel that are not totally incorporated into the clay may form other phases.[2,5,8] Once the gel transforms into clay, subsequent glass dissolution does not form new clay. Instead, components released as the glass dissolves form clay. The observation of the gel layer in Hawaiian basalt glass provides important evidence for the similar corrosion paths of basalt and nuclear waste glasses. The reason that gel layers have not been observed previously is probably because the corrosion of most samples recovered from natural environments have progressed to the point where the gel has transformed to other phases.

Table 1. Results of Analysis of Glass and Gel Layer

Oxide	ICP Glass, wt%	EDS Glass, wt%	EDS Gel, wt%	Oxide	ICP Glass, wt%	EDS Glass, wt%	EDS Gel, wt%
Al_2O_3	14.4	15.3	17.0	MnO	0.13	0.13	0.17
CaO	9.68	10.4	0.18	Na_2O	2.81	3.70	0.00
Fe_2O_3	12.0	9.39	54.8	SiO_2	51.5	51.8	13.8
K_2O	0.50	0.69	0.17	TiO_2	2.05	1.61	2.07
MgO	6.69	6.90	0.00	others	0.24	0.08	11.8

To further investigate the nature of the gel layer, EELS spectra were generated with GIF for the Si-K edges of both the gel layer and the glass. The results were then used to calculate the momentum transfer-weighted scattering factor, $k^2\chi(k)$, and the radial distribution function (RDF), both of which provide information on local environments of Si in the glass or in the gel layer. Figure 2 shows the EELS spectra of background-subtracted Si-K edges acquired from the glass and the gel layer. Both curves feature similar energy-loss near-edge structure for Si atoms in the layer and in glass. There is no indication of extra pre-edges or substantial peak shifts, which suggests that the valence states of Si atoms are similar in the glass and gel layer. Figure 2b and 2c shows the $k^2\chi(k)$s and RDFs of

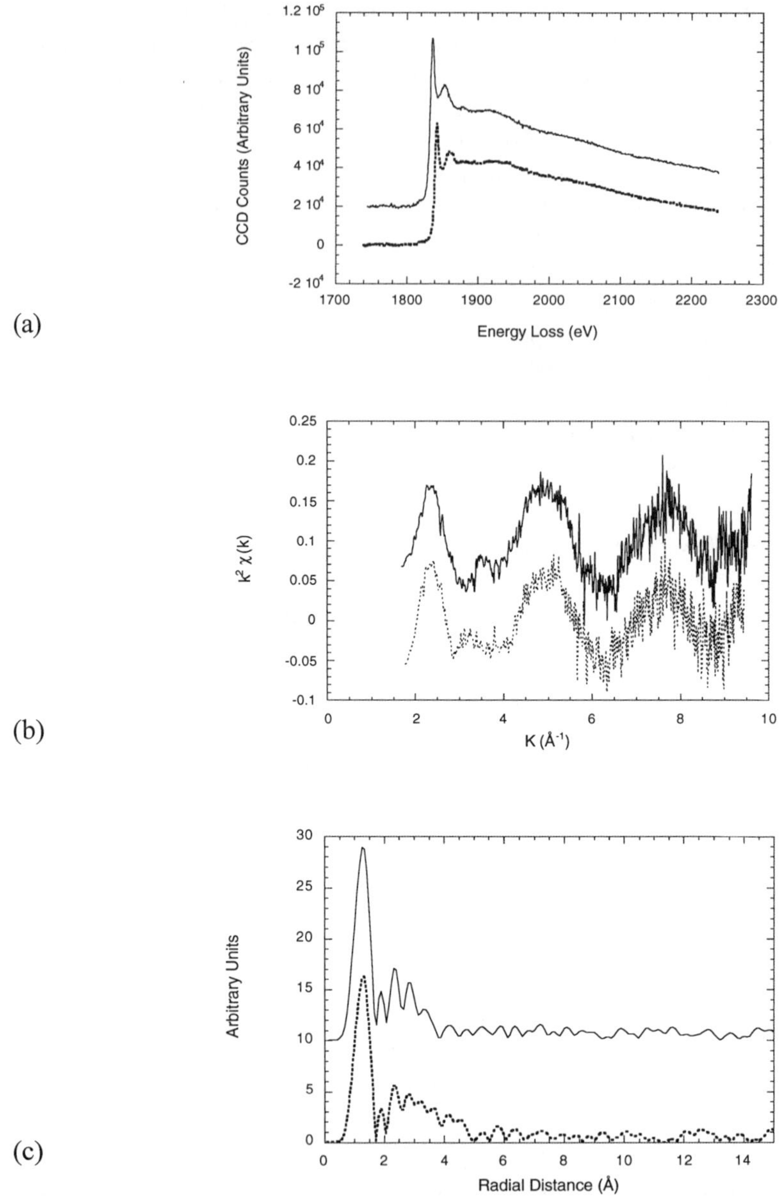

Fig. 2. (a) Si-K ionization edges, (b) EXELFS, and (c) RDF from the Si-K edges in gel layer (upper spectra, solid lines) and in glass (lower spectra, dotted lines).

Si-K edges in the glass and in the layer, respectively. Both of these functions provide information regarding the local order around the Si atoms. Again, the spectra have very similar features; in particular, both first peaks of RDFs, which represent the Si-O bond lengths, are located at nearly identical positions (1.62Å). These results reflect that both the glass and the layer have comparable Si-O structural arrangements, indicating that the gel layer preserves the Si-O network of the precursor glass and has not yet transformed into a clay.

CONCLUSIONS

We have analyzed the interface between palagonite and glass in 700-year-old subaerially altered Hawaiian basalt glasses with high-resolution TEM, EDS, and GIF. Different AEM samples prepared from basalt recovered from the same vicinity displayed two significantly different interfacial microstructures. In the most commonly seen type of interface, the glass and palagonite were physically separated, the glass surface was smooth and unaltered, and the palagonite was crystallized primarily as goethite. The other type of interface showed a gel layer between the glass and the crystalline phases. The gel layer was depleted of Na, Ca, and Mg relative to the glass and had a Si environment similar to that in the glass. To our knowledge, this is the first time that a gel layer has been observed on a naturally reacted basalt glass. This observation provides an important link in the corrosion mechanism of basalt in natural environments and nuclear waste glasses in laboratory tests.

ACKNOWLEDGMENT

The authors thank Drs. J. Fortner and A. Bakel for their critical reading of the manuscript and J. Andrew for editorial assistance. The work at ANL was sponsored by the U.S. Department of Energy, Office of Environmental Management, under Contract W-31-109-ENG-38.

REFERENCES

[1]R.C. Ewing, "Natural glasses: analogues for radioactive waste forms," in *Scientific Basis for Nuclear Waste Management*, Edited by G.J. MaCarthy et al., Plenum Press, New York, pp. 57-68 (1978).

[2]J.L. Crovisier, J. Honnores, and J.P. Eberhart, "Dissolution of basaltic glass in seawater: Mechanism and rate," *Geochim. Cosmochim. Acta*, **51** pp. 2977-2990 (1987).

[3]H. Furnes, "Element mobility during palagonitization of a subglacial hyaloclastite in Iceland," *Chem. Geol.*, **22** pp. 249-264 (1978).

[4]B. Grambow, M.J. Jercinovic, et al., "Weathered basalt glass: A natural analogue for the effects of reaction progress on nuclear waste alteration," *MRS Symposium Proceedings,* **50** pp. 263-272 (1985).

[5]M.J. Jercinovic, K. Keil, M.R. Smith, and R.A. Schmitt, "Alteration of basaltic glasses from north-central British Columbia, Canada," *Geochim. Cosmochim. Acta*, **54** pp. 2679-2696 (1987).

[6]J.S. Luo, W.L. Ebert, and J.K. Bates, "Formation and evolution of palagonites formed on subaerially altered Hawaiian basaltic glass: An analytical electron microscopy study," submitted to Chemical Geology.

[7]J.S. Luo, S. Wolf, W. Ebert, and J.K. Bates, "Quantitative SEM/EDS analysis of high-level waste glasses," in *Proceedings of the 29th Annual Conference of the Microbeam Analysis Society*, Edited by E.S. Etz, VCH Publishers Inc., New York, p375 (1995).

[8]Z. Zhou and W.S. Fyfe, "Palagonitization of basaltic glass of DSDP Site 335, Leg 37: Texture, chemical composition, and mechanism of formation," *Am. Miner.*, **74** pp. 1045-1053 (1989).

[9]W.L. Ebert, J.K. Bates, et al., "The long-term alteration of Borosilicate waste glasses," Ceram. Trans., **39** pp. 333-340 (1993).

AN APPROACH TO THERMOCHEMICAL MODELING OF NUCLEAR WASTE GLASS

Theodore M. Besmann and Edward C. Beahm
Oak Ridge National Laboratory
Oak Ridge, TN 37831

Karl E. Spear
Pennsylvania State University
University Park, PA 16802

ABSTRACT

This initial work is aimed at developing a basic understanding of the phase equilibria and solid solution behavior of the constituents of waste glass. Current, experimentally determined values are less than desirable since they depend on measurement of the leach rate under non-realistic conditions designed to accelerate processes that occur on a geologic time scale. The often-used assumption that the activity of a species is either unity or equal to the overall concentration of the metal can also yield misleading results. The associate species model, a recent development in thermochemical modeling, will be applied to these systems to more accurately predict chemical activities in such complex systems as waste glasses.

INTRODUCTION

High-level nuclear and transuranic wastes are currently foreseen as being incorporated in a host glass for permanent disposal. A large number of glasses have been explored, with borosilicate glass as the typical base composition. Prediction of the long-term stability of these waste forms requires an accurate understanding of the chemical activity of the constituent species. These are necessary for determining volatility, leaching behavior, and corrosion reactions. However, accurate thermochemical values for the constituents of glasses require a global model for the glass phase. To date, such a model is lacking, requiring the use of experimental simulation or oversimplistic assumptions. Figure 1 is a schematic of the leaching process from glass. It can be seen that any assumption regarding the activity of a species in the glass will directly influence the eventual concentration in the leachate. This illustrates the dependence of leachate concentrations on the assumption of activity for a metal oxide constituent of the glass and the need for an accurate understanding of species chemical potentials. Unfortunately, because of the lack of a suitable model for the thermochemistry of glasses, it has been necessary to either assume a metal oxide is present at unit activity or that it has an activity equal to the overall concentration of the metal in the glass. These approaches can lead to highly misleading results. Accurate predictions of leaching from a chemically complex nuclear waste glass requires accurate thermodynamic activities of its component oxides.

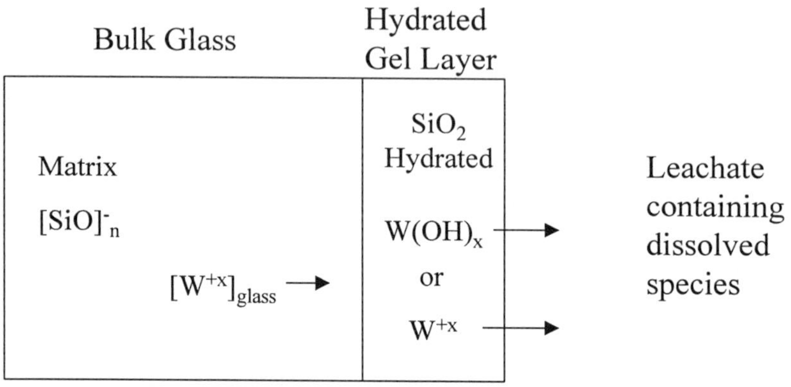

Fig. 1. Schematic of the leaching process from glass. The concentration of the leached waste species is dependent on the activity of the ion, $[W^{+x}]$, in the glass.

The work presented in this paper covers initial efforts in developing a comprehensive model for waste glass compositions. The effort is focussed on producing an accurate, easy to understand and use thermochemical representation for describing the stability of waste glasses. The model must be able to generate accurate values for thermodynamic activities of glass components as a function of temperature and glass composition.

The approach used here utilizes the common description of a glass as a supercooled liquid solution phase. It therefore requires developing an accurate model for the complex oxide liquid phase. This requires a critically assessed thermodynamic database consistent with the model for binary and ternary combinations of the major components in a typical waste glass. These systems can then be combined to represent the thermodynamic behavior of complex, multicomponent glass systems.

SOLUTION MODEL

An associate model developed in the 1980s was used to represent the thermochemical behavior of liquid oxide solutions. With an accurate model of the liquid solution, the supercooled liquid therefore represents the chemically complex nuclear waste glass. The model was initially utilized for complex solutions by Hastie, Bonnell, and co-workers (2-5). A review of much of their research incorporating this model is given in Bonnell and Hastie (5). Pantano, Spear and co-workers also used the model for predicting and explaining interface reactions in fiber-reinforced glass-matrix composite systems (6-8), and the behavior of tin and iron in float glass systems (9-10). Other thermodynamic models have been used to represent nuclear waste glass, most notably the modified quasichemical model of Pelton, Blander and co-workers (11-12) who initially used it to represent molten slag phases, and later nuclear materials, including waste glass (13). The associate model, however, is substantially easier for non-specialists to understand and use, and yet the model still accurately represents the limiting thermodynamic activities of components in these metastable glass phases.

Energies of interaction between end-member component oxides beyond those of ideal mixing often exist. These are considered explicitly in representations such as the quasichemical model (14). The use of interaction energies in such an approach results in a significant multiplication of terms making it cumbersome to develop relationships for complex systems. These energies are included in the associate model by adding "associate species," with their respective formation energies, to the solution. Thus the complex interaction terms are simply embodied in additional species in the solution. For example, in using the associate model for the Na_2O-Al_2O_3 binary oxide system examined in this paper, an ideal liquid solution phase was created from the liquid components $Na_2O(l)$ and $Al_2O_3(l)$ along with a $NaAlO_2(l)$ associate liquid species. In calculating the equilibrium state of the liquid phase containing the three species, the minimization of the total free energy determines the relative mole fractions of the species. As the system is treated as an ideal solution, the activities are by definition equivalent to the species mole fraction. The resulting activities of these species can be used to predict leaching behavior, corrosion reactions, and other important phenomena.

THERMODYNAMIC DATA AND CALCULATIONS

The thermodynamic computer program ChemSage™ (15), thermodynamic data from the SGTE database, 1996 update (16), estimated thermodynamic data, and phase diagrams from Phase Diagrams for Ceramists (17) and other sources (18-20) were used in our assessments and calculations. Unavailable data which had to be estimated included fusion enthalpies and entropies, enthalpies of formation at 298 K, absolute entropies at 298 K, and heat capacity (Cp) equations for temperatures from 298 to ~3000 K. Trends in chemical reaction values were used as described by Spear (21) for entropies and heat capacities. Enthalpies of formation were derived by adjusting values until the computed phase diagram stability regions were consonant with those of accepted, published diagrams. All values were assessed to determine if they would yield reasonable thermodynamic and phase relations over wide ranges of temperature and composition.

Data for the Na_2O-Al_2O_3 System

A search of the SGTE database (16) for the Na_2O-Al_2O_3 system revealed that it did not contain values for the fusion of $NaAlO_2$, nor any data for β''-Al_2O_3 [Na_2O -6•Al_2O_3] and β-Al_2O_3 and [Na_2O -11•Al_2O_3]. These data were estimated, including fusion enthalpies and entropies, enthalpies of formation at 298K, absolute entropies at 298K, and Cp equations for temperatures from 298 to ~3000K. A liquid phase which produced a good thermodynamic fit of the binary phase diagram (17,18), however, required only three liquid species: Na_2O, Al_2O_3, and $NaAlO_2$. Additional associate species when included in the solution phase did not improve the fit to the phase diagram.

Data for the Na_2O-SiO_2 System

Thermodynamic data for crystalline phases in the Na_2O-SiO_2 system were available from the SGTE database (16), but a liquid solution phase had to be developed and all of the system data refined by manually optimizing it with respect to the binary temperature-composition phase diagram (17,19). Species in the liquid solution included: Na_2O, Si_2O_4, $Na_{1.6}Si_{0.4}O_{1.6}$, $Na_{1.33}Si_{0.67}O_2$, $NaSiO_{2.5}$. (Note that these formulae have the same atom ratios as the typically denoted crystalline phase formulae: Na_2O, SiO_2, Na_4SiO_4, Na_2SiO_3, $Na_2Si_2O_5$.) Each associate species in the liquid phase was written so that

$$a \text{ Na} + b \text{ Si} = 2/\text{formula unit}$$

where the number of metals per formula unit equals 2. This general rule was followed for each system studied since it gives equal weight to all liquid species in the solution with regard to the entropy of mixing, and resulted in the best fits of the phase diagrams. The fundamental reasons for why better phase diagram fits were obtained when the sum of the non-oxygen atoms equaled 2/formula unit rather than 1/formula unit is not understood.

Data for the SiO_2-Al_2O_3 System

The SiO_2-Al_2O_3 system data were taken from the SGTE database (16), but the data for the mullite liquid associate required adjustment to give the best fit of the phase diagram (17,20). The liquid species included Si_2O_4, Al_2O_3, and $Al_{1.5}Si_{0.5}O_{3.25}$.

RESULTS AND DISCUSSION

Na_2O-Al_2O_3 System

Figure 2(a) is the calculated phase diagram for this system after fitting the liquid associate thermodynamic data to reproduce the melting temperatures of its stoichiometric crystalline phases. Figure 2(b) contains plots of the calculated activities of the liquid species (Na_2O, Al_2O_3, $NaAlO_2$) as a function of composition at 1200°C. The values were computed by suppressing the formation of crystalline phases and only then computing the thermochemical equilibrium state of the system. In this way the activities of the associates in the supercooled liquid can be represented. Since the associate solution model assumes ideal mixing behavior, the activities are numerically equal to the mole fractions of the respective species in the liquid solution phase. From Fig. 2(b) it is apparent that the associate $NaAlO_2$(l) dominates the liquid phase at Al_2O_3 mole fractions from about 0.25 to 0.75, and is almost unity at its stoichiometric composition in the glass. This indicates that the liquid solution energy in this composition range is dominated by the formation energy of the associate species, with very little contribution from the ideal mixing term. The strong association energy of $NaAlO_2$(l) causes the Na_2O(l) activity to be lowered from about 10^{-1} to 10^{-6} as the composition of the liquid changes from the Na_2O-rich side to the Al_2O_3-rich side of the $NaAlO_2$ liquid composition. A similar, but opposite effect occurs with the Al_2O_3(l) activity as this composition region is traversed.

Na_2O-SiO_2 System

Figure 3(a) is the calculated phase diagram for this system after fitting the liquid associate thermodynamic data to reproduce the melting temperatures of its stoichiometric crystalline phases. The calculated activities of the liquid species (Na_2O, Si_2O_4, $Na_{8/5}Si_{2/5}O_{8/5}$, $Na_{4/3}Si_{2/3}O_2$, $NaSiO_{5/2}$) are plotted in Fig. 3(b) as a function of composition at 1200°C. Again, the activities are calculated at 1200°C with the formation of crystalline phases suppressed so that these activities represent those in a supercooled liquid phase. The liquid associates dominate the liquid phase at SiO_2 mole fractions of from 0.2 to 0.8, with two of the associates attaining almost unit activities at liquid compositions corresponding to their respective stoichiometric compositions. The third associate, $NaSiO_{5/2}$, is not as stable as the other two, and thus does not dominate the liquid solution energy at its composition. As is seen from Fig. 3(b), the strong association energies of the liquid associate species in this system cause the component activities of Na_2O and Si_2O_4 to remain at very low values except near the compositional extremes.

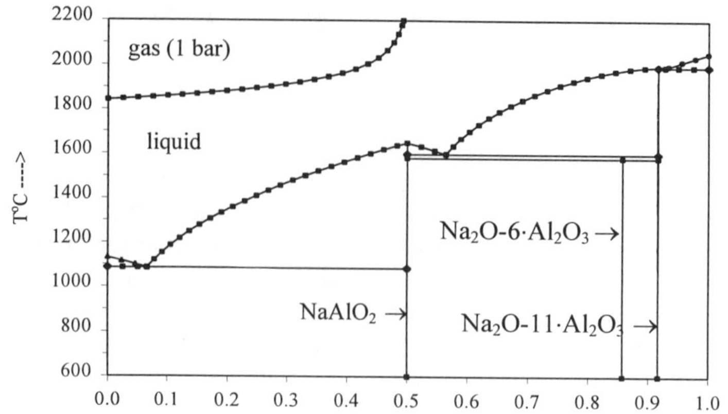

Fig. 2(a). Calculated phase diagram for the Na_2O-Al_2O_3 system using ChemSage™ (15), optimizing the thermodynamic data for the liquid phase to reproduce the melting temperatures (17,18) of the stoichiometric compound phases.

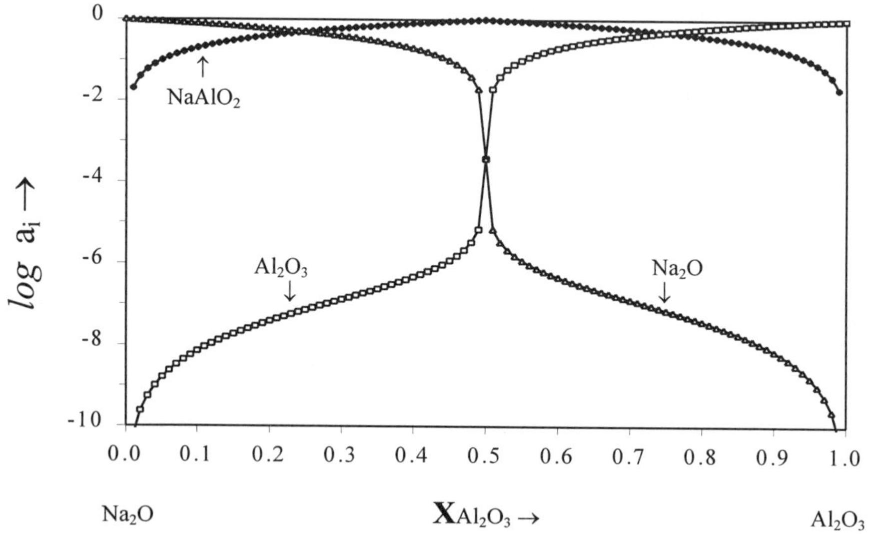

Fig. 2(b). Calculated activities of Na_2O, Al_2O_3, and $NaAlO_2$ liquid species at 1200°C for the Na_2O-Al_2O_3 system using ChemSage™ (15) using the thermodynamic data which produced the phase diagram in Fig. 2(a). The formation of equilibrium crystalline phases was suppressed.

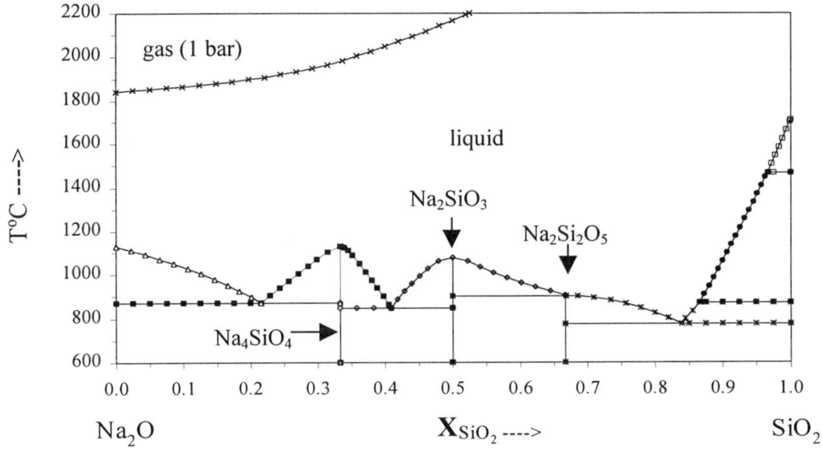

Fig. 3(a). Calculated phase diagram for the Na_2O-SiO_2 system using ChemSage™ (15), optimizing the thermodynamic data for the liquid phase to reproduce the melting temperatures (17,19) of the stoichiometric compound phases.

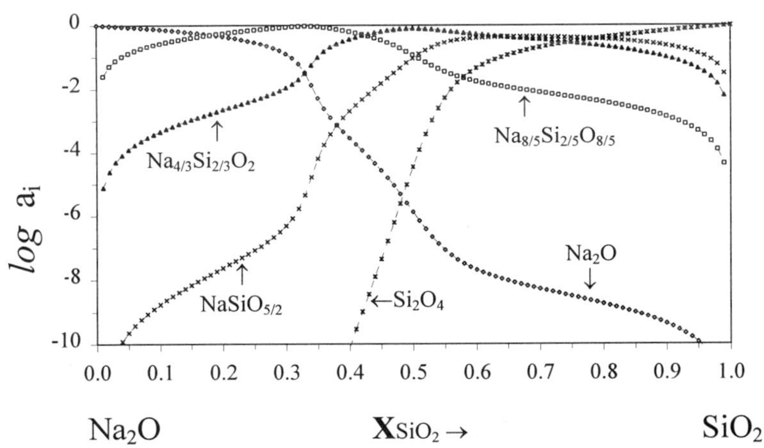

Fig. 3(b). Calculated activities of Na_2O, Si_2O_4, $Na_{8/5}Si_{2/5}O_{8/5}$, $Na_{4/3}Si_{2/3}O_2$, and $NaSiO_{5/2}$ liquid species at 1200°C for the Na_2O-SiO_2 system using ChemSage™ (15) using the thermodynamic data which produced the phase diagram in Fig. 3 (a). The formation of equilibrium crystalline phases was suppressed.

SiO_2-Al_2O_3 System

Figure 4(a) gives the calculated phase diagram for this system after fitting the liquid associate thermodynamic data to reproduce the melting temperatures of its stoichiometric crystalline phases. Figure 4(b) is a plot of the calculated activities of the liquid species (Si_2O_4, Al_2O_3, $Al_{1.5}Si_{0.5}O_{3.25}$) at 1200°C. Note the difference in the activity scale (3 orders of magnitude) as compared to the other two systems discussed in this paper (10 orders of magnitude). The liquid species Si_2O_4 and Al_2O_3 dominate the liquid phase, even at the stoichiometric composition of mullite. The weak association energy of $Al_{1.5}Si_{0.5}O_{3.25}$(l) results in appreciable mixing energy in the liquid at all compositions. This may be the result of the need to decrease the stability of the liquid mullite species in order to obtain an accurate melting temperature.

CONCLUSIONS

As demonstrated in the current work, as well as in previous studies, the associate model is a simple and accurate means for representing the activities of the constituent species of oxide melts and glasses. The approach of developing a critically assessed thermodynamic database for binary and ternary combinations of the major components in a typical waste glass is continuing with assessments of other binary alkali and alkaline earth silicates, aluminates, and borates. At the same time, the database being created will be used for producing datasets for assessments of ternary and higher order systems so that the importance of ternary associate species on the liquid solution stability can be determined. These data will then be combined to predict the thermodynamic behavior of the more complex glass systems used for high-level nuclear and transuranic wastes. In the case of leaching from a waste form, the activity of the each of the more important (higher activity) species, whether single oxide and associate species, should be used in the calculation of the equilibrium hydration reaction. This procedure should give the most accurate result with regard to equilibrium concentrations of leaching constituents.

ACKNOWLEDGEMENTS

Research supported by the DOE Environmental Management Science Program funded by the Office of Environmental Management's Office of Science and Technology, and administered jointly with the Office of Energy Research under contract DE-AC05-96OR22464 with Lockheed Martin Energy Research Corporation.

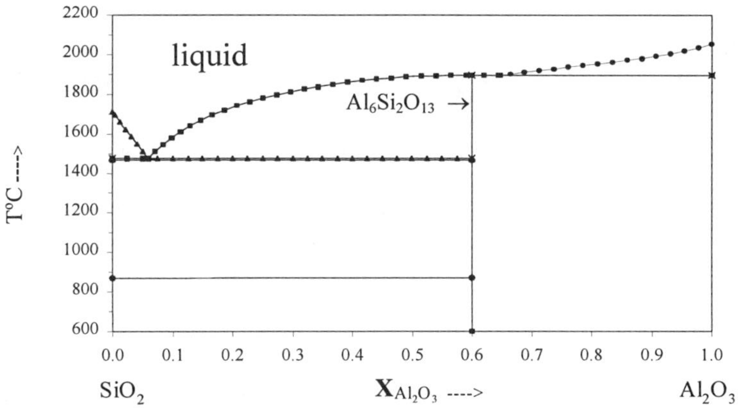

Fig. 4(a). Calculated phase diagram for the SiO_2-Al_2O_3 system using ChemSage™ (15), optimizing the thermodynamic data for the liquid phase to reproduce the melting temperatures (17,20) of the stoichiometric mullite phase.

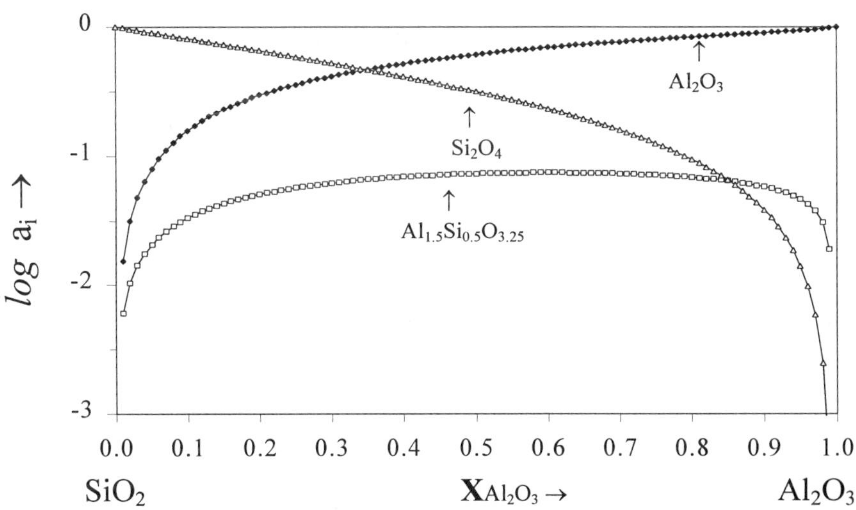

Fig. 4(b). Calculated activities of Si_2O_4, Al_2O_3, and $Al_{1.5}Si_{0.5}O_{3.25}$ liquid species at 1200°C for the SiO_2-Al_2O_3 system using ChemSage™ (15) and the thermodynamic data which produced the phase diagram in Fig. 4 (a). The formation of equilibrium crystalline phases was suppressed.

REFERENCES

1. A.J.G. Ellison and A. Navrotsky, *Scientific Basis for Nuclear Waste Management XIII* (eds. V.M. Oversby and P.W. Brown) MRS Sympos. Proc. Vol. 176, (1990) pp. 193-207.
2. J.W. Hastie and D.W. Bonnell, *High Temp. Sci.* 19, 275-306 (1985).
3. J.W. Hastie, *Pure and Appl. Chem.* 56, 1583 (1984).
4. J.W. Hastie, E.R. Plante, and D.W. Bonnell, *Vaporization of Simulated Nuclear Waste Glass, NBSIR 83-2731*, NIST, Gaithersburg, MD (1983).
5. D.W. Bonnell and J.W. Hastie, "A Predictive Thermodynamic Model for Complex High Temperature Solution Phases XI," *High Temp. Sci.* 26, 313-334 (1990).
6. K.E. Spear, P. Benson, and C.G. Pantano, *High Temperature Materials Chemistry IV* (Z.A. Munir, D. Cubicciotti, and H. Tagawa, eds.) The Electrochemical Society, Pennington, NJ (1988), pp. 345-354.
7. P.M. Benson, K.E. Spear, and C.G. Pantano, *Ceram. Eng. Sci. Proc.* 9(7-8), 663-670 (1988).
8. C.G. Pantano, K.E. Spear, G. Qi, and D.M. Beall, *Advances in Ceramic-Matrix Composites, Transactions, Vol. 38* (N. Bansal, ed.) The American Ceramic Society, Westerville, OH (1993), pp. 173-198.
9. T.E. Paulson, K.E. Spear, and C.G. Pantano, *High Temperature Materials Chemistry IX, Proceedings 97-39*, The Electrochemical Society, Pennington, NJ, 1997, pp 194-202.
10. Thomas E. Paulson, Karl E. Spear and Carlo G. Pantano, "Thermodynamic Analysis of the Tin Penetration Profile in High-Iron Float Glass," To be published in proceedings of *International Congress on Glass*, Summer 1998.
11. A.D. Pelton and M. Blander, *Met. Trans. B*, **17B**, 805 (1986).
12. M. Blander and A.D. Pelton, *Geochim. Cosmochim. Acta*, **51**, 85- (1987).
13. Arthur D. Pelton, *Pure and Applied Chem.* 69(11) 2245-2252 (1997).
14. C. H. P. Lupis, *Chemical Thermodynamics Of Materials*, p. 446, North-Holland Pub., New York, 1983.
15. Gunnar Eriksson and Klaus Hack, *Met. Trans. B*, **21B**, 1013-1023 (1990); *ChemSage™, Version 4.0*, GTT Technologies, Kaiserstrasse 100, 52134 Herzogonrath, Germany, (1998).
16. *SGTE Pure Substance Database, 1996 Version*; produced by the Scientific Group Thermodata Europe, and obtained through GTT Technologies (see ref. 14.)
17. *Phase Diagrams for Ceramists, Volumes 1-12*, (1964-1996) The American Ceramic Society, Westerville, OH.
18. G. Eriksson, Ping Wu, and A.D. Pelton, *Calphad* **17(2)**, 189-205 (1993).

19. Ping Wu, G. Eriksson, and A.D. Pelton, *J. Am. Cer. Soc.* 76(8), 2059-2064 (1993).
20. Gunnar Eriksson and Arthur D. Pelton, *Met. Trans. B*, **24B**, 807-816 (1993).
21. K.E. Spear, Chapt. 3 in *Treatise on Solid State Chemistry, Vol. 4* (N. B. Hannay, ed.), Plenum Publishing Corp., New York (1976) pp. 115-192.

LEACHING BEHAVIOUR OF TOXIC CERAMIC WASTE IMMOBILIZED IN CEMENT MATRICES

Monfort, E.; Enrique, J.E.; Cantavella, V.; Sánchez, E.
Instituto de Tecnología Cerámica
Asociación de Investigación de las Industrias Cerámicas
Universitat Jaume I. Castellón (Spain).

KEYWORDS: Immobilization, toxic ceramic waste, leaching kinetics.

ABSTRACT

A study was undertaken of the leaching kinetics of a toxic ceramic waste produced in glazed ceramic tile manufacture, which had been immobilized in a cement matrix.

The study allowed characterizing the efficiency of the stabilization/solidification process by simulating the behaviour of the material at a disposal site. The leaching test used was the method proposed by the American Nuclear Society for nuclear waste, subsequently adapted by Côté and Isabel for studying toxic waste.

Successive leachant renewals are used in the adapted test, followed by calculation of the coefficient of effective diffusivity to quantify the degree of immobilization. In these computations, mass transfer is assumed to be diffusion controlled, so that the coefficient of effective diffusivity should be independent of leachant renewal frequency.

However, on analyzing the resulting data for the various toxic elements found, the coefficient of effective diffusivity was observed to depend on leachant renewal frequency. A kinetic model is proposed in this study, which allows obtaining a coefficient of effective diffusivity that is independent of leachant renewal frequency.

1 INTRODUCTION

Leaching kinetics studies on immobilized solid waste are aimed at determining the rate at which the process takes place, and at obtaining information on the process-controlling mechanisms. The shape of the curve representing the amount leached of a given element can provide data on the mechanism (or series of mechanisms) that produce leaching, while also offering some insight into how leaching proceeds as time elapses.

The study does not undertake to derive or propose a model explaining the microstructural phenomena arising in the leaching process. All the models considered in this study make the assumption that the solid is a continuous, homogenous medium.

2 TEST USED IN THE LEACHING KINETICS STUDY

The ANS method is a standard test put forward by the American Nuclear Society[1], used by many workers in kinetics studies on radioactive waste immobilized in cement matrices.

The method proposed by Côté and Isabel[2] involves an adaptation of the ANS procedure for studying immobilized toxic wastes. Demineralized water is used as a leachant, and the assumption is made that pollutant diffusivity in water is much greater than in the solid, which justifies not stirring the leachant.

The main difference between this method and the ANS procedure lies in the leachant renewal times. Côté and Isabel suggested using three series of leachant renewal times, as against the single series proposed by the ANS method, since in the case of immobilized products obtained from toxic and hazardous wastes, the study of different chemical species with varying effective diffusivities is typically of interest.

The range of concentrations between analytical detection and saturation determines the range of effective diffusivities that may be determined for a given leachant renewal frequency.

A study elsewhere[3] reviewed and analyzed the various test methods used for studying the leaching kinetics of heavy metals in immobilized solid waste[4-10].

3 METHOD AND EXPERIMENTAL TECHNIQUE

3.1 Test specimen preparation

The material to be immobilized was a sludge produced at a facility that simultaneously manufactured glazed stoneware floor tile and porous wall tile. In both cases, manufacturing was by the single-firing process. Sampling was programmed so as to obtain a representative sludge sample. Leaching tests run to characterize the material revealed that the two most significant elements from an environmental point of view were lead and zinc. The sludge used contained 4.30 wt% PbO and 4.32 wt% ZnO.

The optimum mixture for immobilizing this kind of sludge in a cement matrix was determined in a previous study. The mixture was made up of 77.5% sludge and 22.5% Portland cement. Mixing took place by adding 43% water (by weight) to the starting dry materials[11].

Besides the sludge-cement test specimens, a series of specimens just containing cement was also prepared to determine the behaviour of the cement matrix in the leaching tests.

3.2 Experimental procedure

The leachant used was demineralized water, which underwent no stirring during leaching intervals. The leachant volume used at each renewal was 1 l, which met the ANS requirement for the actual dimensions of the test specimens.

All the test specimens underwent leaching for 28 days according to the three leachant renewal schedules. In the standard renewal schedule (STD), the leachant was renewed at: 2, 7, 24, 48, 72, 96, 120, 192, 264, 336, 504 and 672 hours. The double (DBL) and half (HLF) renewal frequencies were derived from the standard leachant renewal schedule.

3.3 Chemical analysis of the leachates

Lead and zinc concentrations were determined on a Perkin-Elmer, model 2380, atomic absorption spectrophotometer fitted with a Perkin-Elmer HGA-400 graphite camera. In the testing conditions used, detection limits were 1 and 7 ppb ($\mu g\ L^{-1}$) for lead and zinc respectively.

4 RESULTS AND DISCUSSION

4.1. Experimental results

The leachate samples were analyzed, determining the concentrations of the studied elements. The concentrations in the leachate were generally quite low.

The parameter F_j, will be used to characterize the leaching process, defined as:

$$F_j = \frac{\Delta M_j}{M_{j0}} \frac{V_p}{S_p}$$

F_j is therefore a parameter proportional to the mass of the component leached and possesses length dimensions. F_j is readily calculated from the experimental data.

If the process-controlling mechanism were purely diffusional, on plotting F_j as a funtion of $t^{1/2}$, the plots would be linear. However, effective diffusivity was observed to depend on the leachant renewal schedule. On raising the leachant renewal frequency, diffusivity rose. No references to any studies on this phenomenon emerged in the literature surveyed. For this reason, it was decided to further analyze the variation of effective diffusivity as a function of the leachant renewal schedule.

4.2 Proposed kinetic model

None of the models reviewed in the literature considered whether the quantity of leached material might depend on the leachant renewal frequency, although some studies[2] did remark this phenomenon.

A model that takes into account the influence of leachant renewal time is proposed on the basis of these results.

It will be assumed that the variation of the concentration of A_j with time is given by Fick's second law. The solid is therefore assumed to be of semi-infinite thickness, with mass transfer just occurring in a single direction (at right angles to the surface).

To completely determine the concentration profile as a function of time, i.e., to solve the problem, the starting concentration distribution (initial condition) and boundary conditions need to be specified. The scheme of the proposed model is shown in Fig. 1.

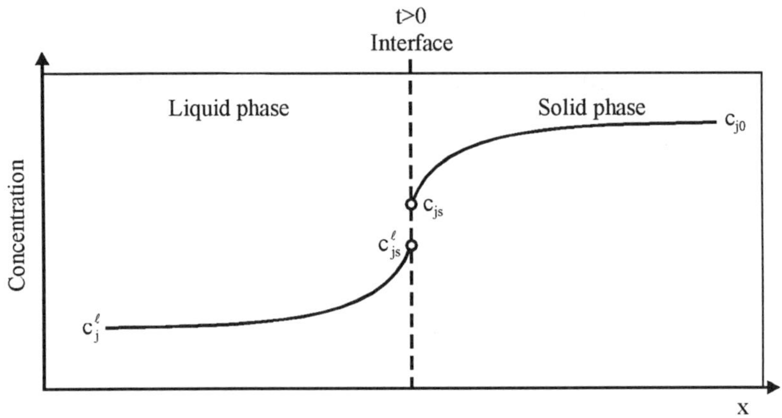

Equilibrium relation	$K_{ej} = \dfrac{c_{js}^{\ell}}{c_{js}}$		
Equation (Fick's 2nd Law)	$\dfrac{\delta c_j}{\delta t} = D_j \dfrac{\delta^2 c_j}{\delta x^2}$		
Starting conditions	$t = 0$	$0 \le x < \infty$	$c_j = c_{j0}$
Boundary conditions	$x = 0$	$t > 0$	$n_j = -D_j \dfrac{\delta c_j}{\delta x} = k_j\left(K_{ej} \cdot c_{js} - c_j^{\ell}\right)$

Fig 1. Proposed kinetic model.

In order to take into account the effect of the initial rinse, the term F_{j0} (quantity of instantly leached material) has been added.

To solve the equations of the proposed kinetic model a numerical method needs to be used. The method chosen was that of finite differences (using the implicit form)[12]. The calculations were carried out using the Mathematica (version 2.2) software, on a HP9000/730 workstation.

4.3 Results of the fit of the experimental points to the proposed kinetic model

Table I shows the results obtained on fitting the experimental points to the proposed model.

Table I. Parameters obtained with the proposed kinetic model

Component	Series	F_{j0} (m)	D_j ($m^2\ s^{-1}$)	K_{ej}	k_j ($m\ s^{-1}$)
Zn	DBL	4.0×10^{-7}			
	STD	1.0×10^{-7}	120×10^{-20}	5.68×10^{-7}	10^5
	HLF	0.7×10^{-7}			
Pb	DBL	2.0×10^{-7}			
	STD	1.0×10^{-7}	11×10^{-20}	15.7×10^{-7}	10^5
	HLF	1.9×10^{-7}			

Figs. 2 and 3 depict the plots of the experimental values of F_j, together with the values of F_j calculated according to the proposed model (solid line) and according to a diffusional model in which effective diffusivity was assumed to be independent of leachant renewal frequency (dashed line).

As Figs. 2 and 3 show, the curves obtained with the proposed kinetic model fit the experimental results well. There are only some values at very long times, which exceed the values predicted by the model. One of the most interesting features of the proposed model is that it predicts a rise in the leaching rate after leachant renewal. The increase becomes more striking as the period between two successive renewals is lengthened.

It should finally be highlighted that the values of D_j allow establishing which element will be most effectively retained. The diffusivity of zinc is around 10 times greater than that of lead.

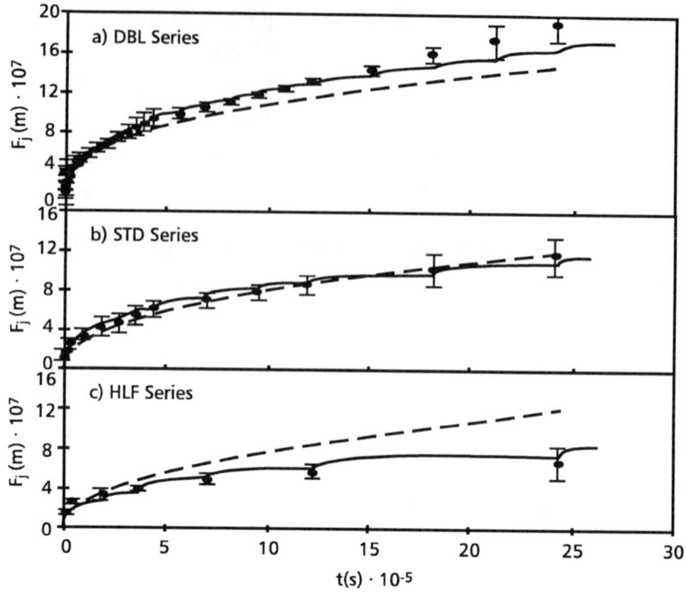

Fig 2. Values of Fj for zinc as a function of time.

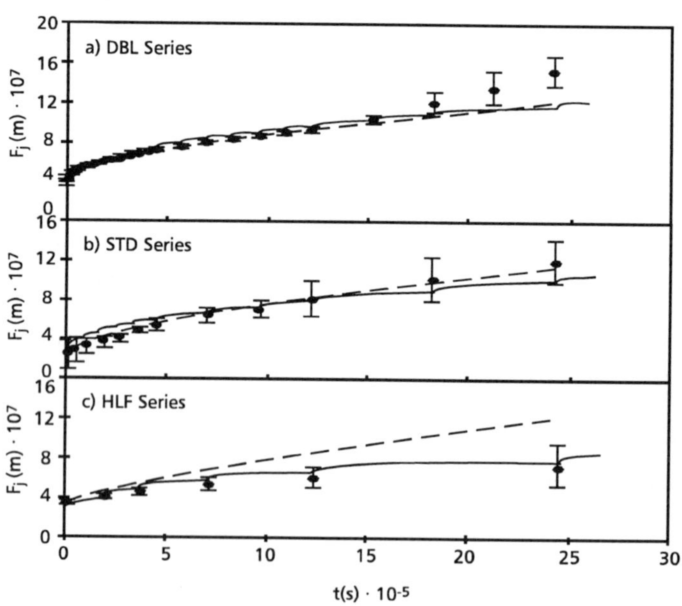

Fig 3. Values of Fj for lead as a function of time.

5 CONCLUSIONS

In order to describe the variation of effective diffusivity with leachant renewal frequency, an alternative kinetic model to the models encountered in the literature is proposed. This model satisfactorily describes the observed leaching behaviour, in which the amount of leached material depends on the leachant renewal frequency.

The proposed kinetic model yields an effective diffusivity value for each of the two studied components, lead and zinc. This effective diffusivity is independent of the leachant renewal frequency under the studied operating conditions. The other independent parameters of the renewal frequency are the equilibrium constant (K_{ej}) and the individual coefficient of mass transfer (k_j).

NOMENCLATURE AND UNITS

c_{jS}^{l}: concentration of A_j in the liquid phase in contact with the interface (kg m^{-3})
c_j^{l}: concentration of A_j inside the liquid phase (kg m^{-3})
c_{jS}: concentration of A_j in the solid phase in contact with the interface (kg m^{-3})
c_j: concentration of A_j inside the solid phase (kg m^{-3})
D_j effective diffusivity (m s^{-2})
F_j parameter proportional to the mass of the component leached (m)
F_{j0} parameter proportional of the quantity of instantly leached material (m)
k_j: individual coefficient of mass transfer (m s^{-1})
K_{ej} is the equilibrium constant, ratio between c_{jS}^{l} and c_{jS}.
ΔM_j: cumulative mass of A_j leached (kg)
M_{j0}: mass of A_j initially present in the solid (kg)
n_j: mass flow density (kg m^{-2} s^{-1})
S_p: test specimen surface area (m^2)
t time (s)
V_p: test specimen volume (m^3)
x distance from the solid surface (m)

ACKNOWLEDGEMENT

This paper includes part of the results obtained in project AMB94-0219, granted by the Comisión Interministerial de Ciencia y Tecnología (CICYT), Spain.

6 REFERENCES

[1] ANSI/ANS-16.1, 1986. American National Standard Measurement of the Leachability of Solidified Low-Level Radioactive Wastes by a Short-Term Test Procedure. Le Grange Park, IL: American Nuclear Society, 1986.

[2] Côté, P.L.; Isabel, D. Application of a dynamic leaching test to solidified hazardous wastes. In: Hazardous and industrial waste management and testing:

third symposium. Philadelphia: American Society for Testing and Materials, 1984, p. 48-60

[3] Enrique, J.E.; Monfort, E.; Cantavella, V. et al. Leaching kinetics of toxic elements immobilized in cement matrices (I). Selecting a test method. Materials Engineering,8 (02), 151-169, 1997.

[4]Cioffi, R.; De Vito, B.; Palumbo, M., et al. Building materials: a study on the kinetics of toxic metals release. Materials engineering, 4(3), 379-386, 1993.

[5] Medici, F., et al. Release of toxic elements from solidified wastes: a mathematical model. In: Stabilization and solidification of hazardous, radioactive and mixed wastes. Philadelphia: American Society for Testing and Materials, 1992, vol. 2, p. 171-181.

[6] Gilliam, T.M.; Sams, T.L.; Pitt, W.W. Testing protocols for evaluating monolithic waste forms containing mixed waste. In: Nuclear waste management II. Westerville: American Ceramic Society, 1986, p. 305-311.

[7] De Angelis, G.; Marchetti, A.; Balzamo, S. Leach studies: influence of various parameters on the leachability of cesium from cemented BWR evaporator concentrates. In: Stabilization and solidification of hazardous, radioactive and mixed wastes. Philadelphia: American Society for Testing and Materials, 1992, vol. 2, p. 182-192.

[8] Zamorani, E.; Sheikh, I.A.; Serrini, G. Physical properties measurements and leaching behavior of chromium compounds solidified in a cement matrix. *Nucl. chem. waste manage.*, 8, 239-245, 1988.

[9] Buil, M.; Revertegat, E.; Oliver, J. A model of attack of pure water or unsaturated lime solutions of cement. In: *Stabilization and solidification of hazardous, radioactive and mixed wastes*. Philadelphia: American Society for Testing and Materials, 1992, vol. 2, p. 227-241.

[10] Groot, G.J. de; Sloot, H.A. van der. Determination of leaching characteristics of waste leading to environmental product certification. In: *Stabilization and solidification of hazardous, radioactive and mixed wastes.* Philadelphia: American Society for Testing and Materials, 1992, vol. 2, p. 149-170.

[11] Enrique, J.E.; Monfort, E. Situación actual y perspectivas de futuro de los residuos de la industria cerámica. *Cerámica Información,* 221, 20-32, 1996.

[12]Carnahan, B.; Luther, H.A.; Wilkes, J.O. Cálculo numérico: métodos, aplicaciones. Madrid: Rueda, 1979.

THE DISSOLUTION RATE OF CHEMICALLY DURABLE MATERIALS: ZIRCON

Katheryn B. Helean, The University of New Mexico, Department of Earth and Planetary Sciences, Northrop Hall, Albuquerque NM 87131
Werner Lutze, The University of New Mexico, Department of Chemical and Nuclear Engineering, Farris Hall, Albuquerque NM 87131
Rodney C. Ewing, The University of Michigan, Department of Nuclear Engineering and Radiological Sciences, The University of Michigan, 2355 Bonisteel Blvd., Ann Arbor MI 48102

ABSTRACT

A high temperature Soxhlet extractor was designed to measure the forward rate of dissolution of the highly durable material, zircon ($ZrSiO_4$), in the range 120 to 250°C. Zircon has been proposed as an immobilization phase for the disposition of excess weapons Pu. Because of its high chemical durability, a leaching technique was designed to measure the forward rate of zircon dissolution in the absence of solubility-limiting phases (e.g. zirconia). The forward dissolution rates for zircon based on the Si release rates, using the high temperature Soxhlet extractor, are: 4.1×10^{-4} g m^{-2} day^{-1} at 250°C, 1.7×10^{-4} g m^{-2} day^{-1} at 200°C and 7.1×10^{-5} g m^{-2} day^{-1} at 120°C. The activation energy for zircon dissolution is 23 kJ/mol. The forward rate of dissolution of crystalline zircon was extrapolated to 90°C: 4.6×10^{-5} g m^{-2} day^{-1}. Thus, at 90°C in an open aqueous system, a 100 μm crystal of zircon would require 150,000 years for complete dissolution. Under static conditions, the dissolution rate would be substantially reduced. These data are consistent with the high chemical durability of zircon in a wide range of geologic environments.

INTRODUCTION

Zircon, a highly refractory mineral, is ubiquitous in the Earth's crust and has been proposed as a ceramic host phase for the immobilization and disposition of excess weapons-grade plutonium [1,2,3,4]. Zircon is an extremely durable accessory mineral found in igneous and metamorphic rocks. Zircon's physical and chemical durability are such that individual grains can undergo many cycles of weathering and erosion followed by transport and deposition with limited physical abrasion or chemical dissolution. More than 4,000 studies have been conducted on zircon since 1970 (as cited in Georef). However, none of these studies have measured the dissolution rate of zircon in water in the temperature range, 100-250°C. Static dissolution tests are ineffective due to the estimated very low (0.1 to 1.0 ppb) saturation concentrations of Si and Zr over zircon. Thus, we have developed an alternative experimental technique, a modified Soxhlet extractor. The sample is exposed to distilled water under flowing conditions such that dissolution proceeds at a forward rate and measured solution compositions are not solubility limited. A Soxhlet extractor was designed to withstand temperatures up to 350°C and the corresponding pressures.

DESIGN AND FUNCTION OF THE SOXHLET EXTRACTOR

In order to prevent the saturation of Zr and Si over zircon in solution, a Soxhlet extraction technique was used. Soxhlet devices have been commonly used in dissolution studies of nuclear waste glasses. Glass, however, dissolves more readily than zircon at all temperatures. Thus, commercially available Soxhlet extractors cannot be operated at temperatures greater than 100°C. They are not typically designed for pressures exceeding 1 bar.

A Soxhlet extractor was designed for use in this study to operate at temperatures up to 350°C and at the respective water vapor pressures. The device is made of stainless steel (Fig. 1). The base of the extractor sits in a heating mantle with a temperature control. A sample cup, with a volume of 20 ml, is suspended by stainless steel tabs in the center of the base. The sample cup receives water droplets from the water cooled condensing coil which sits, suspended, in the top of the extractor. When the sample cup is filled with water, a siphon drains the sample cup. In this manner the zircon sample is exposed to a quasi-continuous flow of distilled water, and Zr and Si concentrations are kept as low as possible above the zircon sample.

The base of the vessel is coated with gold to prevent leaching of Si from the stainless steel of the Soxhlet extractor. A flow monitor is connected to the water source and shuts off the electricity to the heating mantle in the event that the flow of water to the cooling coil is interrupted. This is a necessary safety measure that allows long-term experiments to be conducted without risking the loss of the samples.

The synthetic zircon powder had a BET surface area of 0.048 m^2 /g and an average grain diameter of 50 μm. The zircon powder is placed in the sample cup in a Teflon bag with a pore size of 11 μm, which is smaller than the minimum particle size as confirmed by particle size measurements. The leachate along with any dissolved Zr or Si species is siphoned off and concentrates in the solvent reservoir at the base of the vessel. Zr and Si are expected to saturate in the volume (150 ml) of liquid in the base. This would limit the measurement of kinetic data beyond the saturation concentration of Zr and Si which are expected to be in the range of 0.1 to 1.0 ppb. In order to prevent the saturation and subsequent precipitation of ZrO_2 or $ZrSiO_4$, silicic acid, H_4SiO_4, was prevented from reaching saturation by adding NaOH to the water in the vessel which maintains a pH of 12. At this pH, silicic acid is dissociated, and the solubility limit of zircon cannot be reached. This experimental design allows us to measure the irreversible dissolution rate of zircon. This is the forward rate as given by: $k=1/\nu\ dc/dt\ V/S$ [g m^{-2} d^{-1}], where c is the concentration of Si in solution, t = time, ν= stoichiometric factor for Si in $ZrSiO_4$, V= solution volume and S= surface area of zircon. The quasi-continuous flow of distilled water over the zircon sample in the Soxhlet extractor ensures that the rate of dissolution does not depend on the concentrations of the dissolving Zr and Si. If a static test were employed, such that the reaction was first order with respect to Zr or Si, then the extremely low solubility of these elements over zircon would prevent the measurement of dissolved species in solution and thereby prevent quantification of the forward rate.

Two Soxhlet extractors were used to dissolve a crystalline synthetic zircon powder in the temperature range 120 to 250°C. Samples were drawn from the solvent reservoir of the Soxhlet extractor over a period of approximately 300 hours. The solutions were analyzed for Si using the molybdate blue method. A linear regression of the Si concentration over time gave the forward rate of zircon dissolution at the temperatures studied. These data were used to calculate the activation energy for the dissolution of crystalline zircon.

SAMPLE DESCRIPTION

A commercially prepared crystalline synthetic zircon powder was obtained from Cerac, Inc., Milwaukee, WI., lot # 96017-A-1. The manufacturer's spectrographic analysis reported a Zr wt. % value which is 1.15 wt. % lower than the theoretical value of 49.76 wt.%. A complete structural and chemical analysis of the zircon confirmed that the material is only zircon and the elements accounted for are minor impurities, such as Hf, in the zircon phase.

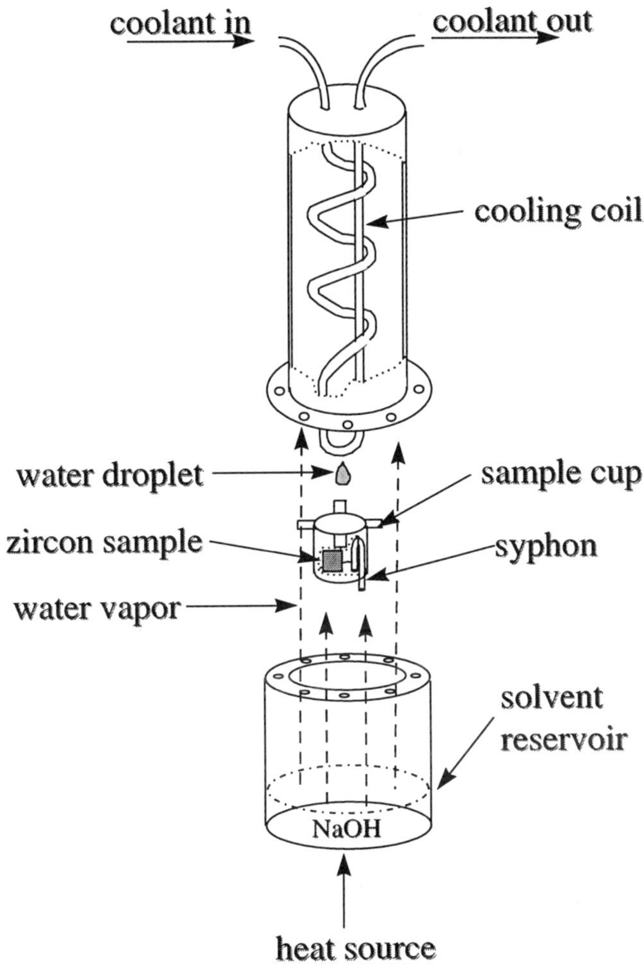

Fig. 1. A schematic representation of the Soxhlet extractor designed to work at temperatures up to 350°C and the corresponding water vapor pressures. The top cover houses the cooling coil. The middle section contains the sample cup with the siphon. The base serves as a reservoir and is gold plated to prevent corrosion of the stainless steel and contamination of the solvent.

XRD was used to check for the presence of crystalline phases other than zircon such as SiO_2 or ZrO_2. X-ray diffraction analysis (XRD) was completed on a Scintag diffractometer operated at 20kV using Cu radiation (λ-1.5418 Å). The XRD pattern matches the JCPDS reference standard zircon pattern (06-0266). No non-zircon peaks were identified. TEM was used to determine if any unreacted amorphous silica, which would not appear in the XRD pattern, was present in the sample. Transmission electron microscopy (TEM) was done on a JEOL 2000FX operated at 200kV with a point to point resolution of 2.5 Å. No amorphous SiO_2 was observed. A major element chemical analysis was completed using the SEM-EDS. Energy dispersive X-ray spectroscopy (EDS) was completed on a JEOL JSM-5800LV scanning electron microscope (SEM) operated at 15 kV, a focused beam diameter of 1 micron and a counting time of 60 s. The results of six analyses are given in Table I. The ideal composition weight percent values for Zr, Si and O are listed for comparison. All totals were low suggesting the presence of approximately 3 weight percent impurity elements. Oxygen was calculated, not directly measured.

Table I. Major element EDS analyses. The standard used was a pure zircon. All values are reported in weight percent. The values for the ideal zircon composition are reported in the last column for comparison.

element	#1	#2	#3	#4	#5	#6	theor.
Zr	47.52	48.66	48.90	47.11	47.93	49.26	49.76
Si	14.73	15.01	13.97	14.61	14.89	14.77	15.32
O	34.64	34.83	34.16	34.97	34.51	34.81	34.92
total	96.89	98.50	97.03	96.69	97.33	98.84	100

Trace element analysis was conducted using a quadrapole-type VG 5800 inductively coupled plasma mass spectrometer (ICP-MS). The zircon sample was dissolved in a lithium metaborate, $LiBO_2$, flux at 1450°C and then dropped into a 5% HNO_3 solution. The zircon solution was stirred for several hours until all particles were dissolved. Blank solutions were prepared using the $LiBO_2$ flux. The blank concentrations were then subtracted from the sample concentrations for the trace elements analyzed. The results, reported as ppm except for Hf which is given in wt.%, are given in Table II. Twenty impurity elements were detected and analyzed including Y, La, Ce, REE, Ti, Hf, Th, U, and Fe. All trace elements detected are common constituents in zircon samples. The large number of impurity elements was deemed acceptable, because all can be incorporated into the structure of zircon and the effect of their presence on the matrix dissolution of zircon, as measured from Si release rates, was presumed to be negligible.

The zircon particle size distribution was measured using a Coulter LS230 instrument, which uses the scattering of laser light by the particles to determine particle size. The laser light has a wavelength of 750 nm and can detect particles from 0.4 μm to 2000 μm in diameter. The synthetic zircon powder had a BET surface area of 0.048 m^2 /g and an average grain diameter of 50 μm as measured by the Coulter particle size analyzer. The zircon powder was placed in a Teflon sample bag with a pore size of 11 μm, which is smaller than the minimum particle size as confirmed by particle size measurements.

Table II. ICP-MS analyses of crystalline synthetic zircon. All values are reported as ppm except Hf, which is reported in wt.%. The isotope measured is indicated after the element in each column. The error corresponds to 2σ.

Y-89	**La-139**	**Ce-140**	**Pr-141**	**Nd-146**	**Eu-151**	**Sm-152**
939.7±61	13.9±0.6	46.8±2.7	2.0±0.1	15.0±0.9	0.7±0.1	4.1±0.3
Gd-158	**Tb-159**	**Dy-161**	**Ho-165**	**Er-166**	**Tm-169**	**Yb-174**
23.3±1.5	4.5±0.2	76.5±2.5	29.9±1.3	142.2±7.3	28.8±1.5	257.8±11.6
Lu-175	**Ti-48***	**Hf-178#**	**Th-232**	**U-238**	**Fe-56***	
42.6±1.2	2101±8.3	1.22±0.1%	720.3±7.0	275.0±29	1303±22	

*high background in the blank for Ti and Fe
#Hf in unknowns are above highest standard

EXPERIMENTAL

Approximately 1 g of crystalline synthetic zircon powder was placed in a Teflon mesh bag with a pore size of 11 μm. The bag was closed using Teflon tape, and the sample bag was placed into the sample cup of a Soxhlet extractor. Approximately 150 ml of a NaOH solution with a pH=12 was added to the solvent reservoir. The Soxhlet extractor was sealed and placed into a heating mantle set for the desired temperature (120-250°C). The Soxhlet was removed from the heating mantle, cooled and a 2 ml sample was drawn from the solvent reservoir periodically for up to 300 hours. As the samples were collected, the temperature was recorded by a digital thermocouple and the pH of the solution in the sample cup was measured. The samples were then analyzed for Si using the molybdenum-blue method. The reagents used were ammonium molybdate ($(NH_4)_6Mo_7O_{24} \cdot 4H_2O$) dissolved in an aqueous solution of H_2SO_4; an oxalic acid ($HO_2CCO_2H \cdot 2H_2O$) solution, which was used to suppress phosphorous interference; the reducing agent was hydroxylamine hydrochloride ($NH_2OH \cdot 2HCl$) / stannous chloride ($SnCl2 \cdot 2H_2O$) dissolved in dilute H_2SO_4. Soluble silica species react with molybdate at approximately 25°C and pH=1.2 to form a yellow silicamolybdate complex. Molybdenum is reduced in stannous chloride and a blue complex of heteropoly acid forms which has an absorbance maximum at 820 nm. The intensity of the color is proportional to the concentration of silica. A Bausch and Lomb Spectronic 88 spectrophotometer was used to measure the transmittance at 820 nm for each sample and for a series of known standards. The detection limit was 5 ppb.

A volume correction of the Si concentrations was necessary due to the reduction in total solution volume when solvent samples were taken. The correction was: $(V_n/V_i)C_n = C_f$ where V_n is the total solution volume just prior to collecting sample n, V_i is the initial solution volume, C_n is the Si concentration of sample n; C_f is the final, volume corrected Si concentration.

RESULTS

THE FORWARD RATE OF ZIRCON DISSOLUTION

Three dissolution experiments were completed at 250, 200 and 120°C. A series of blanks were collected over five days for each experiment to measure background Si concentrations. A linear regression analysis of the background data revealed no significant contribution from the stainless steel of the vessel to the measured Si-release rate, i.e. the least-squares fitted lines had slopes indistinguishable from zero at a 95% confidence level. The volume corrected Si concentrations were plotted against time and a linear regression model of these data was used to quantify the rate of zircon dissolution at each temperature (Fig. 2a,b,c). The dissolution rates are proportional to the slope of the least-squares fitted lines. For the experiment at 250±15°C, the average pH in the sample cup was 9±0.9 and the dissolution rate as measured by the Si release profile was $4.14 \times 10^{-4} \pm 2.8$ g m^{-2} day^{-1}. For the experiment at 200±7°C, the average pH in the sample cup was 9.3±0.4 and the dissolution rate measured by the Si-release rate was $1.66 \times 10^{-4} \pm 1.9$ g m^{-2} day^{-1}. For the experiment at 123±9°C, the average pH in the sample cup was 8.8±0.5 and the dissolution rate measured by the Si-release rate was $7.10 \times 10^{-5} \pm 0.9$ g m^{-2} day^{-1}. The errors reported correspond to the standard error. The pH increase in the sample cup is explained by carry over of small droplets from the reservoir. The dissolution rate of zircon reported in this study relates to a pH of about 9. The pH dependence of zircon dissolution was not studied.

THE ACTIVATION ENERGY OF ZIRCON DISSOLUTION

The rate of zircon dissolution was measured at three temperatures, 250, 200 and 120°C. These three rates were used to calculate the activation energy of zircon dissolution from the Arrhenius equation: $k(T) = Ae^{-Ea/RT}$ where k is the measured rate at temperature, T in Kelvin, A is the pre-exponential factor, E_a is the activation energy in J, and R is the gas constant, 8.3145 J mol^{-1} K. The activation energy, E_a was calculated from a plot of ln(k) against 1/T (Fig. 2d). The calculated Ea is 23.2 kJ/mol.

DISCUSSION

The Soxhlet extractors used in this study worked effectively for the duration of these experiments. However, after several weeks of continuous use the Si background concentrations were too high to conduct experiments. The increased background Si was caused by degradation of the gold coating in the solvent reservoir and subsequent leaching of Si from the stainless steel. This complication can be avoided in the future by modifying the Soxhlet design to allow for a removable Teflon liner in the solvent reservoir.

The results of this study are similar to previously reported results which were obtained under very different conditions [5]. The extrapolated value for the dissolution rate of zircon at 90°C from this study is 4.6×10^{-5} g m^{-2} day^{-1}. Ewing et al. (1982) reported a value of 9.2×10^{-4} g m^{-2} day^{-1}. Their rate is one order of magnitude higher, which may be due to the different leachant, and to the fact that the surface area of the zircon was less well know than in this study.

The Arrhenius activation energy for the zircon hydration reaction, 23 kJ/mol, is consistent with the reported activation energies for zirconolite, $CaZrTi_2O_7$, another proposed Pu-host phase. At pH=9, zirconolite has an Arrhenius activation energy in the range 19-25 kJ/mol [7]. However, the dissolution rate of zirconolite based on Ca-release at 90°C is 5×10^{-2} g m^{-2} day^{-1} under static flow conditions. This rate is three orders of magnitude larger than the dissolution rate of zircon.

CONCLUSIONS

1) The high-temperature Soxhlet extractor is an effective tool for measuring the dissolution rates of extremely inert materials, in this case zircon. An improved design with a Teflon liner will better control the background concentrations of dissolved species at temperatures less than 200°C.

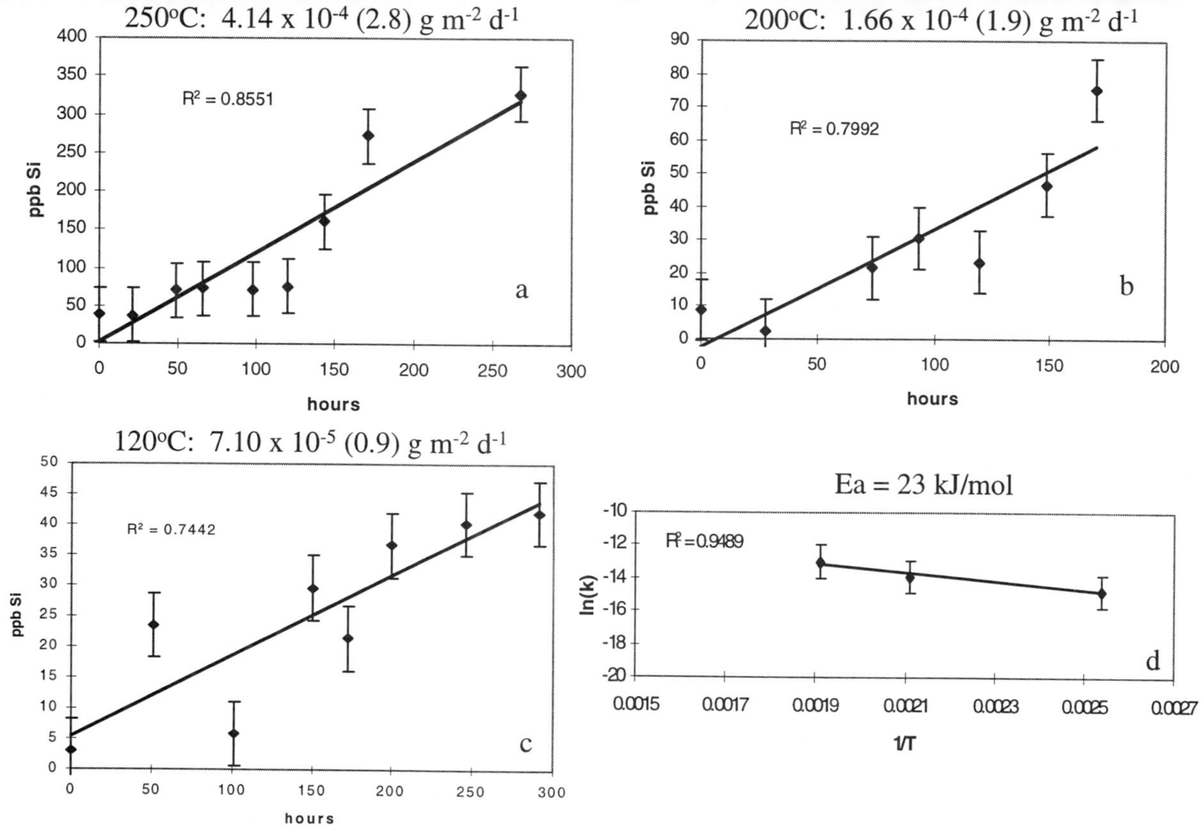

Fig. 2. The Si concentrations as a function of time for the zircon dissolution tests at a) 250, b) 200 and c) 120°C. The dissolution rates calculated from the Si concentration are reported for each temperature. A plot of ln(k) vs. 1/T is shown (d). The slope is -Ea/R for the zircon dissolution reaction, where Ea is the activation energy. The R-square values are given. All error bars correspond to the standard error.

2) The forward rates of zircon dissolution are low. As a function of temperature, the rates are 4.1×10^{-4} g m^{-2} day^{-1} at 250°C, 1.7×10^{-4} g m^{-2} day^{-1} at 200°C and 7.1×10^{-5} g m^{-2} day^{-1} at 120°C.

3) Zircon dissolution is a thermally activated process that can be described by the Arrhenius equation. The Arrhenius activation energy for zircon dissolution is 23 kJ/mol.

REFERENCES

[1] B.E. Burakov, "A Study of High-Uranium Technogenous Zircon (Zr,U)SiO4 from Chernobyl "Lavas" in Connection with the Problem of Creating a Crystalline Matrix for High-Level Waste Disposal"; pp. 19-28 in *SAFE WASTE -93*, **2**, 1993.

[2] Burakov, B.E, E.B. Anderson, B. Ya. Galkin, V.A. Starchenko and V.G. Vassiliev, "Crystalline Host-Phases for Immobilization of Weapons Plutonium and Waste Actinides"; pp. 85-89 in *Disposal of Weapons Plutonium*, Edited by E.R. Merz and C.E. Walter, Kluwer Academic Publisher, Dordrecht, 1995.

[3] R.C. Ewing, W. Lutze and W.J. Weber, "Zircon: A Host-Phase for the Disposal of Weapons Plutonium," *Journal of Materials Research*, **10**[2], 243-246 (1995).

[4] R.C. Ewing, W. Lutze and W.J. Weber, "Crystalline Ceramics: Waste Forms for the Disposal of Weapons Plutonium"; pp. 65-83 in *Disposal of Weapons Plutonium*, Edited by E.R. Merz and C.E. Walter, Kluwer Academic Publisher, Dordrecht, 1995.

[5] R.C. Ewing, R.F. Haaker and W. Lutze, "Leachability of Zircon as a Function of Alpha Dose"; pp. 389-397 in *Scientific Basis for Nuclear Waste Management-V*, Edited by W. Lutze, Elsevier Science Publishing, 1982.

[6] M.P. Tole, "The Kinetics of Dissolution of Zircon ($ZrSiO_4$)," *Geochimica et Cosmochimica Acta*, **49**, 453-458 (1985).

[7] A.E. Ringwood, S.E. Kesson, K.D. Reeve, D.M. Levins and E.J. Ramm, "Synroc"; pp. 233-334 in *Radioactive Waste Forms for the Future-XI*, Edited by W. Lutze and R.C. Ewing. North-Holland Press, 1988.

SOLUTION EXCHANGE CORROSION TESTING WITH A GLASS-ZEOLITE CERAMIC WASTE FORM IN DEMINERALIZED WATER AT 90°C

L. J. Simpson, C. W. Vander Kooi, and V. N. Zyryanov
Chemical Technology Division
Argonne National Laboratory
9700 South Cass Avenue
Argonne, IL 60439

ABSTRACT

A ceramic waste form of glass-bonded zeolite is being developed for the long-term disposition of fission products and transuranic elements in wastes from the U.S. Department of Energy's spent nuclear fuel conditioning activities. Solution exchange corrosion tests were performed on the ceramic waste form and its potential base constituents of glass, zeolite 5A, and sodalite as part of an effort to qualify the ceramic waste form for acceptance into the Civilian Radioactive Waste Management System. Solution exchange tests were performed at 90°C by replacing 80 to 90% of the leachate with fresh demineralized water after set time intervals. The results from these tests provide information about corrosion mechanisms and the ability of the ceramic waste form and its constituent materials to retain waste components. The results from solution exchange tests indicate that radionuclides will be preferentially retained in the zeolites without the glass matrix and in the ceramic waste form, with respect to cations like Li, K, and Na. Release results have been compared for simulated waste from candidate ceramic waste forms with zeolite 5A and its constituent materials to determine the corrosion behavior of each component.

INTRODUCTION

The Electrometallurgical Treatment Program at Argonne National Laboratory (ANL) is developing conditioning processes for the potential treatment of a diverse inventory of U.S. Department of Energy's (DOE's) spent nuclear fuel that may not be amenable to direct geologic disposal [1]. The Electrometallurgical Treatment Program uses molten anhydrous LiCl-KCl salt in an electrorefining process that accumulates Rb, Cs, Sr, Ba, Y, La, Ce, Pr, Nd, Pm, Sm, Eu, Ru, Pd, and Ag fission products and transuranics. The anode residue, which contains the metal hulls and noble metal fission products, will be incorporated into a metal waste form [2]. Fission products that accumulate in the salt will be loaded into zeolites through an ion exchange and salt occlusion process [1]. This salt-loaded zeolite material will be mixed with glass frit, packed into Type 304 stainless steel canisters, and hot isostatically pressed to produce the ceramic waste form of glass-bonded zeolite (Table I) [3-7].

The materials used in this independent evaluation of the glass-zeolite ceramic waste form include Bayer glass 57, nonradioactive salt-loaded zeolite 5A, salt-loaded sodalite, and "baseline" ceramic waste form samples with equal amounts by mass of salt-loaded zeolite 5A and glass 57 (Table I). The samples used in these experiments were provided by the Ceramic Waste Form Development Group of the ANL Electrometallurgical Treatment Program [5,6].

Table I. Nominal Elemental Compositions in Mass% of the Glass 57, Salt-Loaded Zeolite 5A, Salt-Loaded Sodalite, and Ceramic Waste Form with Zeolite 5A.

Elements	Li	B	Na	Al	Si	Cl	K	Ca	Y	Sr	Cs	Ba	La	Ce	Nd	O
Glass 57		4.9	5.0	3.8	29		0.5	0.8		0.01						56
SLZ	1.4		4.3	15	16	13	5.4	7.5	0.01	0.05	0.4	0.1	0.07	0.2	0.5	36
SLS	0.66		15	17	18	6.1	2.5		0.01	0.02	0.2	0.07	0.03	0.1	0.2	40
CWF	0.7	2.5	4.6	9.5	22	6.4	2.9	4.2	0.01	0.03	0.2	0.07	0.04	0.1	0.2	46

SLZ: salt-loaded zeolite 5A; SLS: salt-loaded sodalite; CWF: ceramic waste form with zeolite 5A.

To be accepted into the DOE Civilian Radioactive Waste Management System, the ceramic waste form must be reviewed and deemed "acceptable" based on the requirements set forth in the Waste Acceptance System Requirements Document (WASRD) for spent nuclear fuel and high-level waste [8,9]. We have started qualification testing of the ceramic waste form to meet the WASRD requirements by following the testing methodology outlined in the American Society for Testing and Materials (ASTM) document C1174-91 [10], which will be used to identify corrosion mechanisms and develop a database that may be used for predictive modeling. The ultimate goal of qualification testing is to provide appropriate information such that the effects of the ceramic waste form behavior and radionuclide release on the overall repository performance can be determined. Glass corrosion, zeolite corrosion, ion-exchange effects, and zeolite partition coefficient effects are a few of the processes that need to be investigated to understand the total corrosion behavior of the ceramic waste form.

The most important consequence of ceramic waste form corrosion is the release of radionuclides. Since the vast majority of the radionuclides in the ceramic waste form should be contained within the zeolites, groundwater accessibility to the zeolites and zeolite corrosion and ion exchange mechanisms will control the radionuclide release. A conceptual model of ceramic waste form corrosion was developed based on test results obtained over the past two years [3-7].

Some salt may have been released from the zeolites during the hot isostatic pressing process and resides in the waste form as free salt. Therefore, upon initial contact of the ceramic waste form by an aqueous solution, accessible free salt is quickly released. Over the next few weeks, some of the alkali metals and chloride may be preferentially released (as compared to the alkaline earth and rare earth elements) from the zeolites at the exposed surface regions [3-7]. However, since the zeolitic material is dispersed throughout the entire glass matrix, the glass provides a barrier between the leachant and the vast majority of the zeolites.

Additional zeolites become exposed to the leachant and additional ions are released from the zeolites as the glass corrodes, more fractures form or become interconnected, or as the leachant traverses through interconnected zeolites. Within a month, in tests conducted at 90°C in demineralized water, the overall corrosion behavior of the baseline ceramic waste form is governed primarily by glass dissolution. It should be noted that the chemical processes associated with the dissolution of the glass part of the waste form may play less of a role for the reference ceramic waste form material, being produced for the Experimental Breeder Reactor II (EBR-II) Spent Fuel Treatment Demonstration Project, which will have only 25 mass% glass as compared to the baseline material used in these tests that contains 50 mass% glass.

Solution exchange tests have been designed and performed to help identify the relative effects of free salt content, glass and zeolite matrix dissolution, zeolite ion exchange and preferential release, solubility, and water flux [3,4,7] on the release of radionuclides from the ceramic waste form. Where as solution exchange tests are usually used to study diffusion-controlled reactions [11], the tests performed here studied release mechanisms. For example, solution exchange removes the effect of the initial pulse of released free salts. To aid in interpreting the results of the tests with the ceramic waste form, separate tests were performed with the constituent materials of the waste material, namely glass 57 and salt-loaded zeolite 5A. Also, solution exchange tests were performed with salt-loaded sodalite, which will be the main crystalline phase used in the EBR-II demonstration waste form. Once ceramic waste form samples with sodalite become available, solution exchange tests will be performed with this material.

The results from solution exchange tests with nonradioactive salt-loaded zeolite 5A and sodalite, glass 57, and crushed baseline ceramic waste form samples will be discussed here. The primary goal of this investigation is to provide information about the release mechanisms of specific radionuclides and the corrosion behavior of the ceramic waste form to supplement previous observations from materials characterization and product consistency tests [3-7].

EXPERIMENTAL METHODS AND DEVELOPMENT

The ceramic waste form samples were cut from their hot isostatically pressed containers with a coring drill; ethanol was used as a coolant. Samples were crushed and sieved (-100 to +200 mesh; 75 to 150 μm) according to the product consistency test protocols [9], except that ethanol was used in place of demineralized water for rinsing fines from the crushed material [3,4,9]. This change was imposed to limit alkali metal and chloride loss from the ceramic waste form material during sample preparation. Analysis of ethanol rinse solutions indicates that less than 1% of the total salt in the crushed material was released during the rinsing process to remove fines [7].

Salt-loaded zeolite 5A samples were prepared by blending 21 g of simulated waste salt with 79 g of zeolite 5A in an argon atmosphere at 500°C. The salt-loaded sodalite sample was made with 90 g of zeolite 4A mixed with 10 g of simulated salt at 500°C, then converted to sodalite by increasing the temperature to 900°C. The glass 57 material, procured directly from Bayer, had an average particle diameter of approximately 40 μm.

Solution exchange tests were initiated with the baseline ceramic waste form and with its constituent materials in demineralized water to investigate the release of specific components. In each test approximately 80 to 90% of the leachate was exchanged with an equal amount of fresh demineralized water after 1 and 4 hr, and 1, 3, 7, 14, 28, and 56 d. The removed leachate was analyzed for anion and cation content. The tests all maintained pH values between 8 and 9 after the initial solution exchange at 1 hr (Fig. 1).

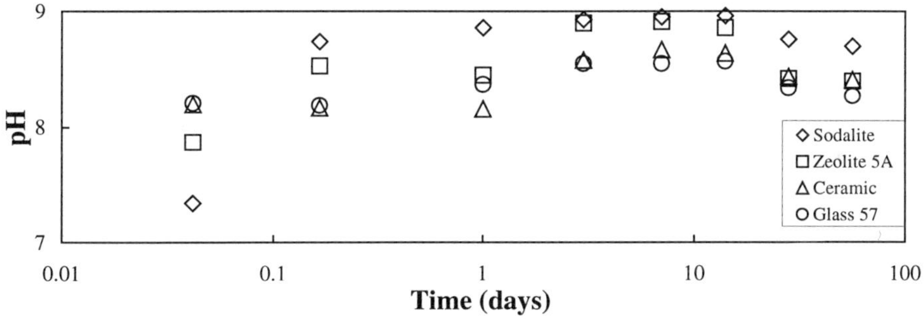

Fig. 1. Leachate pH measurements for solution exchange tests with the ceramic waste form and constituent materials in demineralized water at 90°C. pH values measured at room temperature.

The tests were performed in Teflon vessels at 90°C. The solution volumes used in the tests were adjusted to the geometric surface areas of the samples to provide a sample surface area to leachant volume ratio (S/V) of 2000 m^{-1} for the tests with each material. Demineralized water was used as the leachant in these tests. The surface areas were estimated for the samples based on density (~2200 kg/m^3 for the zeolite and ~2500 kg/m^3 for the glass) and average sieve opening or mean particle diameters. The surface areas were determined purely by geometry; no attempt was made to account for surface roughness, fractures, or internal surfaces of the zeolites.

Leachate analyses were performed with a Fison Instruments Plasma Quad inductively coupled plasma mass spectrometer (ICP/MS) for the elemental cation analysis, ion chromatography (IC) for anion analysis, and a Brinkmann 605 pH meter. The leachate cation and anion analyses are accurate to within ±10% for elements present above the level of quantitation. The pH measurements are accurate to within ±0.05 pH units. The analyses were performed at room temperature, and the solutions collected for cation analysis were passed through a 0.45 μm filter to ensure complete separation from test material and were acidified to prevent precipitate formation.

The leachate analyses provide information about the amount of a specific element in solution. The leachate results are reported as normalized concentration (NC_i, g/m^3) or cumulative "elemental release percent" ($ER_i\%$), defined as $NC_i = C_i / f_i$ and $ER_i\% = (100\ C_i\ V_l)/ (M\ f_i)$, where C_i is the leachate concentration of element i (mg/L), f_i is the mass fraction of element i in the sample, V_l is the leachant volume, and M is the initial sample mass. The NC_i provides a measure of the amount of material that dissolved, based on the release of a particular element during each interval. Cumulative $ER_i\%$ adds the $ER_i\%$ results for each replacement interval to provide a measure of the total sample mass that has reacted.

RESULTS AND DISCUSSION

The NC_i results of solution exchange tests with baseline ceramic waste form samples in demineralized water at 90°C are plotted as a function of time for several elements found in the leachate (Fig. 2). Since boron is only present in the glass phase, the normalized boron concentration in solution is indicative of the extent to which the glass has corroded [12]. In the solutions removed after 1 and 4 hr, the normalized release of alkali metals and Cl is greater than the release of boron. This is consistent with accessible free salt being released very quickly. The normalized concentrations of Li, Na, K, and Cl in solution being greater than or equal to that of boron in the solution exchange tests after 1 hr (Fig. 1) suggest that the occluded salt is being released from the zeolite 5A particles that are exposed to the leachant. After the 4 hr solution exchange time, the normalized concentration of all the elements was the same as or less than the normalized concentration of boron. This is consistent with the glass providing a barrier between the leachate and the unexposed zeolites. Thus, the glass must corrode or crack before additional zeolites are contacted by water and additional material can be released to solution. The fact that the normalized concentrations of all the elements become nearly constant after the fourth solution exchange at seven days is consistent with the corrosion of the baseline ceramic waste form becoming governed by the solubility limited dissolution of specific solution species [3-7]. For silicate-based glasses, the orthosilicic acid concentration controls the rate of glass reaction with water [12]. Similarly, the orthosilicic acid concentration appears to control the rate of dissolution of the glass matrix in the ceramic waste form [7].

For the sake of clarity, data for several elements were not included in the plots. Unless otherwise stated, the elements within specific chemical groups behave in a similar fashion. For example, Li is plotted in Fig. 2 to 4 but is representative of the alkali metals Li, Na, and K; Sr is used to represent the alkaline earths (Sr and Ba were always very close); and Nd or Ce is used to represent the rare earth elements (i.e. La, Ce, Pr, Nd, Sm, and Eu).

The release of rare earth elements from the baseline ceramic waste form in tests at 90°C in demineralized water follows the same trends as the glass corrosion described above, but the normalized concentrations are always one to two orders of magnitude lower than the normalized concentration of Si or Boron [3-7]. In these tests, the normalized concentrations of the alkaline earth elements were also low. However, the normalized concentration of the alkaline earth elements was significantly higher than that of the rare earth elements in the initial three solution exchange tests, but the same or somewhat lower in the final four solution exchange tests (compare the release of the representative elements of Sr with that of Ce and Nd in Fig. 2). This suggests that significantly more alkaline earth elements may be present as free salts than the rare earth elements, but the alkaline earth elements are probably held more strongly by the zeolite 5A.

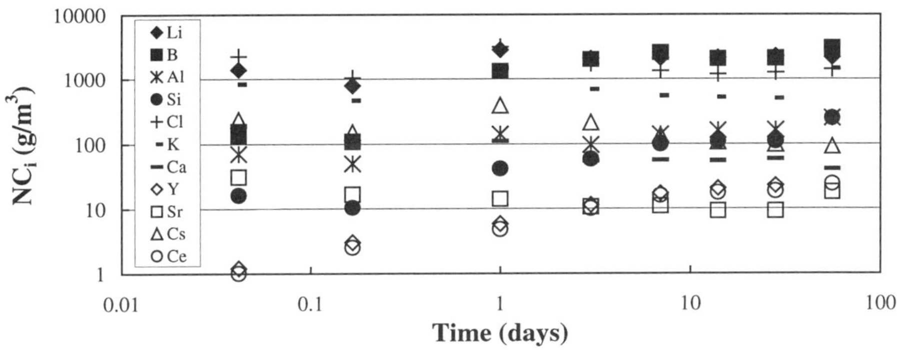

Fig. 2. Leachate NC_i measurements for solution exchange tests with the ceramic waste form in demineralized water at 90°C.

The zeolite 5A material's preferential retention of alkaline earth elements over rare earth elements is also observed in solution exchange tests with the salt-loaded zeolite 5A material without the glass binder (compare the release of the representative elements of Sr with that of Ce and Nd in Fig. 3). The normalized concentration of alkaline earth elements in solution is two orders of magnitude lower than the normalized concentration of Si and Al after the 1 d solution exchange. This is probably due to the dissolution rates being lower for the specific phases that contain the alkaline earth elements. Except for the free salt released during the 1 hr solution exchange test, the alkaline earth elements should be contained in the zeolite 5A structure. The presence of Si and Al in the leachates indicates that the zeolite structure is corroding. So either the specific zeolites or phases that contain alkaline earths are more corrosion resistant than are the zeolites without alkaline earths, or else the alkaline earths are being ion exchanged into uncorroded zeolites if and when they are released due to dissolution of the zeolite matrix [3,4].

The results displayed in Fig. 3 indicate a preferential release of alkali metals and Cl from the zeolite 5A. The cumulative ER_i% values were slightly greater than 100% for some alkali metals, probably because of cumulative and systematic uncertainties involving spectral analytical interferences in the measured concentrations and perhaps contamination. The data indicate that approximately 100% of Li, Na, K, and Cl is released to solution from the zeolite 5A material by the 14 d solution exchange test. The rare earth elements and Y are released, but with normalized concentration values an order of magnitude lower than those of Si and Al; this suggests that some preferential retention of the rare earth elements and Y by the zeolite 5A occurred. These observations are consistent with previously reported results from modified product consistency tests that used only the salt-loaded zeolites without the glass binder [3-5]. The release and adsorption of specific elements to and from the leachate was explained by the preferential exchange of ions between the zeolite and the leachant [3,4]. In this previous work with salt-loaded zeolites [3-5], the release order was correlated with ionic radius and charge.

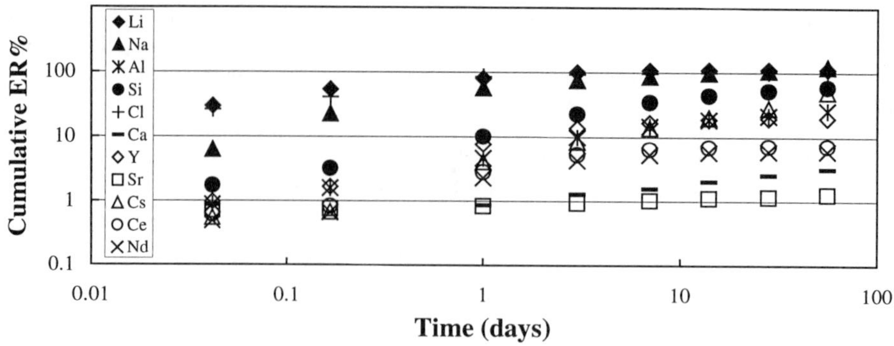

Fig. 3. Cumulative $ER_i\%$ values for solution exchange tests with salt-loaded zeolite 5A material without the glass binder in demineralized water at 90°C.

The elemental release from sodalite without the glass binder in demineralized water at 90°C (Fig. 4) is significantly different from that of zeolite 5A. After the 1 hr solution exchange tests, in which the accessible free salt was removed, the release of all the elements was lower than the release of Si to solution. This indicates that sodalite matrix dissolution must occur before any of the radionuclides can be released to solution. The release order of the alkali metals from the sodalite is Li > Na > K. Furthermore, as was observed in solution exchange tests with the zeolite 5A, alkali metals are released to a greater extent than the alkaline earth or rare earth elements. However, the release of the alkaline earth elements exceeded that of the rare earth elements from the sodalite material with the release of Ce being significantly lower than that of the other rare earth elements (Fig. 4). Since Ce is often used as a surrogate for Pu, the lower Ce release observed from the sodalite material suggests that the release of Pu from sodalite might also be low.

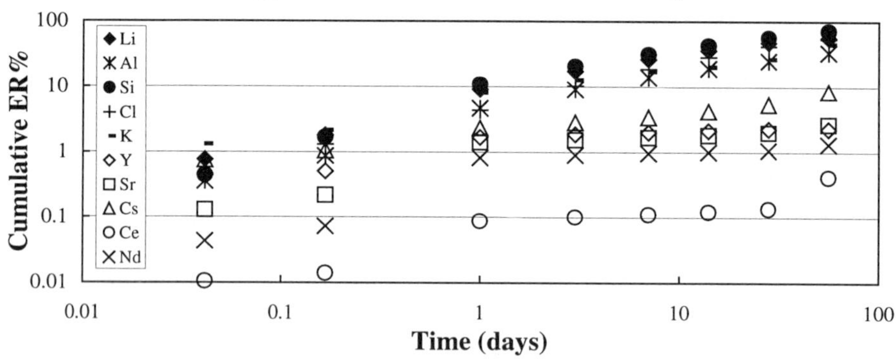

Fig. 4. Cumulative $ER_i\%$ values for solution exchange tests with salt-loaded sodalite material without the glass binder in demineralized water at 90°C.

The incongruent release of Cs, alkaline earth elements, and rare earth elements in solution exchange tests with sodalite and demineralized water at 90°C indicates that sodalite or other phases that may contain these elements are more corrosion resistant than is the sodalite with the alkali metal salts. In the appropriate conditions, the corrosion resistance of a phase may be associated with the solubility of the specific phase. This suggests that the corrosion of the phases that contain alkaline earth and rare earth elements may be solubility controlled [13].

The salt-loaded zeolite 5A and sodalite have cubic morphologies with edge lengths of up to 4 μm. Thus, the salt-loaded zeolite and sodalite powders used in these solution exchange tests have extremely high specific surface areas that are not representative of the actual material to be disposed. Thus, the results from the tests with the 4 μm powders are not directly comparable with tests conducted with materials of larger sizes. The results from solution exchange tests with 4 μm powders provide information about the relative release rates and mechanisms of different salt-loaded zeolite 5A and sodalite components.

The effects of particle size can be observed by comparing the cumulative ER_i% data in Fig. 3 through 5. For these solution exchange tests, the initial S/V ratio of 2000 m^{-1} required significantly less sample to be used in the tests where the particle size of the material was smaller and, thus, the ratio of sample surface area to sample volume of the material was larger. Even if the corrosion mechanisms and rates are the same, the corrosion results could be affected by the diminishing availability of material. This is demonstrated by the cumulative ER_i% data in Fig. 3 and 4, where the dissolution of most of the zeolitic material had occurred by the 28 and 56 d tests. Similarly, while only 15% of the glass from the ceramic waste form samples, which had an average particle size of 110 μm, had corroded after 56 days (as given by the boron release), approximately 35% of the material in the tests with glass 57 (average particle size ~40 μm) had corroded (Fig. 5). The smaller amounts of material used in some tests enhance the effects of diminishing particle sizes and sample availability, which may significantly affect the test results.

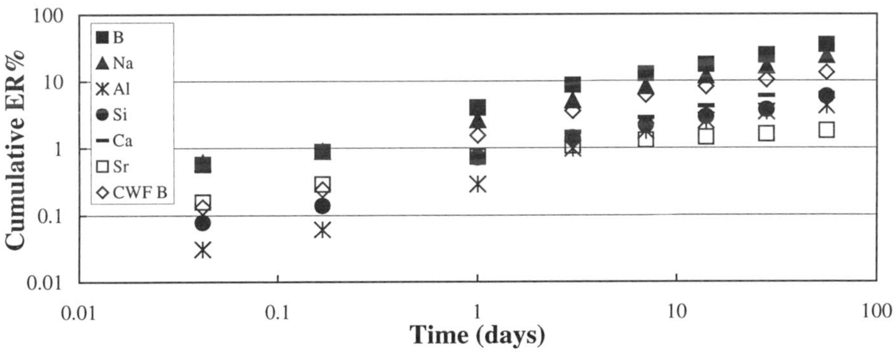

Fig. 5. Cumulative ER_i% values for solution exchange tests with glass 57 (~ 40 μm diameter average) and boron values for the ceramic waste form material (CWF B in legend, ~110 μm diameter average) in demineralized water at 90°C.

CONCLUSIONS

The ceramic waste form has a two-component structure of salt-loaded zeolite in a glass matrix that incorporates a relatively large amount of diverse fission products and waste materials associated with DOE's nuclear fuel conditioning activities. The solution exchange tests provided a substantial amount of information about the corrosion and release mechanisms of specific elements from the different components in the ceramic waste form. Release of free salts, preferential release of elements, ion exchange, and matrix dissolution were all distinguishable with the solution exchange procedure used for these tests.

In general, singly charged ions are preferentially released, relative to the divalent and trivalent ions in demineralized water, from the salt-loaded zeolite 5A by itself or in the glass matrix of the ceramic waste form. The order of cation release from the salt-loaded zeolites is affected by the solubility and zeolite 5A retention preference of specific ions. After an initial release of free salt, the results from the solution exchange tests with salt-loaded sodalite indicate that dissolution of the sodalite matrix must occur before occluded and ion-exchanged elements, including fission products, are released to solution. Furthermore, in the solution exchange tests with sodalite, phases that contain Ce were less soluble than phases that contained other rare earth elements; these in turn were less soluble than phases that contained alkaline earth elements followed by sodalites that contained alkali metals (i.e. solubility of phases that contain Ce << other rare earths < alkaline earths < Cs < K < Na < Li). The order of the solubility of the phases that contain the different elements may be related to corrosion resistance. Detailed studies will be performed to determine the phases that contain the rare earth and alkaline earth elements. However, the detectability of the relatively low amount of rare earth and alkaline earth elements in the reference ceramic waste form could impede the identification of the phases that contain these elements.

ACKNOWLEDGMENTS

This work was supported by the U.S. Department of Energy, Nuclear Energy Research and Development Program, under Contract No. W-31-109-Eng-38. Valuable discussions with Drs. J. K. Bates, D. M. Strachan, D. J. Wronkiewicz, W. L. Ebert, S. F. Wolf, M. A. Lewis, C. C. Pereira, and J. P. Ackerman are greatly appreciated.

REFERENCES

[1]C. C. McPheeters, J. P. Ackerman, and G. K. Johnson, "Electrometallurgical Treatment of Spent Nuclear Fuels for Repository Disposal," *1st Seminar on Molten Salts in Nuclear Technologies*, Dimitrovgrad, Russia (June 19-23 1995).

[2]D. P. Abraham, S. M. McDeavitt, and J. Y. Park, "Microstructure and Phase Identification in Type 304 Stainless Steel-Zirconium Alloys," Met. and Mat. Transactions A, **27A**, 2151 (1996).

[3]L. J. Simpson and D. Wronkiewicz, "Evaluation of Standard Durability Tests Towards the Qualification Process for the Ceramic Waste Form," *Mat. Res. Soc. Symp. Proc.* **465** (1997).

[4]L. J. Simpson, D. J. Wronkiewicz, and J. A. Fortner, "Development of Test Acceptance Standards for Qualification of the Glass-Bonded Zeolite Waste Form," ANL Report No. ANL-98/5 (1998).

[5]M. A. Lewis, D. F. Fischer, and L. J. Smith, "Salt-Occluded Zeolites as an Immobilization Matrix for Chloride Waste Salt," *J. Am. Ceram. Soc.* **76**(11), 2826 (1993).

[6]M. A. Lewis, D. F. Fischer, and C. D. Murphy, "Properties of Glass-Bonded Zeolite Monoliths," *Ceramic Trans.* **45**, 277 (1994).

[7]L. J. Simpson and D. J. Wronkiewicz, "Development of Test Acceptance Standards for Qualification of the Glass-Bonded Zeolite Waste Form," ANL Final Annual Report (1998).

[8]"Waste Acceptance System Requirements Document," U.S. Department of Energy, Office of Civilian Radioactive Waste Management, Report No. DOE/RW-0351, Rev. 2 (1996).

[9]Amer. Soc. for Testing and Mat., "Annual Book of ASTM Standards," C1285-94 (1996).

[10]Amer. Soc. for Testing and Mat., "Annual Book of ASTM Standards," C1174-91 (1987).

[11]"American National Standard Measurement of the Leachability of Solidified Low-Level Radioactive Wastes by a Short-Term Test Procedure", ANSI/ANS-16.1 (1986).

[12]"High-Level Waste Borosilicate Glass, A Compendium of Corrosion Characteristics (Vol. 2)", U.S. Department of Energy, Office of Waste Management, DOE-EM-0177 (1993).

[13]P. Aagaard and H. C. Helgeson, "Thermodynamic and Kinetic Constraints on Reaction Rates among Minerals and Aqueous Solution I," *Amer. J. of Sci.*, **282**, 237 (1982).

CHARACTERIZATION AND DURABILITY TESTING OF A GLASS-BONDED CERAMIC WASTE FORM

S. G. Johnson*, S. M. Frank, T. P. O'Holleran, M. Noy, T. DiSanto, K. M. Goff
and K. Bateman
Argonne National Laboratory-West
P.O. Box 2528
Idaho Falls, ID 83403

ABSTRACT

Argonne National Laboratory is developing a glass bonded ceramic waste form for encapsulating the fission products and transuranics from the conditioning of metallic reactor fuel. This waste form is currently being scaled to the multi-kilogram size for encapsulation of actual high level waste. This paper describes characterization and durability testing used to evaluate the product of the waste treatment process. The emphasis is on results from application of glass durability tests such as the Product Consistency Test and characterization methods such as X-ray diffraction and scanning electron microscopy. The information presented forms the basis for a suite of tests utilized for assessing product quality during scale-up and parametric testing.

BACKGROUND

Argonne National Laboratory has developed an electrometallurgical process for conditioning nuclear spent fuel prior to emplacement into a geologic repository [1]. A demonstration of this technology is taking place using spent fuel from the Experimental Breeder Reactor II (EBR-II). This technology involves the use of a molten salt electrorefiner that separates the active fission products, i.e., Cs, Sr, Ba, Rb, the rare earth elements, and the transuranics, i.e., Pu, Np, Am, Cm, from the uranium and noble metal fission products when processing spent metallic alloy fuel. The active fission products and the transuranics remain in the salt which is a eutectic mixture of KCl and LiCl. The high chloride content of this waste stream precludes the use of a more conventional borosilicate glass for immobilization. A glass-bonded ceramic waste form has been developed for this purpose that incorporates the salt in a crystalline (zeolitic) phase.

The salt from the electrorefiner is removed and prepared for encapsulation after the conditioning of the fuel. Contacting the electrorefiner salt at elevated temperatures with zeolite 4A results in the capture and encapsulation of the salt with the fission products into the alumino silicate framework of the zeolite [2]. The resulting zeolite is then mixed with a small amount of glass binder (25 wt%) and consolidated at high pressure and temperature in a specially designed steel can [3]. This processing takes place in a Hot Isostatic Press (HIP) at temperatures in the 973-1173 K (700-900 °C) range and pressures of 103-172 MPa (15-25 kpsi). During the HIP process the zeolite undergoes a phase transformation to sodalite. This process has been extensively researched and developed by the team at Argonne in Illinois[2]. The scale-up of this process has been underway for two years at Argonne in Idaho [4]. The physical and material aspects of this scale-up have been addressed in the literature [4] and will not be discussed further here. However, there is another issue that deserves attention, that of devising the appropriate criteria for judging the process itself. This issue has been raised with other High Level Waste (HLW) producers [5]. However, since Argonne's waste form is a multiphase ceramic a different set of criteria is needed to demonstrate the level of control of the process. Traditional HLW glass producers have been reliant upon the sole application of the Product Consistency Test for demonstration of their product quality; this approach is not appropriate for a multiphase ceramic waste form such as Argonne's.

This paper will present results from several tests and analyses of the 2 kg size ceramic waste form currently being produced. The goal of these tests are to support scale-up and production of the actual HLW products. To that end this paper will describe the tests or methods being applied, present representative results from those tests and finally make a recommendation as to the role that these tests or methods will play in the production of the ceramic waste form.

EXPERIMENTAL

This section is divided into two parts. Very briefly, the apparatus used for production of the ceramic waste form is described. More conventionally, the motivation and scientific apparatus applied are described.

I. Sample Production

Hot Isostatic Press (HIP)

The HIP uses high temperatures and pressures to produce consolidated ceramic waste forms for the disposal of salt from the electrorefiner. The HIP, model MIH-9, was bought commercially from ABB Autoclave (Columbus, OH). The HIP has a working cavity of 15 cm (6 inch) diameter with a 30 cm (12 inch) height. The HIP has a maximum operating temperatures of 1450 °C and a maximum operating pressure of 206 MPa (30,000 psi). The HIP is controlled using a computerized operator control station and uses the computer software FIX-32. The various functions of the HIP are controlled from the computer console. The product is a ceramic waste form sealed in a stainless steel HIP can, 11 cm (dia.) by 15 cm in length.

II. Characterization and Testing

The group of tests presented in this paper are used to provide a basis for control of the actual demonstration/production process through product testing. In order to be an effective tool for showing the production process is "in control" the battery of tests must be as complete as possible while providing for reasonable turn around time. The tests utilized are summarized in Table I along with a brief statement of the motivation for applying each method. A description of the apparatus and/or method follows.

Table I. Tests and techniques to be applied to ceramic waste products as a measures of product quality.

Test or Technique	Information obtained	Purpose
1) Density	the degree of consolidation	Indicator of process upset during filling of HIP can or process cycle
2) XRD	phase composition	Indicator of material blending error or process cycle upset
3) SEM/TEM	microstructure and phase composition	Indicator of blending error, segregation of feed materials or process cycle upset
4) 7 day Crushed Sample Durability Test	a relative performance measure	Indicator of the failure of feedstream preparation or process upset

The phase process cycle upset in Table I means an improper temperature, pressure and/or time cycle in the HIP. The purpose section in Table I is meant to be a brief synopsis and cause and effect and not an inclusive listing.

A) Density Measurements

The pycnometer (Quantochrome Co., Boynton Beach, FL) uses helium gas to obtain an accurate volume determination of an object and combines that with a known mass to determine the density.

B) X-ray Diffractometer (XRD)

The X-ray diffractometer used is a Scintag X1 Powder X-Ray diffractometer (Scintag Inc., CA) with a theta-theta goniometer, Cu K-alpha X-Ray tube with normal focus and top loading powder sample holder.

C) Scanning Electron Microscopy (SEM) with Energy Dispersive Spectroscopy (EDS)

The SEM work was performed with a Zeiss DSM 960A digital scanning electron microscope. The energy dispersive detector was provided by Oxford Instruments.

D) Durability Testing

The samples were subjected to a crushed leaching test. This was conducted in manner consistent with ASTM C1285-94, the Product Consistency Test (PCT) [6]. The test was conducted for 7 days at 363 K (90 °C) using a surface area to volume of leachant ratio of 2000 m^{-1}. The test material was the -100 to +200 mesh size fraction. ASTM type I water was used as a leachant.

RESULTS AND DISCUSSION

I. Density

The density of the waste form was determined using the technique of helium pycnometry. Typical density for this waste form is 2.30-2.40 g/cm^3 and indicate acceptable consolidation during the HIP process. Several waste forms were analyzed using 7-13 samples in different locations of the cans and no trend in the density was observed. This implies that no segregation of material occurs within the can, or if it does, the HIP cycle then homogenizes the product. This is important since the zeolite and glass used are slightly different in size and density and could lead to segregation. There is no supporting evidence for this potential problem either in the density analyses or in the XRD results.

II. X-ray Diffraction

Figure 1 presents the XRD pattern for a representative ceramic waste form. Sodalite ($Na_6Al_6Si_6O_{24} \cdot 2NaCl$) is a major phase with nepheline ($NaAlSiO_4$) and Halite (NaCl) being present as minor phases. Samples from various areas of the HIP can revealed no variation in the pattern was observed. The lack of a zeolite pattern indicates the relatively complete transformation of the zeolite 4A to sodalite during the HIP cycle (within the detection limits of the technique of XRD). This is important because our previous experiments have shown that the behavior of the sodalite waste form is superior to that of a waste form consisting of unconverted zeolite based on a 7 day PCT. The presence of nepheline indicates that a small portion of the sodalite was converted to nepheline, i.e., the conversion process was "over-driven". This has not been shown to be detrimental to the waste form performance as measured using the 7 day PCT and thus is of minimal concern. The small amount of sodium chloride present, which is a by-product of the conversion process, has not been shown to be deleterious to the waste form performance based on a 7 day PCT.

III. Scanning Electron Microscopy

Figure 2 features a backscattered electron SEM micrograph of material from the same HIP can as that displayed in Fig. 1. Three phases are commonly observed using SEM in the ceramic waste form: a "zeolitic" region, a glass region and a rare earth oxide or oxychloride or some combination of the two. These are identified in Fig. 2. Occasional sodium chloride crystals are also identified but these are very few in number and well dispersed throughout the waste form. The zeolitic region is positively identified only through the XRD results, however, it is readily identified by the distinctive ratio of silicon to aluminum and the presence of chloride as observed when applying EDS. The rare earths tend to react with the small amount of residual moisture present in the zeolite 4A precursor and form rare earth oxychlorides during the salt contacting step and then are converted to oxides, at least partially, when processed in the HIP. The exact stiochiometry of the typical rare earth phase observed varies in oxygen content, chloride content and the mixture of rare earth cations present. The remainder of the surrogate fission products present in the salt are below detection limits, but presumable are dispersed within the sodalite phase.

IV. Durability Test

The results from a conventional 7 day PCT-B test are presented in Table II which shows normalized release rates for a typical ceramic waste product (CWP) for the pertinent elements. The elements analyzed for include those of the major matrix phases present (Si, Al, B, Na), eutectic salt (Li, K, Cl) and the encapsulated surrogate fission products (Cs, Rb) of most interest. Data is also presented for selected elements of the glass standard used as an internal quality control standard, ARM-1. The results for the ARM-1 standard are within the acceptable range for this material and this test based on those values commonly cited in the literature [7]. The materials were freshly prepared in accordance with ASTM C-1285-94. Please note that the rare earths were also analyzed for, however, they were present in concentrations below the limits of our analytical techniques, inductively coupled plasma atomic emission spectroscopy and graphite furnace atomic absorption spectroscopy.

Table II. Normalized release rates for a 7 day PCT conducted at 90 C. All NRR are given in g/m^2day and are normalized for the mass fraction of the elements present in the final waste form. Values are given for the ARM-1 standard only for pertinent elements.

Element	C W P	ARM-1
Li	4.6e-2	3.5e-2
Na	9.7e-3	3.0e-2
K	1.1e-2	--
Rb	9.5e-3	--
Cs	5.8e-3	--
Si	9.5e-3	1.8e-2
Al	9.4e-3	--
B	2.4e-2	2.9e-2
Cl	2.1e-2	--

CONCLUSIONS

This paper presents results of several characterization and testing methods applied to a demonstration scale ceramic waste form from Argonne National Laboratory's electrometallurgical treatment process. This group of tests were chosen to provide a basis for control of the actual production process. The cadre of tests presented here forms the basis for our "Measures of Product Quality" for our ceramic waste form process.

ACKNOWLEDGMENTS
Argonne National Laboratory is operated for the US Department of Energy by the University of Chicago. This work was supported by the US Department of Energy, Reactor Systems, Development and Technology, under contract W-31-109-ENG-38. The following individuals are acknowledged for their valuable assistance: M. Simpson (preparation of ceramic waste form materials), S. A. Barker (GF-AAS of leachates), and D. G. Cummings (ICP-AES of leachates).

REFERENCES

[1] Prog. Nucl. Energy, **31** (1997).

[2] Lewis, M. A., Hash, M., Glandoff, D., "Effect of Different Glass and Zeolite A Compositions on the Leach Resistance of Ceramic Waste Forms", *Scientific Basis for Waste Management XX*, W. Gray and I. Triay, ed., MRS, Pittsburgh, vol. **465**, 433-440, (1997).

[3] Goff, K. M., Simpson, M. F., Johnson, S. G., Bateman, K. J., Battisti, T. J., Frank, S. M., paper contributed to the ANS topical conference on Spent Fuel, Charleston, SC (1998).

[4] Goff, K. M., Benedict, R. W., Bateman, K., Lewis, M. A., Periera, C., Musick, C. A., "Spent Fuel Treatment and Mineral Waste Form Development at Argonne National Laboratory-West", *Spectrum '96*, ANS, La Grange, IL, 2436-2443, (1996).

[5] Bibler, N. E. and Bates, J. K., "Product Consistency Leach Tests of Savannah River Site Radioactive Waste Glasses", *Scientific Basis for Waste Management XIII*, V. Oversby and P. W. Brown, ed., MRS, Pittsburgh, vol. **173**, 327-338, (1990).

[6] ASTM C1285-94, ASTM, Philadelphia (1995).

[7] Jantzen, C. M., Bibler, N. E., Beam, D. C., Crawford, C. L., Pickett, M. A., "Characterization of the Defense Waste Processing Facility (DWPF) Environmental Assessment (EA) Glass Standard Reference Material (U)", WSRC-TR-92-346.

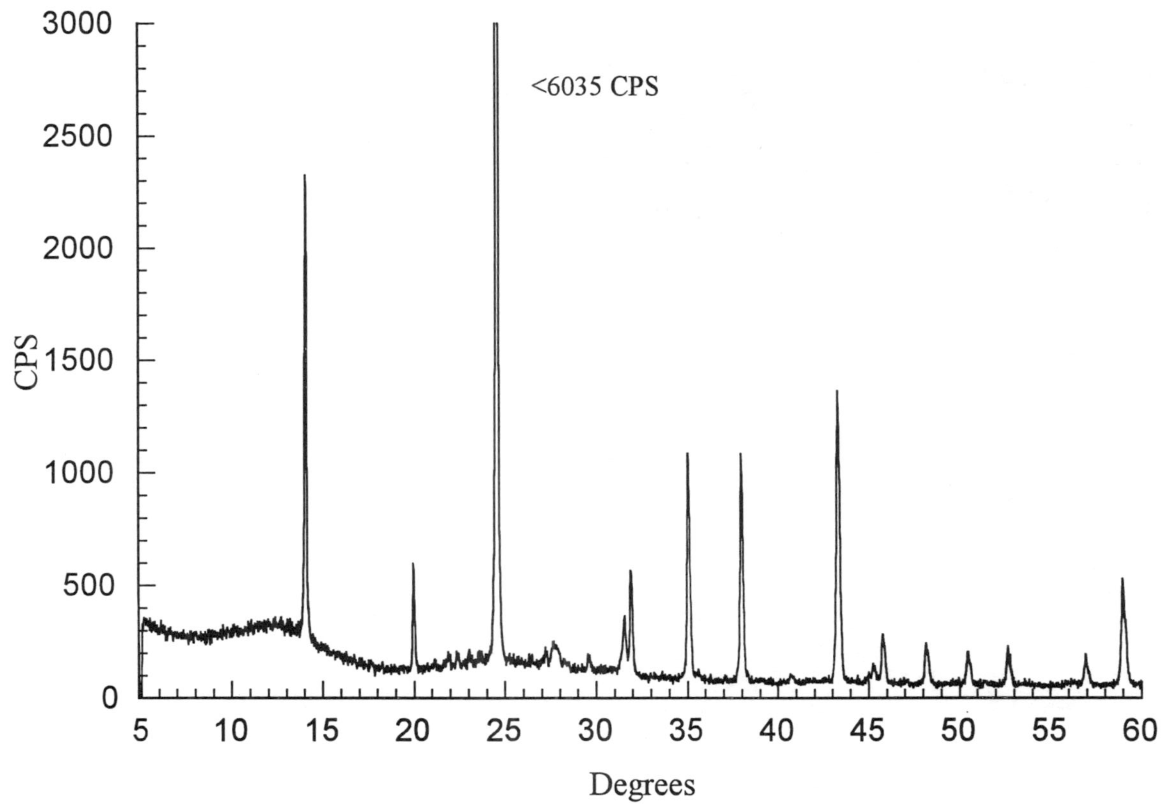

Figure 1. XRD pattern for a typical ceramic waste form product. The major phase is sodalite ($Na_6Al_6Si_6O_{24}\cdot 2NaCl$). Two minor phases are also observed: nepheline ($NaAlSiO_4$) and halite ($NaCl$).

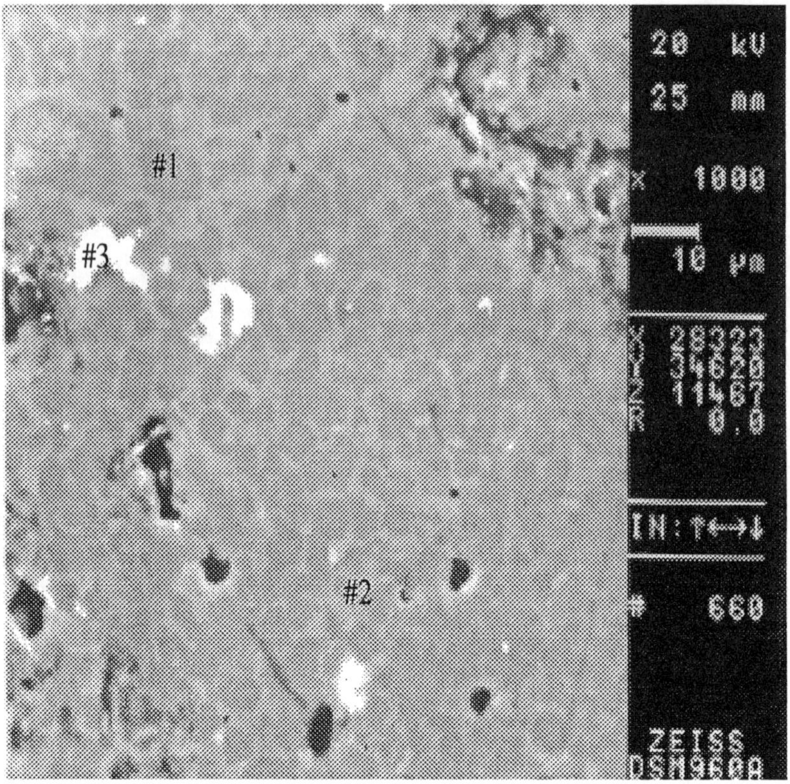

Figure 2. Back scattered SEM micrograph of a the same product featured in the Fig. 1. The major phases, #1 (Sodalite), #2 (Glass) and #3 (Rare earth oxide or oxychloride), are identified.

Actinide Material Treatment and Stabilization

SYNROC DERIVATIVES FOR EXCESS WEAPONS PLUTONIUM

E. R. Vance, A. Jostsons, S. Moricca, M. W. A. Stewart, R. A. Day, B. D. Begg, M. J. Hambley, K. P. Hart and B. B. Ebbinghaus*
Materials Division, ANSTO, Menai, NSW 2234, Australia
*Lawrence Livermore National Laboratory, Livermore, CA 94550, U.S.A.

ABSTRACT

Zirconolite-rich (80 wt%) ceramics containing ~ 12 wt% Pu and 17 wt% of Hf and Gd as neutron absorbers were prepared by the oxide and alkoxide routes, and consolidated by sintering in air at 1275-1400°C and hot isostatic pressing at 1280°C/150 MPa in a stainless steel container respectively. Pyrochlore-rich ceramics containing ~ 10 and 20 wt% of U and Pu respectively, ~ 15 wt% of neutron absorbers, and 2-15 wt% of process chemicals were produced by oxide and alkoxide routes, and sintered in argon at 1300 or 1350°C. Pu leach rates at 70 or 90°C in both the zirconolite- and pyrochlore-rich ceramics were typically 10^{-4} g/m^2/day after 7 days, with the rare earth and Hf leach values being ~ 3 times greater and ~ 10 times less respectively than those for Pu. Leach rates of other elements were generally comparable with those for Synroc-C.

INTRODUCTION

Zirconolite-rich Synroc has been developed for actinide-rich wastes in general [1] and particularly since 1994 as a candidate ceramic host for the immobilisation option for the final disposition of excess weapons Pu [2,3]. Experimentally, zirconolite is the most durable Synroc phase in aqueous media, and there is an abundance of data on natural zirconolites to reinforce this conclusion [4].The incorporation of Pu at levels of 10-20 wt% in pure zirconolite utilising processing atmospheres ranging from oxidising to reducing is reasonably well understood on the basis of detailed compositional, X-ray absorption and X-ray diffraction data [2,3]. The solubility in zirconolite of the neutron absorbers Hf, Gd and Sm has been shown to be extensive (1, 0.7, and 0.7 formula units respectively [3]). Gd and Sm occupy the Ca and Zr sites in zirconolite, and Hf occupies the Zr site [3].

There are advantages in using a multi-phase zirconolite- or pyrochlore-rich Synroc derivative. A multi-phase material assists in minimising the grain size to improve mechanical properties and reduce microcracking effects attendant upon radiation damage. Also, modest changes in the waste/precursor ratio and impurities generally only affect the abundances of the phases present and not their identities.

In the present work, zirconolite-rich titanate ceramics containing ~ 12 wt% of high-fired PuO_2 plus significant amounts of Hf, Gd and Sm were made by an oxide route followed by sintering in air at 1275-1400°C. Similar materials were also made by the alkoxide route, in which the Pu was introduced as a nitrate solution, with final consolidation by hot isostatic pressing at 1280°C/150MPa. Microstructural and leaching data on these materials are presented.

The main "impurity" in the Pu of interest in the remainder of the present work is U. At least some of the (depleted or natural) U would be added as a criticality control agent for ^{235}U which forms from the radioactive decay of ^{239}Pu. The aim here was to co-incorporate ~ 10 and 20 wt% of Pu and U respectively in a Synroc-type ceramic. To maximise the actinide content of a Synroc-type ceramic, it is necessary to employ the pyrochlore phase rather than the closely-related zirconolite. Though the experimental evidence on the resistance of synthetic pyrochlore to aqueous dissolution is limited, natural pyrochlores can retain actinides for billions of years even when extensive secondary alteration due to contact with fluids at temperatures < 100°C has occurred [6].

EXPERIMENTAL

Samples were made in a glovebox line by both the oxide and alkoxide/nitrate [1] routes. After calcination at 750°C in air or argon, the powders were ball-milled and consolidated by (a) pelletising, followed by sintering in air or argon, (b) uniaxial hot-pressing in metal bellows or (c) hot isostatic pressing. Scanning electron microscopy, X-ray diffraction and X-ray absorption spectroscopy were carried out as described previously [2,3].

Leaching was performed by modified MCC-1 methods, mostly in deionised water at 70°C (polythene container) or 90°C (teflon container). Elemental extractions in solution and the vessel walls were measured to yield total (normalised) differential rates. The samples were contained in Pt crucibles inside the leaching vessels to prevent alpha particles from the solid samples from contacting the vessel walls. Leaching at 200°C was carried out in inconel Parr bombs. Pu and U in the leachates were measured by α-counting and Ca, Gd, Sm, Ca and Ti by atomic absorption spectrophotometry or ICP/MS or ICP/AES methods as appropriate.

RESULTS AND DISCUSSION

Oxide-route, Sintered Zirconolite-rich Ceramics Containing 12 wt% of PuO_2 and 17 wt% of Neutron Absorbers

Pu was targeted towards the Ca site of zirconolite as Pu^{3+}. Also incorporated in zirconolite were the neutron poisons Gd and Sm, each in a 1:2 molar ratio with respect to Pu, and designed to be partitioned equally between the Ca and Zr sites. Hf, in a 1:1 molar ratio with respect to Pu, was substituted on the Zr site. The nominally 80 wt% of zirconolite in the ceramic was designed to have the following composition: $Ca_{0.595}Pu_{0.27}Gd_{0.135}Sm_{0.135}Zr_{0.595}Hf_{0.27}Ti_{1.73}Al_{0.27}O_7$. Ten wt% each of Ba hollandite and rutile made up the rest of the ceramic. In some samples, the Sm was replaced by an equimolar amount of Gd.

The samples were sintered in air at 1275-1400°C. A typical microstructure is given in Fig. 1a, for a sample sintered for 4h at 1375°C. Zirconolite was the majority phase incorporating nearly all the Pu; minor hollandite and rutile were also present, together with < 2 % of the Pu inventory remaining as unincorporated PuO_2. The average grain size was ~ 4 μm and the density ~ 93% of theoretical. Hot-pressing the same calcined precursor at 1300°C in a metal bellows allowed incorporation of all the Pu in the zirconolite and yielded a finer-grained (~ 1 μm), fully-dense ceramic (see Fig. 1b).

These data, though still somewhat incomplete, show that the Pu leach rates are very low, in the range of 10^{-4} -10^{-5} g/m²/day at 70°C. For the neutron absorbers, the Gd values exceed the Pu leach rates by factors of 2-10, whereas the Hf values are generally lower than the Pu values, by at least a factor of 10. The Ba and Ca leach rates were within a factor of ~ 3 of those for Synroc-C, and the rare earth and Pu leach rates also showed order-of-magnitude agreement with the corresponding rates in Synroc-C [7]. The leach rates decreased with increasing leaching time as normally observed for Synroc-C, and the increase in leach rate at 200°C was generally only about a factor of 10.

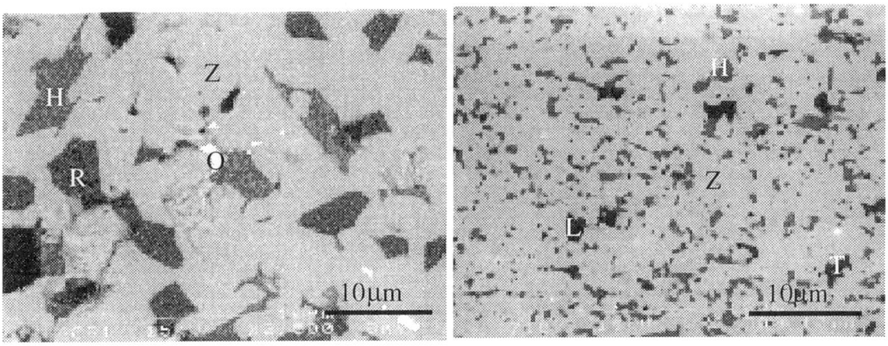

(a) (b)

Figure 1. Backscattering scanning electron micrographs from zirconolite-rich ceramics:- (a) oxide route, sintered 4h at 1375°C; (b) hot-pressed at 1300°C/20MPa/3h. Z = zirconolite, H = hollandite, O = PuO_2, R = rutile, L = alumina and T = aluminium titanate.

The sintered ceramics were leached at 70°C and 200°C in deionised water and the results are summarised in Table I.

Table I. Leaching rates ($g/m^2/day$) of sintered (4h at 1375°C in air) zirconolite-rich ceramics

Leach period (days)	0-7	7-28	28-84	0-7	7-28	28-84
Temperature (°C)		70			200	
Element						
Pu	6×10^{-5}	2×10^{-5}	2×10^{-5}	9×10^{-6}	2.5×10^{-6}	na
Ba	0.017	na	6×10^{-4}	na	na	0.02
Ca	0.15	na	0.02	na	na	na
Gd	7×10^{-4}	na	5×10^{-5}	na	na	2×10^{-4}
Hf	$<2 \times 10^{-5}$	na	$<2 \times 10^{-6}$	na	na	$<3 \times 10^{-5}$
Ti	4×10^{-4}	na	1×10^{-4}	na	na	1.5×10^{-3}
Zr	$<4 \times 10^{-5}$	na	$<2 \times 10^{-5}$	na	na	$<1 \times 10^{-4}$

na: data not available

Alkoxide-route, HIPed Zirconolite-rich Synroc Containing 12 wt% of Pu and 17 wt% of Neutron Absorbers

A 360 g ceramic sample containing 50 g of Pu oxide and having the same overall composition as the sintered sample described in the previous section, was made by the alkoxide/nitrate route, followed by calcination in argon, and hot isostatic pressing at 1280°C/150MPa for 2h in a metal can as a final consolidation step. The microstructure (see Fig. 2) contained the expected phases, having approximately the predicted composition, and no uncombined PuO_2 remained. X-ray absorption data on the Pu L_{III} edge at 18.06 keV showed that the Pu was present as Pu^{3+}[3], as per the original design.

Leaching data at 70°C gave the values shown in Table II.

Table II. Modified MCC-1 leach rates ($g/m^2/day$) at 70°C for HIPed Pu-doped Synroc sample

Element	0-7 days	7-14 days	14-28 days
Ba	0.09	0.04	0.02
Ca	0.02	<0.01	0.02
Gd	1.3×10^{-4}	7×10^{-5}	5×10^{-5}
Hf	$<10^{-5}$	$<10^{-5}$	$<10^{-5}$
Sm	1.3×10^{-4}	5×10^{-5}	3×10^{-5}
Ti	1.2×10^{-4}	1.7×10^{-4}	6×10^{-5}
Zr	1.1×10^{-4}	1.1×10^{-4}	2×10^{-5}
Pu	7×10^{-5}	2×10^{-5}	1×10^{-5}

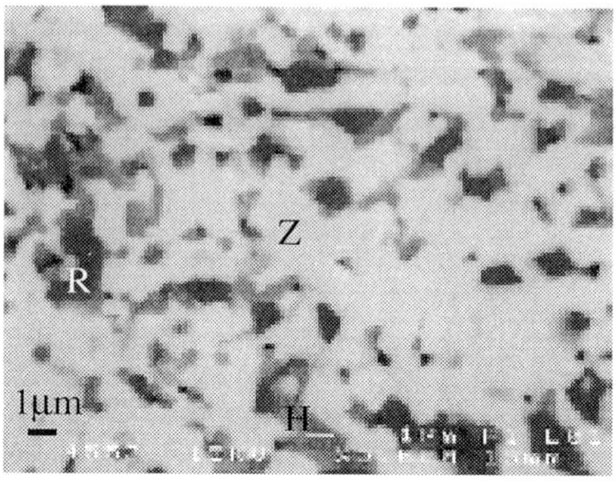

Figure 2. Backscattering scanning electron micrograph from hot isostatically pressed zirconolite-rich ceramic. Z = zirconolite, R = rutile and H = hollandite.

The rare earth leach rates were factors of 2-5 higher than the Pu rates, whereas the Hf rates were substantially lower than the Pu rates. All leach rates were in essential agreement with those of Synroc-C[7] and the sintered zirconolite-rich ceramic.

Incorporation of 10 and 20 wt% Respectively of Pu and U, plus 15 wt% of Neutron Poisons, in Pyrochlore-rich Synroc Ceramics.

Following the basic strategy of Synroc-F[8], a ceramic containing 10 and 20 wt% of Pu and U was designed to contain 95 wt% of a pyrochlore-structured phase having the following composition: $Ca_{0.89}Gd_{0.22}Pu_{0.22}U_{0.44}Hf_{0.23}Ti_2O_7$, plus 5 wt% of rutile containing 25 wt% of HfO_2. i.e. $Ti_{0.91}Hf_{0.09}O_2$. Here Pu and U were targeted as +4 ions, with Gd being equally partitioned between the Ca and actinide/Hf sites.

The Pu/U oxide was fired for 2h in air at 1000°C, and the ceramic was fabricated by an oxide-route, using either wet-or dry-milling, with final consolidation by pelletising and sintering for 4h at 1275-1400°C in an argon atmosphere. The final material contained ~ 90 wt% of pyrochlore, plus small amounts of rutile, brannerite $(U,Pu)Ti_2O_6$ and $(U,Pu)O_{2+x}$. The actinide oxide phase abundance ranged from ~2 % at the lower firing temperatures to <1% at the higher temperatures. Densities in excess of 90% of theoretical were obtained at the higher firing temperatures.

The effects of various additional processing impurities (see Table III) were also studied. These impurity compositions were chosen as representative of likely impurity contents across a range of impure Pu stockpiles. The samples corresponding to waste compositions 3A and 5A were made by the oxide route, whereas those corresponding to the 7A, 8A and 9A waste compositions were made by the alkoxide route. Samples 3A and 5A were sintered at 1350°C; the rest were sintered at 1300°C to avoid warping. All samples were sintered for 4h in argon.

Table III. Waste additions (%) added to baseline* pyrochlore-rich Synroc.

Sample	3A	5A	7A	8A	9A
Waste compound					
Al_2O_3	0.22	0.11	0.32	1.80	0.53
MgO	0.24	0.46	0.13	1.00	0.46
$CaCl_2$			0.16	2.52	0.70
Ga_2O_3	1.30		0.14		
Fe_2O_3	0.14	0.16	0.08	0.58	0.16
Cr_2O_3	0.02		0.02	0.15	0.08
NiO	0.09		0.04	0.38	0.14
CaF_2			0.12	1.49	0.46
K_2O	0.04		0.07	1.21	0.34
Na_2O			0.06	0.54	0.15
MoO_2		0.30	0.11	0.54	0.30
SiO_2			0.19	1.72	0.48
Ta_2O_5	0.15		0.06	0.74	0.20
B_2O_3				0.39	0.18
WO_2	0.06			1.89	0.52
ZnO			0.01		0.07

*composition of ceramic without process chemical impurities (wt%): CaO (9.95); TiO_2 (35.86); HfO_2 (10.65); Gd_2O_3 (7.95); UO_2 (23.69); PuO_2 (11.89).

In each case, the major phase in the final ceramic was pyrochlore, but the minor phases (Table 4) varied, with traces of zirconolite, brannerite, rutile, silicate phases (if B and Si were present) and loveringite being formed, depending on the actual inventory of impurities. The titanate phases should be reasonably durable and have been observed in Synroc and tailored ceramics previously [7]. The silicate phase was sometimes present as partly connected intergranular material, but it also appeared enclosed in the titanate phases. Less than 1% of the Pu inventory was present in the silicate phases. Chloride and fluoride were deduced from X-ray fluorescence measurements to be almost completely volatilised during sintering.

Table IV. Additional phases in the various pyrochlore-rich samples

Sample	Phases				
	$(Pu,U)O_{2+x}$	Brannerite	Zirconolite	Rutile	Silicate
3A	trace	very common	common	minor	
5A	trace	common	minor	common	
7A		common		minor	trace
8A*			very common		common
9A		very common	common	minor	minor

*trace of loveringite also observed

The 0-7 day leach rates are given in Table V and they agreed quite well with those for the zirconolite-rich ceramics.

Table V. Normalised leach results from pyrochlore-rich ceramics (7 days at 90°C, modified MCC-1 method).

Sample	3A	5A	7A	8A	9A	dry-mill*	wet-mill*
Element							
Pu	1.3×10^{-4}	2.3×10^{-4}	2.4×10^{-4}	9.5×10^{-5}	1.1×10^{-4}	1×10^{-4}	2×10^{-4}
U	3.9×10^{-4}	4×10^{-4}	3×10^{-4}	5×10^{-4}	5×10^{-4}	4×10^{-4}	5×10^{-4}
Gd	6×10^{-4}	8×10^{-4}	5×10^{-4}	2.2×10^{-4}	3×10^{-4}	3×10^{-4}	5×10^{-4}
Ca	0.032	0.029	0.03	0.043	0.025	0.05	0.03
Hf	$<4 \times 10^{-5}$	$<4 \times 10^{-5}$	$<4 \times 10^{-5}$	$<4 \times 10^{-5}$	$<4 \times 10^{-5}$	$<4 \times 10^{-5}$	2×10^{-4}
Ga	4.8×10^{-3}		2.8×10^{-3}		5.9×10^{-3}		
Ta	$<2 \times 10^{-3}$		$<4 \times 10^{-3}$	$<4 \times 10^{-4}$	$<2 \times 10^{-3}$		
W	$<8 \times 10^{-3}$			$<2 \times 10^{-4}$	$<8 \times 10^{-4}$		
Mo			<0.02	$<6 \times 10^{-3}$	$<6 \times 10^{-3}$		

* baseline ceramic

Finally, the pyrochlore-rich ceramics were heated quickly to 1000°C and slow-cooled to room temperature at 1-2°C/min.; such a thermal cycle simulates the can-in-canister approach in which 10-20% of the volume of a US disposal canister would be occupied by Pu-bearing wasteforms and the canister would be filled with glass from the Defense Waste Processing Facility at Savannah River to constitute a radiation barrier yielding a field of ~1000 R/h at 1 m after 30 yr (equivalent to that of canisterised used fuel). The samples did not exhibit any changes in microstructure or phase constitution. Hence no significant changes other than possible cracking are expected upon filling the canister, already partly filled with cans of pyrochlore-rich Synroc, with DWPF glass.

CONCLUSIONS

Zirconolite- and pyrochlore-rich ceramics show a high potential for the immobilisation of excess weapons Pu which contains significant amounts of process chemicals. While sintered ceramics with very low Pu and neutron absorber leach rates can be readily produced, fabrication by uniaxial or isostatic hot-pressing can give full densification, minimum grain size and maximum leach resistance.

ACKNOWLEDGMENTS

This work was funded in part by the US Department of Energy. We thank J. Foy, R. Stevjanovic, J. Weir, F. Cordaro and E. Loi for analysis of the leachates, and N. Blagojevic for carrying out X-ray fluorescence measurements.

REFERENCES

[1]E.R. Vance, P.J. Angel, B.D. Begg and R.A. Day, "Zirconolite-Rich Titanate Ceramics for High-Level Actinide Wastes", pp. 293-298 *in Scientific Basis for Nuclear Waste Management XVII*, Edited by R. Van Konynenburg and A.A. Barkatt. Materials Research Society, Pittsburgh, PA, USA, 1994.

[2]E.R. Vance, C.J. Ball, R.A. Day, K.L. Smith, M.G. Blackford, B.D. Begg and P.J. Angel, "Actinide and Rare Earth Incorporation in Zirconolite", *J. Alloys and Compounds*, **213/214**, 406-09 (1994).

[3]B. D. Begg, E. R. Vance, R. A. Day, M. Hambley and S. D. Conradson, "Plutonium and Neptunium Incorporation in Zirconolite", pp. 325-332 *in Scientific Basis for Nuclear Waste*

Management XX, Edited by W. J. Gray and I.R. Triay. Materials Research Society, Pittsburgh, PA, USA, 1997.

[4]G. R. Lumpkin, K. P. Hart, P. J. McGlinn, T. E. Payne, R. Giere and C. T. Williams, "Retention of Actinides in Natural Pyrochlores and Zirconolites", *Radiochim. Acta* **66/67**, 469-74 (1994).

[5]E. R. Vance, A. Jostsons, R. A. Day, C. J. Ball, B. D. Begg and P. J. Angel, "Excess Pu Disposition in Zirconolite-rich Synroc", pp. 41-48 in *Scientific Basis for Nuclear Waste Management XIX*, Edited by W. M. Murphy and D. A. Knecht. Materials Research Society, Pittsburgh, PA, USA, 1996.

[6]G. R. Lumpkin and R. C. Ewing, "Geochemical alteration of pyrochlore group minerals: Pyrochlore sub-group", *Amer. Mineral.* **80**, 732-43 (1996).

[7]A. E. Ringwood, S. E. Kesson, K. D. Reeve, D. M. Levins and E. J. Ramm, "Synroc", pp. 233-334 in *Radioactive Waste Forms for the Future*, Edited by W. Lutze and R. C. Ewing, Elsevier, Netherlands, (1988).

[8]S. E. Kesson and A. E. Ringwood, "Safe Disposal of Spent Nuclear Fuel",, *Radioactive Waste and the Nuclear Fuel Cycle*, **4**, 259-74 (1983).

A VERSATILE TWO-STEP PROCESS FOR IMMOBILIZING EXCESS PLUTONIUM

Thomas P. O'Holleran and Stephen G. Johnson
Argonne National Laboratory-West
P. O. Box 2528
Idaho Falls, ID 83403-2528

Peter C. Kong and Bruce A. Staples
Idaho National Engineering and Environmental Laboratory
P. O. Box 1625
Idaho Falls, ID 83415-2210

ABSTRACT

As a consequence of weapon stockpile reduction and the associated shutdown of weapons production facilities, approximately 50 metric tons of plutonium (both weapons-grade and non-weapons-grade) has been declared excess by the United States. Recent experiments demonstrated the feasibility of using high-level waste stored at the Idaho Chemical Processing Plant to immobilize plutonium. The most effective plutonium host phase identified in these experiments was a plutonium zirconate solid solution. Results of recent experiments are reported that show the feasibility of using the highly durable plutonium zirconate host phase as a feed material for high and low temperature encapsulation processes, thereby increasing the potential applications of this material for plutonium dispositioning.

INTRODUCTION

With the end of the cold war, and reductions in the sizes of nuclear arsenals, the United States abruptly ceased production of new nuclear weapons. This resulted in approximately 50 metric tons of surplus plutonium requiring some form of treatment to prepare it for final dispositioning. About 38 metric tons of that inventory is considered weapons grade material. This material is high purity plutonium, in the form of surplus weapons components, or other high purity metallic or oxide forms that could be fabricated into nuclear explosives with relative ease. The Department of Energy (DOE) recently issued a Record of Decision (ROD) electing to burn some of the excess plutonium as mixed oxide (MOX) fuel, and dispose of the remainder along with high-level waste (HLW) glass[1]. The latter approach, known as the "can-in-canister" approach, would immobilize the excess plutonium in a Synroc type ceramic, which would be encapsulated in HLW glass at the Defense Waste Processing Facility. The HLW glass provides a self-protecting radiation field to meet the spent fuel standard for proliferation resistance, as recommended by the National Academy of Sciences[2]. The remaining 12 metric tons is in a variety of forms with diverse chemical and physical properties. This material is plutonium from the numerous steps in the weapons production process, reflecting the sudden cessation of weapons production. Some of these materials are chemically unstable, and most require some form of stabilization for safe interim storage and ultimate disposal. The Defense Nuclear Facilities Safety Board (DNFSB) recommended expeditious treatment of these materials to reduce management and storage risks[3].

Materials for plutonium immobilization have been developed while exploring potential weapons material dispositioning missions for the Idaho National Engineering and Environmental Laboratory (INEEL)[4,5]. In developing these materials, the unique chemistry of the HLW at the INEEL was used to partition plutonium into crystalline host phases to increase plutonium loading and chemical durability (and hence proliferation resistance). Both glassy and crystalline materials were tested. Some of the crystalline materials were formed by devitrifying glasses prepared using conventional glass melting methods. The most effective plutonium host phase identified in these experiments was a plutonium zirconate solid solution with a face-centered cubic (fluorite type) crystal structure. This material is easily prepared by direct solid state reaction of oxide powders in air at 1650° C, and can accommodate appreciable quantities of rare earth neutron absorbers for criticality control[6]. Plutonium release rates were on the order of 10^{-5} to 10^{-6} $g/m^2/d$ as measured using the Product Consistency Test (PCT-A)[7].

The ease with which this material could be prepared suggested a two-step process that could be used to immobilize plutonium from a variety of sources. The first step is to react plutonium (oxide) with zirconia and, if desired, a neutron absorber. Depending on the plutonium source (e. g. an organic-based source stream whcih may generate considerable off-gas), an oxidation step may be desirable prior to reaction with zirconia. The resulting plutonium zirconate powder is then encapsulated in a suitable matrix to make it non-dispersible. Encapsulants evaluated in this work include glass and magnesium phosphate cement. The proposed process is illustrated in Figure 1.

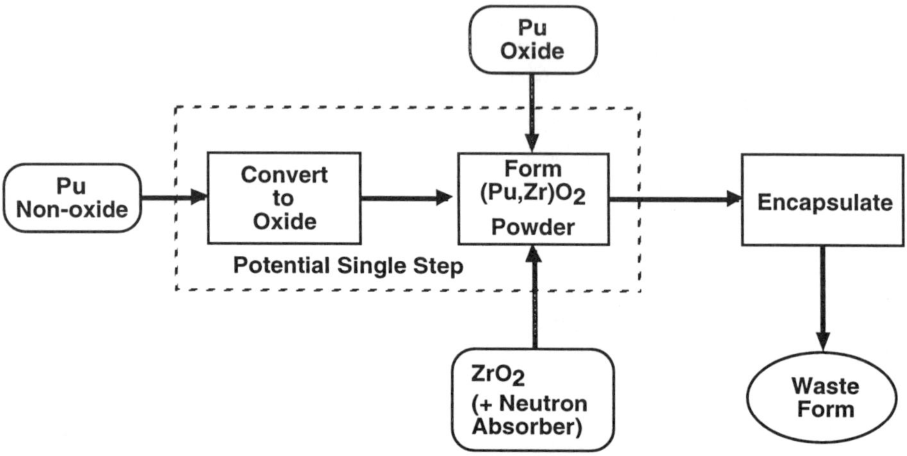

Figure 1: Two-step process for stabilizing plutonium. Depending on the plutonium source, the "Convert to Oxide" step shown inside the dashed box could be accomplished at the same time the material is reacted with zirconia and an optional neutron absorber.

In order to establish the feasibility of the second step of the proposed two step process for plutonium dispositioning (encapsulation of the durable host phase in a HLW glass), it was necessary to show that neither the plutonium host phase nor the matrix is altered by the encapsulation process in any way that would compromise the ability of the composite material to isolate radionuclides from the environment. Glass was selected for evaluation as a matrix material because of its wide acceptance, particularly by "stakeholder" groups, as a durable material for immobilizing toxic and radioactive elements. Hydraulic cement was selected as a second candidate matrix material because of process simplicity and low capital costs.

EXPERIMENTAL

Plutonium Zirconate Preparation

The process to prepare plutonium zirconate (with or without neutron absorbers) is described elsewhere[6]. Briefly, mixtures of PuO_2 and ZrO_2 powders (with or without rare earth oxide powder) are mechanically blended and placed in an alumina crucible. Molar ratios of plutonium to zirconium up to 1.36 have been produced, with up to 8.5 mole % Sm_2O_3. The starting powders are fired for 48 hours at 1650° C in an electric furnace with air atmosphere. A scanning electron micrograph of material resulting from this process step is shown in Figure 2.

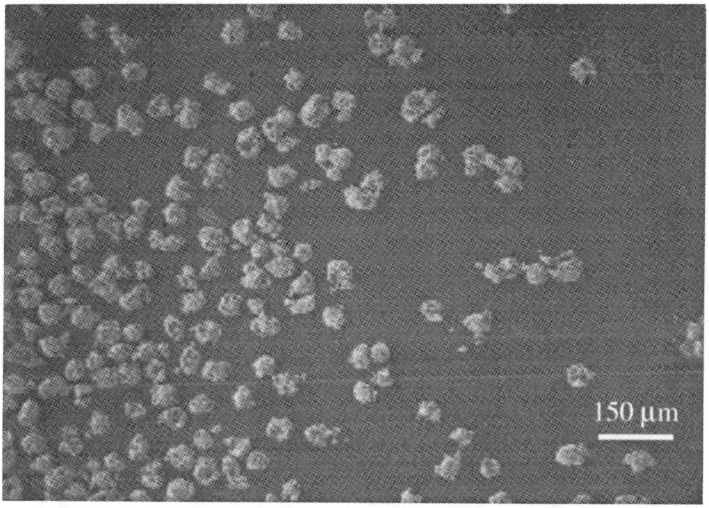

Figure 2: Secondary electron image of plutonium zirconate powder from the high-temperature reaction step.

Glass Encapsulation

A simulated (non-radioactive) HLW glass was selected to demonstrate the feasibility of glass encapsulation. The formulation used in these experiments was developed to immobilize calcined, zirconium-containing HLW stored at the Idaho Chemical Processing Plant[8]. This glass formulation was selected to demonstrate the ability to provide a self-protective radiation field, if necessary, for the immobilized plutonium, and because the highly corrosive nature of the molten glass provides a severe test of the ability of the plutonium zirconate powder to retain plutonium during the encapsulation step. The glass was prepared by mechanically mixing simulated zirconia calcine and "Frit 127" additive in a 3/7 weight ratio, and melting at 1100° C for 4 hours in an alumina crucible using an electric furnace with an air atmosphere. The compositions of the starting materials are given in Table I. The corrosivity of the molten glass results mainly from the high fluoride content of the simulated zirconia calcine.

Table I. Compositions of Run 80 simulated zirconia calcine and Frit 127 glass additive

Run 80 Calcine Compound	wt. %	Frit 127 Compound	wt. %
Al_2O_3	14.4	SiO_2	70.3
B_2O_3	0.3	Na_2O	12.8
CaO	25.9	Li_2O	6.2
Ce_2O_3	0.1	B_2O_3	8.5
Cr_2O_3	0.7	CuO	2.1
Cs_2O	0.3		
CaF_2	34.1		
Fe_2O_3	0.2		
Na_2O	0.3		
NiO	0.1		
SrO	0.8		
ZrO_2	22.8		

After cooling, sufficient plutonium zirconate powder was spread evenly over the surface of the simulated HLW glass (still in the crucible) to give a 20 weight percent loading of plutonium oxide based on glass mass. The crucible was returned to the furnace and heated to 1050° C, held for four hours, cooled to 500° C for a 15 minute annealing hold, then furnace-cooled to room temperature. The crucible with the resulting HLW glass/plutonium zirconate composite was sectioned and polished for scanning electron microscope (SEM) examination.

Cement Encapsulation

Magnesium phosphate cement was selected as a low-temperature encapsulant because of its potential for forming other actinide host phases (such as monazite). The cement matrix must bond well to the plutonium zirconate particles without extracting plutonium from the particles. Magnesium phosphate cement encapsulant was prepared by mechanically mixing the components in the proportions shown in Table II.

Table II. Composition of the magnesium phosphate cement encapsulant

Compound	wt. %
MgO	9.2
KH_2PO_4	31.8
fly ash*	58.9

* Coal fly ash from the Jim Bridger power plant in Rock Springs, WY

Deionized water was mixed with the cement at a water to cement ratio of 0.20 by weight. Sufficient plutonium zirconate was mixed with the slurry to give a 20 wt. % plutonium oxide loading. The mixture was allowed to cure for 20 days at room temperature in a closed polyethylene container (initial set occurs in about 4 hours), after which the sample was removed, sectioned, and polished for SEM examination.

RESULTS AND DISCUSSION

Glass Encapsulation

The dense plutonium zirconate had settled to form a glass/plutonium zirconate layer about half the depth of the melt with a well defined boundary layer. No plutonium or other variation in composition could be detected in the glassy phase, either above or within the settled layer. The volume ratio of HLW glass to plutonium zirconate in the settled layer was estimated at about 0.5, which would give an average plutonium concentration in the composite layer of about 43 weight percent. This gravity-driven concentration of plutonium shows that plutonium loadings in excess of the targeted 20 weight percent can be easily achieved.

Zircon grains from the crucible were incorporated into the melt near the glass/crucible interaction zone as a result of attack by the corrosive HLW glass melt. Zircon grains appear as the medium gray phase in Figure 3.

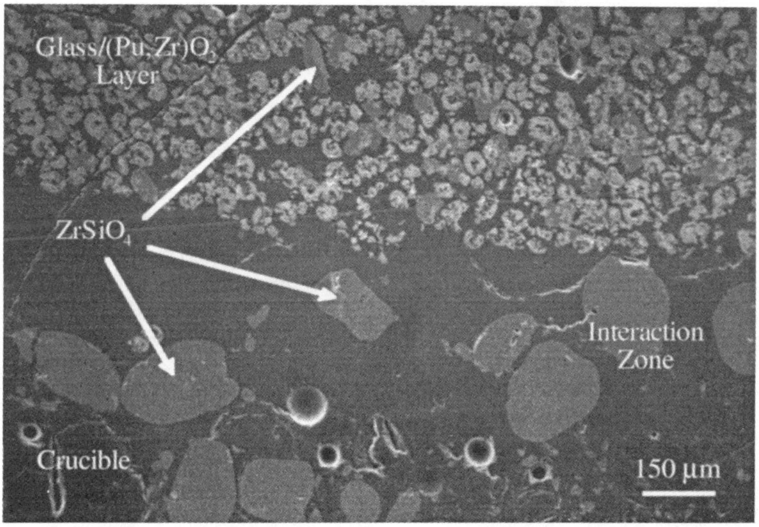

Figure 3: Secondary electron micrograph of the glass/crucible interaction zone showing zircon grains from the crucible incorporated into the glass/plutonium zirconate composite layer.

Energy dispersive X-ray spectroscopy (EDXS) analysis of zircon grains incorporated into the glass melt showed that these grains had absorbed about 3 to 4 mole percent PuO_2, while zircon grains in the crucible showed no detectable plutonium. EDXS analysis of matrix glass phase in the immediate vicinity of plutonium zirconate phases and zircon near the interaction zone showed no detectable plutonium in the glass phase.

A likely explanation for this observation is that plutonium is sufficiently soluble in the glass at the melt temperature (1050° C) to allow diffusion from the plutonium zirconate phase into zircon grains from the crucible. If there is any residual plutonium in the glass in excess of about

0.5 atom percent (estimated detection limit), as the glass cools and plutonium solubility decreases it most likely partitions into the more thermodynamically stable zirconium-containing crystalline phases. Had zircon not been introduced into the melt from the crucible, any plutonium dissolved into the glass from the plutonium zirconate at elevated temperature (in excess of about 0.5 atom percent) would have returned to the zirconate phase during cooling, leaving both materials essentially unaltered. In either case, the net amount of plutonium transported to the zircon grains was small, resulting in no detectable changes in either the plutonium zirconate or the simulated HLW glass, thus demonstrating the feasibility of encapsulating plutonium zirconate in a HLW glass.

The criticality implications of the observed settling of plutonium zirconate in the glass melt were investigated by modeling the behavior of plutonium zirconate powder being fed into a Defense Waste Processing Facility (DWPF)-type canister along with molten 127-based HLW glass at typical DWPF feed rates. Settling velocities were calculated using Stokes law, assuming equilibrium between drag force and gravitational force adjusted for buoyancy according to equation 1.

$$v = \frac{2r_e^2 g(\rho_s - \rho_f)}{9\eta} \qquad (1)$$

Where: r_e is the effective particle radius (taken here as the actual particle radius, 22.5 μm, since the particles are roughly spherical),

g is the acceleration due to gravity,

ρ_s is the density of the solid (9.38 g/cm^3),

ρ_f is the density of the fluid (2.2 g/cm^3), and

η is the viscosity

Viscosity data on 127-based HLW glass[9] were fit to the Vogel-Fulcher-Tamman equation to derive an expression for viscosity as a function of temperature. Worst case cooling data from a DWPF canister instrumented with internal thermocouples (canister B-55, thermocouple 7)[10] was fitted to an exponential cooling rate model to convert the viscosity vs. temperature relation to a viscosity vs. time relation. The viscosity vs. time relation allowed estimation of particle settling during cooling according to equation 2.

$$x = \frac{2r_e^2 g(\rho_s - \rho_f)}{9} \int_0^t \frac{dt}{\eta(t)} \qquad (2)$$

Where x is particle displacement, and

$$\eta(t) = \exp\left[A + \frac{B}{m_1 + \frac{m_2}{\exp(m_3 t)}}\right] \qquad (3)$$

The limits of integration in equation 2 were set at t = 90 minutes by evaluating settling rate as a function of temperature to determine when the settling rate became vanishingly small. Constants

A and B in equation 3 were obtained by regression fitting viscosity vs. temperature data to the Vogel-Fulcher-Tamman equation, and constants m_1, m_2, and m_3 were obtained by regression fitting canister cooling data to an exponential cooling rate function. Evaluating equation 2 gives an estimated particle displacement of 0.001 cm (+0.002, -0.0005 cm). The error estimate is based on twice the standard deviation of the observed particle size distribution (9 μm). This leads to approximately a 0.1% increase in plutonium concentration as a result of settling, which does not have significant implications for criticality control.

Cement Encapsulation

In contrast to the glass encapsulant, no settling of the plutonium zirconate was observed in the magnesium phosphate cement matrix. Figure 4 shows a representative plutonium zirconate particle encapsulated in the cement matrix. Good bonding between the particle and matrix material to the lower left is evident. EDXS analysis detected no plutonium contamination in either the matrix material within 2 μm of the plutonium zirconate particle, or the fly ash (round objects in Figure 4). No evidence of deleterious interactions between plutonium zirconate and any of the cement constituents were observed.

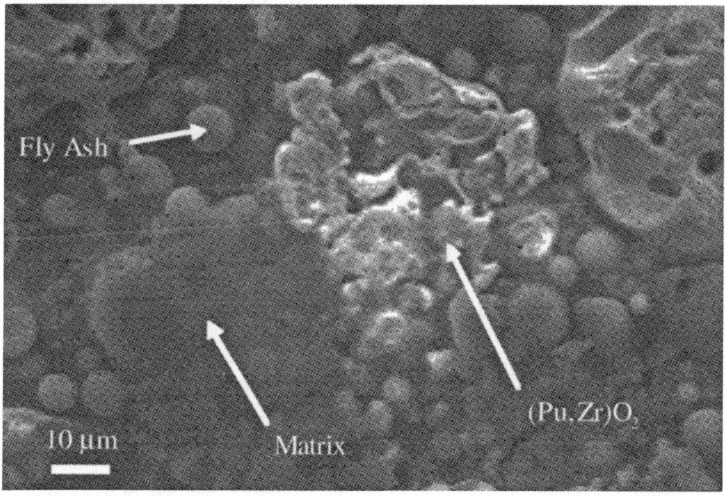

Figure 4: Secondary electron image of a plutonium zirconate particle encapsulated in magnesium phosphate cement.

CONCLUSIONS

The feasibility of a two step process for immobilizing plutonium has been demonstrated. The first step, production of plutonium zirconate powder, is a straightforward process resulting in a highly durable plutonium host phase which can also accommodate neutron absorbers for criticality control. The second step, encapsulation of the plutonium zirconate powder, was demonstrated for a high-temperature process (glass encapsulant) and a low temperature process (cement encapsulant). While significant settling of the plutonium zirconate in the glass matrix was observed in these laboratory-scale experiments, calculations suggest that settling would not be significant if the plutonium zirconate powder is added to a glass canister during glass pouring. If the plutonium zirconate is mixed with glass in a melter, some stirring of the melt would likely be necessary. While some evidence of plutonium diffusion into the glass at high temperatures was observed, the plutonium apparently partitions back into the zirconate phase upon cooling. Otherwise, no significant interactions between the glass and plutonium zirconate were observed. No settling of the plutonium zirconate in the cement matrix was observed, nor where any significant interactions between the cement and plutonium zirconate observed. Both the glass and cement formulations tested performed acceptably as encapsulants for the plutonium host phase.

ACKNOWLEDGMENTS

The authors gratefully acknowledge the contributions of M. Hansen, S. Frank, E. Wood, and D. Barber. Argonne National Laboratory is operated for the U. S. Department of Energy by the University of Chicago. This work was supported by the U. S. Department of Energy, Reactor Systems, Development and Technology, under contract W-31-109-ENG-38.

REFERENCES

[1] U.S. Department of Energy, "Record of Decision for the Storage and Disposition of Weapons-Usable Fissile Materials Final Programmatic Environmental Impact Statement," published at 62 FR 21, January, 1997.

[2] Committee on International Security and Arms Control, National Academy of Sciences, "Management and Disposition of Excess Weapons Plutonium," National Academy Press (1994).

[3] Defense Nuclear Facilities Safety Board, "Recommendation 94-1," published at 59 FR 03, June, 1994.

[4] T. P. O'Holleran, S. G. Johnson, S. M. Frank, M. K. Meyer, M. Noy, E. L. Wood, D. A. Knecht, K. Vinjamuri, and B. A. Staples, "Glass-Ceramic Waste Forms for Immobilizing Plutonium," *Scientific Basis for Waste Management XX*, Ed. W. Gray and I. Triay (Mat. Res. Soc. Proc., pp. 1251-1258 (1997)

[5] D.A. Knecht, K. Vinjamuri, S.V. Raman, B.A. Staples, J.D. Grandy, S. Johnson, T.P. O'Holleran, and S. Frank, " Summary of HLW Glass-Ceramic Waste Forms Developed in Idaho for Immobilizing Plutonium," *Waste Management '98*, Waste Management Symposia, Inc., in press, available on-line at http://wmsym.org/.

[6] T. P. O'Holleran, S. G. Johnson, P. C. Kong, and B. A. Staples, " Plutonium Host Phases Derived From High-Level Waste At The Idaho Chemical Processing Plant," *Ceramic Transactions,* **87**, in press.

[7] Annual Book of ASTM Standards, 12.01, pp. 797-814, ASTM, Philadelphia, PA (1995).

[8] B. A. Staples, D. A. Pavlica, and H. S. Cole, "Properties of Formula 127 Glass Prepared with Radioactive Zirconia Calcine," pp. 125-134 in *Scientific Basis for Nuclear Waste Management II*, Edited by D. G. Brookins. Materials Research Society, Pittsburgh, PA, 1979.

[9] H. S. Cole, D. Gombert II, J. R. Berreth, "Properties of Vitrified ICPP Zirconia Calcine," ENICO-1083, prepared for the U. S. Department of Energy, Idaho Operations Office 1980

[10] R. E. Edwards, "SGM Run 8 - Canister and Glass Temperatures During Filling and Cooldown," USDOE Report DPST-87-801, Savannah River Laboratory, E. I. DuPont de Nemours and Co., Aiken, SC, 1987

HEAT CAPACITY, THIRD LAW ENTROPY, AND FORMATION ENERGETICS OF ZIRCONOLITE, $CaZrTi_2O_7$.

Robert L. Putnam, Alexandra Navrotsky
Thermochemistry Facility
4440 Chemistry Annex
University of California, Davis
Davis, CA 95616

Brian F. Woodfield, Jennifer L. Shapiro, Juliana Boerio-Goates
Department of Chemistry and Biochemistry
C100 BNSN
Brigham Young University
Provo, UT 84602

Abstract:

High temperature oxide melt solution calorimetry combined with low temperature adiabatic calorimetry have been used to characterize zirconolite, $CaZrTi_2O_7$, a mineral that is being considered as a host phase for the disposal of weapons plutonium and uranium. The standard formation energies, $\Delta_f H^o$ and $\Delta_f G^o$, as well as the third law entropy $\Delta^{298.15}_{0}S$ are (-3713.8 ± 4.7) kJ•mol^{-1}, (-3514.5 ± 4.7) kJ•mol^{-1}, and 193.3 J•K^{-1}•mol^{-1} respectively. The heat capacity, third law entropy, and formation energies of zirconolite are presented from 298.15 K to 1500 K. The further application of these calorimetric techniques to the issue of nuclear waste materials is discussed.

Introduction:

The United States has chosen to utilize a ceramic waste material for the disposition of excess weapons plutonium and uranium. The waste form is expected to contain phases closely related to the naturally occurring minerals zircon ((Act,Zr)SiO_4)[1], zirconolite ((Act,Ca)(Act,Zr)Ti_2O_7), and/or pyrochlore (CaActTi_2O_7)[2] where *Act* is an actinide element, primarily U^{+4} and Pu^{+4}. These minerals are extremely refractory and have been shown to have retained their natural loadings of actinide elements over periods of up to billions of years in a variety of weathering conditions [1,2]. Extensive solid solubility exists between each of these minerals and the oxides UO_2 and PuO_2. However nothing is known about the thermodynamics of actinide substitution in the target mineral compositions.

A database of thermodynamic quantities for each of these systems needs to be obtained before the durability and stability of zirconolite or other ceramic waste forms can be properly assessed for long-term geologic disposal. Unfortunately, the very property that makes these materials desirable for nuclear waste disposal, that is their highly refractory nature resulting in low solubility, makes obtaining much of the thermochemical data problematic. Recent developments in calorimetric techniques[3] have made it possible to overcome this difficulty. Pure zirconolite, $CaZrTi_2O_7$, is a starting point for such thermodynamic analysis and represents the first step toward establishing a thermodynamic database for the UO_2 - PuO_2 - $CaZrTi_2O_7$ system.

Experimental:

We have applied two different calorimetric techniques, low-temperature adiabatic calorimetery[4,5,6] and high-temperature oxide melt solution calorimetry [3,7,8] to obtain the formation energetics, heat capacities, and third law entropies of zirconolite, $CaZrTi_2O_7$, from T= 0 K to T= 1500 K. It is beyond the scope of this paper to describe in detail the calorimetric techniques used in this study. Readers who seek detailed descriptions of these two well-established calorimetric techniques are referred to the works cited above.

Results and discussion:

Table I contains thermodynamic quantities obtained for zirconolite in our studies [3, 6] at selected temperatures from T= 298.15 K to T= 1500 K.

Table I. Molar heat capacity, third law entropy, enthalpy of formation and Gibbs free energies of formation for zirconolite, $CaZrTi_2O_7$, from T= 298.15 K to T= 1500 K.

T (K)	C_p ($J \cdot K^{-1} \cdot mol^{-1}$)	$\Delta_0^T S_m^o$ ($J \cdot K^{-1} \cdot mol^{-1}$)	$\Delta_f^T H_m^o$ ($kJ \cdot mol^{-1}$)	$\Delta_f^T G_m^o$ ($kJ \cdot mol^{-1}$)
298.15	211.9	193.3	-3713.8	-3514.6
500	255.5	314.9	-3708.7	-3380.9
700	278.0	404.6	-3700.6	-3251.2
900	296.4	476.8	-3692.1	-3123.8
1100	314.7	538.0	-3682.5	-2998.5
1300	334.2	592.1	-3689.0	-2872.1
1500	355.4	641.4	-3670.8	-2747.8

The heat capacity[6] of zirconolite from T= 0 K to T= 1500 K can be represented by equation 1 where $D(\Theta_D/T)$ and $E(\Theta_E/T)$ are Debye and Einstein functions respectively.

$$C^o_{p,m} = 3R\cdot[m\cdot D(\Theta_D/T)+n\cdot E(\Theta_E/T)]+A_1\cdot T + A_2\cdot T^2 \quad (1)$$

The coefficients A_1, A_2, m, n, Θ_D, and Θ_E are obtained from a non-linear fit of the measured heat capacity data (T= 20 K to T= 400 K) and are given in Table II. The Debye and Einstein terms are approximations to the harmonic contributions to the heat capacity from lattice phonons and internal vibrations while the terms which are linear and quadratic in T provide for the anharmonic contributions.

The heat capacity curve of zirconolite over the extended temperature range T= 0 K to T= 1500 K can be found in figure 1. Filled symbols in Figure 1 represent easured C_p values. The C_p data point at T= 1023 K was obtained through heat content measurements on zirconolite for temperature intervals between room temperature (T= 298 K) and T= 973 K, 1043 K, and 1073 K and was not used in fitting the heat capacity equation. This data point is used to check the extrapolation of the fitted data beyond T= 400 K. The extrapolated C_p agrees with the measured C_p at T= 1023 K to within 1.5 $J\bullet K^{-1}\bullet mol^{-1}$ or less than 1 per cent of measured C_p and we conclude that the extrapolation of our heat capacity equation to higher temperatures is valid.

Table II. Parameters for the heat capacity equation for zirconolite, $CaZrTi_2O_7$. Coefficients are used in equation 1.

Parameter	Value	Parameter	Value
m	5.5902	Θ_E	725.66
Θ_D	417.33	A_1	22.550
n	4.883	A_2	0.028109

Table I represents a piece of the thermodynamic data base that must be established for the UO_2 - PuO_2 - $CaZrTi_2O_7$ system. Key also to this data base, novel waste treatment processes[9], and the studies of other waste forms, especially nuclear waste glasses, is the thermodynamics of the binary oxides in dissolution environments.

Navrotsky [10] has shown that the dissolution enthalpies of key oxides in both molten oxide solvents and in water are nearly linear with the molar volume of the parent oxide. Furthermore, the trend observed is relatively solvent-independent and a relative offset may be applied to characterize the solution

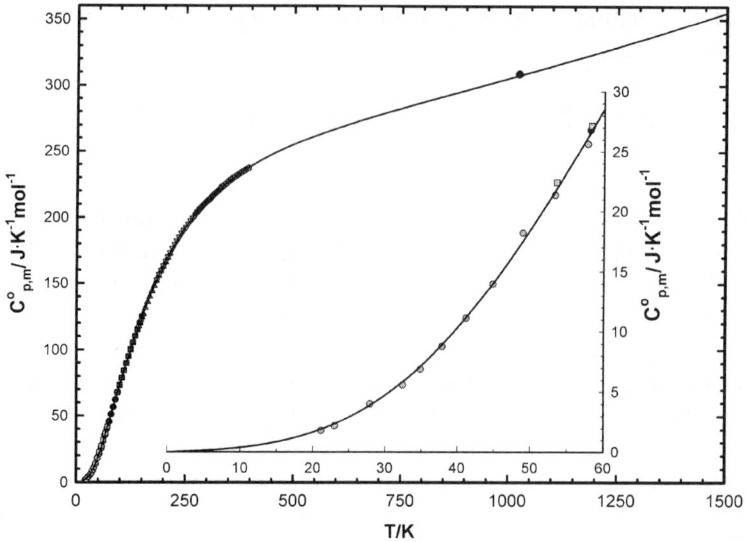

Figure 1. The molar heat capacity of zirconolite, $CaZrTi_2O_7$, from T= 0 K to T= 1500 K. Filled symbols represent measured values. The solid line is the fitted curve as defined by Equation 1.

energetics in differing solvents. The actinides appear to behave similarly[10].

Seeking a better understanding of the dissolution energetics of the binary oxides of some tri- and tetravalent actinides and lanthanides, Navrotsky[10] has attempted to predict the solution thermodynamics of the oxides of several lanthanides and actinides related to nuclear waste. These predictions have been based on crystal chemical concepts, systematic trends, and available thermochemical data for M^{+4} cations in solutions ranging from water to borate glasses. Predictions such as these are first approximations and involve complex correlations with differing solvents, starting crystal structures, and chemical properties. More specific data are clearly needed.

We have recently obtained additional thermochemical data and have begun to notice a previously unobserved trend that can be applied to provide potentially more accurate and a consistent picture of the fundamental thermodynamic behavior of the actinide elements in nuclear waste glasses.

Figure 2 shows the measured heats of solution in lead borate melt ($2PbO_2 \bullet B_2O_3$) at T= 973 K for three series of MO_2 materials. Each series has a different crystal structure, quartz, rutile, and fluorite, and represents a metal cation (M) coordination environment of 4, 6, and 8 respectively. It is notable that, within each series, the measured data (filled symbols ■, ▲, and • respectively) are linear with respect to the ionic potential (cation charge over the radius of the cation[11], Z/r) in the given coordination environment . Linear correlations (r^2) for the data representing the fluorite, rutile, and quartz structured MO_2 materials are 1.0 , 0.98, and 1 respectively. (Though the quartz data set contains only two data points.)

With such a trend it becomes possible to predict the enthalpies of solution in lead borate for the actinide MO_2 materials that share the cubic fluorite structure with CeO_2, ZrO_2, and HfO_2. Our predictions can be found in Table III and are identified graphically as the open triangles (△) in Figure 2.

Table III. Cation ionic radii (CN8) and estimated enthalpies of solution in molten lead borate ($2PbO \bullet B_2O_3$) at T= 973K for MO_2 (cubic fluorite structure), where M is Th, Pa, U, Np, Pu, Am, Cm, Bk , or Cf. Values calculated from the linear regression of the measured enthalpies of solution for cubic fluorite structured ZrO_2, HfO_2, and CeO_2 as shown in Figure 2.
(ΔH_{sol} = -99.6067 (Z/r) + 486.7768 $r^2 = 0.9999$)

Actinide MO_2	Ionic Radius r (Å)	$\|Z/r\|$ ($e \bullet Å^{-1}$)	ΔH_{sol} ($kJ \bullet mol^{-1}$)
Th	1.05	3.81	107.09
Pa	1.01	3.96	92.17
U	1.00	4.00	88.20
Np	0.98	4.08	80.38
Pu	0.96	4.17	71.30
Am	0.95	4.21	67.32
Cm	0.95	4.21	67.32
Bk	0.93	4.30	58.35
Cf	0.92	4.35	53.70

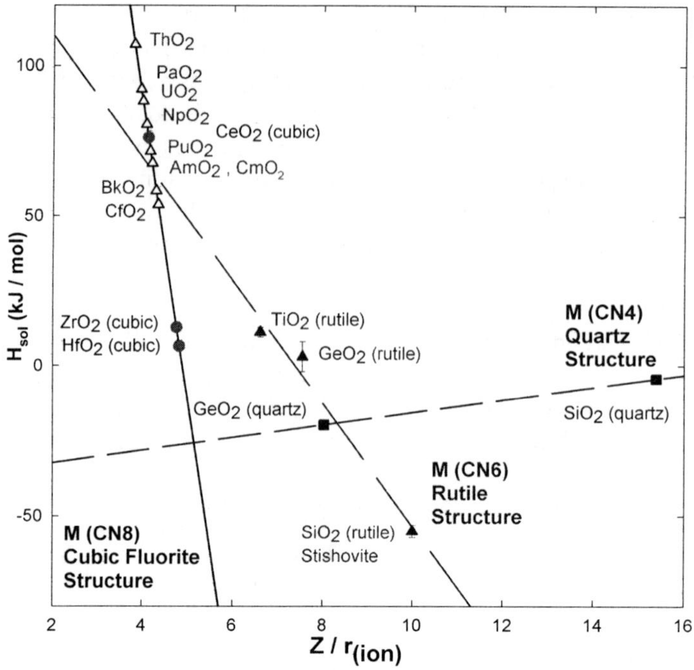

Figure 2. The enthalpy of solution of MO_2 materials in lead borate[3,12,13,14,15,16] ($2PbO{\bullet}B_2O_3$) at T= 973K. Measured enthalpies of solution for cubic fluorite(CN8), • , rutile(CN6), ▴, and quartz(CN4), ■ , crystal structured MO2 materials.. Open triangles, △ , represent predicted values for actinide MO_2. The lines represent mathematical fits to the measured data. For the fluorite data the fitted line is:
ΔH_{sol} = -99.6067 (Z/r) + 486.7768 (r^2 = 0.9999).

The correlations in Figure 2 are also interesting from a crystal chemical point of view and may have broader application. Comparison of the three different coordination environments shows that with increasing coordination number the slope of the line representing the solution thermodynamics becomes more negative. This is most likely related to the metal cation's decreasing tolerance of variations in the M-O bond length as the M-O bond becomes shorter and stronger. It is also interesting that the lines intersect at values of Z/r where polymorphism (at either atmospheric or high pressure) has been observed. It is possible that similar trends can be found in other oxides such as the M_2O_3 polymorphs (A,B, and C type) of the lanthanides. Additional work is needed to further clarify these observations.

In conclusion, we have successfully measured the formation energetics of a highly refractory ceramic material, zirconolite, through the only means available at present, high temperature oxide melt solution calorimetry. We have utilized low temperature adiabatic calorimetry to define the heat capacities and third law entropies of this ceramic. Studies are continuing with an emphasis on establishing a solid thermodynamic database for materials related to nuclear waste disposal. Application of such data is key to waste form process refinement, the modeling of waste form behavior in a geologic environment over geologic time, waste form optimization and characterization, and material dissolution studies.

We acknowledge B. Ebbinghaus and R. VanKonynenburg at Lawrence Livermore National Laboratory for synthesis of the zirconolite sample. This work was supported by a contract from Lawrence Livermore National Laboratory and by Department of Energy grant number DE-FG07-97ER45673.

References

1. Ewing, R. C.; Lutze, W.; Weber, W. J.; "Zircon: A host-phase for the disposal of weapons plutonium" *J. Mater. Res.*; **1995**, 10, 2, 243-246.

2. Lutze, W.; Ewing, R. C. (eds) *Radioactive Waste forms for the Future*, North Holland Publishing, New York, NY **1988**.

3. Putnam, R.L.; Navrotsky, A.; Woodfield, B.F.; Boerio-Goates, J.; Shapiro, J.L; "Thermodynamics of Formation of Zirconolite, $CaZrTi_2O_7$, From T= 298 K to T= 1500 K.". Submitted to *J. Chem. Thermodyn.*

4. Boerio-Goates, J.; Woodfield, B.F.; "Calorimetric studies of the phase transition in iodoform" *Can. J. Chem* **1988,** 66, 645-650.

5. Battley, E.H.; Putnam, R.L.; Boerio-Goates, J.; "Heat capacity measurements from 10 to 300 K and derived thermodynamic functions of lyophilized cells of

Saccharomyces cerevisiae including the absolute entropy and the entropy of formation at 298.15 K." *Thermochim. Acta* **1997,** 298, 37.

6. Woodfield, B. F.; Boerio-Goates, J.; Shapiro, J. L.; Putnam, R. L. ; Navrotsky, A.; Ebbinghaus, B.; "Molar Heat Capacity and Thermodynamic Functions of zirconolite, $CaZrTi_2O_7$." Submitted *J. Chem. Thermodyn.*

7. Navrotsky, A.;"Recent Progress and New Directions in High Temperature Calorimetery" *Phys. Chem. Min.*, **1977**, 2, 89-104.

8. Navrotsky, A.;"Progress and New Directions in High Temperature Calorimetry Revisited" *Phys. Chem. Min.*, **1997**, 24, 222-241.

9. Forsberg, C. W.; Beahm, E. C.; Rudolph, J. C.;"Direct conversion of halogen-containing wastes to borosilicate glass." *Mat. Res. Soc. Symp. Proc.*; **1997**, 465, 131-137.

10. Navrotsky, A. "Thermochemistry of Crystalline and Amorphous Phases Related to Radioactive Waste." Sterne, P. A. eds. *Actinides and the Environment*, 1998, 267-297.

11. Shannon, R.D.; Prewitt, C.T.; *Acta Cryst.* **1969**, A32, 751-762.

12. Navrotsky, A; "Thermodynamic Relations among Olivine, Spinel, Phenacite Structures in Silicates and Germanates. III The system CuO-MgO-GeO_2" *J. of Solid State Chem.*, **1974**, 11, 10-16.

13. Navrotsky, A; "Enthalpies of transformation among the tetragonal, hexagonal, and glassy modifications of GeO_2." *J. Inorg. Nucl. Chem.*, **1971**, 33, 1119-1124.

14. Akaogi, M.; Navrotsky, A.; "The quartz-coesite-stishovite transformations: new calorimetric measurements and calculation of phase diagrams." *Physics of the Earth and Planetary Interiors*, **1984**, 36, 124-134.

15.Molodetsky, I.; Navrotsky, A.; Lajavardi, M.; Brune, A.; *Zeits. Phys. Chem.* In press for March 1998.

16. Putnam, R. L., Navrotsky, A.; Unpublished data. Thermochemistry Facility, University of California, Davis.

CERAMIC FORMS FOR IMMOBILIZING PU USING ZR, Y, AL METAL ADDITIVES

B. E. Burakov and E. E. Anderson
V. G. Khlopin Radium Institute, 28,
2-nd Murinskiy Ave.,
St. Petersburg, 194021, Russia
fax: +7-812-346-1129,
E-mail: burakov@riand.spb.su

D. A. Knecht
Lockheed Martin Idaho Technologies Co.
P.O. Box 1625, Idaho Falls, ID 83415
USA
E-mail: dieter@inel.gov

ABSTRACT

Plutonium residue wastes can contain a larger amount and variety of impurities than wastes from weapons dismantling and can contain >50% impurities including such elements as and oxides of Am, Al, Ca, Mg, Ga, Fe, K, La, Na, Mo, Nd, Si, Ta, Ce, Ba, B, W, Zn, Zr, C, and Cl. For some of these residues, direct conversion to glasses or ceramics will be difficult. Previous work has demonstrated that zirconia-zircon and garnet-perovskite-based ceramics appear to be promising hosts for plutonium-containing wastes and for the impurities. Garnet solid solutions formed from oxides of Y, Al, Ga, and Gd have a structure that allows incorporation of a broad range of elements, including Fe, Zr, V, Ca, Cr and rare earths. Using Ce as a Pu simulant, the garnet system produced a garnet-perovskite mixture with improved waste loading capability. This paper will give the characteristics of these polyphase ceramic forms for immobilizing Pu.

INTRODUCTION

The immobilization of Pu is connected not only with weapons Pu from metallic pits. There are also Pu-residue wastes containing a large amount and variety of impurities up to 70 wt.% including the following ranges in wt.%: Pu - 30-88; U – 0-20; Am – 0-5; Th – 0-5; Al – 0-2; C – 0-2; Ca – 0-5; Cl – 0-30; Cr – 0-2; Cu – 0-2; Ga – 0-2; Fe – 0-2; K – 0-5; Mg – 0-5; Na – 0-5; Ni – 0-2; Si - 0-2; Sr – 0-2; Ta – 0-1; W – 0-1; Zn – 0-2. These wastes consist of complex chemical compositions including different uncharacterized phases, some of them, for example actinide oxides, are chemically inert. Therefore, direct conversion of these wastes into ceramic or glass can be difficult.

Our approach to the immobilization of these wastes is based on the following issues:

1) Because of complex chemical and phase compositions and non-homogeneity of the wastes, the stoichiometry of their starting precursor for the synthesis of a final waste form cannot be provided.
2) Ceramic or glass-ceramic forms are more favorable for the immobilization. It is necessary to fully incorporate the actinides in the lattice of durable crystalline host-phases in the form of a solid solution.
3) The primary consideration must be given to preparation of the starting precursor and increasing its chemical reactivity.

This paper considers the application of two systems: zircon-zirconia ($ZrSiO_4$-ZrO_2) and garnet-perovskite [$(Y,Gd)_3(Al,Ga)_5O_{12}$ – $(Y,Gd)(Al,Ga)O_3$] for the immobilization of Pu-residue wastes of complex chemical composition. The selection of these systems is based on previous studies of durable crystalline host-phases developed to accommodate actinides.[1–9]

It is proposed that Pu and other actinides are incorporated either separately from other impurities in the lattices of zircon and zirconia or together with most of the impurities in the garnet and perovskite lattices. The application of the metal powder additives Zr, Y, Al compared to the inert oxides ZrO_2, Y_2O_3, Al_2O_3 is generally more effective, especially if other components of the wastes are of complex chemical and phase composition. In previous work the use of metallic Zr powder in the starting precursor for zircon synthesis resulted in increased chemical reactivity and high actinide incorporation into the zircon lattice. Similarly, the use of Y metal powder improved the synthesis of actinide-doped garnet.[1,8] Potential problems due to precursor nonstoichiometry can be addressed by using excess metal additives because the oxidation of unreacted metals during ceramic synthesis forms inert oxide phases with no impact on the chemical and mechanical durability of the final ceramic waste form.

This paper presents experimental results obtained in the framework of co-operation between Lockheed Martin Idaho Technologies Co. and V. G. Khlopin Radium Institute for the immobilization of 94-1 plutonium residue wastes.

EXPERIMENTAL

A simulant of Pu residue waste with composition as shown in Table I was prepared using commercial chemicals. Uranium dioxide and cerium dioxide were used as simulants for plutonium in the residue waste. Waste forms were prepared using the Pu residue waste simulant at 10 wt.% and precursor chemicals to form zirconia-zircon and garnet-perovskite waste forms. Additional waste forms were made to determine the behavior of Ce or U (at 10 wt.%) in pure garnet without the other components of the residue waste.

Table I. Phase and chemical composition (in wt.%) of blend simulating Pu-residue wastes

UO_2	25	U	22.00	O	3.00
CeO_2	25	Ce	20.30	O	4.70
Al_2O_3	2	Al	1.06	O	0.94
CaC12	10	Ca	3.60	Cl	6.40
Cr_2O_3	2	Cr	1.37	O	0.63
CuO	2	Cu	1.60	O	0.40
Ga_2O_3	2	Ga	1.49	O	0.51
Fe_2O_3	2	Fe	1.40	O	0.60
KCl	5	K	2.60	Cl	2.40
NaCl	10	Na	3.90	Cl	6.10
NiO	2	Ni	1.57	O	0.43
SiO_2	2	Si	0.94	O	1.06
$SrCl_2$	2	Sr	1.10	Cl	0.9
Ta_2O_5	1	Ta	0.82	O	0.18
WO_3	1	W	0.79	O	0.21
ZrO_2	2	Zr	1.60	O	0.40
MgO	5	Mg	3.00	O	2.00

The zircon-zirconia precursor doped with 10 wt.% of simulant (substituting for zirconium) was prepared using an excess of metal zirconium powder as the only source of Zr to form the estimated final composition 80-90 wt.% of zircon and the rest as zirconia. The precursor was ball milled in acetone for 6 hours. The resulting blend was sintered in two stages: to 500°C with heating rate 10°C per minute and then to 1450°C with heating rate 20°C per minute. Sintering was completed in 30 minutes at 1450°C.

The garnet precursor $Y_3Al_5O_{12}$–$Gd_3Ga_5O_{12}$ (YAG-GGG with mole ratio 1:1) doped with 10 wt.% simulant of Pu residue waste (Table I) was prepared using excess of Al metal powder (about 5-10 wt.% based on garnet stoichiometry) as the only source of Al. Other chemicals were used in the form of commercial oxide powders. The resulting blend was ball milled in acetone for 6 hours and then sintered in two stages: to 300°C with a heating rate of 10°C per minute and to 1300°C with a heating rate of 20°C.

To determine the behavior of Ce and U (as simulants for Pu) with no additional residue components, the stoichiometric precursor garnents $Gd_3Al_5O_{12}$ (GAG) and $Gd_3Ga_5O_{12}$ (GGG) were each doped separately with 10 wt.% Ce or 10 wt.% U and were prepared using commercial oxide powders. All four samples were melted in air for a short time (20-30 sec) at a temperature of about 1900-2000°C.

All samples were examined by XRD and ESEM methods.

RESULTS AND DISCUSSION

Zircon-Zirconia System

The resulting material consists of two visually distinct layers: a top layer (about 80 vol.%) and bottom (about 20 vol.%). XRD analysis shows that top layer contains about 5-10 and 10-15 wt.% amorphous phase in the upper and lower parts, respectively. The bottom layer of the sample contains about 35-40 wt.% amorphous phase. The main crystalline phases in the top layer are zircon (from 45 to 85 wt.%), monoclinic zirconia (from 30 to 10 wt.%) and crystobalite (from 20 to 5 wt.%). The main crystalline phase in the bottom of the sample is monoclinic zirconia (about 95 wt.%).

Because of the small particle size (less than one micron), it was not possible to accurately determine the microprobe analysis of the crystalline phases, however, it was found that Ce and U did not form separate phases (detectable on XRD or at ESEM resolution) and were concentrated in zircon (up to 0.9-1.1 wt.% Ce and to 0.4 wt.% U) and zirconia (up to 3.3 wt.% Ce and to 2.0 wt.% U) matrices. The concentration of Ce and U in the silica (crystobalite and amorphous glass phase) matrix was less than detection limits (0.05 wt.%).

Garnet-Perovskite System

It was found that the formation of Pu residue simulant-doped garnet and perovskite phases depends on the type of precursor garnet stoichiometry (Table II). Doping the GAG precursor by U or Ce causes the formation of mainly perovskite, which well accommodates these admixtures. Doping the GGG precursor by the same amount of U or Ce does not prevent the formation of the garnet (GGG) phase. However, GGG accommodates only Ce but not U (Table II). Therefore, the incorporation of U in the garnet lattice requires the charge compensation what was assumed earlier.[9] Preliminary results were obtained using Sn oxide additive in the GGG precursor. The use of 4 wt.% Sn (estimated on metal) allows to increase the U incorporation in GGG up to 5.5 wt.%.

Table II. Influence of Pu simulant admixtures in the starting precursor on the formation of final crystalline phases

Stoichiometry of starting precursor	Main crystalline phase in melted sample	Maximal content of Pu simulant in main phase
$Gd_3Al_5O_{12}$ doped with 10 wt.% Ce	Perovskite (Gd,Ce)AlO_3	~ 8 wt.% Ce
$Gd_3Al_5O_{12}$ doped with 10 wt.% U	Perovskite (Gd,U)AlO_3	~ 7 wt.% U
$Gd_3Ga_5O_{12}$ doped with 10 wt.% Ce	Garnet $(Gd,Ce)_3Ga_5O_{12}$	~ 6 wt.% Ce
$Gd_3Ga_5O_{12}$ doped with 10 wt.% U	Garnet $Gd_3Ga_5O_{12}$	less than 0.1 wt.% U

As a result of melting of the YAG-GGG precursor doped with 10 wt.% Pu-residue simulant (Table I) a polyphase crystalline ceramic was obtained. The SEM micrographs are shown in Figures 1a-b. XRD analysis confirms that the main phase is garnet (gray phase in Figure 1) about 70 wt.%. Other phases with the structures close to Ga_2O_3, Al_2O_3-corundum and Perovskite are possible phases which were not positively identified. Garnet composition varies in different grains and can be correlated with two types of cathodoluminescence—red and green (Table III). In these samples the garnet compositions are described by formulas: $(Y_{1.7},Gd_{1.3})(Al_{4.5}Ga_{0.5})O_{12}$ and $(Y_{1.6},Gd_{1.4})(Al_{4.3}Ga_{0.7})O_{12}$.

Table III. Chemical composition of Garnet from microprobe analysis (in wt.%)

Gd	Ce	Al	Ga	U	Y	O	Total
red luminescence							
29.4	0.2	17.1	5.1	0.3	22.3	27.6	102.1
29.3	0.4	17.2	5.3	0.2	23.1	27.9	103.4
29.1	0.1	17.2	5.3	0.2	20.8	27.2	99.8
green luminescence							
29.5	0.5	16.7	6.4	—	19.7	27.0	99.8
30.6	0.4	16.2	6.8	—	20.8	27.1	101.8
30.8	0.3	15.9	7.3	—	18.3	26.4	99.0

It is interesting to note the difference between Ce and U accommodation in garnet: the higher content of Ce and the lower content of U are correlated with the higher content of Gd and Ga. The same behavior of Ce and U is found for the second phase (unidentified dark phase in Figure 1) of this sample with the following average composition in wt.%: Gd – 9.9; Ce – 3.7; Al – 35.4; Ga – 11.0;

U – 0; Y – 0.7; and O – 37.6. The third phase, probably with a perovskite structure (white inclusions in Figure 1) contains in wt.%: Gd – 22.8; U – 31.5; Y - 7.4; Zr – 17.3; Ce – 1.8; Al – 1.5; and O – 17.5 and is responsible for all Zr and most of the U accommodation. The most important result is that no separate Ce and U phases were found.

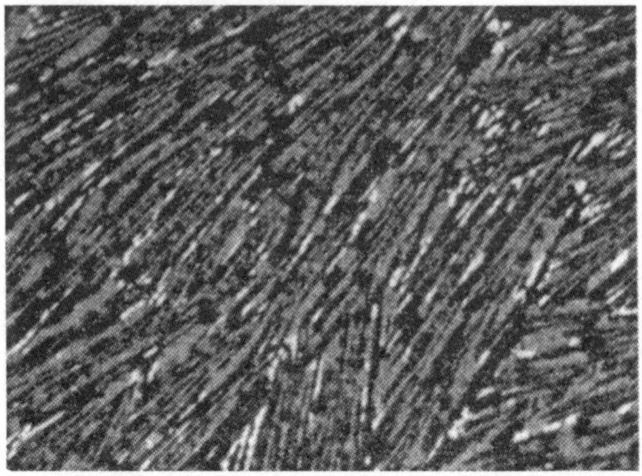

1a. 1000 µm

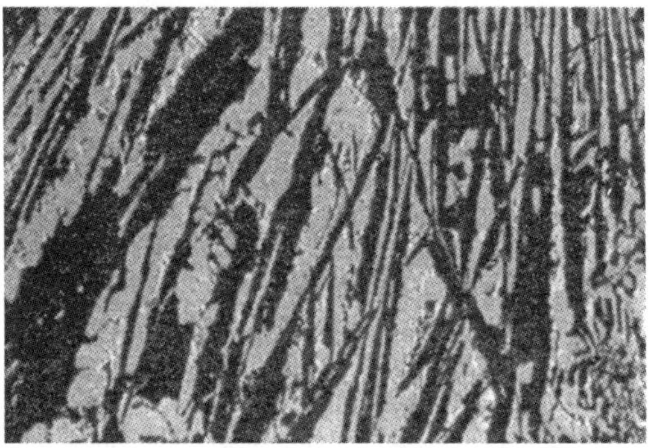

1b. 250 µm

Figure 1. SEM back-scattered image of polyphase ceramic:
$(Y_{1.6-1.7},Gd_{1.4-1.3})(Al_{4.3-4.5},Ga_{0.70.5})O_{12}$-Garnet (gray);
$(Gd,Ce)(Al,Ga)O_x$ -(dark);
$(U,Gd,Y)ZrO_x$ – Perovskite (white).

CONCLUSIONS

These results allow us to make the following preliminary conclusions:

1. The use of the metallic additives Zr, Y, and Al in the starting precursors is now more attractive for the immobilization of Pu-residue wastes of complex chemical compositions in a ceramic matrix, based on zircon-zirconia or garnet-perovskite (aluminate-gallate) systems.
2. Formation of host-phases doped with actinide simulants from a precursor with metallic additives is not accompanied by formation of separate simulant phases.
3. The zircon-zirconia system incorporates only actinide simulants and rare-earth elements into the lattice of crystalline host-phases. Other elements from the Pu residue waste simulant are concentrated in the silicate glass phase.
4. The garnet-perovskite system incorporates most elements of the simulated Pu residue waste into the lattice of the crystalline host-phases.
5. It is assumed that the actinide simulant incorporation into the garnet lattice can be improved in some cases using charge compensated elements. The ratio yield between garnet and perovskite was assumed also to be controlled by some additives in the starting precursor.

ACKNOWLEDGMENT

The authors are very grateful to Mrs. M. Zamoryanskaya for the help in microprobe analysis and Mrs. M. Yagovkina for XRD work.

REFERENCES

[1]E. B. Anderson, B. E. Burakov, V. G. Vasiliev, “A Creation of Crystalline Matrix for Actinide Waste in Khlopin Radium Institute,” *Proceedings of the International Conference SAFE WASTE-93, Avignon, France,* Vol. 2, pp. 29-33 (June 13-18, 1993).

[2]C. Prunier, M. Salvatores, Y. Guerin, A. Zaetta, “First Results and Future Trends for the Transmutation of Long-Lived Radioactive Wastes,” *Proceedings of the International Conference SAFE WASTE-93, Avignon, France,* Vol. 2, pp. 298-311 (June 13-18, 1993).

[3]E. B. Anderson, B. E. Burakov, V. A. Starchenko, V. G. Vasiliev, N. Shulyak, R. G. G. Holmes, W. Weaver, D. Goddard, R. Clegg, S. Richardson, “UK-Russian Collaboration High Level Waste Immobilization Studies,” *Proceedings of the International Conference GLOBAL-95, Versailles, France,* Vol. 1, pp. 216-223 (September 11-14, 1995).

[4]K-I. Kuramoto, Y. Makino, T. Yanagi, S. Muraoka, Y. Ito, "Development of Zirconia- and Alumina-Based Ceramic Waste Forms for High Concentrated TRU Elements," *Proceedings of the International Conference GLOBAL-95, Versailles, France,* Vol. 2, pp. 1838-1845 (September 11-14, 1995).

[5]R. C. Ewing, W. Lutze, W. J. Weber, "Zircon: a Host-Phase for the Disposal of Weapons Plutonium," *Journal of Material Research,* **10**, pp. 243-246 (1995).

[6]R. C. Ewing, W. J. Weber, W. Lutze, "Crystalline Ceramics: Waste Forms for the Disposal of Weapons Plutonium," pp. 65-83, *Disposal of Weapon Plutonium*, Eds. E. R. Merz and C. E. Walter, Kluwer Academic Publisher, Dordrecht, 1996.

[7]B. E. Burakov, E. B. Anderson, B. Ya. Galkin, V. A. Starchenko, V. G. Vasiliev, "The Crystalline Host-Phases for Immobilization of Weapons Plutonium and Waste Actinides," pp. 85-89, *Disposal of Weapon Plutonium,* Eds. E. R. Merz and C. E. Walter, Kluwer Academic Publisher, Dordrecht, 1996.

[8]B. E. Burakov, E. B. Anderson, S. I. Rovsha, S. V. Ushakov, R. C. Ewing, W. Lutze, W. J. Weber, "Synthesis of Zircon for Immobilization of Actinides," *Scientific Basis for Nuclear Waste Management XIX, Material Research Society Symposium Proceedings*, W. M. Murphy and D. A. Knecht, Eds., v. 412, pp. 33-39 (1996).

[9]B. E. Burakov, E. E. Strykanova, "Garnet Solid Solution of $Y_3Al_5O_{12}$-$Gd_3Ga_5O_{12}$-$Y_3Ga_5O_{12}$ (YAG-GGG-YGG) as a Prospective Crystalline Host-Phase for Pu Immobilization in the Presence of Ga," *Proceedings of the International Conference WASTE MANAGEMENT'98,* Tucson, Arizona, USA (March 1-5, 1998), (in press).

SOLID SOLUTIONS OF Ce, U, AND Th IN ZIRCON

S.V. Ushakov[1], W.Gong[2], M.M. Yagovkina[3], K.B. Helean[2], W. Lutze[2] and R.C. Ewing[4]
[1]V.G. Khlopin Radium Institute, 28, 2nd Murinskiy Ave, 194021 St. Petersburg, Russia
[2]Center for Radioactive Waste Management, Univ. of New Mexico,
Albuquerque, NM87131, USA
[3]JSC "Mekhanobr-Analyt", 8a, 21 Linija, 199026, St. Petersburg, Russia
[4]Dept. of Nuclear Engineering&Radiological Sciences, University of Michigan, MI48109, USA

ABSTRACT

Zircon ($ZrSiO_4$) has been proposed as a candidate host phase for storage and disposal of excess weapons plutonium. Zr can be replaced by a variety of elements including Hf, Am, Pu, Np, U, Pa, and Th. Little to nothing is known about miscibility in most of these systems. A variety of solid solutions of $(Zr,A)SiO_4$ where A = Ce, U, and Th using the sol-gel technique have been synthesized. Though more experiments need to be conducted to refine our results, it is clear that the solubility of these elements are limited to less than 10 at%. The solubility limits determined by analytical electron microscopy and X-ray diffraction data are in at.%: 7±1 (Ce), 4±1 (U), and 2±0.5 (Th). Ce (as Ce(IV)) was used to simulate Pu. The measured solubility limit is an approximate value. The presence of some Ce(III) in the solid solution could not be excluded. The solubility of Pu in $ZrSiO_4$ is unknown. The highest Pu concentration reported in the literature is 9.2 at.%.

INTRODUCTION

Zircon ($ZrSiO_4$) has been proposed as a candidate host phase for storage and disposal of excess weapons plutonium[1,2]. The waste form could be a single phase zircon ceramic or a dual phase zircon/zirconia ceramic[3]. Zircon is a ubiquitous mineral and contains small amounts of uranium and thorium naturally. Long-term chemical durability of zircon has been evidenced by geochronologists who found that zircon alteration is small even under extreme conditions in nature[4]. Over 4,000 references were found (georef since 1960) on zircon dating. Zircon is one of the most durable minerals and the oldest dated mineral on earth and in the universe (about 3.4 billion yr.).

Incorporation of plutonium into the zircon structure requires information on solubility limits not only for Pu but also for its daughter product uranium. A variety of compositions of $ASiO_4$ in which A^{4+} = Hf, Zr, Am, Pu, U, Pa, and Th have been synthesized. $HfSiO_4$ (hafnon) $ZrSiO_4$ (zircon), $ThSiO_4$ (thorite), and $USiO_4$ (coffinite) occur naturally. Complete miscibility has been reported between hafnon and zircon[5,6]. Limited miscibility has been reported between zircon and coffinite and zircon and thorite[7].

We have synthesized various solid solutions of $(Zr,A)SiO_4$ where A^{4+} = Ce, U, Th by the sol-gel method. These compositions and the observed phase assemblages were used to calculate solubility limits of the respective A cations in zircon. Uranium and thorium were studied to compare our experimental approach and the results with reported data by Mumpton and Roy[7]. Ce as Ce^{4+} is a good substitute for Pu. The ionic radii of Pu^{4+} and Ce^{4+} are close: 0.96 and 0.97 Å, respectively (VIII coordination) [8]. The solubility limit of Ce^{4+} could be used as guidance to limit the number of experiments when working with Pu.

EXPERIMENTAL

The sol-gel method used to synthesize zircon precursor material is similar to that reported by Haaker and Ewing[9]. Tetraethoxysilane ($Si(OC_2H_5)_4$) and zirconyl oxynitrate ($ZrO(NO_3)_2{\cdot}2H_2O$) were mixed in a methanol-aqueous solution. The uranium and cerium were added as uranyl nitrate ($UO_2(NO_3)_2{\cdot}6H_2O$) and cerium ammonium nitrate ($(NH_4)_2Ce(NO_3)_6$). Thorium was added as thorium chloride ($ThCl_4{\cdot}xH_2O$). After gelation, the precursors were dried at ~90°C and calcined at ~900°C.

SYNTHESIS AND ANALYTIC METHODS

Calcined gels with nominal compositions $Zr_{0.9}Ce_{0.1}SiO_4$ and $Zr_{0.5}Ce_{0.5}SiO_4$ were used as formed for zircon synthesis with or without milling. The material was cold-pressed into cylindrical pellets, 25 grams each. The samples were sintered for up to 12 hours. There was a small beneficial effect of milling on the zircon solid solution yield. The sintering temperatures varied between 1400°C and 1700°C.

Samples of U- and Th-doped zircon gel, 1-5 g, were milled and then sintered in a vacuum furnace (10^{-5} atm.) at 1400-1550°C for 2 hour. The stoichiometry of the gels are given in Table 1. Milling of the dried gels was required with these materials to get any reaction even at 1500°C.

X-ray powder diffraction (XRD) measurements were executed using CuKα and CoKα radiation on a Rigaku Denki diffractometer. Semi-quantitative estimates of the phase contents were conducted using Rigaku software. Lattice constants refinement was carried out using quartz as an internal standard. Analytical electron microscopy (AEM) examination of the synthesized phases was performed using a JEM-2010 transmission electron microscope equipped with an Oxford Link ISIS EDS system. The microscope was operated at 200 keV. Chemical compositions were calculated assuming the stoichiometry of oxides and normalizing to 100 wt.%. The analytical errors were estimated to be ±5%. Bright field imaging was used to document the microstructure of the zircon ceramic containing Ce.

RESULTS

XRD

The results of XRD measurements have been compiled in Table 1. In addition to zircon solid solutions, cristobalite, monoclinic ZrO_2, these crystalline phases were detected: i) tetragonal $(Zr,A)O_2$ solid solutions in all samples, except the sample doped with 50 at.% Ce; ii) the cubic solid solutions, $(A,Zr)O_2$ in the gel doped with 10 at.% Ce with high zircon yield and in the samples doped with 50 at.% U and Ce, respectively; iii) uranium oxide, $UO_{2.25}$, in the samples with 50 at.% U; iv) thorium oxide, ThO_2, and thorite, $ThSiO_4$ at 1400°C in the samples of initially 4 at.% Th; $ThSiO_4$ with monazite structure (huttonite) in the sample sintered at 1550°C.

Figure 1 (left) shows yields of zircon solid solutions as a function of the sintering temperature, obtained from unmilled samples of $Zr_{0.9}Ce_{0.1}SiO_4$ gel. The yield increased with temperature from 1400 to 1600°C. The zircon yield at 1700°C was slightly lower than that at 1600°C. This may indicate partial zircon decomposition. Zircon is known to decompose at temperatures higher than 1650°C. The cell parameter of Ce-doped zircon (Fig 1-right) decreased with increasing sintering temperature. This effect can be attributed to the reduction of Ce^{4+} to Ce^{3+} which is expected to increase with temperature.

Table 1. Zircon composition, final products, and cell parameters of zircon solid solutions.

Designed zircon compositions (gel)	Sintering conditions T,°C/time,h	Final products** Zrn(ss) = zircon solid soln. decreasing yield →	Zircon solid solutions cell parameters, (Å) *a*	*c*
$Zr_{0.9}Ce_{0.1}SiO_4$	*air** 1700/1	Zrn(ss), CZ, Cr, MZ, TZ	6.606±0.001	5.980±0.001
$Zr_{0.9}Ce_{0.1}SiO_4$	*air* 1600/1	Zrn(ss), MZ, Cr, TZ, CZ	6.610±0.001	5.983±0.001
$Zr_{0.9}Ce_{0.1}SiO_4$	*air* 1500/1	Zrn(ss), MZ, TZ, Cr, CZ	6.609±0.001	5.979±0.002
$Zr_{0.9}Ce_{0.1}SiO_4$	*air* 1400/1	MZ, TZ, Zrn(ss)	6.618±0.002	5.987±0.002
$Zr_{0.9}Ce_{0.1}SiO_4$	*air* 1400/8	Zrn(ss), CZ, TZ, Cr		
$Zr_{0.5}Ce_{0.5}SiO_4$	*air* 1400/8	CZ, Zrn(ss), Cr		
$Zr_{0.9}Ce_{0.1}SiO_4$	*air* 1400/12	Zrn(ss), CZ, TZ, Cr		
$Zr_{0.92}U_{0.08}SiO_4$	*vac.* 1500/2	MZ, Zrn(ss), TZ, Cr	6.621±0.001	5.993±0.001
$Zr_{0.92}U_{0.08}SiO_4$	*vac.* 1400/2	MZ, Zrn(ss), TZ, Cr	6.617±0.001	5.995±0.001
$Zr_{0.5}U_{0.5}SiO_4$	*vac.* 1500/2	CZ, TZ, Zrn(ss), Cr, U	6.621±0.001	6.000±0.002
$Zr_{0.5}U_{0.5}SiO_4$	*vac.* 1400/2	CZ, TZ, Zrn(ss), Cr, U	6.617±0.001	5.993±0.001
$Zr_{0.96}Th_{0.04}SiO_4$	*vac.* 1400/2	MZ, Zrn(ss), TZ, T, Thr, Cr	6.614±0.002	6.000±0.002
$Zr_{0.96}Th_{0.04}SiO_4$	*vac.* 1550/2	Zrn(ss), MZ, TZ, T, Thr, Ht, Cr	6.614±0.001	5.987±0.001

**vac.*=vacuum (10^{-5} atm); **Cr=cristobalite (SiO_2); CZ=cubic $(A,Zr)O_2$; TZ=tetragonal $(Zr,A)O_2$, MZ=monoclinic ZrO_2 or (Zr,A)O2 zirconia solid solutions; U=$UO_{2.25}$; T=ThO_2; Thr=$ThSiO_4$ thorite; Ht=$ThSiO_4$ (huttonite).

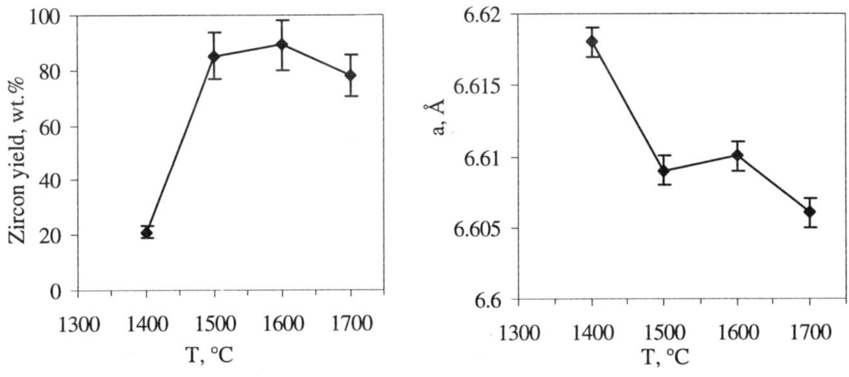

Figure 1. Yield of zircon solid solutions prepared from $Zr_{0.9}Ce_{0.1}SiO_4$ gel (gel not milled, sintered at 1400°C in air; 1 hour (left); respective *a*-cell parameter (right).

AEM RESULTS

Figure 2 shows the microstructure of a zircon ceramic obtained after sintering a milled and cold pressed gel with 10 at.% Ce at 1400°C for 8 hours. There are three crystalline phases, $(Zr,Ce)SiO_4$, $(Ce,Zr)O_2$, and SiO_2 (cristobalite). The occurrence of $(Zr,Ce)O_2$ was observed prior to the the formation of zircon at shorter sintering time. This phase then reacted with SiO_2 to form the zircon solid solution. During sintering for 12 hours zircon crystals grew in size, whereas $(Ce,Zr)O_2$ and SiO_2 particles decreased both in abundance and in size. No significant difference in the average Ce concentration was found in the zircon solid solution after 8 hours (Fig. 2-left) and 12 hours (Fig. 2-right), indicating that the maximum Ce concentration in zircon was reached before eight hours. Further solid state reaction increased only the yield of the phase.

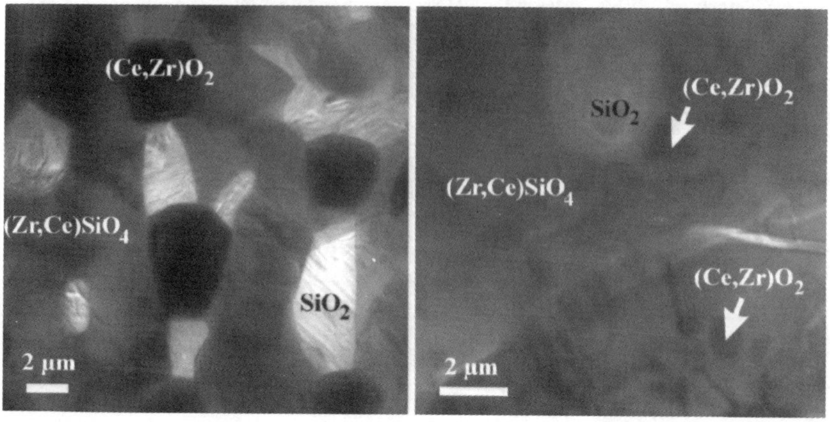

Figure 2. TEM Micrographs showing two stages of reaction progress during the formation of $(Zr,Ce)SiO_4$ solid solution from $(Ce,Zr)O_2$ and SiO_2: after 8 h (left), after 12 h (right) at 1400°C. Nominal composition of the gel was $Zr_{0.9}Ce_{0.1}SiO_4$.

Table 2. The results of AEM analyses of zircon solid solutions with Ce and U

Designed zircon composition (gel)	Sintering conditions T(°C) time(h)	Composition of zircon solid solution after sintering (NA = number of analyses)				
		SiO_2, data/aver. wt.%	ZrO_2, data/aver. wt.%	CeO_2, data/aver. wt.%	"$CeSiO_4$" data/aver. mole%	NA
$Zr_{0.9}Ce_{0.1}SiO_4$	1400	29.2-32.6	60.4-64.0	4.3-7.8	4.6-8.6	19
	8 and 12	31.8	62.1	6.2	6.7	
$Zr_{0.5}Ce_{0.5}SiO_4$	1400	30.7-33.7	59.4-62.9	5.8-7.6	6.3-8.2	8
	8	32.1	61.4	6.5	7.1	
		SiO_2, wt.%	ZrO_2, wt.%	UO_2 wt.%	$USiO_4$ mole%	
$Zr_{0.92}U_{0.08}SiO_4$	1500	30.9-35.0	60.0-62.1	4.8-7.3	3.2-5.1	3
	2	32.9	60.8	6.3	4.1	

The results of AEM chemical analyses of zircon solid solutions with Ce and U are given in Table 2. The measured cerium content in zircon varies between crystals from 4.3 to 7.8 wt.% CeO_2. Solid solutions with Ce were synthesized at 1400°C from gels with 10 and 50 at.% Ce, respectively, with the nominal zircon composition. The average of all analyses corresponds to 7±1 mole% "$CeSiO_4$" dissolved in zircon. Solid solutions with U were synthesized at 1500 °C.

Analyses of uranium show that zircon contains between 4.8 and 7.3 wt.% UO_2, corresponding to 3.2 to 5.1 mole% $USiO_4$.

DISCUSSION

CERIUM

Two sets of experiments, one with a $Zr_{0.9}Ce_{0.1}SiO_4$ and the other with a $Zr_{0.9}Ce_{0.1}SiO_4$ gel composition yielded the same average Ce concentration in solid solution with zircon: 7±1 mole% at 1400°C. The appearance of other reaction products indicates that that the solubility of "$CeSiO_4$" was exceeded. Experiments at temperatures above 1400°C were not evaluated. These higher temperature experiments should be conducted under controlled oxygen fugacity in order to maintain Ce as Ce(IV).

URANIUM AND THORIUM

Assuming a straight-line relation of unit cell parameters between two end-member compounds, $ZrSiO_4$ and $ASiO_4$, as a function of A, Mumpton and Roy[7] estimated zircon compositions from the change in the (200) *d* spacing. We used the lattice constant *a* because it shows a stronger change between the end-member compositions $ZrSiO_4$, $ThSiO_4$ and $USiO_4$. The calculation yielded (4±1) mole% $USiO_4$ and (2±0.5) mole% $ThSiO_4$. For the samples with uranium, chemical analyses were conducted using AEM-EDS of the solid solution compositions and the result (3.2 to 5.1 mole%) is in fair agreement with the value obtained from calculations using the measured *a* cell parameter (Table 2).

Studies of zircon crystals from the melt in the destroyed nuclear reactor at Chernobyl showed that uranium was dissolved in zircon. As previously reported[1], these zircon crystals formed at temperatures higher than 1500°C. They contained numerous inclusions of U(IV) phases, indicating the presence of excess uranium during zircon formation. Microprobe analyses and cell parameter calculations[10] showed that zircon contained 4±2 mole% $USiO_4$. These results are also in good agreement with ours and with the results of Mumpton and Roy[7]. Thus, the uranium solubility in zircon is less than 6 mole%.

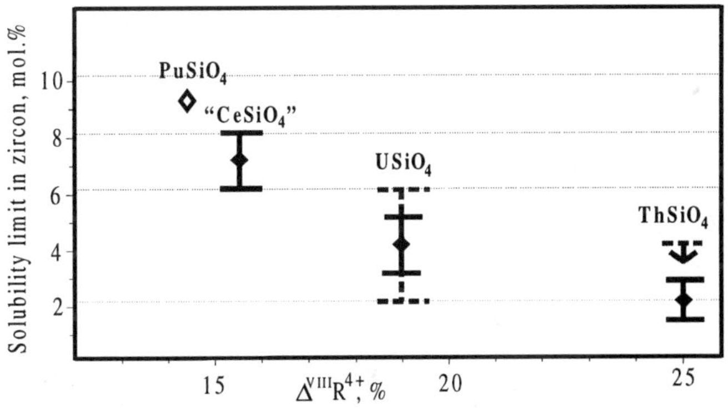

Fig. 3. Solubility limits of "$CeSiO_4$" (1400°C), $USiO_4$ (1400-1500°C) and $ThSiO_4$ (1400-1550°C) in zircon. $\Delta^{VIII}R(IV)=(R_i-R_{Zr})/R_{Zr})\cdot 100$, where R - ionic radius, *i* - Pu, Ce, U, Th. Dotted lines indicate results by Mumpton and Roy for $USiO_4$ at 1350°C and $ThSiO_4$ at 1175-1400°C.

It was reported recently[11] that the miscibility in the $ZrSiO_4$-$ASiO_4$ systems would increase with temperature. The previously reported solubility limits of $USiO_4$ and $ThSiO_4$ in zircon by Mumpton and Roy[7] were obtained at temperatures below 1400°C. Our results were obtained at 1500°C and the crystals found in the Chernobyl lava formed at even higher temperature. It appears that there is little or no increase in the solubility of uranium in this temperature range. The value for the solubility of Th is a factor of two lower than Mumpton and Roy's[7] though our temperature was higher than theirs.

PLUTONIUM

The solubility of Pu in zircon is unknown. The highest concentration of Pu in zircon reported in the literature[2] is 9.2 at.%. Using the dependence of the solubility of A-cations in zircon on their ionic radii in the tetravalent state, the solubility of Pu is expected to be higher but close to that of Ce (Fig. 3). The solubility of Pu in zircon is high enough to make it an attractive waste form. Increasing Pu content above a few eight percent requires the addition of neutron absorbers. Criticality concerns are more likely to limit waste loading than the thermodynamic solubility would allow.

ACKNOWLEDGMENTS

This research was supported by the University of New Mexico Center for Radioactive Waste Management. The analytical electron microscopy study was completed in the Transmission Electron Microscopy Laboratory of the Department of Earth and Planetary Science, University of New Mexico, which is supported by NSF, NASA and State of New Mexico. The authors thank Dr. Burakov (V.G. Khlopin Radium Institute, St. Petersburg, Russia) for help in the experimental part.

REFERENCES

[1]B.E. Burakov, "A Study of High-Uranium Technogenous Zircon (Zr,U)SiO_4 from Chernobyl "Lavas" in Connection with the Problem of Creating a Crystalline Matrix for High-Level Waste Disposal"; pp. 29-33 in *SAFE WASTE -93*, **2**, 1993.

[2] R.C. Ewing, , W. Lutze, , W.G. Weber, "Zircon: A Host-Phase for the Disposal of Weapons Plutonium," *Journal of Materials Research*, **10**[2], 243-246 (1995).

[3] B.E. Burakov, E.B. Anderson , B.Ya. Galkin, , V.A. Starchenko, V.G. Vasiliev, "The Crystalline Host-Phases for Immobilization of Weapons Plutonium and Waste Actinides"; pp. 85-89 in *Disposal of Weapons Plutonium*, Edited by E.R. Merz and C.E. Walter, Kluwer Academic Publisher, Dordrecht, 1995.

[4] R. C. Ewing, W.G. Weber, W. Lutze, "Crystalline Ceramics: Waste Forms for the Disposal of Weapons Plutonium, pp. 65-83 in *Disposal of Weapons Plutonium*, Edited by E.R. Merz and C.E. Walter, Kluwer Academic Publisher, Dordrecht, 1995.

[5] J.A. Speer, "The Actinide Orthosilicates"; pp. 113-135 in *Reviews in Mineralogy-Orthosilicates*, **5**, 2nd ed. Edited by Paul H. Ribbe, Mineralogical Society of America, Chelsea, MI, 1982.

[6] J.A. Speer and B.J. Cooper, "Crystal Structure of Synthetic hafnon, $HfSiO_4$, comparison with Zircon and the Actinide Orthosilicates," *American Mineralogist*, **67**, 804-808 (1982).

[7] F.A. Mumpton and R. Roy, "Hydrothermal Stability Studies of the Zircon-Thorite Group," *Geochemica et Cosmochemica Acta*, **21**, 217-238 (1961).

[8] R.D Shannon and C.T Prewitt, "Effective Ionic Radii in Oxides and Fluorides," *Acta Crystallographica*, **A32**, 751-767 (1976).

[9] R.F. Haaker and R.C. Ewing, "Solution-Gelation Method for Preparing Polycrystalline Zircon," *American Ceramic Society*, **64**[11], C-149 (1981).

[10] S.V. Ushakov, PhD thesis, St.Petersburg State University (*in russian*).

[11] V.A. Kurepin, "Preliminary Evaluation of Thermodynamic Mixing Properties and Miscibility Limits for Cubic Dioxide $(Zr,M)O_2$ and Zircon $(Zr,M)SiO_4$ Solid Solutions ($M=An^{4+}$, Ln^{4+}), *Abstracts*, 21st International Symposium on the Scientific Basis for Nuclear Waste Management, "MRS-97", 919-920 (1997).

IMMOBILIZATION OF ACTINIDES IN TITANATE CERAMIC WASTEFORMS

Allen W. Apblett* and Edwin H. Walker, Jr. Department of Chemistry, Oklahoma State University, Stillwater, OK, 74078-3071.

ABSTRACT

Synthetic sodium trititanate has been demonstrated to be highly effective for the removal of Th^{4+} ions from aqueous solution. The maximum loading was found to be 14.9 % by weight of Th^{4+}. The effectiveness of this material appears to result both from hydrolysis/absorption reactions and ion-exchange reactions. Sintering of the thorium-loaded titanate yielded a mixture of thoria and a phase that is believed be a thorium-substituted hexatitanate, $Na_2Ti_{6-X}Th_xO_{13}$. An ion-exchanger that could be converted to a dense titanate ceramic was synthesized by preparation of an ammonium titanate gel. This material was found to have a very high ion-exchange capacity for thorium and the loaded material yielded an anatase/ $ThTi_2O_6$ composite upon sintering to 900°C. Alternatively, a thorium titanate ceramic could be prepared directly from solid thorium nitrate by dissolution in a liquid titanium carboxylate followed by pyrolysis at 800°C.

INTRODUCTION

A successful waste form for radioactive metals must not only have the ability to chemically incorporate the elements but it must also be extremely stable in the geological environment. Therefore, many scientists have adopted a "naturalistic" approach to waste disposal materials whereby the choice of waste form is based upon the natural geological repository for these elements. Thus, ceramic waste forms are sought which mimic those minerals that have sequestered the hazardous metals for billions of years. Such an approach has culminated in the development of highly successful materials such as zircon ($ZrSiO_4$) for plutonium immobilization [1] and titanate ceramics such as Synroc for nuclear waste disposal [2].

Ion-exchange materials are particularly valuable in hazardous waste treatment because of their ability to remove pollutants such as heavy metals and radionuclides from contaminated water. Typically, ion exchangers are used to concentrate toxic metals and are subsequently regenerated by chemical treatment for reuse. This results in the release of the pollutants into solution and requires subsequent steps for their eventual incorporation in a ceramic waste form. Our approach, shown schematically in Figure 1, is to design the ion exchanger so that once it is loaded with radionuclides it may be converted directly to the desired waste form. From the point of view of economics and simplicity, the ion exchangers for titanate-based waste forms should consist of a titanate phase that has exchangeable cations. The material should also be commercially-available and inexpensive or readily prepared from economical starting materials.

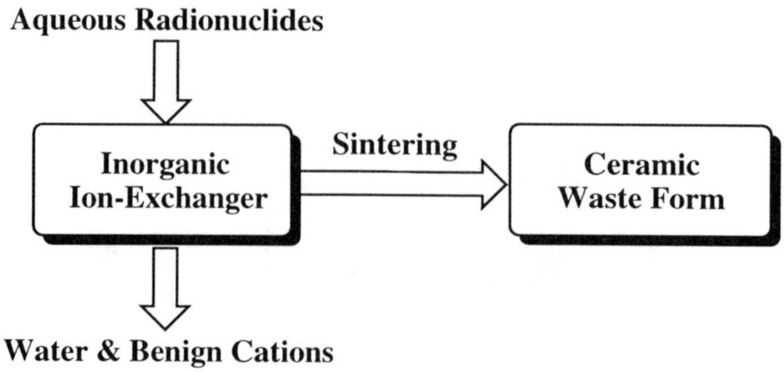

Figure 1. Schematic of Ion-Exchange/Ceramic Waste Form Process

Sodium titanates are promising candidates as ion-exchangers for the removal of actinides from waste streams in a form that is amenable to direct conversion to a titanate waste form. For example, the sodium ions in $Na_2Ti_3O_7$ undergo ion-exchange with other metals and most alkali metal titanates are ionic conductors [3]. As well, sodium titanates undergo rapid hydrolysis in water leading to a dramatic increase in pH that may be utilized to effect the hydrolysis of actinide metal ions [4]. Previously, sodium titanate has been successfully used for removal of soluble strontium and plutonium from aqueous solution in a process developed by the Savannah River Laboratory for vitrification of high level waste [5, 6]. Furthermore, Apblett *et. al.* have demonstrated that a commercially-available sodium titanate (a mixture of approximately 1:1 $Na_2Ti_3O_7$ and $Na_2Ti_6O_{13}$) served well as an ion exchanger for removing thorium and uranium from aqueous solution, absorbing up to 8.5% of their weight of the actinides and reducing the concentrations of these ions to below 10 ppb in solution [4].

Herein, the results of an investigation of the effectiveness of a synthetic sodium trititanate and ammonium titanate gels for application in an ion-exchange/ceramic waste form process for remediation of actinides are reported.

EXPERIMENTAL

. $Ti(MEEA)_4$ was prepared according to a literature procedure [7]. Water was deionized and distilled before use. All other reagents were commercial products and were used without further purification. Metals analyses were performed by inductively-coupled plasma atomic emission spectroscopy (ICP-AES) on a Perkin Elmer Optima instrument or by X-ray fluorescence spectroscopy on a Spectro X-Lab energy-dispersive XRF spectrometer. Thermogravimetric studies were performed using 10-20 mg samples in a Seiko TG/DTA 220 instrument under a 100 ml/minute flow of dry air. The temperature was ramped from 25°C to 825°C at a rate of 2°C per minute while the off-gases were analyzed on a Fisons Thermolab mass spectrometer. Bulk pyrolyses at various temperatures were performed in ambient air in a temperature-programmable muffle furnace using ca. 4 g samples, a temperature ramp of 5°C/minute and a hold time of 12 hours. X-ray powder diffraction patterns were obtained on a Scintag XDS 2000 diffractometer using copper K_α radiation. Microanalysis for hydrogen and nitrogen content was performed by Tulane University's Coordinated Instrument Facility.

Preparation of $Na_2Ti_3O_7$ Sodium nitrate (0.121 g) was added to $Ti(MEEA)_4$ (0.948 g) and the mixture was gently heated with a heat gun until the sodium nitrate completely dissolved to give a pale yellow liquid. The liquid was subsequently pyrolyzed at 900°C for one hour to yield $NaTi_3O_7$ (0.214 g) as a fluffy white solid.

Preparation of Ammonium Titanate A mixture of concentrated ammonium hydroxide (9.95 g) in absolute ethanol (25 ml) was cooled to -25°C and was then added to a cooled mixture (-25°C) of titanium isopropoxide (42.47 g) in 25 ml of absolute ethanol. A white precipatate formed quickly but the material was allowed to age for 12 hours in a sealed flask at 0°C. The material was then isolated by filtration using a fine sintered glass Buchner funnel and was washed twice times with 500 ml of water. The white solid was the air dried in a an oven at 50°C to give a yield of 15.16 g. Nitrogen content was determined to be 3.14 %.

RESULTS AND DISCUSSION

Previously, it had been demonstrated that a commercial sodium titanate mixture of $Na_2Ti_3O_7$ and $Na_2Ti_6O_{13}$ was capable of removing 8.7% by weight of thorium from aqueous solution [4]. Sintering of the thorium-loaded exchange material at 800°C yielded a ceramic in which the main phase appeared to be $Na_2Ti_6O_{13}$. Since the hexatitanate had already been present in the ion exchanger, its observance in the final product could be attributed either to amorphization of the trititanate or conversion of thorium-exchanger trititanate to a hexatitanate-like phase. Therefore, in this investigation, significantly purer trititanate was prepared and tested as a sinterable ion-exchanger for titanate waste forms. The synthesis of this material took advantage of novel precursors for titanate materials that have recently been developed based on metal carboxylates that are room-temperature liquids. This recently discovered family of metal carboxylates has provided novel approaches for the preparation of refractory ceramic bodies and the synthesis of ceramic films by liquid phase processing. These carboxylates, salts of 2-[2-(2-methoxy)ethoxy]ethoxy acetate, MEEA, have polyether linkages which undoubtedly contribute to their unusual characteristics by chelating and "solvating" the metal ions.

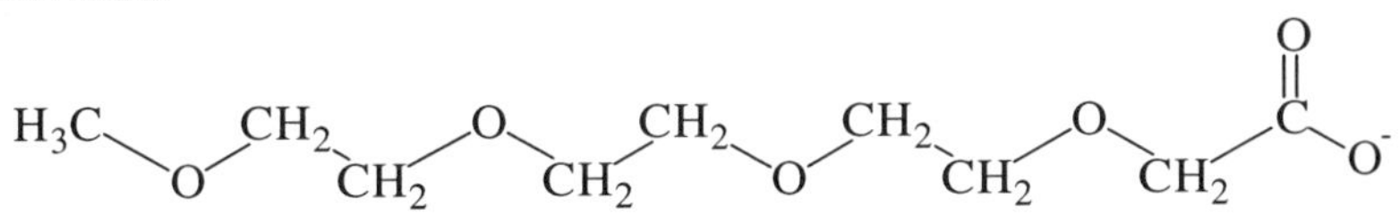

MEEA

Furthermore, these liquids are excellent solvents for other metal salts so that precursor solutions for multi-metallic ceramic materials are readily prepared. Previously, a homogeneous precursor solution for barium titanate was synthesized using a liquid metal carboxylate precursor that was prepared by dissolving barium acetate in $Ti(MEEA)_4$ [7]. Similarly, precursors for nickel ferrite, $MgAl_2O_4$, and yttrium aluminum garnet were synthesized by dissolution of $Ni(NO_3)_2$ in the liquid iron carboxylate, $Fe_3O(MEEA)_7 \cdot 5.5(H_2O)$ [8], $Mg(NO_3)_2$ in $Al(MEEA)_2OH$ [9], and $Al(OH_2)_6(NO_3)_3$ in $Y(MEEA)_3$ [10], respectively. These precursor solutions yield crystalline phase-pure ceramics at exceptionally low temperatures (400-800°C) due to the extremely homogeneous mixing of the metal ions. Similarly, fairly pure $Na_2Ti_3O_7$ was prepared at 900°C from a liquid precursor which was obtained by dissolving sodium nitrate into $Ti(MEEA)_4$. The fluffy white solid prepared in this manner displayed an X-ray powder (XRD) diffraction pattern that corresponded to fairly pure sodium trititanate with only a trace of hexatitanate present (Figure 2). Inductively-coupled plasma atomic emission spectroscopic (ICP-AES) analysis for sodium and titanium confirmed the correct composition.

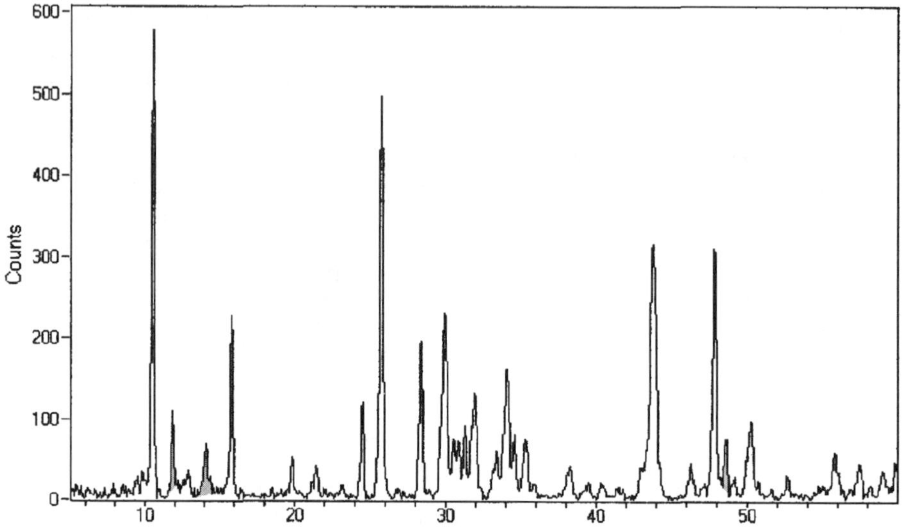

Figure 2. XRD Pattern of Synthetic $Na_2Ti_3O_7$. Shaded reflections are attributable to $Na_2Ti_6O_7$

The reaction of the sodium trititanate with a solution containing slightly more than a half molar equivalent of Th^{4+} (i.e. more than a sufficient amount for replacement of all the sodium atoms according to Eq. 1) leads to a significant release of sodium ions results and the absorption of 14.9 weight percent of thorium, as demonstrated by X-ray fluorescence spectroscopy. The latter amount can, therefore, be taken as the maximum loading with thorium (and, by analogy, other tetravalent actinides such as Pu^{4+}). This corresponds to about 0.74 milliequivalents per gram, approximately half of the theoretical capacity of $Na_2Ti_3O_7$ for Th^{4+} of 1.66 mEq/g based on Eq. 3. The uptake of thorium by the pure sodium trititanate is significantly higher than that reported previously for the commercial mixture of titanate phases (0.40 mEq/g) [4] indicating that $Na_2Ti_3O_7$, with it's more open structure (Figure 3), is much more active in ion-exchange than the $Na_2Ti_6O_{13}$ phase.

$$2\ Na_2Ti_3O_7 + Th^{4+} \rightarrow ThTi_6O_{14} + 4\ Na^+ \qquad (1)$$

The thorium-loaded sodium titanate was heated as a powder to 400°C, was ball-milled, and was then cold-pressed at 15 tons and sintered at 900°C. XRD analysis (Figure 4) indicated that the resulting phases were primarily $Na_2Ti_6O_{13}$ and thorium dioxide. The latter result is surprising since thorium dioxide was not observed in similar experiments with the commercial sodium titanate mixture. Nevertheless, our previous investigation concluded that part of the reduction of thorium in solution was due to the hydrolysis of Th^{4+} caused by the increase in pH that results from reaction of sodium trititanate with water. The observance of thoria in this investigation confirms this conclusion and the fact that it was not observed in the previous investigation may be attributed to the higher content of $Na_2Ti_3O_7$ in the synthetic ion exchanger leading to a greater pH increase. The formation of $Na_2Ti_6O_{13}$ from thorium-exchanged $Na_2Ti_3O_7$ and the leach-resistance of the thorium (no thorium is observed by ICP-AES analysis of a 4 M hydrochloric acid solution that has been stirred with sintered thorium-loaded powder for one week) suggest that the thorium has been incorporated into the oxide network of $Na_2Ti_6O_{13}$ (Figure 3), possibly by substitution for titanium ions.

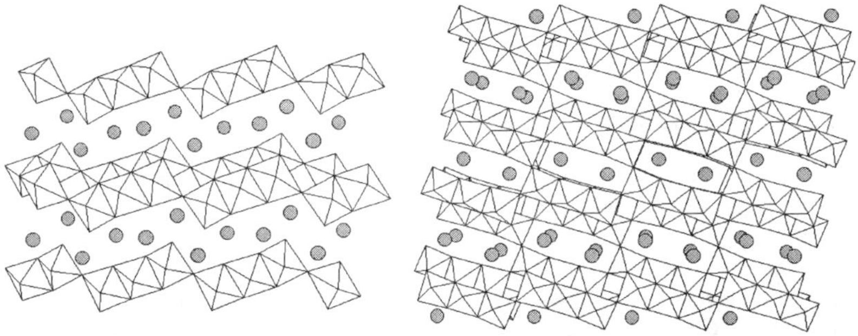

$Na_2Ti_3O_7$ $\qquad$ $Na_2Ti_6O_{13}$

Figure 3. Structures of Sodium Trititanate [11] and Sodium Hexatitanate [12]. Polyhedra represent TiO_6 Octahedra and Circles Represent Sodium Ions

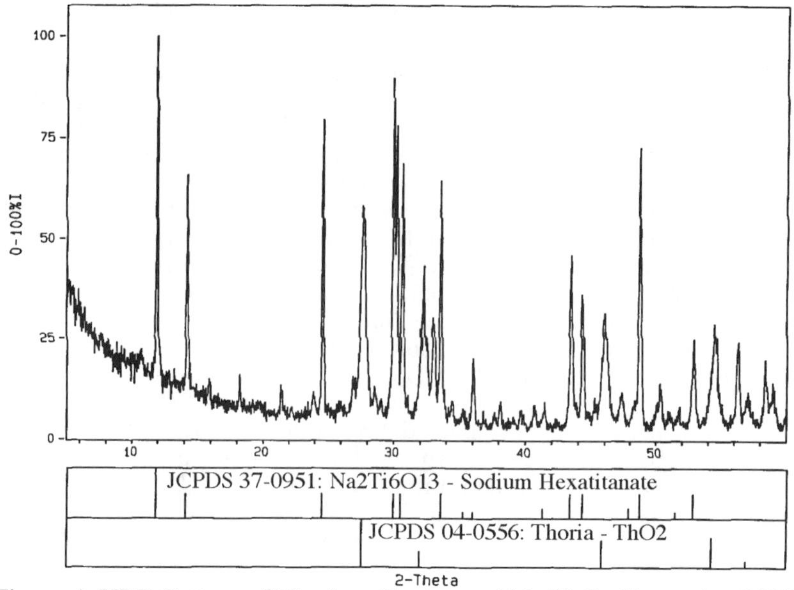

Figure 4. XRD Pattern of Thorium-Exchanged $Na_2Ti_3O_7$ Sintered at 900°C

The inability to achieve complete exchange of sodium ions with Th^{4+} ions results in formation of a ceramic phase, $Na_2Ti_3O_7$, that has an open structure that is unlikely to be stable and leach-resistant over the long term. An appropriate solution to this problem is to prepare layered titanates with cations such as ammonium ions that may be burned-out in the sintering step. Layered titanates, $[RNH_3]_2Ti_3O_7$, have previously been prepared by conversion of $Na_2Ti_3O_7$ to $H_2Ti_3O_7$ by acid-washing followed by reaction of the resulting titanic acid with alkylamines [13]. However, in this investigation it was found that a highly effective amorphous ammonium titanate ion-exchange material could be prepared expediently by reaction of titanium ethoxide with concentrated ammonium hydroxide. This material is amorphous to X-rays and contains 3.5% by weight of ammonium ions (1.9 mEq/g). The hydrolysis of titanium ethoxide under basic conditions yields negatively charged surfaces on the hydrosylate that promote the absorption of cations [14]. Reaction of the ammonium titanate with an excess of thorium nitrate (5 mmoles/gram), results in a material that contains 29 weight percent of thorium, as determined by XRF spectroscopy. This result corresponds to an exchange-capacity of 41 % by weight (1.8 mEq/g) of the ion exchange material for thorium. Thus, it appears that almost every anionic site that is occupied by ammonium ions can act as a binding site for a Th^{4+} ion. This leads to a remarkably high ion-exchange capacity for thorium as compared to other inorganic ion exchangers such as zeolites which have capacities of 0.083 mEq/g [15]. Sintering of the thorium-loaded ion exchanger at 800°C yields thorium titanate, $ThTi_2O_6$ and anatase (Figure 5). The resulting ceramic is extremely resistant to leaching of thorium. No thorium was detected in the leachate that resulted from stirring the sintered powder in 4 M nitric acid (1 g exchanger / 10 ml acid) at room temperature for one week.

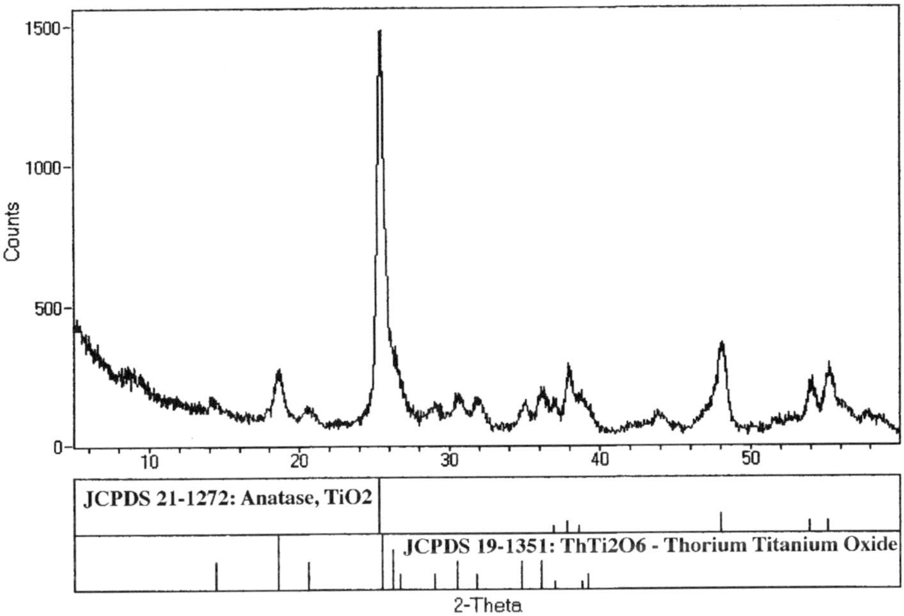

Figure 5 XRD Pattern of Thorium-Loaded Ammonium Titanate Sintered at 800°C

Many actinides such as plutonium have already been isolated in concentrated form and their incorporation into a ceramic waste form does not require an ion-exchange step. Since $Na_2Ti_3O_7$ could be readily prepared from a solution of sodium nitrate in $Ti(MEEA)_4$, it was likely that this approach could also be used to form titanate waste forms. Therefore, a 0.5 molar equivalent $Th(NO_3)_4 \cdot 6H_2O$ was added to the titanium liquid carboxylate. Heating to 80°C and stirring resulted in dissolution of the thorium nitrate and yielded a viscous yellow liquid. Heating this liquid to 800°C yielded a white solid that had an XRD pattern that corresponded to phase-pure thorium titanate, $ThTi_2O_6$.

CONCLUSION

Sodium titanate is a highly effective reagent for removal of thorium (and, by analogy, other actinides) from aqueous solution. However, residual sodium ions in the ion-exchanger prevent the sintering of the thorium-loaded material to a durable waste form. This problem can be circumvented by resorting to an amorphous ammonium titanate ion-exchange material that has extremely high capacity for Th^{4+} and can be sintered directly to a dense thorium titanate ceramic. A liquid titanium metal carboxylate provides an alternative for preparation of titanate waste forms via homogeneous precursor solutions.

ACKNOWLEDGEMENTS

We thank the Louisiana Board of Regents for support of this research through Grant # LEQSF(1997-00)-RD-B-13.

REFERENCES

[1] W. Lutze, R. C. Ewing, and A. Prinja, "Stabilization and Disposal of Surplus Weapons Plutonium in Zircon," Presented at Emerging Technologies in Hazardous Waste Management VIII, Birmingham, AL, 1996.

[2] W. Lutze and R. C. Ewing, Eds., "Radioactive Wasteforms for the Future" North-Holland, Amsterdam, 1988.

[3] S. Kikkawa, F. Yasuda, and M. Koizumi, "Ionic Conductivities of $Na_2Ti_3O_7$, $K_2Ti_4O_9$ and Their Related Materials," *Mater. Res. Bull.,* **20**, 1221-1227 (1985).

[4] A. W. Apblett, A. Ahmadi, and E. H. Walker, "Incorporation of Actinides into Leach Resistant Phases Via Ion-Exchange Materials" *Ceram. Trans.,* **72**, 189-195 (1996).

[5] R. F. Schumacher, "DWPF Batch 1 Waste Glass Investigations"*Ceram. Trans.,* **23**, 453-463, (1991).

[6] M. J. Plodinec, "Vitrification Chemistry and Nuclear Waste," *J. Non-Crystalline Solids,* **84**, 206-214, (1986).

[7] A. W. Apblett, G. D. Georgieva, L. E. Reinhardt, and E. H. Walker, "Precursors for Aqueous and Liquid-Based Processing of Ferroelectric Thin Films"; pp. 95-105 in *Synthesis and Characterization of Advanced Materials*, , Edited by M. Serio, D. M. Gruen, and M. Ripudaman, American Chemical Society, Washington, D.C, 1998.

[8] A. W. Apblett, M. L. Breen, and E. H. Walker, "Synthesis of Nickel Ferrite Using Liquid Metal Carboxylates" *Mater. Chem.* (in press).

[9] A. W. Apblett, L. E. Reinhardt, and E. H. Walker, "Novel Liquid Precursors for Spinel and Alumina Ceramic Coatings" *Proceedings of Unified International Technical Conference on Refractories,* **Vol III**, 1503-1507 (1997).

[10] A. W. Apblett, L. E. Reinhardt, and E. H. Walker, "Liquid Metal Carboxylate Precursors for Yttrium Aluminum Garnet" *Proceedings of Unified International Technical Conference on Refractories,* **Vol III**, 1525-1529 (1997).

[11] S. Andersson and A. D. Wadsley, "The Crystal Structure of $Na_2Ti_3O_7$" *Acta. Cryst.,* **14**, 1245-1249 (1961).

[12] S. Andersson and A. D. Wadsley, "The Structures of $Na_2Ti_6O_{13}$ and $Rb_2Ti_3O_7$ and the Alkali Metal Titanates" *Acta. Cryst.,* **15**, 194-201 (1962).

[13] H. Izawa, S. Kikkawa, and M. Koizumi, "Formation and Properties of N-Alkylammonium Complexes with Layered Tri- and Tetra-Titanates" *Polyhedron,* **2**, 741-744 (1983).

[14] J. R. Bartlett and J. L. Woolfrey, "Preparation of Multicomponent Powders by Alkoxide Hydrolysis. 1. Chemical Processing"*Chem. Mater.,* **8**, 1167-1174, (1996).

[15] P. Misaelides, D. Charistos, A. Filippidis, and I. Anousis, "Interaction of HEU-Type Zeolite Crystals with Thorium Aqueous Solutions" *Chem. der Erd - Geochem.,* **56**, 143-156 (1996).

SOLID SOLUTIONS $Th_{4-x}M_x(PO_4)_4P_2O_7$ ($M^{(IV)}$ = U, Np, Pu) : SYNTHESES, CHARACTERIZATION AND LEACHING TESTS

Nicolas Dacheux, Anne C. Thomas, Benoît Chassigneux, Vladimir Brandel, Michel Genet
Radiochemistry Group, Nuclear Physics Institute, University of Paris Sud, Bat 100, ORSAY cedex, F-91406, FRANCE

ABSTRACT

In the framework of radwaste storage, the thorium phosphate-diphosphate $Th_4(PO_4)_4P_2O_7$ (TPD) has been synthesized in the ThO_2 - P_2O_5 system. This material is very resistant to aqueous corrosion. In this paper, we report the conditions of substitution of thorium by tetravalent actinides in the TPD structure. Th^{4+} can be replaced up to 75 mole % by U^{4+} , 52 mole % by Np^{4+} and 41 mole % by Pu^{4+}. Leaching tests of $Th_{4-x}Pu_x(PO_4)_4P_2O_7$ in distilled water lead to a leaching rate equal to 1.2 10^{-6} g/(m^2·d) which confirms the good properties of the TPD in comparison to other ceramics. The conditions required to get pellets are also discussed : the TPD is easily sintered after pressing at 500 MPa at room temperature then heating at 1250°C for 10 hours.

INTRODUCTION

Considering that phosphate matrices could be potential candidates for the immobilization of actinides coming from an advanced reprocessing or for the final disposal of the excess plutonium from dismantled nuclear weapons, the chemistry of thorium phosphates has been reexamined. In this field, we already reported the synthesis and the characterization of the thorium phosphate-diphosphate $Th_4(PO_4)_4P_2O_7$ (TPD) which is a material very resistant to aqueous corrosion [1, 2]. It can be synthesized by wet and dry chemical processes.

This compound was specially tailored for the immobilization of ^{239}Pu (α, $T_{1/2}$ = 24 500 yr) and the radiogenic ^{235}U (α, $T_{1/2} = 7.04 \times 10^8$ yr) coming from the breaking up of nuclear weapons. We are also interested in the behavior of ^{237}Np (α, $T_{1/2} = 2.14 \times 10^6$ yr) for the long term storage. Thus, we studied the substitution of thorium by tetravalent uranium, neptunium or plutonium in the TPD structure in order to synthesize $Th_{4-x}U_x(PO_4)_4P_2O_7$ (TUPD), $Th_{4-x}Np_x(PO_4)_4P_2O_7$ (TNPD) and $Th_{4-x}Pu_x(PO_4)_4P_2O_7$ (TPPD) solid solutions.

For the eightfold coordination number, the ionic radius of U^{4+}($^{[8]}r_U$ =

1.00 Å), Np^{4+}($^{[8]}r_{Np}$ = 0.98 Å) and Pu^{4+}($^{[8]}r_{Pu}$ = 0.96 Å) are close to that of Th^{4+}($^{[8]}r_{Th}$ = 1.05 Å). Then it should be possible to replace large amounts of thorium by these three tetravalent actinides in the TPD lattice. Nevertheless, for neptunium, the chemistry is very complicated due to the presence of several oxidation states. From the synthesis of TUPD, TNPD and TPPD solid solutions, the maximum substitution of Th^{4+} by U^{4+} , Np^{4+} or Pu^{4+} in the TPD structure was determined.

SYNTHESES

In our previous work concerning the thorium phosphate chemistry, we showed that for a mole ratio r = Th/PO_4 in the range $0.5 < r < 1$, only the thorium phosphate-diphosphate (r = 2/3) is well defined at high temperature. For all other ratios, the system obtained is polyphase [1]. For this reason, all syntheses were performed using the mole ratio r = 2/3.

Solid solutions of thorium-neptunium (IV) phosphate-diphosphate (TNPD) or thorium-plutonium (IV) phosphate-diphosphate (TPPD) were prepared by wet methods of synthesis from a mixture of thorium nitrate, neptunium (or plutonium) nitrate and phosphoric acid solutions. For thorium-uranium (IV) phosphate-diphosphate (TUPD), solutions were prepared in hydrochloric acid to avoid the oxidation of U^{4+} into UO_2^{2+}. Details concerning these solutions have been already published [3-5]. They were obtained by dissolution of actinide salts (nitrate, chloride or oxide) in nitric (or hydrochloric) acid. The concentrations were equal to 2.13 ± 0.02 M (thorium nitrate), 1.00 ± 0.02 M (thorium chloride), 1.67 ± 0.03 M (uranium chloride), 0.209 ± 0.002 M (neptunium nitrate), 0.528 ± 0.001 M (plutonium nitrate) while that of the phosphoric acid solution used was 4.97 ± 0.02 M.

In order to synthesize solid solutions $Th_{4-x}M_x(PO_4)_4P_2O_7$, actinide solutions were mixed with phosphoric acid for various mole ratios Th/M. For all the solids, the mole ratio r = $(Th+M)/PO_4$ was fixed to 2/3 with 1.5% of excess of phosphoric acid.

The mixtures were slowly evaporated between 100°C and 200°C in porcelain crucibles for at least 4 hours until the residue was dried. They were ground in water, evaporated and heated in alumina nacelles with a rate of 3°C/min at 300°C for 2 hours then at 1000°C in air for 14 hours in an Adamel furnace (in glove box for TNPD and TPPD solid solutions) and in inert atmosphere in a Pyrox furnace (for TUPD) [3-5]. After heating the samples were green (TUPD), turquoise-green colored (TNPD) and pink (TPPD).

CRYSTALLOGRAPHIC STUDY OF VARIOUS SYSTEMS

Solid solutions were characterized by X-ray diffraction. X-ray powder diffraction diagrams were collected with a Jobin-Yvon CGR apparatus using Cu $K_{\alpha 1}$ ray in glove box. Silicon (JCPDS file number 27-1402) was added as internal reference in order to calibrate the angular positions of the diffraction lines. Patterns were collected from 10° to 50° (2θ) each 0.01° using an acquisition time equal to 1.2 s/step.

As the crystal structure of TPD is now well known, we have studied the deformation of the lattice for solid solutions $Th_{4-x}M_x(PO_4)_4P_2O_7$ (M^{IV} = U, Np, Pu) when thorium is partly replaced by a smaller tetravalent actinide.

The ThO_2 - MO_2 - P_2O_5 (M = U, Np, Pu) systems

In our previous study concerning the TUPD solid solutions, we used an electron probe to verify whether the solids are single or polyphase [3,5]. After heating at 1000°C, we checked that the powdered samples synthesized by wet methods of syntheses were homogeneous and single phase.

For each TUPD, TNPD and TPPD solid solution, the unit cell parameters and volume were refined in order to verify if U^{4+} , Np^{4+} and Pu^{4+} take the place of Th^{4+} in the TPD lattice. The values obtained are reported in Table I.

Table I. Unit cell parameters of solid solutions $Th_{4-x}M_x(PO_4)_4P_2O_7$

Compound	*a* (Å)	*b* (Å)	*c* (Å)	*V* (Å^3)
$Th_4(PO_4)_4P_2O_7$ [1]	12.8646(9)	10.4374(8)	7.0676(5)	949.0(1)
$Th_{3.6}U_{0.4}(PO_4)_4P_2O_7$	12.851(3)	10.410(3)	7.061(2)	944(1)
$Th_3U(PO_4)_4P_2O_7$	12.816(4)	10.400(2)	7.043(2)	939(1)
$Th_2U_2(PO_4)_4P_2O_7$	12.772(4)	10.366(3)	7.016(3)	929(1)
$Th_{1.6}U_{2.4}(PO_4)_4P_2O_7$	12.758(6)	10.373(8)	6.998(4)	926(2)
$ThU_3(PO_4)_4P_2O_7$	12.705(3)	10.334(5)	6.996(3)	918(2)
$Th_{3.6}Np_{0.4}(PO_4)_4P_2O_7$	12.822(4)	10.419(3)	7.046(2)	941.3(9)
$Th_{3.2}Np_{0.8}(PO_4)_4P_2O_7$	12.797(2)	10.394(2)	7.034(1)	935.7(4)
$Th_{2.4}Np_{1.6}(PO_4)_4P_2O_7$	12.751(3)	10.359(3)	7.006(2)	925.4(8)
$Th_{3.6}Pu_{0.4}(PO_4)_4P_2O_7$	12.829(5)	10.410(4)	7.049(3)	941(1)
$Th_3Pu(PO_4)_4P_2O_7$	12.779(6)	10.365(3)	7.027(3)	930(1)

For all the solid solutions, we observed the decrease of the unit cell parameters when x increases which is in good agreement with the substitution of Th^{4+} by a smaller tetravalent cation. More, the variations are linear for a large x range value. The variation of the unit cell parameters in terms of x were obtained by linear regression from experimental data. These variations are reported in Table II for TUPD, TNPD and TPPD solid solutions and in Figure 1 for TNPD. All are consistent with the ionic radii values reported in literature for each tetravalent actinide. Indeed, for a given substituent amount, smaller is the radius of the tetravalent actinide substituted to thorium in the TPD structure, greater is the deformation of the lattice.

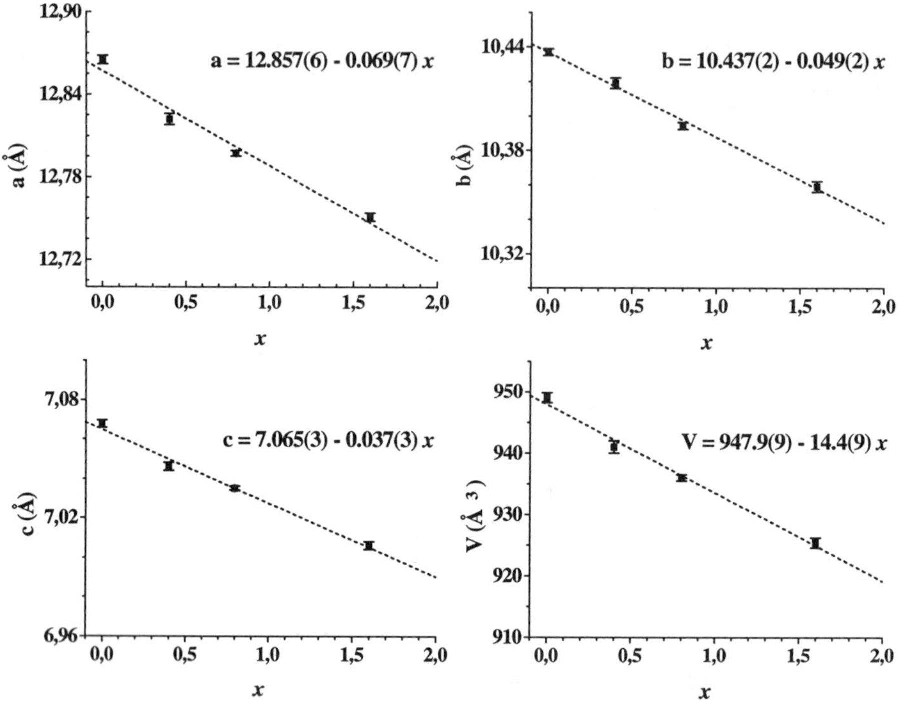

Figure 1. Variation of the unit cell parameters and volume of TNPD

We showed that it is possible to determine the maximum substitution of thorium by another tetravalent actinide (x_{max}) from the study of TUPD solid solutions [3]. The calculated x_{max} values are equal to 3.00 , 2.14 and 1.67 for U^{4+},

Np^{4+} and Pu^{4+} respectively. These values were also determined from the variation of the unit cell volume. The experimental values obtained : 3.00 , 2.09 and 1.63 are in good agreement with the calculated. They correspond to the replacement of 75 mole % of Th^{4+} by U^{4+}, 52 mole % by Np^{4+} and 41 mole % by Pu^{4+}.

Table II. Variations of the unit cell parameters for $Th_{4-x}M_x(PO_4)_4P_2O_7$

	U	Np	Pu
a (Å)	12.871(6)-0.053(3) x	12.857(6)-0.069(7) x	12.864(1)-0.085(1) x
b (Å)	10.431(7)-0.030(3) x	10.437(2)-0.049(2) x	10.438(1)-0.072(1) x
c (Å)	7.069(4)-0.026(2) x	7.065(3)-0.037(3) x	7.067(2)-0.042(2) x
V (Å^3)	949.0(6)-10.0(4) x	947.9(9)-14.4(9) x	949.0(1)-19.0(1) x

The ThO_2 - UO_2 - NpO_2 - PuO_2 - P_2O_5 system

From the studies of TUPD, TNPD and TPPD solid solutions, we can predict the field of existence of $Th_{4-x_U-x_{Np}-x_{Pu}}U_{x_U}Np_{x_{Np}}Pu_{x_{Pu}}(PO_4)_4P_2O_7$ solid solutions (TUNPPD) making the assumption that no more redox reactions occur between uranium (IV), neptunium (IV) and plutonium (IV) in phosphoric acid during the synthesis.

Considering the variation of the a parameter, solid solutions can be theoretically synthesized for :

$$5\,x_U + 7\,x_{Np} + 9\,x_{Pu} \leq 15 \qquad (1)$$

while the maximum substitution of thorium by the three other tetravalent actinides (uranium, neptunium or plutonium) should be obtained for :

$$5\,x_U + 7\,x_{Np} + 9\,x_{Pu} = 15. \qquad (2)$$

These solid solutions of thorium - uranium (IV) - neptunium (IV) - plutonium (IV) phosphate-diphosphates (TUNPPD) could be synthesized by dry chemical processes. For wet ways of synthesis, the use of nitrate solutions is not possible because of the oxidation of tetravalent uranium into uranyl UO_2^{2+} [5]. Moreover, the heating treatment should be performed under inert conditions as for TUPD solid solutions in order to prevent the formation of $U_{1-x}Th_x(UO_2)(PO_4)_2$ in air as already reported [5].

The MO_2 - P_2O_5 (M = U, Np, Pu) systems

In our investigation, we also studied the MO_2-P_2O_5 system in order to verify whether $M_4(PO_4)_4P_2O_7$ (M = U, Np, Pu), analogous to TPD were synthesized. Considering an initial mole ratio M/PO_4 equal to 2/3, the system is polyphase for the three actinides studied. It is composed of :

- $U(UO_2)(PO_4)_2$ {triclinic, a = 8.8212(4) Å, b = 9.2173(5) Å, c = 5.4772(3) Å, α = 102.622(4)°, β = 97.748(4)°, γ = 102.459(5)°} and α-UP_2O_7 {cubic, a = 8.606 Å} for tetravalent uranium when heated in air, $U_2O(PO_4)_2$ {orthorhombic, a = 7.0878(5) Å, b = 9.0362(8) Å, c = 12.702(1) Å} and α-UP_2O_7 in inert atmosphere ;

- $Np_2O(PO_4)_2$ {orthorhombic, a = 7.0334(2) Å, b = 9.024(3) Å, c = 12.587(6) Å} and α-NpP_2O_7 {cubic, a = 8.586(1) Å} for tetravalent neptunium ;

- $PuPO_4$ {monoclinic, a = 6.73(2) Å, b = 7.00(2) Å, c = 6.42(2) Å, β = 103.8(4)°} and α-PuP_2O_7 {cubic, a = 8.562(4) Å} for tetravalent plutonium.

SINTERING

The sintering of TPD took place through the preparation of a precursor which was obtained after the slow evaporation of the initial acid mixture of thorium and phosphoric acid solutions at about 150°C as previously detailed. The size of the particles provided by this dry residue was widely spread from 10 to 140 microns and the specific area was around 5.5 m^2/g. This precursor was firstly shaped into pellet forms via a room temperature pressing at 100 - 1500 MPa depending on the density required for the sintered material. It was thermally treated at 400°C for 5 hours then at 1250°C for 10 hours with a heating rate of 2°C/min. Variations of the sample density and shrinkage were determined in terms of the pressure used. From 200 to 800 MPa, the density varies from 83 to 93% of the calculated value while the shrinkage decreases from almost 15 to 10%. For P = 500 MPa, the hardness of the ceramized TPD was measured at 430 HV.

LEACHING TESTS

Leaching tests were performed for all TUPD, TNPD and TPPD solid solutions. Samples of 200 mg of powdered sample was put into 10 mL of distilled water (or acidic media) under agitation at room temperature and at 90°C. The results obtained for TPPD solid solutions at room temperature in distilled water are reported in Figure 2. They confirm the low solubility of the pure TPD structure. It is noticeable that the leaching rate of the matrix is independent on the plutonium amount loaded in TPD.

For instance, for $Th_{3.6}Pu_{0.4}(PO_4)_4P_2O_7$ (10 mole % of thorium replaced by plutonium), the equivalent mass of dissolved matrix is equal to 338 $\mu g/m^2$ after 300 days of contact between the solid and the solution. That corresponds to 21.7 $\mu g/m^2$ of plutonium released in solution. Considering that the dissolution rate becomes almost constant after 200 days ($1.2\ 10^{-6}$ $g/(m^2{\cdot}d)$), 1 mg of plutonium is released in solution after about 46 years of contact (surface of contact = 1 m^2). This value is good in comparison with other ceramics like zircons, apatites or synroc.

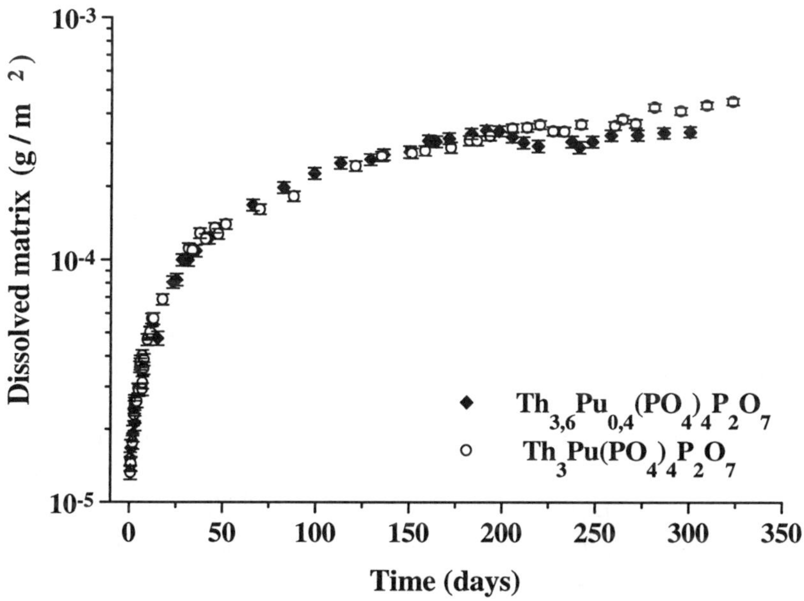

Figure 2. Leaching tests of TPPD solid solutions

CONCLUSION

Solid solutions $Th_{4-x}M_x(PO_4)_4P_2O_7$ were synthesized for several x values by substitution of thorium by tetravalent uranium, neptunium and plutonium. For each actinide we determined the maximum substitution of thorium according to the variations of the unit cell parameters and volume. This value is equal to 3.00 , 2.09 and 1.63 for uranium, neptunium and plutonium, respectively. That corresponds to the replacement of 75 mole % of Th^{4+} by U^{4+}, 52 mole % by Np^{4+} and 41 mole % by Pu^{4+}. These results show that the TPD may be used as a host material for large amounts of tetravalent actinides.

Leaching tests of TUPD, TNPD and TPPD solid solutions are under progress in distilled water or in acidic media. From these results, the TPD appears very resistant to corrosion. Only 0.14% of the matrix is dissolved after 300 days of leaching in distilled water (this value is one hundred times lower in the Volvic french granitic mineral water).

In the MO_2-P_2O_5 systems (M = U, Np, Pu), attempts to synthesize analogues of the $Th_4(PO_4)_4P_2O_7$: $U_4(PO_4)_4P_2O_7$, $Np_4(PO_4)_4P_2O_7$ and $Pu_4(PO_4)_4P_2O_7$ were unsuccessful. The systems obtained were always polyphase.

REFERENCES

[1]P. Bénard, V. Brandel, N. Dacheux, S. Jaulmes, S. Launay, C. Lindecker, M. Genet, D. Louër and M. Quarton,"$Th_4(PO_4)_4P_2O_7$, a New Thorium Phosphate : synthesis, characterization and crystal structure determination", *Chemistry of Materials*, **8** 181-188 (1996).

[2]N. Dacheux, "Matrices à base de phosphate d'uranium et de thorium : synthèses, caractérisations et lixiviation", Thèse, Université de Paris XI, Orsay, IPNO-T 95-04, March 31, 1995.

[3]N. Dacheux, R. Podor, V. Brandel and M. Genet, "Investigations of systems ThO_2-MO_2-P_2O_5 (M = U, Ce, Zr, Pu). Solid solutions of thorium-uranium (IV) and thorium-plutonium (IV) phosphate-diphosphates", *Journal of Nuclear Materials*, **252** **[3]** 179-186 (1998).

[4]N. Dacheux, A.C. Thomas, V. Brandel and M. Genet, "Investigation of the system ThO_2-NpO_2-P_2O_5. Solid solutions of thorium-neptunium (IV) phosphate-diphosphate", *Journal of Nuclear Materials*, In Press (1998).

[5]N. Dacheux, V. Brandel, M. Genet, K. Bak and C. Berthier, "Solid solutions of uranium and thorium phosphates : synthesis, characterization and X-ray photoelectron spectroscopy", *New Journal of Chemistry*, **[20]** 301-310 (1996).

AMERICIUM/CURIUM MELTER 2A PILOT TESTS

Michael E. Smith, Andrew P. Fellinger, Timothy M. Jones, Carey B. Miller, Donald H. Miller, Theresa K. Snyder, Michael E. Stone, and Douglas C. Witt

Westinghouse Savannah River Company

Savannah River Technology Center

Aiken, SC 29808

ABSTRACT

Isotopes of americium (Am) and curium (Cm) were produced in the past at the Savannah River Site (SRS) for research, medical, and radiological applications. These highly radioactive and valuable isotopes have been stored in an SRS reprocessing facility for a number of years. Vitrification of this solution will allow the material to be more safely stored until it is transported to the DOE Oak Ridge Reservation for use in research and medical applications. To this end, the Am/Cm Melter 2A pilot system, a full-scale non-radioactive pilot plant of the system to be installed at the reprocessing facility, was designed, constructed and tested. The full-scale pilot system has a frit and aqueous feed delivery system, a dual zone bushing melter, and an off-gas treatment system. The main items which were tested included the melter feed system, the dual zone bushing melter, the drain tube with dual heating and cooling zones, glass compositions, and the off-gas system which used for the first time a film cooler/shorter melter plenum. Most of the process and equipment were proven to function properly, but several problems were found which will need further work. A system description and a discussion of test results will be given.

INTRODUCTION

Isotopes of americium (Am) and curium (Cm) were produced in the past at the Savannah River Site (SRS) for research, medical, and radiological applications. They have been stored in a nitric acid solution in a SRS reprocessing facility for a number of years. Vitrification of the americium/curium (Am/Cm) solution will allow the material to be more safely stored until it is transported to the DOE Oak Ridge Reservation. Oak Ridge is responsible for marketing radionuclides for research and medical applications. To this end, the Am/Cm Melter 2A bushing melter pilot system was installed and tested at the SRS.

A bushing melter was chosen because of its small size and its shape (which could be needed for future criticality requirements if other radioisotopes such as plutonium were subsequently vitrified in the same system). Platinum/rhodium bushing melters are commonly used in the glass industry, especially in the manufacture of fiber glass. In these industrial applications, however, either molten glass or dry feed is delivered to the bushing. There has never been a bushing melter which was designed for the additional requirements of slurry feeding. Slurry feeding greatly increases the level of difficulty for glass melting as higher power requirements are needed at the upper portion of the bushing. An off-gas system is also required to treat any volatile or entrained components coming out of the melter. In addition, the numerous remote operation requirements for this radioactive feed had never been attempted to be met before.

PROCESS DESCRIPTION

The Melter 2A system is a full-scale pilot plant of the actual system to be installed at the SRS. It comprises a feed delivery system, a dual zone bushing melter, a split power supply and an off-gas treatment system (see Figure 1). The main differences between this and the previous Am/Cm Melter 2 system[1] are a re-designed dual zone melter, a shorter melter plenum, an off-gas film cooler, and PC control/monitoring of the process.

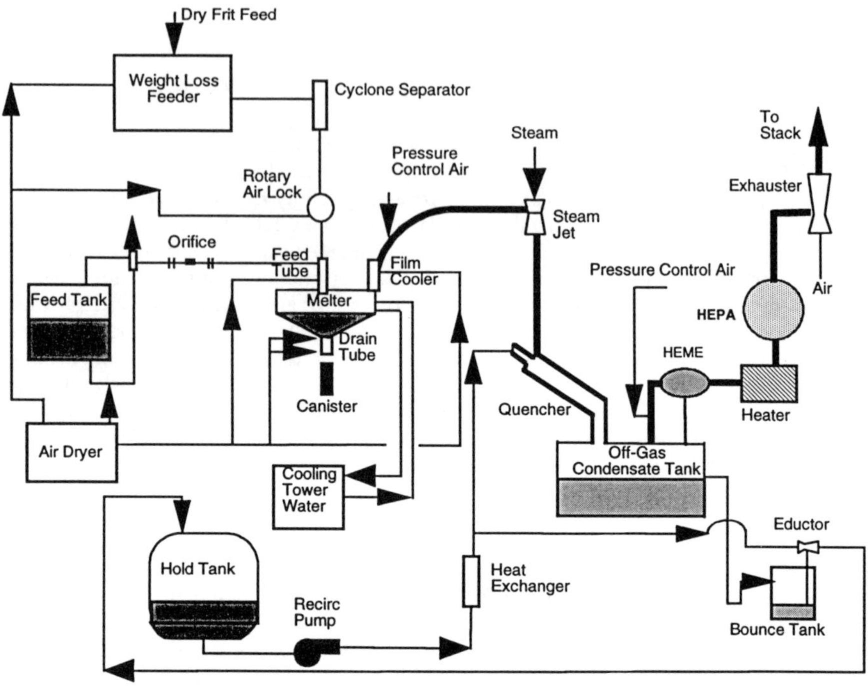

Figure 1. Am/Cm Melter 2A bushing melter pilot system schematic

A surrogate (non-radioactive) Am/Cm nitric acid feed solution, along with glass forming frit, are fed to the melter in two separate streams in a carefully controlled mass flow ratio. The feed is introduced into the vapor space of the bushing melter which is equipped with a top plenum to contain the melter off-gasses. The glass melt flows from the bottom of the melter by gravity through a heated drain tube into a stainless steel canister[2].

The melter vessel is a 90% platinum/10% rhodium alloy, bottom-sloped, rectangular vessel which acts like a resistance heater when current is passed through it. The melter is about 28 cm tall, 25.4 cm wide, and 7 cm deep. Two sets of platinum heater screens are located just below the melt line to aid in melting the feed. The melter is operated at about 1350° to 1500°C. The glass flows by gravity from the bottom of the melter through a 80% platinum/20% rhodium heated drain tube (nominal 0.5 cm inner diameter) into a stainless steel canister. Glass pouring is stopped by applying cooling air to the upper and lower portions of the drain tube.

As the melter is fed and the liquid is vaporized, the expanding gasses are drawn through a film cooler into the off-gas treatment system by a steam jet. The steam jet, coupled with a control air purge that is introduced into the off-gas system between the film cooler and the steam jet, maintains a target melter plenum pressure that is negative with respect to ambient pressure. The vapor exiting the steam jet enters a quencher where it is cooled and scrubbed of particulate entrained into the off-gas system during melter feeding. The condensate drains to the off-gas condensate overflow tank (COT). Overflow from the COT goes to the Bounce Tank. The vapor exits the top of the COT and is drawn through a High Efficiency Mist Eliminator (HEME), a High Efficiency Particulate Airfilter (HEPA) pre-heater and a HEPA by an air jet. The pre-heater raises the off-gas temperature above the dew point prior to entering the HEPA filter to prevent condensation on the HEPA. The air jet discharges the treated off-gas to the atmosphere through a stack.

TEST PLAN DESCRIPTION

Testing was planned in three phases as described below. Melter 2A performance was planned to be evaluated using Frit 25SrABS-F (the frit developed that has a lower liquidus than the original reference lead based Frit B2000)[1] which enabled operating the melter at a lower temperature and correspondingly provided a greater temperature delta between operating temperature and the temperature at which the integrity of the platinum begins to be potentially compromised. However, during Phase 2 testing severe foaming was encountered when a nitric acid solution was fed to the melter which had been initially inventoried with 50SrABS Hybrid glass (a surrogate glass representing the product of vitrification of the Frit 25SrABS-F and a surrogate feed). As a consequence, it was necessary to depart from the original experimental plan and consequently several of the Phase 2 and Phase 3 test objectives were not met. The compositions of the various feed streams are shown in Tables I and II. B1000 is a leaded glass that represents the vitrification product of Frit B2000 and a surrogate feed.

Table I. Major components of Am/Cm surrogate feed solution

Component	Equivalent oxide weight percent
Lanthanum	11.23
Cerium	12.90
Praseodymium	12.90
Neodymium	27.35
Samarium	6.34
Europium	1.25
Gadolinium	3.22
Erbium	12.90
Iron	2.19
Manganese	8.34

Table II. Compositions of various Am/Cm frits and glasses
(Note: all values in weight percent oxide -major constituents only)

Oxide	B2000	25SrABS-F	B1000	50SrABS-H
SiO_2	44.0	33.7	26.6	21.9
B_2O_3	9.2	13.5	3.3	8.8
Al_2O_3	6.0	24.9	14.1	16.4
La_2O_3	7.1	25.0	16.5	20.2
PbO	24.3	---	11.0	---
BaO	9.0	---	3.6	---
SrO	---	2.9	---	1.9
MnO_2			---	2.9
CeO_2			6.8	4.5
Pr_2O_3			---	4.5
Nd_2O_3			17.6	9.5
Sm_2O_3				2.2
Er_2O_3				4.5

Phase 1 covered the startup of the melter and focused on characterizing the performance of the zoned power supply. Temperatures of the glass pool, the platinum vessel and external surfaces of the melter assembly were measured during steady state and transitional periods of melter operation. Phase 2 testing was designed to characterize the response of the zoned power supply to aqueous feeding, obtain some measure of the rate of glass production achievable with the power supply and evaluate the capability of the off-gas system to control the pressure in the melter plenum. Initially, only a nitric acid solution was to be used during feed simulations. Ultimately, it was planned to feed 50SrABS Hybrid frit along with the nitric acid solution. Phase 3 testing was to determine nominal and maximum rates of vitrification while feeding the reference frit (25SrABS-F) and a surrogate feed solution (a nitric acid solution containing lanthanides and actinide stand-ins) compositionally identical to the feed solution projected for actual radioactive operation.

TEST RESULTS

Melter 2A Startup - Phase 1

The melter was initially charged with 50SrABS Hybrid glass cullet prior to heatup. With both zones in manual control, the melter was energized and a nominal heatup rate corresponding to 2°C/min was initiated. Upon reaching 1350°C, both zones were placed in automatic temperature control. Zone 1 (upper) and Zone 2 (lower) required 249 volts/44.3 amps (primary) and 231 volts/12.5 amps respectively to maintain the control temperature setpoint in both zones at 1350°C. Because of the foaming problem (and splattering of solids) encountered once nitric acid feeding was initiated, it was difficult to sustain any significant feed rates and correspondingly not possible to meet the test objective of quantifying vitrification rate capacity of the system. Consequently, the remainder of the original experimental plan was abandoned and subsequent efforts were directed at trying to determine the cause of the foaming (see Foaming section).

Glass Pouring

One of the test goals during the performance assessment of Melter 2A was to thoroughly characterize the glass draining process and develop the process operating parameters. The three main objectives of the glass pouring experiments were to (1) demonstrate the effectiveness of the two zone resistance drain tube heater on starting and stopping the flow of glass; (2) determine the reproducibility of the starting and stopping process; and (3) identify the values of process parameters required for acceptable glass pouring. Twenty-one glass pours were done with Melter 2A during this testing. Of those twenty-one pours, nine pours were with a glass of the 50SrABS Hybrid glass composition, seven of the 25SrABS-F/surrogate feed composition, and five pours of the Frit B2000/surrogate glass composition. The melter temperature was adjusted as required, based upon the temperature-viscosity relationship for the composition of the glass being poured.

Glass pouring began with gradual movement of the cold glass plug at the extreme tip end of the drain tube upon removal of the tip cooling air. Melter Zone 2 temperature setpoints were 1350°C to 1380°C for SrABS glass (1450°C to 1480°C for Frit B2000 glass), and pours transitioned from glass droplets to a steady stream of glass within seconds. Temperatures of Melter Zone 2 and the drain tube were found to be critical. For example, when the melter Zone 2 temperature was maintained at 1380°C (versus 1350°C) for 50SrABS glass, a steady glass stream immediately followed the cold glass plug as it disengaged from the drain tube without any glass droplets. This type of pour initiation is needed to minimize the chance of glass pluggage or the formation of glass stringers (thin fibers) during pouring. In each case, the drain tube was heated to 1350°C or higher prior to initiation of glass pouring.

To stop the glass pouring, 2800 l/hr cooling air was applied to the upper portion of the drain tube for 80 seconds, with 450 l/hr air being applied to the lower tip end of the tube 60 seconds after the initial application of the upper cooling air. Power input was also reduced to the drain tube heater assembly when the cooling air was applied to stop the glass pouring. The lower tip cooling air was continuously applied at 450 l/hr flow to maintain the cold glass plug until the next glass pour was initiated. The upper drain tube cooling air remained off until needed to stop the next glass pour. In general, the glass pouring tests showed that the dual zone heating and cooling performed quite well. Optimization of the process (such as drain tube temperature and time of upper and lower cooling air during poor stoppage) is still needed and is being done.

FOAMING

Severe foaming was encountered during Phase 2 testing when the melter began to be fed with a nitric acid solution and SrABS frit. The foaming problem rendered completion of the original experimental plan of little value. Subsequent testing focused on developing an understanding of the cause for the foaming and compositionally characterizing the foam layer using the reference frit, 25SrABS-F. In an effort to develop an understanding of the cause of the foaming, a series of feed cycle simulations using 25SrABS-F frit under the following conditions were conducted:

- Surrogate Feed of Varying Nitric Acid Concentrations (0.16M to 3.2M)
- Surrogate Feed Without Manganese
- Surrogate Feed Without Metal Oxides (Mn, Fe, Cr, Ni)
- Deionized Water as the Feed Solution

These cycles were chosen based on past SRS experience with nitric acid feed and redox conditions which can cause foaming. Twenty-seven feeding cycles were made with 25SrABS-F frit. A 3.2 Molar nitric acid surrogate solution was prepared and fed at 2.6 l/hr with the 25SrABS-F frit at 0.5 kg/hr to build the glass inventory within the melter. As the liquid feed rate varied between 2.4 and 2.8 l/hr for the different solutions fed, the frit feed rate was adjusted to balance the surrogate components being fed with the liquid. With the exception of feeding deionized water only, all combinations using liquid surrogate feeds foamed and splattered.

Analysis of samples of the foamy, porous cold cap formed by feeding the 25SrABS-F frit with liquid nitric acid surrogate feed indicated the presence of rare earth oxides. Mass spectrometry analysis of gas trapped in the pores of the cold cap samples detected nitrogen and oxygen. The foaming mechanism postulated from this information was the decomposition of the rare earth nitrates to an oxide. This decomposition liberates NO_x gas which is trapped in the viscous cold cap. This foam layer then acts as an insulating blanket which in turn reduces the melting efficiency of the bushing. If feeding is continued, the bushing cannot melt the additional feed and the entire melt pool becomes covered with a cold cap. This theory was backed by the finding that the addition of just nitric acid to an SrABS glass melt pool produced a thin, non-porous oxide layer due to the extraction of the lanthanides from the glass. This extraction first converts the rare earths to a nitrate form and then quickly decomposes these nitrates to an oxide layer on the glass surface. This decomposition results in the liberation of gases which then attempt to rise out of the melt pool/cold cap. Without a cold cap, however, there was no viscous layer to trap these gases, and therefore these rare earth oxides were re-dissolved into the glass fairly rapidly.

The next logical step was to determine compositional differences in the two glass systems that would cause one to foam and the other to not foam. This comparison of the 25SrABS-F frit and the B2000 frit (see Table II) showed that the La_2O_3 weight percent was much higher in the 25SrABS-F frit (25% versus 7%). This higher lanthanide content would contribute to the foaming problem in two ways. First, there would be more lanthanide nitrates to decompose. Secondly, the lanthanides in the SrABS glass are one of the main fluxes to lower the viscosity of the glass (B2000 frit uses lead as its main flux). Therefore, extraction of the lanthanides would cause the viscosity of the cold cap/upper melt pool to increase much more for the SrABS glass than the leaded glass. The high viscosity cold cap would then make it much more difficult for the gases formed to escape.

Following the decision to abandon the SrABS glass system and resume testing with the original leaded glass, the melter was inventoried with Frit B1000 glass (a glass simulant compositionally similar to the vitrified product of the proper blend of Frit B2000 and the surrogate feed). This glass matrix has a liquidus of around 1350°C and accordingly requires a glass processing temperature in the 1425°C to 1450°C range. As previously experienced with Frit B1000 tests with the first generation Am/Cm melter (Melter 1), no foaming was observed while feeding Frit B1000 with 2.8 M nitric acid. The cold cap produced by the frit and liquid feed melted into glass within 25 minutes after stopping feed. Feeding 0.87 M nitric acid surrogate solution by itself onto the Frit B1000 molten glass surface for 10 minutes also showed no evidence of foaming.

Frit B2000 together with surrogate feed was next fed onto the pool of Frit B1000 glass. After feeding for 30 minutes at 0.5 kg/hr frit addition and 2.7 l/hr of 0.87 M nitric acid surrogate, the B2000 composition feed was observed to form a normal cold cap which eventually spread across the entire glass surface. Many small bubbles could be seen breaking the upper surface of the crusty cold cap, but no foaming was detected. The upper zone of the melter was increased to 1470°C and the lower zone increased to 1500°C. Additional feeding of B2000 frit and surrogate liquid feed produced a normal appearing cold cap which allowed gasses to pass through without creating foam. The cold cap was incorporated into the molten glass within 30 to 45 minutes after feeding was stopped. Subsequent assessment of melter 2A with the B2000 system was halted because of pluggage problems with the off-gas HEME filter. Initial assessment of the causes of the HEME pluggage were lead volatilization during vitrification and the inability of the scrubber to effectively scrub out the lead aerosols and HEME design deficiencies (see next section).

EQUIPMENT PROBLEMS

Film Cooler Pluggage

The film cooler is designed to cool the off-gas stream and walls of the off-gas piping to minimize the amount of deposit formation at the inlet to the off-gas system. Process air flows through slots in the film cooler into the off-gas stream. The slots are designed to provide a laminar flow of cooling air along the walls of the film cooler. The film cooler was operated with air flows of 39-41 kg/hr during operation with the SrABS frit. The airflow was increased during operation with B2000 frit to 59 kg/hr (because of the previous pluggage experience and higher melter operating temperatures). The airflow to the film cooler was sufficient to prevent damage to the Inconel insert, but not its ceramic fiber reinforced mullite gasket and Inconel spacer. During operation at 1350°C, the spacer showed signs of corrosion along the bottom edge. The spacer was slightly deformed when the operating temperature was increased to 1450°C, and required repair prior to re-installation. The ceramic gasket was heavily damaged during operation with leaded glass (Frit B1000) at 1450°C. Large sections along the bottom edge were found missing. The exit temperature of the film cooler ranged from 80°C while idling with SrABS glass to a high of 220°C during feeding with surrogate feed and B2000 frit.

Deposition occurred on all exposed surfaces of the film cooler insert, spacer, ceramic gasket, and downstream piping. The piping downstream of the film cooler contained a 0.16 cm layer of deposits and was cleaned on two different occasions. Deposits on the film cooler insert and downstream piping were friable and easily removed by mechanical scraping and/or water rinse. Deposits on the film cooler gasket and metal spacer were not friable. The ceramic gasket could not be cleaned and was replaced each time that it was removed. The metal spacer was cleaned with a water rinse each time. During SrABS operation, the deposits were rare earth borates and oxides which were caused by splatter from the cold cap. The deposits formed during B2000 frit feeding were lead borates and oxides due to volatilization.

HEME Pluggage / HEPA Pluggage

The quencher/scrubber is designed to condense the majority of condensables from the off gas and to scrub entrained particles from the off-gas line gas. This prolongs the life expectancy of the downstream HEME and HEPA filters. HEME and HEPA performance were acceptable during operation with SrABS glass. However, during operation with the leaded glass (Frit B2000 and B1000 glass), the HEME plugged six times and the HEPA plugged twice (HEPA plugged only after HEME was removed). Analysis of the HEME flushes and condensate indicate that a majority of the lead volatilized from the melter was collecting on the HEME filter with relatively small amounts collecting in the condensate. HEME flush sample analyses indicated that the original HEME plugged after accumulating 20 grams of lead, while replacement HEME filters plugged with less than 5 grams.

Lead oxide volatilization creates sub-micron particles entrained in the off-gas. Sub-micron particles are not effectively removed by the existing quencher, therefore these particles pass through the quencher and collect on the HEME. When the HEME was removed, these particles collected on the HEPA filter. The quantity of lead measured on the HEME was less than that expected to cause pluggage. Particle size distribution and chemical composition of the solids exiting the melter's film cooler and entering the HEME have been determined for the leaded glass. The main findings are as follows:

- During melter idling (non-feeding), virtually all of the particulate exiting the melter and just before the HEME is in the submicron range (0.25 microns) and is mostly PbO.
- Reduction of the idling melter glass temperature from 1450°C to 1150°C resulted in a particulate evolution rate of only 2% as that at 1450°C. In addition, lead volatility was reduced to a level below the detection limit of the test.
- Feeding the melter at 1450°C (as compared to idling) resulted in a eight fold particulate increase at the melter exit. Most of the particles were in the 0.91 to 9.2 micron range and had about the same composition as the feed. The B and Pb were still in the submicron range.
- Melter feeding at 1450°C (compared to idling) resulted in a 36% reduction in Pb volatility.

CONCLUSIONS/RECOMMENDATIONS

The following is a list of the conclusions and recommendations based on the M2A testing to date:

- The dual zone Melter 2A worked quite well and was able to achieve acceptable melt rates when used with the non-foaming leaded glass.
- SrABS glass tended to foam to the point that operation with this glass is not practical. The Frit B1000 leaded glass did not foam. Subsequent work was done to understand the foaming mechanism. The reference glass composition is now the leaded glass.
- The current film cooler design did not prevent off-gas pluggage when used with air flows of 41-59 kg/hr. The film cooler deposits consisted mostly of lanthanide oxides and borates. Lead oxide deposits occurred when leaded glass was present in the bushing melter.
- The dual zone drain tube heater with dual cooling air locations worked quite well for the initiation and stoppage of glass pours. Optimization of this process, however, is still needed. Process parameters which will be optimized include drain tube temperature and timing of cooling air flows during initiation and stoppage of glass pouring.
- The off-gas HEME and HEPA plugged several times during the tests. It is recommended that further testing be done to solve this problem. This testing should include possible reconfiguration of the off-gas piping arrangement near these filters to determine if pluggage can be minimized solely by mechanical means.

REFERENCES

[1]R.F. Schumacher, W.G. Ramsey, F.M. Johnson, T.M. Jones, D.H. Miller, and B.J. Hardy, "Development of a Remote Bushing System for Americium-Curium Vitrification"; pp. 3-12 in *Ceramic Society Transactions Vol. 87 - Environmental Issues and Waste Management Technologies in the Ceramics and Nuclear Industries III,* Edited by D. K. Peeler and J. C. Marra, American Ceramic Society, Westerville, OH, 1998.

[2]T.M. Jones, B.J. Hardy, and M.E. Smith, "Americium/Curium Bushing Melter Drain Tests (U)," WSRC-MS-97-0168, presented at the *21st Annual Actinide Separations Conference*, Charleston, SC, June 26, 1997.

AMERICIUM/CURIUM VITRIFICATION PILOT TESTS - PART II

J. E. Marra,* M. A. Baich, A. P. Fellinger, B. J. Hardy, T. M. Jones, C. B. Miller, D. H. Miller, D. K. Peeler, T. K. Snyder, M. E. Stone, J. C. Whitehouse, and D. C. Witt
Westinghouse Savannah River Company
Savannah River Technology Center
Aiken, SC 29808

ABSTRACT

Isotopes of americium (Am) and curium (Cm) were produced in the past at the Savannah River Site (SRS) for research, medical, and radiological applications. These highly radioactive and valuable isotopes have been stored in an SRS reprocessing facility for a number of years. Vitrification of this solution will allow the material to be more safely stored until it is transported to the DOE Oak Ridge Reservation for use in research and medical applications. A previous paper described operating results from the Am-Cm Melter 2A pilot system, a full-scale non-radioactive pilot facility. This paper presents the results from continued testing in the Pilot Facility and also describes efforts taken to look at alternative vitrification process operations and flowsheets designed to address the problems observed during Melter 2A pilot testing.

INTRODUCTION

Approximately 15,000 liters (3,600 gallons) of solution containing isotopes of americium (Am) and curium (Cm) are currently stored in F-Canyon Tank 17.1. These isotopes were recovered during plutonium-242 production campaigns in the mid- and late-1970's. The continued storage of these isotopes was identified as an item of primary concern in the Defense Nuclear Facility Safety Board's (DNFSB) Recommendation 94-1. Currently, there are no existing SRS facilities that can be used to stabilize this material for safe interim storage and transportation to the heavy isotopes program at the Oak Ridge National Laboratory (ORNL). An analysis of several alternatives resulted in the recommendation to stabilize the Am-Cm solution in a high-lanthanide glass. The Multi-Purpose Processing Facility (MPPF) in the F-Canyon at SRS will be used for the vitrification process. Pretreatment operations will be performed in existing canyon vessels to separate actinides and lanthanides from other impurities (primarily iron, aluminum, and sodium) subsequent to the vitrification operation.

In 1994, a study was performed to determine the most attractive disposition option for the solution of Am and Cm that is currently stored in the F-Canyon. Vitrification in the form of a high-lanthanide glass was selected as the most attractive option. The Savannah River Technology Center (SRTC) was charged with developing the vitrification flowsheet, designing the prototype equipment, and demonstrating integrated system operations. The vitrification operation was originally planned to be performed in a platinum-rhodium (Pt-Rh) bushing melter that will be fed an Am-Cm liquid-frit combination. A Pt-Rh bushing melter was chosen because of its small size and its shape (which could be needed for future criticality requirements if other radioisotopes such as plutonium were subsequently vitrified in the same system). Platinum-rhodium bushing melters are commonly used in the glass industry, especially in the manufacture of fiberglass. In these industrial applications, however, either molten glass or dry feed is delivered to the bushing. There has never been a bushing melter that was designed for the additional requirements of slurry feeding. Slurry feeding greatly increases the level of difficulty for glass melting, as higher power requirements are needed at the upper portion of the bushing. An off-gas system is also required to treat any volatile or entrained components coming out of the melter. In addition, the numerous remote operation requirements for this radioactive feed had never been attempted before. The SRTC effort has demonstrated proof of concept with over 250 hours of integrated melter system operation. Significant progress has been made on the development program, however, several technical issues have arisen that require additional development in order to ensure confidence for reliable remote operation. These issues deal primarily with demonstrating reliability of the feed system and with design and operation of the bushing melter. A previous paper[1] discussed initial development activities in the Am-Cm Melter 2A bushing melter pilot system. This system was installed and operated at SRS in 1997. As a result of the issues associated with remote operation, a decision was made in early 1998 to focus the development on a batch process operated in an induction-heated vessel. This paper discusses the development of the batch process and compares operation in the induction melter to the original bushing melter system.

DEVELOPMENT AND OPERATING ISSUES

Bushing Melter System

The Melter 2A system is a full-scale pilot plant of the actual system to be installed at the SRS. The basic operation involves feeding a surrogate Am-Cm nitric acid feed solution, along with glass forming frit, in a carefully controlled mass flow ratio into the vapor space of a resistance heated Pt-Rh bushing melter operating at approximately 1350-1450°C (depending on the glass composition used). The glass flows from the bottom of the melter by gravity through a drain tube and into a stainless steel canister. The canister will then be sealed for later transport to ORNL.

As the melter is fed and the liquid is vaporized, the expanding gases are drawn into the off-gas treatment system through a film cooler by a steam jet. The steam jet, coupled with a control air purge that is introduced into the off-gas system between the film cooler and the steam jet, maintains a target melter plenum pressure that is negative with respect to ambient pressure. The vapor exiting the steam jet enters a quencher where it is cooled and scrubbed of particulate entrained into the off-gas system during melter feeding. The condensate drains to the off-gas Condensate Overflow Tank (COT). Overflow from the COT goes into a 55-gallon bounce tank. The vapor exits the top of the COT and is drawn through a high efficiency mist eliminator (HEME), a high efficiency particulate air (HEPA) filter pre-heater and a HEPA by an air jet. The pre-heater raises the off-gas temperature above the dew point prior to entering the HEPA filter to prevent condensation in the HEPA. The air jet discharges the treated off-gas to the atmosphere through a stack. A schematic diagram of the original bushing melter vitrification system is shown in Figure 1.

Significant achievements were made in the development of the bushing melter system, however, the R&D effort has identified systems and components requiring further development prior to initiating final design for reliable and remote operation in the MPPF. Many of these issues were discussed in the preceding paper,[1] that described early pilot facility experience with the Melter 2A system. Additional work was performed in late-1997 to address the issues identified during the early testing. While progress was made during this phase of testing, significant issues remained. These issues are summarized below.

- Feed System
  Reliable performance of this system has not been achieved in the Pilot Facility. In order to achieve the low flow rates (2 liter/hr), while maintaining acceptable melting rates, the liquid feed system contained small orifices and small diameter tubing (0.05"). This system has been susceptible to plugging and required frequent mechanical cleaning, a procedure that will be difficult to adapt to the MPPF environment. Addressing these concerns was likely to require substantial modifications to existing components. Evaluation of alternative technologies was likely to significantly increase the scope of the development program. The

preferred alternative technology, slurry feeding, was greatly complicated by the low feed rate requirement.

- Melter System
  Splatter and deposition observed in the melter plenum in the Pilot Facility was greater than that expected. As the liquid was fed into the melter, a significant amount was ejected into the plenum area of the melter (note: the surface of the melt pool was approximately 3" from the off-gas entrance under normal operating conditions). This liquid deposited in the cooler areas of the plenum where the rare-earth nitrates would decompose to form rare-earth oxides. These oxides were not melted in the cool temperature regions of the plenum. On several occasions, this splatter was sufficient to plug the off-gas system and the liquid feed tube.

  Additionally, both Melters 2A and 2B (which were of identical design) failed prematurely. In both cases, the failure occurred at the base of the lower electrical connection 'ears' indicating a structural deficiency in the design of Melter 2A and 2B. A newly designed melter was procured which addressed some of the structural design issues of Melters 2A and 2B, however, the failure of these melters raised concerns regarding the long-term structural integrity of the bushing in its current slab-like geometrical configuration.

- Off-Gas System
  Splatter and plugging problems were also observed in the off-gas system. The earlier paper[1] discussed plugging of the HEME and HEPA filter and the off-gas sampling performed to better understand this phenomenon. A redesign of the system to provide increased scrubbing of particulates was proposed. A more significant issue was the splatter of material onto the entrance of the off-gas system in the area of the film cooler that resulted in plugging of the film cooler and subsequent pressurization of the melter system. Pressurization of the melter is undesirable due to the potential to contaminate clean areas of the MPPF.

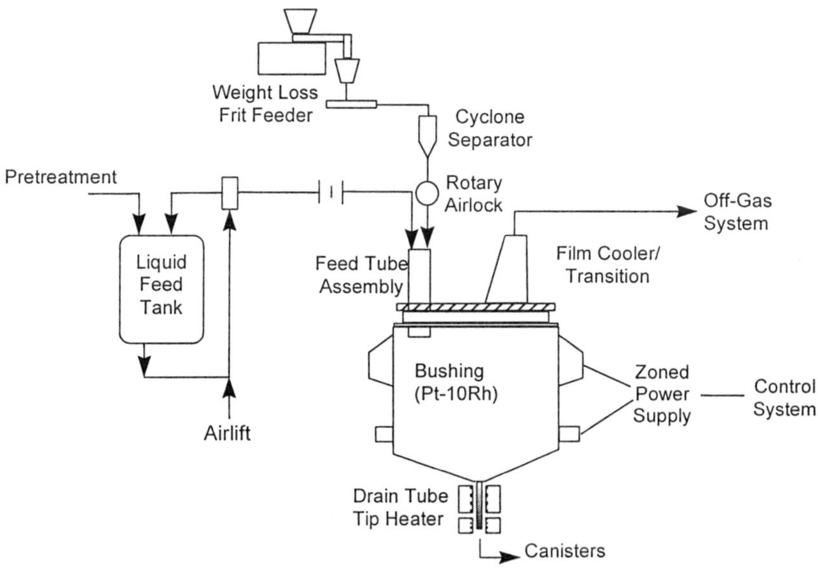

Figure 1 - Schematic Diagram of Bushing Melter System

The combination of the splatter/deposition issue and the melter life concerns was likely to lead to a significant redesign of the current bushing system. Additionally, some of the earlier requirements that led to the design of the bushing melter system were relaxed. For example, the requirement that the equipment be of critically safe geometry (which led to the use of the slab bushing melter) was removed by demonstrating that the Am-Cm solution was infinitely safe. The other primary requirement that led to the bushing system design was the desire to achieve fairly high production rates (several kilograms per hour) in the system. This led to the current liquid feed system, which was originally designed for a continuous feed and pour system. The throughput requirement resulted from the potential to utilize this system for production campaigns other than Am-Cm where significant quantities of glass will be produced. Since the Am-Cm campaign will only generate approximately 330 kg of glass, the throughput requirement was removed.

The problems encountered during the development of the bushing melter system, coupled with removal of some of the original processing requirements, made it prudent to evaluate alternative processes.

Batch Melter Process

A batch melter process is currently under investigation as an alternative to the bushing melter system. Flowsheet development for this process is underway. Additionally, a pilot facility is under construction and will be operational in early-1998. In this process, the nitric acid-based Am-Cm solution or surrogate is transferred into a reaction vessel. Eight-weight percent (8 wt.) oxalic acid is added to the solution to precipitate the rare earths as oxalates. The precipitate is allowed to settle and the excess liquid is decanted. A dilute oxalic acid wash solution (0.10M) is added to remove excess contaminants (e.g., Fe, Cr) and reduce the acidity of the feed. The precipitate is again allowed to settle and the washwater is decanted. The resultant slurry is mixed and then gravity fed to a cylindrical Pt-Rh vessel that is induction heated. The proper amount of frit is added to result in a final glass composition containing 35 weight percent of the Am-Cm solution. Depending on the frit particle size used, the frit addition can be made to the precipitation vessel prior to the transfer to the vessel or can be preloaded in the melter vessel. A simplified schematic of the batch process equipment is shown in Figure 2.

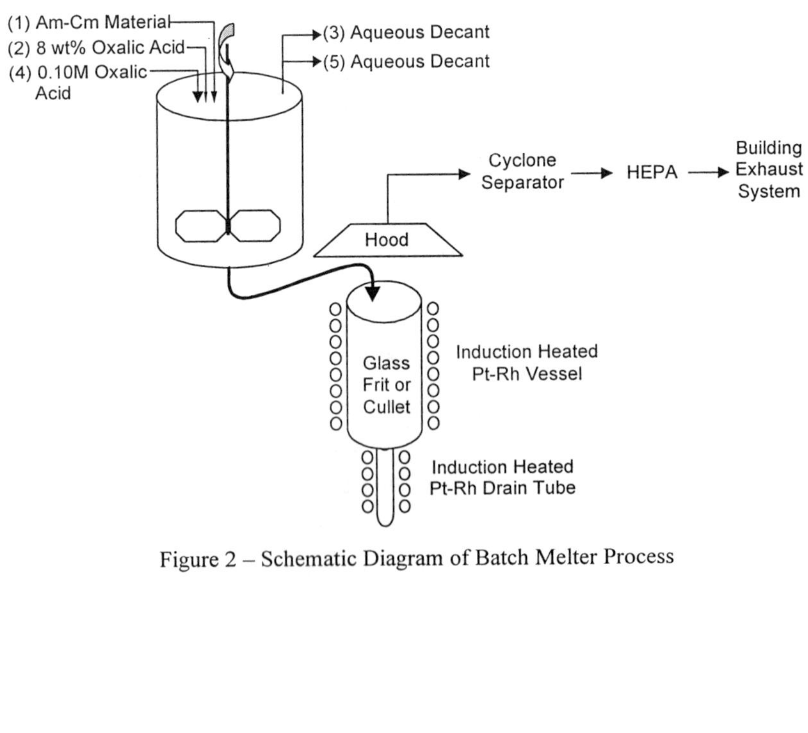

Figure 2 – Schematic Diagram of Batch Melter Process

The oxalate precipitate/frit slurry is added to the melter vessel while holding at a between 100 and 150°C). This temperature is held until the bed has dried. The vessel temperature is then raised at rate of ~4 to 8°C /min to a glass melting temperature of ~1350-1380°C. During this temperature increase, the remaining nitrate is evolved and the oxalate precipitate decomposes to the associated rare-earth oxides. The melt is held at temperature for approximately 1-4 hours to ensure adequate mixing of frit and oxides and is drained from the bottom of the melter by gravity through a drain tube and into a stainless steel canister. The vessel is sized so that a single charge fills a single canister. The canister will then be sealed for later transport to ORNL.

As discussed earlier, efforts are underway to develop a batch processing operation performed in a cylindrical induction-heated melter. This work has been ongoing since November 1997. To date, efforts have focused on developing a process flowsheet and installing a prototypical melter system in the pilot facility. The batch melter system and process addresses many of the issues associated with the bushing melter and greatly simplifies the process, making it desirable for remote operation. A summary of the advantages of the batch melter system is provided in Table 1.

Table 1. Comparison of Slab Bushing Melter Problems/Issues to Batch Melter Potential

System/Equipment/ Process	Slab Bushing Melter	Batch Cylindrical Induction Melter
Melter System	• Melter Life	• Less Stress (Unrestrained, Cylindrical) • Easier/Inexpensive to Replace Vessel
Melter Feed System	• Challenging Low Liquid Feed Rate • System Pluggage	• Batch Process - Higher Flow Rate • Reduced System Pluggage
Off-Gas System	• Film Cooler Entrance Pluggage • Complex Control Scheme	• Simple, Hood Design • 'Decoupled' Offgas System
Melter Pour System	• Drain Tube Pluggage (Devitrification) • Drips & Stringers	• Melt & Drain -- Single Canister – Less Time in Devitrification Zone • Mechanical Drain Valve Possible • Option to Operate as 'In-Can' Melter
Primary Disadvantages/Issues	• *Melter Life* • *Feed Pluggage* • *Off-Gas Pluggage* • *Drain Tube Pluggage*	• *Increased Manipulator Handling* • *Slightly Longer Campaign* • *Potential Slurry Feed/Dry Feed Development*

Development of the batch process and associated equipment is being accomplished by a combination of lab-scale tests and pilot system testing. Lab-scale tests are accomplished in Pt-Rh crucibles in muffle furnaces. Pilot system testing is being accomplished in a cylindrical, resistance-heated vessel that was installed to develop glass drain tube operations (known as the Drain Tube Test Stand, DTTS). This facility is representative of the Cylindrical Induction Melter (CIM) being installed; the primary difference being in the heating mode (resistance heating vs. induction heating). The lab-scale and DTTS tests are being used to refine the flowsheet in preparation for testing in the CIM facility which will qualify equipment for eventual installation in the MPPF.

FLOWSHEET DEVELOPMENT

Significant progress has been made on flowsheet development since beginning this effort. Initial testing has demonstrated the ability to produce glass that meets acceptance standards. The primary issues encountered to date relate to controlling decomposition of the oxalate during the heating cycle. The decomposition reaction generates CO and CO_2 that can result in expansion of the powder bed. Volume expansion is undesirable because it may limit the amount of glass that can be produced in a single batch; the batch height will have to be controlled so that the material is contained within the Pt-Rh vessel at all times. Additionally, initial testing resulted in the formation of bubbles in the glass melt at temperatures in excess of 1200°C. The cause of these bubbles remains under investigation, however, preliminary analysis of the evolved gases indicates that the composition is rich in O_2 and CO_2.

Two possible sources of gas generation at these high temperatures are:

(1) Thermal reduction of CeO2 (or other multivalent oxides), and/or
(2) Decomposition of carbonates.

CeO_2 is known to reduce at high temperatures in a lanthanide borosilicate glass. Recent studies at PNNL indicate an 85% reduction of CeO_2 (+4 to the +3 state) at 1300°C. With this reduction, O_2 would be liberated. Given that CO_2 is one of the high temperature gases detected by the GC/MS during the quartz crucible runs, one possible source of the CO_2 could be the interaction of O_2 liberated from the CeO_2 reduction coupled with residual carbon from oxalate decomposition. Tests are ongoing to confirm this hypothesis. Additional details of the characterization of the melt dynamics in this system will be described in a later paper.

Decomposition of "high temperature" carbonates may provide a second potential source of CO_2. X-ray diffraction patterns of oxalate precipitate calcined at 500°C indicated that presence of "rare-earth" carbonates. Based on this data, one possible mechanism would be the conversion of the rare-earth oxalates to carbonates and ultimate thermal decomposition of the carbonates to oxides liberating CO_2. XRD patterns of the oxalate after calcination at 1300°C for 2 hours indicate only rare earth oxides.

The original glass formulation selected for stabilizing the Am-Cm solution was based on a Loffler glass composition. This glass was chosen based on the fact that the base glass composition is high in lanthanides (La_2O_3) making it a logical choice to stabilize lanthanide materials. The general formulation for this glass (known as Frit 2000) is shown in Table 2.

Frit 2000 met the initial requirements placed on the glass composition as shown below:

- Able to incorporate up to 40 weight percent Am-Cm material,
- Viscosity in the range of 5-15 poise at operating temperature (necessary to pour through the drain tube configuration), and
- Maintain recoverability of the Am-Cm material when dissolved in HNO_3.

However, Frit 2000 required operation at temperatures in excess of 1450°C. After one of the early Am-Cm bushing melters failed due to overtemperature, there was a desire to evaluate frits with lower liquidus/melting temperatures. A strontium aluminum borosilicate glass was identified as a potential candidate. The composition of this glass (known as 25SrABS-F) is also shown in Table 2. Based on successful testing of the 25SrABS-F glass in the laboratory, it was selected as the baseline composition for initial evaluation in bushing Melters 2A and 2B. However, during early testing of Melter 2A, severe foaming was encountered when the melter began to be fed with

Table 2. Simplified Formulations for Glasses Considered for Am-Cm Stabilization

Oxide	Frit 2000	25SrABS-F
SiO_2	44.2	33.7
B_2O_3	9.20	13.5
Al_2O_3	6.05	24.9
La_2O_3	7.12	25.0
PbO	24.38	---
BaO	9.02	---
SrO	---	2.9

a nitric acid solution and SrABS-F. Subsequent testing focused on developing an understanding of the cause for the foaming and compositionally characterizing the foam layer using the reference frit, 25SrABS-F.

Analysis of samples of the foamy, porous cold cap formed by feeding the 25SrABS-F with liquid nitric acid surrogate feed indicated the presence of rare earth oxides. Mass spectrometry analysis of gas trapped in the pores of the cold cap samples detected nitrogen and oxygen. The foaming mechanism that was postulated from this information was the decomposition of the rare earth nitrates. This decomposition liberates NO_x gas, which is trapped in the viscous cold cap. This foam layer then acts as an insulating blanket, which in turn reduces the melting efficiency of the bushing. If feeding is continued, the bushing cannot melt the additional feed and the entire melt pool becomes covered with a cold cap. This theory is backed by the finding that the addition of just nitric acid to an SrABS glass melt pool produced a thin, non-porous oxide layer due to the extraction of the lanthanides from the glass. This extraction first converts the rare earths to a nitrate form and then quickly decomposes these nitrates to an oxide layer on the glass surface. This decomposition results in the liberation of gases, which then attempt to rise out of the melt pool/cold cap. Without a cold cap, however, there was no viscous layer to trap these gases, and therefore these rare earth oxides were re-dissolved into the glass fairly rapidly.

Based on the less than adequate results with the 25SrABS-F in Melter 2A, a decision was made to reevaluate the use of Frit 2000 in the bushing melter system. This frit composition did not produce a porous, cold cap, but the lead in the composition was significantly volatile at the melter operating temperature (1450°C) and caused plugging in the off-gas system. Modifications were being made to the off-gas system to accommodate the increased volatility when other processing problems led to the decision to abandon the slab-bushing melter system.

Due to acceptable processing experience with the Frit 2000 in Melters 2A and 2B, it was selected as the 'reference' frit composition for initial testing of the batch processing flowsheet. Early testing on this flowsheet suggested, however, that the Frit 2000/surrogate system was subject to rapid devitrification in the drain tube of the Pt-Rh vessel. The drain tube functions as a 'freeze-valve' to hold the glass in place by application of forced cooling air on the tip of the tube. As a result, there exists a region where the glass in the drain tube is below the liquid temperature of the glass. Initiation of the glass pour is accomplished by interrupting the cooling airflow allowing the tube to heat up. As the tube heats up, the glass melts and flow begins. If there is a region of substantial devitrification, it is difficult to initiate flow. This was observed when Frit 2000/surrogate glass was held in the melter for periods in excess of 24 hours. Laboratory testing indicated that complete devitrification occurred after periods of 24 hours in the B2000/surrogate system. At the same time, it was recognized that the change in process flowsheet made it possible to reexamine the use of the SrABS glass composition; the new flowsheet removes much of the NO_3^- which was shown to produce unacceptable foaming during the melting operation. Testing of this system had shown that its crystallization properties are more favorable to operation of the drain tube than the B2000/surrogate composition. Testing to date has shown acceptable performance with the SrABS glass composition in the batch flowsheet; it is now the 'reference glass' composition.

SUMMARY

Significant progress has been made in the development of a vitrification process and equipment for stabilization of the Am-Cm solutions at the Savannah River Site. Equipment problems, primarily associated with developing the confidence for operation in a remote environment, have resulted in programmatic changes from a semi-continuous process in a slab Pt-Rh bushing melter to a batch process in a cylindrical induction heated melter. Testing and development of the batch process and equipment is in progress. Initial results described in this paper suggest that this process variation is more readily adaptable to remote processing environments.

ACKNOWLEDGEMENTS

This paper was prepared in connection with work done under Contract No. DE-AC09-96SR18500 with the U.S. Department of Energy.

REFERENCES

[1]M. E. Smith, A. P. Fellinger, T. M. Jones, C. B. Miller, D. H. Miller, T. K. Snyder, M. E. Stone, and D. C. Witt, "Americium/Curium Melter 2A Pilot Tests (U)," Submitted for publication in the *Proceedings of the International Symposium on Waste Management Technologies in Ceramic and Nuclear Industries,* The 100th Annual Meeting of The American Ceramic Society, Cincinnati, Ohio, May 3-6, 1998.

GLASS FRIT DEVELOPMENT FOR ENCAPSULATION OF ROCKY FLATS ASH

J. D. Vienna, J. Luey, P. Hrma, H. Li, D. E. Smith, R. D. Scheele, H. L. Wigent, and J. V. Crum, Pacific Northwest National Laboratory, Battelle Boulevard, PO Box 999, Richland, WA 99352

ABSTRACT

A glass frit composition has been optimized for encapsulation of plutonium-bearing ash stored at Rocky Flats Environmental Technology Site (RFETS). The frit composition was developed with existing glass property data and tested with simulated and real RFETS ash. The frit properties of interest are the low- and high-temperature viscosity and sinterability. The chemical durability and resistance to plutonium recoverability of the encapsulated waste form are also of interest. This paper summarizes the frit development and the densification and recoverability results of simulant and radioactive testing of the waste form. The process and waste form are adequate for treatment of ash residues at RFETS.

INTRODUCTION

Roughly 20 Mg of ash residue is stored at Rocky Flats Environmental Technology Site (RFETS). This plutonium (Pu)-bearing ash must be treated for safe storage and disposal in the Waste Isolation Pilot Plant (WIPP). A set of criteria for the final treated waste forms was developed by RFETS to address concerns of safe storage [1], safeguards termination [2], WIPP waste acceptance [3], and Department of Transportation regulations. The ash must be treated by 2002 in existing RFETS facilities. To meet this aggressive schedule, equipment modifications, cycle times, and process upsets must be minimized. Vitrification was chosen as the baseline technology for ash treatment because of the increasing probability that it can economically meet the product criteria within the process schedule.

The vitrification technology discussed in this paper is a low-temperature agglomeration/ encapsulation process being developed by Pacific Northwest National Laboratory (PNNL) as part of a multi-site team composed of staff from PNNL, Los Alamos National Laboratory (LANL), and Safe Sites of Colorado (SSOC) at RFETS. By this process, ash will be mixed with glass frit and the mixture will be sintered to form a solid waste material. Appropriate proportions of frit and ash will be used to maintain adequate concentrations of Pu for termination of safeguards while ensuring that 200 fissile gram equivalent (FGE) (minus twice the measurement uncertainty) is loaded in each transuranic (TRU) waste drum. The full-size primary container for the waste form will be a 0.178 m (7 in.) diameter by 0.216 m (8.5 in.) tall, steel, Vollrath can. The waste form will be batch-vitrified in a muffle furnace for 4 h, at or below 800°C. Two cans will be sealed in a Rocky Flats pipe component stored for eventual disposal in WIPP.

Ash was generated by the incineration of plutonium-bearing organics at RFETS. The furnaces were small (for safety) and operated at low temperatures (nominally 200°C to 900°C) to facilitate Pu recovery processes. This incineration process led to incomplete combustion of the organics. The composition range frequently reported by RFETS for these ash materials is given in Table I. Although reported analyses of ashes often lie outside of this range, it is speculated that this range is applicable to most of the ash in storage. Ash is highly variable in composition with nearly an order of magnitude variation in SiO_2, B_2O_3, Fe_2O_3, and Na_2O and half an order of magnitude variation in Al_2O_3, CaO, MgO, and C. The carbon/organic content is high, ranging up to 36 mass %.

Table I. Rocky Flats ash composition range (mass % reported as oxides, C, Cl, and F) [4]

	Min.	Max.	Ave.		Min.	Max.	Ave.		Min.	Max.	Ave.
Al_2O_3	0.95	5.7	3.33	K_2O	0.24	1.2	0.72	SnO	0.0	0.25	0.13
Am_2O_3	0.02	0.02	0.02	MgO	0.83	8.3	4.57	Ta_2O_5	0.0	0.73	0.37
B_2O_3	0.32	3.2	1.76	MnO_2	0.03	0.08	0.06	TiO_2	1.0	1.7	1.35
BaO	0.58	1.2	0.89	Na_2O	0.0	2.4	1.2	C	7.5	36.0	21.75
CaO	1.1	7.0	4.05	NiO	0.25	0.64	0.45	SiO_2	14.17	74.10	48.49
Cr_2O_3	0.44	0.88	0.66	P_2O_5	0.23	0.23	0.23	F^-	0.04	1.8	0.6
CuO	0.63	1.3	0.97	PbO	0.58	0.92	0.75	Cl^-	0.8	6.4	2.7
Fe_2O_3	1.1	10.3	5.7	PuO_2	1.8	3.8	2.8				

Extensive analyses of the chemical and physical forms of the ashes were conducted by Behrens et al. [5]. The analyzed ash particles were typically 10 nm to ~1 μm in diameter. Plutonium was found in the oxide, large glassy silicate, and alumino-silicate phases. Other phases identified in the ash samples included $SiTiO_x$, $CaAlSiO_x$, SiO_2, and $CaMgSiO_x$.

Decomposition of organics and loss of water at low temperatures (<250°C) and oxidation of carbon at higher temperatures (400°C to 600°C) can cause foaming while processing with viscous glass. The small particle size of ash assists in ash-frit surface reaction and inhibiting frit contact, further complicating the encapsulation process. Hence, the impacts of glass flow, particle size of ash and frit, and off-gas from ash are the primary concerns in frit development.

WASTE FORM REQUIREMENTS

To develop a frit for encapsulation of ash, primary consideration is given to glass viscosity (η). A frit used for encapsulation must sinter at temperatures below 800°C to be used in the RFETS facilities. The densification and removal of porosity during sintering of the frit/ash combinations can be monitored via dimensional changes of the mixture, e.g., $\Delta L/L_0$, where L is the length of one dimension. For initial-stage sintering by a viscous flow mechanism,

$$-\frac{\Delta L}{L_0} = \frac{3\gamma t}{8a\eta}, \tag{1}$$

where γ is the surface tension of glass, t is the heat treatment time, and a is the glass particle size [6]. As the porosity decreases and pores become isolated during the intermediate and final stages of sintering, the kinetics become more complex. This relationship is used only as a first approximation.

For a given sintering temperature, the frit/ash densification rate can most easily be increased by decreasing the frit viscosity and/or the particle size. In general, closure of open, continuous porosity (i.e., encapsulation) occurs at ≅ 90% theoretical density. Assuming an initial in-can density of ≅ 30% (a reasonable tap density), $\Delta L/L_0$ would have to obtain a value of ≅ 0.3 to reach a final density of 90%. For silicate glasses, γ is roughly 0.3 N/m. For a first approximation, using Equation (1) and assuming that t = 1 h with a target $\Delta L/L_0$ of 0.3, the sintering viscosity would have to be 1.8×10^6 Pa·s for a = 750 μm, and 1.8×10^7 Pa·s for a = 75 μm.

Two additional factors must be considered while discussing glass viscosity. First, reactions of glass with waste may alter the effective η, a, and γ, thus influencing sintering. Second, gas generation may blow glass into foam when the porosity is closed. Gas can be generated by decomposition of organic compounds, atmospheric oxidation of carbon, or redox reactions within the waste. Figure 1 shows a thermogravometric analysis (TGA) from a typical dried RFETS ash. There are two areas of high mass-loss rates depicted (differential thermogravimetric (DTG) trace). The DTG peak from 180°C to 300°C is due to loss of water and decomposition of residual organic material; the peak above 800°C is due to the decomposition of carbonates [8]. In some samples, mass-loss in the range of 400°C to nearly 600°C was identified and associated with the oxidation of residual carbon [8, 9]. For successful encapsulation, the temperature at which foaming occurs must not coincide with the temperature range over which the glass sinters. As the upper processing temperature was defined at 800°C [9, 10] and the off-gas generation, due to incomplete calcination, was found to cause foaming at temperatures below roughly 600°C, the sintering temperature range was defined at 600°C to 800°C. The temperature history of ash and frit mixtures in a full-scale can were measured [9, 10]. Figure 2 shows that 2.5 h of heat treatment at 800°C are required to reach a temperature of 600°C in the center of a typical can.

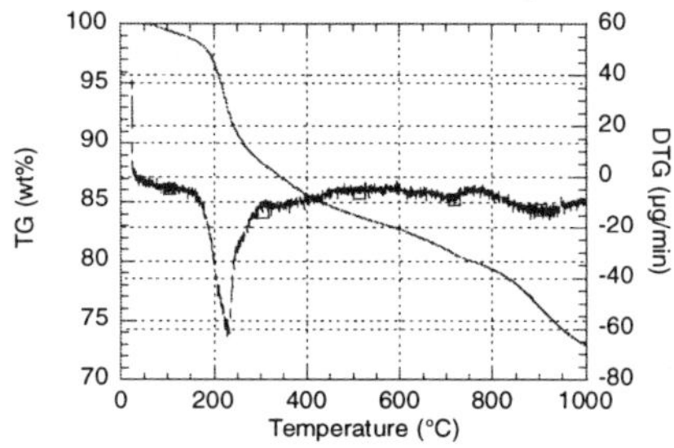

Figure 1. TGA of typical RFETS ash (BASH1H-1).

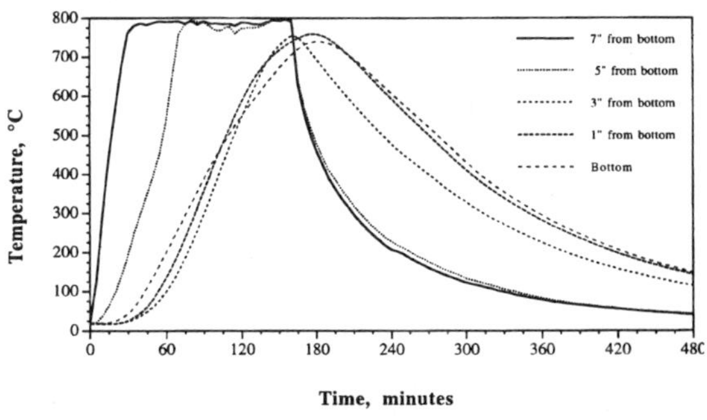

Figure 2. Center-line temperature of 40:60 mixture of ash:frit in a full-scale Vollrathcan heated at 800°C as functions of time and height in the can.

With a heat-treatment cycle-time restriction of ≤ 4 h [10], the densification will have to occur in roughly 1 h at 600°C. To avoid foaming the frit should not seal and trap gas at temperatures below roughly 600°C. We define a frit viscosity criteria such that the time required to achieve a $\Delta L/L_0$ of -0.3 is ≤ 1 h at 600°C and ≥ 3.75 h (maximum time at temperature) at 575°C (see Figure 2). With Equation (1) as a first approximation, the calculated viscosities are ≤ 1.80×10^6 Pa·s at 600°C and ≥ 6.75×10^7 Pa·s at 575°C, assuming that a = 75 μm and γ = 0.3 N/m.

To meet safeguards termination limits with up to 5 mass % Pu in the final waste form, the glass must be resistant to recoverability of Pu from the waste form via acid digestion. A standard Pu recoverability test, developed at RFETS, was used to test in this study [11]. A 20 g sample of waste form was crushed to less than 9.5 mm and digested in 100 mL boiling 9 M HNO_3 with 0.25 moles CaF_2 (excess HF) for 30 min. Less than 15 g Pu / kg of waste form must be solubilized under these conditions for the waste form to pass. The glass frit must be highly resistant to acid attack in order to pass this test.

FORMULATION APPROACH

A glass frit composition must be developed with viscosities ≤ 1.80×10^6 Pa·s at 600°C and ≥ 6.75×10^7 Pa·s at 575°C. To calculate glass compositions which meet the viscosity requirements, a model developed to predict the viscosity of high-level waste glasses was used [7],

$$\ln\eta = \sum_i g_i A_i + \frac{\sum_i g_i B_i}{T - \sum_i g_i T_i}, \tag{2}$$

where g_i is the ith component mass fraction in glass, T is temperature, and A_i, B_i, and T_i are the ith component viscosity coefficients, listed in Table II. This model allows several degrees of freedom in glass composition while meeting the viscosity requirements. These degrees of freedom will be used to develop a glass with high acid leach resistance.

Table II. Component viscosity coefficients [7]

I	SiO_2	B_2O_3	Na_2O	Li_2O	CaO	MgO	Fe_2O_3	Al_2O_3	ZrO_2	Others
A_I	-10.59	-24.41	2.02	5.46	3.95	5.31	-13.63	1.50	-0.36	-2.38
B_I	19,236	15,923	-12,965	-39,177	-18,672	-11,944	14,559	9,524	4,618	1,710
T_I	76.11	263.48	425.72	474.43	1,065.83	752.24	43.64	178.53	540.51	270.74

Although no directly applicable model was found to predict the response of glass to the aggressive leaching test described above, a great deal of knowledge about component effects on acid leach resistance was available. Aluminum oxide is known to greatly increase chemical durability of alkali-silicate glasses [12-14]. This effect is strongest at low Al_2O_3 concentrations and decreases at concentrations near 5 mass % (depending on glass composition) [12,13]. In an acid leaching environment, Al_2O_3 may even decrease leach resistance at higher concentrations [14]. An optimum glass would have roughly 5 mass % Al_2O_3. Zirconia is also known to strongly increase the acid resistance of glass. Like Al_2O_3, the effect of ZrO_2 is strongest at low concentrations and decreases at higher concentrations [14]. To have high durability and meet the recoverability requirement, 5 mass % of ZrO_2 is added to the frit. A combination of B_2O_3, Na_2O, and Li_2O are used as fluxes to maintain adequate η while minimizing impacts on durability. The proposed sodium-borosilicate (NBS) frit composition, 60 mass % SiO_2, 10 mass % B_2O_3, 15 mass % Na_2O, 5 mass % Li_2O, 5 mass % Al_2O_3, and 5 mass % ZrO_2, has predicted viscosity values of 6.75×10^7 Pa·s at 575°C and 1.34×10^7 Pa·s at 600°C. Additionally, the glass is expected to be highly resistant to acid leaching.

RESULTS AND DISCUSSION

Frit viscosity, a primary formulation concern, was measured according to standard procedures with a beam bending viscometer at low temperatures and a spindle viscometer at high temperatures [7,15]. Figure 3 compares the measured viscosity to that predicted from Equation (2), as functions of temperature. The observed frit viscosity values match very closely to those calculated.

The NBS frit was developed to have high acid leach resistance. The standard Pu recovery test was performed on frit:ash simulant mixtures to test this resistance, with Ce as a Pu surrogate. The samples were heat-treated at 650°C or 700°C for 1 h. Recovery values from the resulting samples ranged from 0.2 to 2.1 g Ce/kg of waste form, which is far below the 15 g Pu/kg standard. Comparing these recovery values with those from untreated ash (40.8 g Ce/kg), phosphate frit composition:ash mixtures (4.5 to 8.0 g Ce/kg), and other borosilicate composition:ash mixtures (6.0 to 13.4 g Ce/kg) demonstrates the acid leach resistance of the NBS frit.

Small crucible tests were performed to check the assumptions on viscosity requirements and understand the impacts of ash loading, composition, and particle size on the encapsulation process. To perform these tests, five different ash simulants were mixed with frits of two size ranges in a range of ash loadings. The densification of the mixture was measured after heat treatments of 1 or 2 hours at 600°C, 800°C, or 900°C. Table III reports the subsidence of frit:ash simulant mixtures.

Figure 4 illustrates the effects of ash loading on subsidence of frit:ash mixtures with different ash simulants (lines were added as a guide for the reader). These data show that at 800°C, subsidence decreases with increased ash loading. Although the trends are consistent, differences can be seen between different ash simulants. The most marked difference is that of ash particle size. Ashes Cc (coarse) and Cf (fine) are identical in composition with particle size ranges 88 μm to 354 μm and 1 μm to 5 μm, respectively. These two represent the largest and smallest particle sizes for the ashes tested and show the largest difference in ash loading effects on subsidence. As may be expected, the coarser ashes show higher subsidence at the same ash loading.

Table III. Subsidence[a] results from small crucible tests with ash simulants

Samples	Ash[b]	T (°C)	t (h)	frit[c]	ash loading (mass %)						
					0	10	20	30	40	50	60
CC81M30-60	Cc	800	1	med	-.	-	-	0.53	0.52	0.35	0.10
CC61F0-20	Cc	600	1	fine	0.15	0.25	0.29	-	-	-	-
CC81F0-60	Cc	800	1	fine	0.63	0.50	0.45	0.53	0.57	0.43	0.26
CC82F30-60	Cc	800	2	fine	-	-	-	0.56	0.56	0.48	0.26
CC91F30-60	Cc	900	1	fine	-	-	-	0.63	0.59	0.58	0.43
CF81M30-60	Cf	800	1	med	-	-	-	0.41	0.13	0.09	0.00
CF81F30-60	Cf	800	1	fine	-	-	-	0.58	0.39	0.27	0.21
CF82F30-60	Cf	800	2	fine	-	-	-	0.57	0.52	0.28	0.14
CF91F30-60	Cf	900	1	fine	-	-	-	0.68	0.63	0.57	0.39
P81F30-60	Poz	800	1	fine	-	-	-	0.43	0.30	0.23	0.10
B81F30-60	Black	800	1	fine	-	-	-	0.34	0.29	0.28	0.23
J81F30-60	J Ash	800	1	fine	-	-	-	0.54	0.37	0.17	0.11

Notes: [a] Subsidence is given by decrease in batch height due to heat treatment divided by initial batch height in crucible. [b] Ash simulant descriptions are reported elsewhere [10], Cc and Cf are course and fine ash simulant C. [c] Frit particle sizes are ≤ 420 μm, ≥ 74 μm for med and ≤ 74 μm for fine.

Figures 5 and 6 show the subsidence as functions of ash loading for coarse (Figure 5) and fine (Figure 6) ash C:frit mixtures (lines are added as a guide to the reader). The tests include samples heated at 800°C and 900°C for 1 or 2 hours, with medium or fine frit. As expected higher subsidence occurs in samples with higher temperature, longer time, or smaller frit. These effects are more pronounced in samples with fine ash.

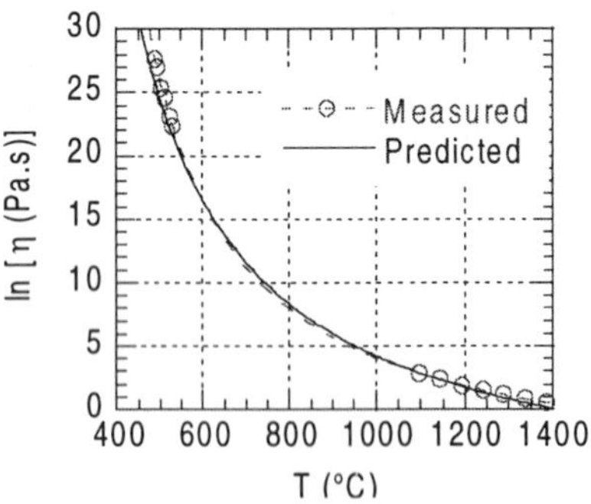

Figure 3. Predicted and measured viscosity of NBS frit as a function of

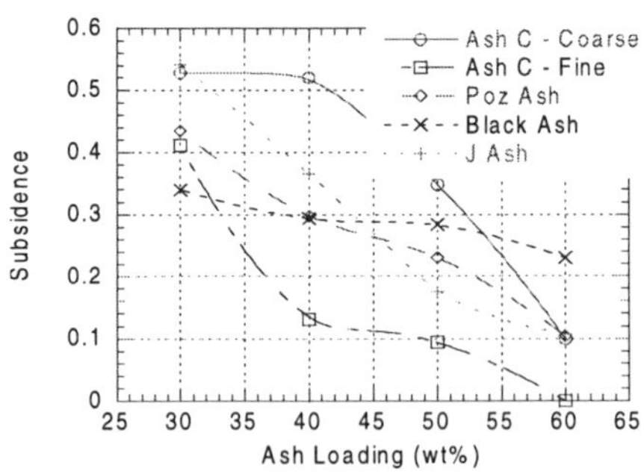

Figure 4. Subsidence after 1 h at 800°C

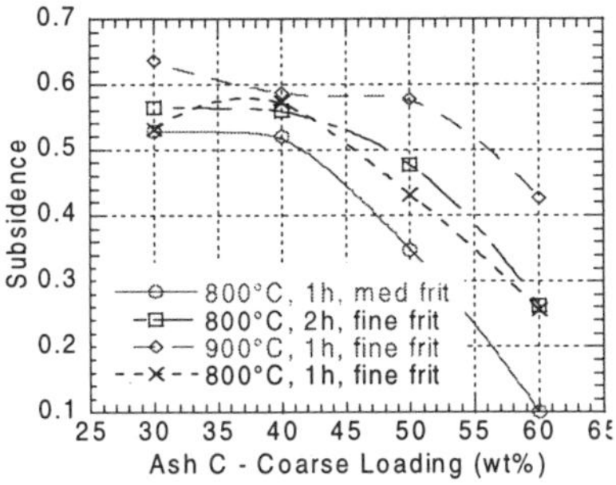

Figure 5. Susidence of ash C-coarse:frit mixtures

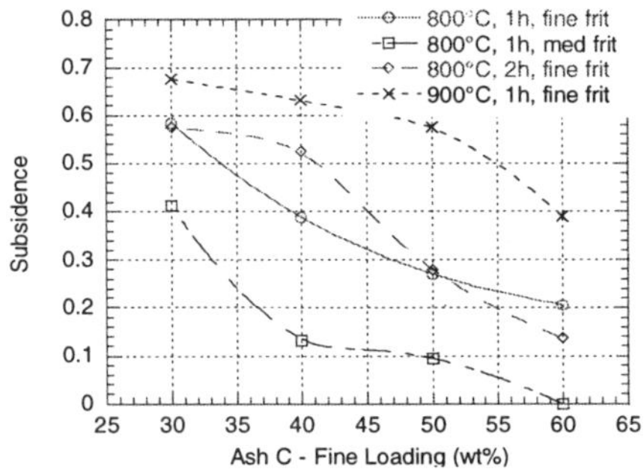

Figure 6. Subsidence of ash C-fine:frit mixtures

Assuming subsidence occurs via a thermally activated process, the activation energies (E_a) were calculated for ash C mixtures. The E_a values increased linearly with ash loading from roughly 15 kJ/mol at ash loading of 30 to 79 kJ/mol at ash loading of 50 for ash Cf (fine) and from roughly 2.5 kJ/mol at ash loading of 40 to 54 kJ/mol at ash loading of 60 for ash Cc (coarse). These values are lower than the activation energy for viscous flow, which is on the order of 250 kJ/mol for the NBS frit at 800°C. These data suggest that viscous flow is not the dominant activated process involved with subsidence of these mixtures at ash loading's of 30 to 60 mass %.

At low ash loadingss (0 to 20 mass %), the samples densified at temperatures as low as 600°C. Low subsidence values shown in Table III are misleading for these samples since radial shrinkage was greater than axial in the conical crucibles. Examination of these samples showed that 1 h at 600°C was adequate to close porosity and obtain > 90% density, as predicted by first approximations. These data suggest that the assumptions regarding viscosity impacts on the sintering process (Equation (1)) are good for high-frit mixtures (ash loading < 30 mass %). At higher ash loading, more sophisticated models are required for glass formulation.

A series of ash simulant samples were fabricated to ensure processability of the ash frit mixture at full-scale. The test parameters and results for the full-scale simulant tests are summarized in Table IV. All full-scale samples were heat treated for 4 hours at 800°C. The ash simulants used in this testing were designed to bound the composition and particle size ranges expected from RFETS ash [10]. These tests confirm the results of small crucible tests and verify the process robustness of this treatment option.

Preliminary tests with radioactive RFETS ash were performed to determine the applicability of non-radioactive surrogate results to real waste, determine Pu recoverability according to the standard test, and prepare samples for waste form characterization. These tests were performed by mixing ashes with fine NBS frit and heating the mixtures in 500 mL steel beakers in a muffle furnace. The current results are summarized in Table V; more studies are being performed and will be reported in a subsequent article. Some mixtures foamed during heat-treatment. The foam subsided by the end of the heat-treatment cycle. The foaming confirms the conclusion of Luey et al. [10] who recommend calcination of the ash before encapsulation. Subsidence and recoverability data are consistent with the non-radioactive test results.

Table IV. Summary of results from full-scale testing with ash surrogates

#	Ash[a]	Frit Size	ash loading	subsidence	recoverability[b]
1	Ash C – coarse	Med	18.47	35.8%	0.788
2	Ash C – coarse	Fine	18.60	41.8%	TBD
3[c]	Ash C – coarse	Med	18.47	34.6%	TBD
4	Ash C – coarse	Fine	40.01	43.9%	0.129
5	Ash B – fine	Med	31.83	N/A	TBD
6	Ash D – fine	Med	28.99	27.1%	TBD
7	Ash A – fine	Fine	19.54	11.6%[d]	2.64
8	U of M	Fine	19.97	16.0%	TBD

Notes: [a] Ash simulant descriptions are reported elsewhere [10]. [b] Recoverability by standard test method in g Ce/kg waste form. [c] Inner container processed inside a larger container. [d] Traces of glass on inside of container indicate 44% expansion at some point during process.

Table V. Summary of results from large-scale testing with radioactive ash

#	Ash[a]	ash loading	g Pu/g glass	total mass	T (°C)	time (h)	subsidence	density (g/cm^3)	recoverability[b]	foam[c]
1-R	RF Ash mix-1	40.9	0.02	450 g	800	1.17	20%	TBD	0.94	++++
2-R	Cal. RF Ash mix-2	40.0	0.01	394 g	760[d]	4.50[d]	0	1.56	2.20	+
3-R	Cal. BASH1H-1	17.6	0.027	390 g	800	2.50	0	2.35	TBD	-
4-R	BASH1H-1	50.0	0.059	130 g	800	2.50	40%	2.60	TBD	++
5-R	BASH1H-1	50.0	0.059	130 g	800	1.00	40%	2.45	TBD	++

Notes: [a] Ash materials are being characterized; RF Ash mixes are mixtures of high-Pu RFETS ash and U of M ash [10], in 30:70 (mix-1) and 12:88 (mix-2) ratios; BASH1H-1 is RFETS ash sent from LANL [5]. [b] Recoverability by standard test method in g Pu/kg waste form. [c] The relative amount of volume expansion is denoted by "+" symbols, a "-" signifies no expansion. [d] The heat-treatment for 2-R was 700°C for 1 h, 0.5 h to 800°C, 800°C for 0.5 h, and 760°C for 4 h.

CONCLUSIONS

A frit, composed of 60 mass % SiO_2, 10 mass % B_2O_3, 15 mass % Na_2O, 5 mass % Li_2O, 5 mass % Al_2O_3, and 5 mass % ZrO_2, was developed for encapsulation of RFETS ash. Formulation was based upon basic glass chemistry knowledge and existing data and models from high-level waste vitrification efforts at Pacific Northwest National Laboratory. A waste form can be fabricated by mixing up to 50 mass % ash with fine NBS frit and heat-treating between 600°C and 800°C for 1 to 4 hours. This final waste form is a monolith, with ash particles encapsulated in a continuous glass phase. The form is highly resistant to leaching with a procedure which subjects the glass to boiling concentrated HNO_3 with HF. The process parameters were developed with non-radioactive ash simulants with varied chemistry and particle sizes. In these laboratory studies the impacts of ash loading, particle size, ash chemistry, and temperature on subsidence, or densification, were measured. Full-scale surrogate tests and preliminary radioactive tests confirm the results of laboratory studies. Foaming occurred in some of the radioactive tests which was attributed to incomplete calcination of the ash. No foaming was observed in samples with completely calcined ash.

REFERENCES

1. "Criteria for Interim Safe Storage of Plutonium-Bearing Solid Materials," Addendum to the Department of Energy Implementation Plan for DNFSB 94-1 (November 1995).
2. "Additional Attractiveness Level E Criteria," Memorandum from E.J. McCallum, DOE NN-51 (July 1996).
3. *Waste Acceptance Criteria for the Waste Isolation Pilot Plant*, Rev. 5, DOE/WIPP-069 (April 1996).
4. J. L. Long, "Analysis of Incinerator Ash & Heel," Internal Rocky Flats Job Report No. PER 1-74, DPA950348 dtd. January 4, 1974.
5. R. G. Behrens, E. C. Buck, N. L. Dietz, J. K. Bates, E. Van Deventer, and D. J. Chaiko, *Characterization of Plutonium-Bearing Wastes by Chemical Analysis and Analytical Electron Microscopy*, ANL-95/35, Argonne National Laboratory, Argonne, IL (September 1995).
6. R.M., *Powder Metallurgy Science*, Metal Powders Industry Federation, Princeton (1984).
7. P. Hrma, G. F. Piepel, M. J. Schweiger, D. E. Smith, D.-S. Kim, P. E. Redgate, J. D. Vienna, C. A. LoPresti, D. B. Simpson, D. K. Peeler, and M. H. Langowski, *Property / Composition Relationships for Hanford High-Level Waste Glasses Melting at 1150°C*, PNL-10359, Vol. 1 and 2, Pacific Northwest Laboratory, Richland, Washington (1994).
8. J. G. Darab, J. D. Vienna, S. T. Autrey, and A. Phillips, Off-Gas from Heat Treatment of Rocky Flats Ash, to be published (1998).
9. J. D. Vienna, H. Li, P. Hrma, J. G. Darab, and J. Luey, *Frit Screening for Rocky Flats Ash and Sand, Slag and Crucible Vitrification*, PNNL-11591, Pacific Northwest National Laboratory, Richland, WA (1997).
10. J. Luey, J. D. Vienna, H. Li, and H. L. Wigent, *Vitrification Process Option for Rocky Flats Incinerator Ash*, PNNL-11770, Pacific Northwest National Laboratory, Richland, WA (1998).
11. T. Burns, Los Alamos National Laboratory, Rocky Flats, personal communication to J. D. Vienna, Pacific Northwest National Laboratory, dtd May 27, 1997.
12. J. D. Vienna, P. Hrma, M. J. Schweiger, and M. H. Langowski, "Compositional Dependence of Elemental Release from HLW Glasses by the Product Consistency Test: A One Component-at-a-Time Study," *Ceramic Transactions* 72, American Ceramic Society, Westerville, OH (1996).
13. X. Feng, P. R. Hrma, J. H. Westsik, Jr., N. R. Brown, M. J. Schweiger, H. Li, J. D. Vienna, G. Chen, G. F. Piepel, D. E. Smith, B. P. McGrail, S. E. Palmer, D. Kim, Y. Peng, W. K. Hahn, A. J. Bakel, W. L. Ebert, D. K. Peeler and C. Chang, Glass Optimization for Vitrification of Hanford Site Low-Level Tank Waste, PNNL-10918, Pacific Northwest National Laboratory, Richland, WA (1996).
14. M. B. Volf, *Chemical Approach to Glass*, Elsevier, New York, NY (1984).
15. R. F. Speyer, *Thermal Analysis of Materials*, Marcel Dekker, New York, NY (1994).

DISTRIBUTION AND SOLUBILITY OF RADIONUCLIDES IN WASTE FORMS FOR DISPOSITION OF PLUTONIUM AND SPENT NUCLEAR FUELS: PRELIMINARY RESULTS

X. Feng, H. Li, L.L. D. L. Li, J.G. Darab, M.J. Schweiger, J.D. Vienna, B.C. Bunker,
Pacific Northwest National Laboratory, Richland, WA 99352
P.G. Allen, J.J. Bucher, I.M. Craig, N.M. Edelstein, and D.K. Shuh, Chemical Sciences Division,
Ernest O. Lawrence Berkeley National Laboratory, Berkeley, CA 94720
R.C. Ewing, and L.M. Wang, Nuclear Engineering and Radiological Sciences, College of Engineering, The University of Michigan, Ann Arbor, MI 48109
E.R. Vance, Australian Nuclear Science and Technology Organization, Menai, NSW2234, Australia

ABSTRACT

Funded by the U.S. Department of Energy (DOE) Environment Management Science Program (EMSP), a multi-institutional research team lead by the Pacific Northwest National Laboratory (PNNL) is currently studying structural aspects of radionuclides, their surrogates, and neutron absorbers in several waste forms (glass, ceramics, and vitreous ceramics) structure and how composition and structure of the waste forms and processing conditions (oxidizing versus reducing conditions) affect their partitioning and solubility limits in these waste forms. In this article, the research methodology is first reviewed. Then, preliminary results are presented, which were obtained from studies of i) the effects of glass composition (in terms of Al_2O_3, B_2O_3, and Na_2O) on solubility limits of HfO_2 and Gd_2O_3, ii) the effects of Pu oxidation states on its solubility limit in borosilicate glasses, and iii) the effects of U oxidation states and composition (in terms of Na_2O) on its solubility limit in soda-lime silicate glasses. The most significant finding from the current study is that Pu^{3+} is shown to be more soluble (by at least 150%) than Pu^{4+} in borosilicate glass.

INTRODUCTION

As a result of the Strategic Arms Reduction Treaties and bilateral offers and agreements between the United States and Russia, the inventory of surplus weapon-grade plutonium (Pu) for disposition is projected to be 350-400 metric tons within the next decade.[1] In addition, more than 13 metric tons of Pu are contained in 110 metric tons of processing wastes throughout the U.S. Department of Energy (DOE) complex. Additionally, the DOE spent nuclear fuel (SNF) inventory consists of approximately 2700 metric tons of heavy metals and also requires disposition.[2,3] The long-term disposition of surplus plutonium and the Pu processing wastes should not only allow for the protection of the public and the environment but should also ensure that the plutonium is not readily recoverable for use in weapons by any nation or organization. One Pu disposition alternative that may be implemented by DOE is immobilization into a glass waste form.[2,3]

Plutonium solubility in glasses with wide compositional ranges has been studied for years, as shown in Table 1. The solubility of Pu oxide varied from less than 0.5 to about 11.4 mass% in glasses processed in oxidizing conditions, but a much higher solubility limit, 22.4 mass% Pu oxide, was reported for silicate glass melted in vacuum.[4-17] However, information is scarce regarding Pu chemistry in the silicate-based glass,[17-19] and no detailed studies, to the best of our knowledge, have been carried out to determine the effect of Pu speciation on its solubility in glass. The speciation of U in silicate glasses has been studied extensively in relatively low concentrations (< 10 mass% UO_3).[20,21] However, uranium solubility information in other glass systems such as borosilicate glasses is limited. Further, the redox chemistry of U, and therefore the effect of redox on its solubility limit in silicate and borosilicate systems, is not well understood.

The objective of this multi-institutional, international research effort is to understand the distributions, solubilities, and releases of radionuclides and neutron absorbers incorporated in waste forms. One approach is by studying their local chemical environments to develop a fundamental understanding of how composition and chemical structure of the host matrix phase as well as processing conditions affect the partitioning of radionuclides and neutron absorbers. Since wastes will vary in form and chemical complexity, glass, ceramic, and glass-ceramic hosts will be considered. The results will provide key knowledge for developing, evaluating, selecting, and matching waste forms for the safe disposal of various wastes associated with Pu, miscellaneous spent nuclear fuels, and other transuranic (TRU) wastes.

The solubilities and distributions of Pu and U in glasses, glass-ceramics, and ceramics are influenced by the oxidation states, ionic size, and coordination of Pu and U in the waste forms. Thus, the following scientific issues need to be understood:

i) the structural roles of Pu and U in the glass as a function of composition and redox chemistry;

ii) the structural roles of Gd (a neutron absorber or potential surrogate for Pu^{3+}) and Hf (a neutron absorber or potential surrogate for Pu^{4+}) in the glass and their effects on the solubility of Pu and U in the glass;

iii) the effects of composition and processing conditions on the redox states of Pu and U, and thus, on their solubility limits in glass, glass-ceramics, and ceramic phases;

iv) the stability of Pu and U in various oxidation states in glass, glass-ceramics, and ceramic phases, which is ultimately related to whether they are released congruently with neutron absorbers from the waste forms in an aqueous environment; and

v) differences between Gd and Pu in site occupancies in lattice structures of solid waste forms, ability of complexation with other species in groundwater, and a tendency to form mineral phases on the surfaces of the corroded waste forms.

This collaborative research program focuses on systematically investigating the solubility limits and partitioning of Gd, Hf, Pu, and U in selected waste forms as a function of waste form composition, temperature, and processing conditions and selectively studying waste form properties in aqueous environments. Both of these issues are linked to determining the local structural environments of radionuclides and neutron absorbers in various phases of the waste forms using various complementary spectroscopic techniques, including x-ray diffraction (XRD), scanning electron microscopy (SEM), transmission electron microscopy (TEM), x-ray absorption fine structure (XAFS), and nuclear magnetic resonance (NMR).

RESULTS

During the initial period of this project, the program focused on determining the solubility limits of Pu, U, Gd, and Hf in silicate and borosilicate glasses. In the coming years the efforts will be expanded to include lanthanide borosilicate glasses, glass-ceramics, and ceramics.

1. Composition Effects on HfO_2 and Gd_2O_3 Solubilities in Borosilicate Glasses

A four-component, sodium-alumino-borosilicate glass system (SiO_2, Al_2O_3, B_2O_3, and Na_2O) has been chosen as a baseline to study the solubility limits of HfO_2 and Gd_2O_3. The component effects on solubility limits were studied by systematically varying the concentrations of Al_2O_3, B_2O_3, and Na_2O, one at a time. The glass systems studied to date, expressed in terms of absolute molar concentrations, include xB_2O_3-$20Na_2O$-$5Al_2O_3$-$60SiO_2$ (x from 5 to 20), $15B_2O_3$-xNa_2O-$5Al_2O_3$-$60SiO_2$ (x from 10 to 30), and $15B_2O_3$-$20Na_2O$-xAl_2O_3-$60SiO_2$ (x from 2 to 15).

Table 2 provides the final target compositions (mass%) of the glasses with HfO_2 and Gd_2O_3 at their solubility limits (with $\pm 1\%$ experimental uncertainty). The solubility limits of HfO_2 and Gd_2O_3 were determined at 1450 and 1560°C, respectively. For HfO_2, its solubility is not affected by B_2O_3 concentration for a fixed Na_2O/Al_2O_3 molar ratio of 4 but increases linearly with the net Na_2O concentration (net moles of Na_2O = moles of total Na_2O - moles of Al_2O_3) at a fixed B_2O_3/Al_2O_3 molar ratio of 3. For Gd_2O_3, its solubility was found to be affected by the Na_2O or B_2O_3 concentrations in the following way. It increases non linearly with the net Na_2O content when the B_2O_3/Al_2O_3 molar ratio is fixed at 3 but linearly with B_2O_3 concentration when the Na_2O/Al_2O_3 molar ratio is fixed at 4. Furthermore, the solubility limits of HfO_2 and Gd_2O_3 decrease significantly as the Al_2O_3 concentration increases, possibly as a result of the decrease in the net Na_2O concentration.

Studies are planned to determine whether a mixture of HfO_2 and Gd_2O_3 would provide an appropriate surrogate for the study of Pu^{4+}/Pu^{3+} in glass systems in terms of their structural roles, solubility limit as a function of processing temperature and conditions (oxidizing and reducing), and their dissolution behaviors in aqueous environments.

2. Effect of Plutonium Redox (Pu^{3+}/Pu^{4+}) on Pu Solubility in Borosilicate Glasses

The redox effects on Pu solubility are being studied systematically for silicate, borosilicate, and lanthanide borosilicate glass systems. Initial work has focused on a borosilicate composition (Pu16MC3) developed at PNNL. The glass composition (mass%) is: Al_2O_3 (4.07), B_2O_3 (13.23), Cs_2O (1.02), Fe_2O_3 (4.74), Gd_2O_3 (11.30), K_2O (1.53), Li_2O (4.09), Na_2O (5.12), P_2O_5 (2.14), and SiO_2 (42.76). The solubility was found to be less than 10 mass% of PuO_2 when the glass was melted under oxidizing conditions in a temperature range from 1300 to 1500°C;[8] PuO_2 crystallites were found in all oxidized glass samples using XRD analysis. Preliminary XAFS analysis suggests that Pu^{4+} was the dominant oxidation state.[8] For the same glass melted under reducing conditions (by melting the glass with graphite powder in a sealed alumina crucible), the solubility of Pu was found to be at least 25 mass% (expressed in an original oxide form, PuO_2) at 1400°C; the reduced glass with 25 mass% PuO_2 was amorphous to XRD.

To determine the local chemical environment of Pu in the glass, Pu L_{III} X-ray absorption spectra were collected on wiggler beamline 4-1 at the Stanford Synchrotron Radiation Laboratory (SSRL) using a Si(220) monochromator. Spectra were collected in transmission mode or in fluorescence mode using a four-pixel Ge detector developed at Lawrence Berkeley National Laboratory (LBNL). The higher harmonic content of the beam was rejected by detuning the monochromator 50%. Spectra were collected from -325 mesh glass powders (except for the oxidized Pu glass that was a thin chip) dispersed in boron nitride, the samples were contained in a quaternary sample containment system. Data reduction was accomplished by standard methodologies[21] using the EXAFSPAK suite of computer codes developed by SSRL. Figure 1 shows Pu L_{III} X-ray absorption spectra of Pu16MC3 glasses (with 10 mass% PuO_2) melted at 1500°C in oxidizing conditions and at 1400°C in reducing conditions. Standard reference spectra for Pu^{4+} and Pu^{3+} are included in the figure for comparison. By comparing with the XAFS spectra of standard reference Pu spectra of Pu^{4+} and Pu^{3+}, the XAFS spectra of Pu in the oxidized and reduced glasses reveal Pu oxidation states that are fully consistent with the melt redox conditions. The proportion of Pu^{3+}, with respect to the total Pu in the reduced glass, was estimated to be about 95%. The proportion of Pu^{4+}, with respect to the total Pu in the oxidized glass, was estimated to be about 90%.[8] Coincidently, under the reducing conditions, the total soluble Pu^{3+} plus Gd^{3+} in the Pu16MC3 glass is close to the Gd^{3+} solubility obtained in the simplified four-component borosilicate glass discussed above (cf. Table 2). This solubility comparison suggests the possibility of using Gd^{3+} as a surrogate to study Pu^{3+} solubility in the glass. Further spectroscopic characterizations continue to quantify local environments of Pu^{3+} and Gd^{3+} in these glasses.

3. Effects of Sodium Content and Redox on Uranium Solubility in Silicate Glasses

A systematic study of U solubility in silicate glass is also being carried out. The initial work has involved a standard soda-lime glass composition, SRM 1830 (obtained from the National Institute of Standards Technology, NIST). The concentration of U, initially in the form of UO_3, was increased from 10 to 40 mass% as shown in Table 3. All glasses (10 grams each) were melted at 1500°C for two hours. XRD measurement showed that glasses with 30 mass% UO_3 (SL–U30) or less were completely amorphous, while U_3O_7 (or $U_2O_5{\cdot}UO_2$) crystallites were detected in glasses with 31 mass% UO_3 or greater. However, in the SL-U40A glass, no uranium-bearing crystals were detected by XRD. The SL–40A glass contains a higher Na_2O concentration than those in the SL–U10 to SL–U40 glasses. The average valence state of U in these glasses, estimated using the XAFS data, are shown in Table 3. It is evident from these results that the valence of U in the oxidized glasses does not change with variation in Na_2O concentration; however, the solubility of U apparently increased from 30 mass% to at least 40 mass% by increasing the Na_2O concentration. This result suggests that oxidized uranium solubility increases with the glass basicity.

When the glass with 40 mass% UO_3 was melted under reducing conditions, i.e., SL–U40R, uranium-bearing crystals with a closer match to UO_2 were found using XRD analysis. U^{4+} is found to be the primary species using XAFS analysis as shown in Table 3. These results suggest that the uranium solubility as U^{4+} is less than 30 mass%. It further suggests that U^{5+} (as implicated by the XRD results of this study) is more soluble than U^{4+}. The local structural environments of U in the silicate glasses are being further studied using the same method applied to the plutonium study discussed above.

Figure 2 shows U L_{III} x-ray absorption spectra obtained from the NIST-based glasses. Initial inspection of the spectra shows that glasses doped with 10-40 wt% UO_3, excluding glasses prepared under reducing conditions, yielded remarkably similar spectra. The oxidation state of U in these glasses more closely resembles that of U_3O_8 than UO_2, characteristic of a mixture of oxidation states in the glasses. In contrast, U in the U40R series glasses is significantly more reduced. The overall line shape, energy position, and presence of the actinyl signature[23] indicate that uranium in the SL-U40RA glass is nearly all tetravalent and in SL-U40R that a significant fraction is tetravalent. Furthermore, the extended XAFS clearly identifies the U^{4+} species in the SL-U40R series glasses from U-U scattering as UO_2[21]. The estimated average valence of U in the oxidizing glass is five, compared with a value about four in the reducing SL-U40R and SL-40AR glasses (cf. Table 3).

SUMMARY

A multi-institutional, international research effort is actively investigating the fundamentals that govern the solubility and distribution of Pu, U, Hf, and Gd in glasses, glass-ceramics, and ceramics. The preliminary results indicate that Pu^{3+} is much more soluble than Pu^{4+} in borosilicate glass, while U^{4+} is less soluble than U^{5+}. The reason for this difference will be further examined using complementary characterization techniques to elucidate additional aspects of their local structural environments.

The solubility of HfO_2 in simplified sodium-alumino-borosilicate glass was found to be independent of B_2O_3 concentration, while the solubility of Gd_2O_3 significantly increases as the B_2O_3 concentration increases. The solubility limits of both HfO_2 and Gd_2O_3 were found to depend on the net Na_2O concentration. This difference will be understood through spectroscopic analysis of the differences in their structural roles in the glass network. The solubility of Gd in a borosilicate glass is similar to that of Pu^{3+} on a molar basis in a borosilicate glass with similar composition, suggesting Gd^{3+} could be a potential surrogate for Pu^{3+} in solubility studies.

ACKNOWLEDGMENT

This work is supported by the Environmental Management Science Program of the U.S. Department of Energy. The Pacific Northwest National Laboratory is operated by Battelle for the U.S. Department of Energy under Contract DE-AC06-76RLO 1830. This research is collaborative with Lawrence Berkeley National Laboratory, which is also operated by the U.S. Department of Energy under contract DE-AC03-76SF00098, the University of Michigan, and the Australian Nuclear Science and Technology Organisation. Part of the work was performed at SSRL, which is operated by DOE, Office of Basic Energy Sciences, Division of Chemical Sciences. L.L. Davis and L. Li acknowledge the Associated Western Universities for their postdoctoral research fellowship appointments at the Pacific Northwest National Laboratory.

REFERENCES

1. S. Lyman. 1993. "The Solubility of Plutonium in Glass." PU/CEES Report No.275. Center for Energy and Environmental Studies, School of Engineering and Applied Science, Princeton University, Princeton, NJ.
2. U.S. Department of Energy, "Record of Decision for Surplus Plutonium Disposition of January 14, 1997." *Nuclear Waste News*, 1997.

3. U.S. Department of Energy. "A DOE-Owned Spent Nuclear Fuel Technology Integration Plan." DOE/SNF/PP-002, Revision 1. U.S. Government Printing Office, Washington, DC, 1996.
4. C.T. Walker and U. Riege, "Compatibility of Actinides with HLW Borosilicate Glass: Solubility and Phase Formation," in *Ceramics in Nuclear Waste Management* ed. by T.D. Chihalla and J. E. Mendel, (1979) p. 198-202.
5. E.S. Prokin, U.S. Kuptsov, T.N. Ananina, and E.E. Ermolaev, "Characteristics of Borosilicate Glass in Modeling of Alpha-radiation and Thermal Conditions of Storage of Glassified Highly Radioactive Wastes," *Radiokhimiya*, 30, 694-698 (1988).
6. Joint U.S./Russian Plutonium Disposition Study, *Investigations of Plutonium Immobilization into the Vitreous Compositions*: A.A. Bochvar All-Russia Research Institute of Inorganic Materials Report UTA96-0330 MOD-1, p53 (1998).
7. J.K. Bates, A.J.G. Ellison, J.W. Emery, and J.C. Hoh, "Glass as a Waste Form for the Immobilization of Plutonium," in *Scientific Basis for Nuclear Waste Management XIX*, Mat. Res. Soc. Symp. Proc., Vol. 412 (MRS, Pittsburgh, PA, 1996) p. 57-64.
8. J.G. Darab, H. Li, M.J. Schweiger, J.D. Vienna, P.G. Allen, J.J. Bucher, N.M. Edelstein, and D.K. Shuh, "Chemistry of Plutonium and Cerium in Vitrified Nuclear Wastes," in *Plutonium Futures - The Science*, Conference Transactions, LA-13338-C, Los Alamos National Laboratory (Los Alamos, New Mexico,1997) p. 143-145.
9. D. Riley, W. Bourcier, J.D. Vienna, T.F. Meaker, D.K. Peeler, and J. Marra, "Dissolution Studies of Plutonium Oxide in LaBS Glass," in *Plutonium Futures - The Science*, Conference Transactions, LA-13338-C, Los Alamos National Laboratory (Los Alamos, New Mexico,1997) p. 119-120.
10. J.D. Vienna, D.L. Alexander, H. Li, M.J. Schweiger, D.K. Peeler, and T.F. Meaker, *Plutonium Dioxide Dissolution in Glass*, Report PNNL-11246, Pacific Northwest National Laboratory, Richland, WA, 1996.
11. M.J. Plodinec, *Development of Glass Compositions for Immobilization of SRP Waste*, Report DP-1517, Savannah River Laboratory, Aiken, NC, 1979.
12. E.S. Prokin, O.A. Alekseev, T.N. Ananina, and E.E. Ermolaev, "Behavior of Plutonium Dioxide in a Molten Phosphate Glass," *Radiokhimiya*, 31, 140-144 (1989).
13. A. S. Nikiforov, A. S. Polyakov, Yu. I. Natyunin, A.V. Demin, M.I. Fedorova, A.A. Bochvar, "Behaviour of Plutonium, Rare Earth Elements and Elements of the Plutonium Group in Vitreous Phosphate Composites During Liquid LHW Vitrification," in *High Level Radioactive Waste Management Proceedings* (Proceedings of the Second Annual International Conference, Las Vegas, Nevada, 1991) p. 738-740.
14. Yu. I. Matyunin, M.N. Astakhov, and M.F. Savin, 1990, Study of Immobilization of α-Radionuclides in Phosphate Glasses," *Radiokhimiya*, 32, 154-158 (1990).
15. A.T. Aldred, "Magnetic Susceptibility of Sodium Disilicate Glasses Containing PuO_2," *J. Non-crystalline Solids*, 40, 347-352 (1980).
16 D.J. Lam, B.W. Veal, and A.P. Paulikas, "X-ray Photo Emission Spectroscopy (XPS) Study of Uranium, Neptunium, and Plutonium Oxides in Silicate-based Glasses," in *Plutonium Chemistry*, ed. by W.T. Carnall and G.R. Choppin, American Chemical Society Symposium Series, Vol. 216, 145-154 (1983).
17. P.G. Eller, G.D. Jarvinen, J.D. Purson, R.A. Penneman, R.R. Ryan, F.W. Lytle, and R.B. Greegor, "Actinide Valences in Borosilicate Glass," *Radiochimica Acta*, 39, 17-22 (1985).
18. D.G. Karraker, "Actinide Valences in Borosilicate Glass," *J. Am. Ceram. Soc.*, 65, 53 (1982).

19. D.P. Karim, D.J. Lam, H. Diamond, A.M. Friedman, D.G. Coles, Bazan, G.L. McVay, "XPS Valence State determination of Np and Pu in Multicomponent Borosilicate Glass and Application to Leached 76-68 Waste Glass Surface," in Scientific Basis for Nuclear Waste Managment, vol. 6, ed. by S.V. Topp (North-Holland, New York, 1982) p. 67.
20. H.D. Schreiber, G.B. Balazs, B.J. Williams, and S.M. Andrew, "Structural and Redox Properties of Uranium in CA-MG-AL-Silicate Glasses," Scientific Basis for Nuclear Waste Management, Vol. 3, ed. by J.G. Moore (Plenum Press, New York, 1981) p.109-114.
21. F. Francois, C.W. Ponader, G. Calas, and G.E. Brown, Jr., "Structural Environments of Incompatible Elements in Silicate Glass/Melt Systems: II. U^{IV}, U^{V}, and U^{VI}," *Geochim Cosmochim. Acta*, 56, 4205-4220 (1992).
22. R. Prins and D. E. Koningsberger, X-ray Absorption: Principles, Applications, Techniques for EXAFS, SEXAFS, and XANES; (Wiley-Interscience, New York: 1998).
23. H.D. Schreiber and M.T. Coolburgh, "Solvations of Redox Ions in Glass-Forming Silicate Melts," *J. Non-Cryst. Solids.*, 181, 225 (1995).

Table 1. Bulk Compositions (mass%) of Host Glasses with PuO_2

System	Alumino-Borosilicate					Lanthanide	Borosilicate		Phosphate				Silicate	
SiO_2	50.5	31.9	46.0 - 46.2	41.5	42.8	25.8	52.5	51.9-58.3					66.7	69.6
TiO_2	4.2			2.02			10	0-11						
ZrO_2				5.3		1.15								
PuO_2			0.41 - 1.23	11.3	10	11.4			1.16	3.6				6.45
B_2O_3	13.6	19.4	2.98 - 3.01	24.1	13.2	10.4	10	9.9-11.1						
Al_2O_3	1.4	16.7	4.95 - 5.01	4.61	4.1	19			21.7	22	15-25	22		
Fe_2O_3		1.1	2.98 - 3.01		4.7					1.45				
Gd_2O_3				6.38	11.3	7.61								
La_2O_3						11								
Nd_2O_3						11.4								
CaO	2.8	15.8					5	4.9-5.6			0-1.0			
NiO										0.1				
SrO						2.22								
Na_2O	27.5	22	20.8 - 21.1	18.3	5.1		18.5	12.5-20.6	23.7	24	23-27	24	33.3	23.9
Li_2O				8	4.1		4	4.9-5.6						
P_2O_5					2.1				53.4	52	49-59	54		
Cs_2O				1.36	0.47									
K_2O				10.3	0.62									
SnO				2.24										
PuO_2 Solubilit	4.4	1.3	0.2 - 0.45	8.1	<10	11.4	4	7	0.2 - 1.1	0.18	0.23	0.24-0.31	22.4 in Vacuum	8.5
Referenc	4	5	6	7	8	8-10	11	11	6	12	13	14	15	16

Table 2. Glass Compositions (mass%) and Preliminary Results on Compositional Effects on Solubility Limits of HfO_2 and Gd_2O_3

HfO_2 Solubility Study												
Systems	xB_2O_3-$20Na_2O$-$5Al_2O_3$-$60SiO_2$				$15B_2O_3$-xNa_2O-$5Al_2O_3$-$60SiO_2$				$15B_2O_3$-$20Na_2O$-xAl_2O_3-$60SiO_2$			
x (mol)	5	10	15	20	10	15	20	30	2	5	10	15
Final Glass Compositions (mass%)												
SiO_2	44.3	42.3	39.4	37.4	51.8	45.1	39.4	34.4	37.9	39.4	41.2	42.3
Al_2O_3	6.3	6.0	5.6	5.3	7.3	6.3	5.6	4.9	2.1	5.6	11.7	17.9
B_2O_3	4.3	8.2	11.4	14.5	15.0	13.1	11.4	10.0	11.0	11.4	11.9	12.3
Na_2O	15.2	14.6	13.6	12.9	8.9	11.7	13.6	17.7	13.0	13.6	14.2	14.5
HfO_2[a]	30.0	29.0	30.0	30.0	17.0	24.0	30.0	33.0	36.0	30.0	21.0	13.0

Gd_2O_3 Solubility Study												
Systems	xB_2O_3-$20Na_2O$-$5Al_2O_3$-$60SiO_2$				$15B_2O_3$-xNa_2O-$5Al_2O_3$-$60SiO_2$				$15B_2O_3$-$20Na_2O$-xAl_2O_3-$60SiO_2$			
x (mol)	5	10	15	20	10	15	20	30	2	5	10	15
Final Glass Compositions (mass%)												
SiO_2	42.4	37.5	32.7	28.3	51.2	36.1	33.8	31.3	31.4	32.7	33.9	42.3
Al_2O_3	6.0	5.3	4.6	4.0	7.2	5.1	4.8	4.4	1.8	4.6	9.6	17.9
B_2O_3	4.1	7.3	9.5	10.9	14.8	10.5	9.8	9.1	9.1	9.5	9.8	12.3
Na_2O	14.6	12.9	11.2	9.7	8.8	9.3	11.6	16.2	10.8	11.2	11.7	14.5
Gd_2O_3[a]	33.0	37.0	42.0	47.0	18.0	39.0	40.0	39.0	47.0	42.0	35.0	13.0

(a) The solubility values were determined within ±1 mass% accuracy.

Table 3. Compositions (mass%) of Soda-lime Silicate Glasses Enriched with UO_3 and Preliminary Results on UO_3 Solubility and Valance of U as Determined by XAFS Analysis

Glass ID	SL-U10	SL-U20	SL-U30	SL-U31	SL-U34	SL-U38	SL-U40	SL-U40A	SL-U40R	SL-U40AR
Al_2O_3	0.11	0.10	0.08	0.08	0.08	0.07	0.07	0.07	0.07	0.07
CaO	7.70	6.85	5.99	5.91	5.65	5.31	5.14	4.71	5.14	4.71
Fe_2O_3	0.11	0.10	0.08	0.08	0.08	0.08	0.07	0.07	0.07	0.07
K_2O	0.04	0.03	0.03	0.03	0.03	0.02	0.02	0.02	0.02	0.02
MgO	3.51	3.12	2.73	2.69	2.57	2.42	2.34	2.15	2.34	2.15
Na_2O	12.38	11.00	9.63	9.49	9.08	8.53	8.25	12.56	8.25	12.56
SO_3	0.23	0.21	0.18	0.18	0.17	0.16	0.16	0.14	0.16	0.14
SiO_2	65.76	58.46	51.15	50.42	48.23	45.30	43.84	40.19	43.84	40.19
TiO_2	0.01	0.01	0.01	0.01	0.01	0.01	0.01	0.01	0.01	0.01
UO_3	9.98	19.96	29.94	30.94	33.93	37.92	39.92	39.92	39.92	39.92
Crystals	N	N	N	Y	Y	Y	Y	N	Y	Y
Average valence	5	5	5	5	5	5	5	5	4	4

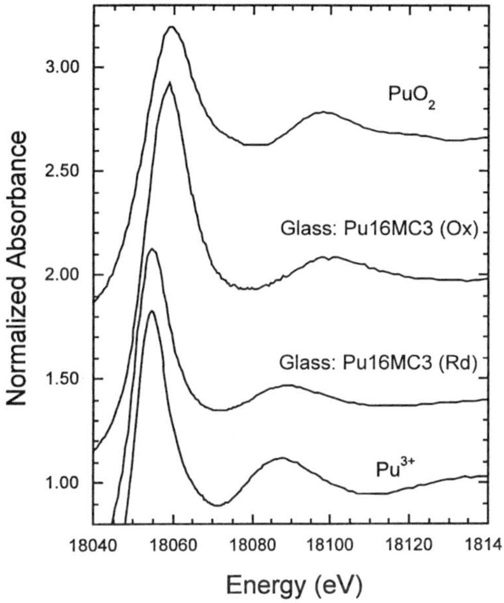

Figure 1. XAFS normalized plutonium L_{III} near edge spectra from a borosilicate glass containing 10 mass% PuO_2 prepared under reducing conditions compared to well-characterized Pu reference spectra for the trivalent and tetravalent oxidation states (the oxidized Pu borosilicate spectrum is reproduced from Reference 8).

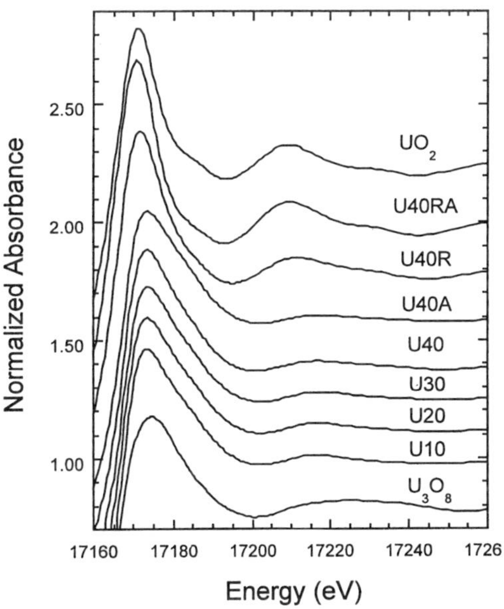

Figure 2. XAFS normalized uranium L_{III} near edges from the soda-lime silicate glasses as a function of uranium oxide loading between 10 and 40 mass% UO_3, glass processing redox conditions, and composition along with appropriate reference uranium oxides. (the spectra are individually labeled with the weight percent of UO_3, special reducing processing conditions by the inclusion of R, and excess alkali (Na_2O) represented by A in the designation, respectively).

KEYWORD AND AUTHOR INDEX